INELASTIC DEFORMATION OF METALS

INELASTIC DEFORMATION OF METALS

MODELS, MECHANICAL PROPERTIES, AND METALLURGY

Donald C. Stouffer
University of Cincinnati

and

L. Thomas Dame
Structural Dynamics Research Corporation

JOHN WILEY & SONS, INC.

New York • Chichester • Brisbane • Toronto • Singapore

This test is printed on acid-free paper.

Library of Congress Cataloging in Publication Data:

Stouffer, Donald C.
Inelastic deformation of metals : models, mechanical properties,
and metallurgy / by Donald C. Stouffer and L. Thomas Dame.
p. cm.
Includes bibliographical references and index.
ISBN 0-471-02143-1 (cloth : alk. paper)
1. Metals—mechanical properties—Mathematical models.
2. Plasticity—Mathematical models. 3. Metals—Metallurgy.
I. Dame, L. Thomas. II. Title.
TA460.S839 1996
620.1′633—dc20 95-4643

Printed in the United States of America

10 9 8 7 6 5 4 3 2 1

Contents

PREFACE

This book is an attempt to bring together the fields of mechanics and metallurgy for the purpose of developing physically based, numerically efficient, and accurate methods for predicting the inelastic response of metals under a variety of loading and environmental conditions. During the middle third of this century (approximately 1930 to 1965) the mathematical foundation of plasticity was developed for modeling metals. A large body of mathematical results had been established for boundary value problems describing structural systems and manufacturing processes. The contributions of researchers in plasticity are responsible for many modern results in continuum mechanics, tensor analysis, and numerical analysis. The underlying assumptions are based on the early experimental observations of material behavior and phenomenological arguments of expected response. Most of the work is based on the assumption that metals do not exhibit any time or loading-rate effects, although methods have been developed to add a time-dependent creep strain. The methods are being used extensively for numerical simulations, but there have not been many direct comparisons to experimental observations. These results have been reviewed, compiled, and presented in books on plasticity. In fact, virtually every book published on plasticity in the last fifty years by authors from the mechanics community has been based almost exclusively on this approach.

During the same period the field of metallurgy has exploded with new results. The concept of a dislocation as a crystalline defect was first postulated in the 1930s. Since then, the early ideas about slip quickly grew into well-informed explanations of how metals really do deform under a wide range of loading and environmental conditions. These results were fueled by the development of a wide range of equipment for mechanical testing and microstructural examination. An elegant mathematical theory of dislocations has also been developed to describe events at the crystal lattice level. However, these results have been largely ignored by the mechanics community. This is, at least in part, because there are just too many dislocations in a structure to attempt structural modeling based on individual motions of dislocations.

As technology evolved since the 1960s it became necessary to model the inelastic mechanical response of metals more accurately for the design of jet engine and rocket components. It became critical to predict time-dependent effects such as creep, recovery, stress relaxation, and strain-rate-dependent properties for use in low-cycle fatigue life analyses. As mechanical testing technology developed, it was recognized that many metals have time-dependent properties at room temperature. AISI type 304 stainless steel is strain-rate dependent at room temperature. Automobile designs are now beginning to incorporate time and rate inelasticity effects. Aircraft aluminum alloys display stress relaxation at room temperature. Television coaxial transmission cables experience plastic deformation due to thermal cycling. Plasticity and low-cycle fatigue design criteria are critical in gas and steam turbines. Every day the role of inelasticity in design becomes more important as we push technology.

In the last twenty years the federal government has spent an enormous amount of money in mechanics and metallurgy looking for alternative approaches. NASA, the Department of Defense, the Department of Energy, and the National Institutes of Health all have (and had) big programs pushing materials and mechanics research. This funding has produced new methods in physically based constitutive modeling for metals. The field has changed, but these methods are not being used by many practitioners in their work. Thus one objective of the book is to bridge the relationship between mechanics and metallurgy for the purpose of constitutive modeling of metals. We synthesize the recent research results in the state-variable approach to constitutive modeling with observations of material behavior and the relevant topics from material science. The objective is to give a continuous and relatively complete connection between material science and constitutive modeling.

Another key objective of the manuscript is to demonstrate how to use the metallurgical information to develop material models for structural simulations and low cyclic fatigue predictions. To achieve this goal, it is necessary to know:

What information is useful and what is not
How to use microscopic observations
The types of variables to use
The steps involved to construct models
How to determine the constants

These topics represent the basics of the state-variable approach. There is a very large body of journal literature on these topics, and perhaps hundreds of state-variable models. In this book we give the principal features of state-variable modeling, include some of the different types of models and discuss attributes, and provide methods to develop models for special situations. The book is not a catalog of models.

Another objective is to provide enough comparisons between data and theory for the reader to make a meaningful judgment about the value and accuracy of a particular model: that is, to show the reader a considerable amount of data, to develop a core knowledge of metals, and to instill an understanding of how metals respond in real service environments. This type of information is necessary to develop a keen engineering judgment.

The final objective is to discuss the numerical methods associated with nonlinear constitutive modeling. Time-independent and time-dependent numerical procedures are different. Time integration schemes, inversion techniques, and subincrementing are all subjects that are necessary for nonlinear finite-element applications.

The manuscript is written primarily for first-year students in a graduate program or practicing engineers. It is felt that the book will fill an important need because plasticity is not part of most undergraduate engineering programs. The book contains new technology that is not easily available, and commercial finite-element codes do not include these models. There are no general equations or single theory for plasticity, so it is necessary to know the types of models that are available and understand their attributes.

The book has three major sections. Part One, "Relationships Between Material and Mechanical Properties," contains the connection between the observed mechanical properties and the corresponding deformation mechanisms. This part is the foundation of the state-variable approach. The experimental results and models in Part One are limited to uniaxial loading. Most of the metallurgical studies and experimental data in the literature are for uniaxial loading. This part of the book contains a discussion of the elements of physical metallurgy that are important for modeling, tensile and fatigue response properties and the associated mechanisms, and creep and creep mechanisms. A number of models for uniaxial tensile, fatigue, and creep response are presented and discussed.

Part Two, "Multiaxial Plasticity and Creep," is the extension from uniaxial to multiaxial loading. The constitutive modeling embodies many of the results in the mechanics literature. This part of the book establishes the mathematical structure and mechanics necessary for the state-variable approach and contains the key elements of the yield surface plasticity approach and classical creep modeling. Solved problems and numerous comparisons with experimental data serve to demonstrate clearly the underlying ideas. The topics in this part are useful for making value judgments about the results of engineering calculations with commercial finite element codes.

Part Three, "State-Variable Approach," brings in the new results. This part of the book will show the reader how to use results from material science to build constitutive models for use in structural design and life prediction. The section begins with a discussion of the philosophy of state variables, the selection of state variables, development of the models, and the evaluation of material parameters for tensile, creep, and cyclic response with and without time-dependent properties. There are topics on multiaxial response, thermomechanical fatigue, single-crystal alloys, numerical implementation, and finite-element applications. The discussion includes many comparison between model and experimental results.

Each chapter of the book is concluded with a summary of the key results and a discussion of the attributes and limitations of the models presented.

Each type of model is discussed for computational efficiency, ease in determining material parameters, and accuracy. A model must be reasonably efficient for numerical implementation. It is not acceptable to develop a model that cannot be used in a finite-element code, where it is expected to be called thousands of times. Another critical factor controlling the usefulness of a model is the ease in finding the material parameters. It is advantageous to be able to find the required parameters from common experimental data rather than special tests. The solution of difficult equations for the material parameters or the development of special tests can make a model useless. Accuracy is a desirable attribute, but it is not always necessary. At certain times it may only be necessary to have an estimate of a particular value; at other times it may be desirable to have a more precise value. The desired level of accuracy has to be balanced by the cost and effort required to obtain the accuracy.

The authors, Drs. Stouffer and Dame, working as a team bring unique expertise to the book. Dr. Stouffer has been active in nonlinear constitutive modeling of materials for 25 years at the University of Cincinnati and has published extensively in the field. The material has been used in courses (classical plasticity, state-variable plasticity, continuum mechanics, life prediction methods and nonlinear finite element modeling) for many years at the University of Cincinnati. Dr. Dame has worked in nonlinear structural analysis methods development at General Electric Aircraft Engines and is currently lead engineer in the development of commercial nonlinear finite-element codes at Structural Dynamics Research Corporation, and is an adjunct professor at the University of Cincinnati. In addition, the authors discussed various sections of the manuscript with experts in the field and wish to thank all for their comments and encouragement. It is this combination of experience that allows the development of a book with such breadth.

June 1995
DCS/LTD

INELASTIC DEFORMATION OF METALS

PART ONE

RELATIONSHIPS BETWEEN MATERIAL AND MECHANICAL PROPERTIES

There are many different types of mechanical response characteristics for metals, depending on the material microstructure, temperature, and method of loading. The response may be time dependent or time independent; there may be a little hardening, a considerable amount of hardening, no hardening, or even softening. It is important to understand how metals really behave under various loading conditions before attempting to construct models for their response. It is equally important to understand how metals behave in order to identify the correct variables for modeling, to know if the results from a particular model are correct, and to know what model to use for a particular application. Part One provides a review of the various types of response characteristics together with a discussion of why the response has the features observed. Since our objective is material modeling, some simple models are included as a first step in inelastic constitutive modeling. Key observations and attributes of the models are presented at the end of each chapter.

CHAPTER 1

Physical Basis of Inelasticity

In this chapter we establish the physical background for understanding the inelastic response of metals. We discuss the properties of dislocations and how they interact with grain boundaries, precipitates, and other dislocations to produce hardening, softening, and time-dependent effects observed in inelastic material response. The concepts presented are key to relating the observed response to the underlying metallurgical phenomena and serve as the basis for the state-variable approach presented in Part Three.

1.1 INTRODUCTION

Most metals are an aggregate of very many small crystals or grains. Most metals are therefore polycrystalline, although single-crystal alloys are currently being used in special aerospace applications as discussed in Chapter 8. Because of the small grain size, microscopes operating in the range of 1000× and higher may be required to examine the structural features of grains. Occasionally, very large grains are visible to the naked eye. Structures requiring magnification are known as microstructures. The basic atomic arrangement inside a grain is a crystal structure. Dislocations are disruptions or irregularities in the crystal lattice structure in a grain. Inelastic flow occurs primarily from the movement of the dislocations through the crystal structure. The grain boundaries, particles, and other dislocations affect the propagation or movement of dislocations. During inelastic deformation the number of dislocations (dislocation density) increases and the features of the microstructure change continuously as the deformation progresses. These events at the microstructural level result in the observed inelastic effects.

A correlation between the microstructure and inelastic response was demonstrated by Moteff (1980) for type 304 stainless steel at 650°C. Figure 1.1.1 shows five microstructures from five specimens at different values of strain during tensile tests. It can be seen that the microstructure evolves and changes during deformation. The number of dislocations (the dark lines) increase from an almost dislocation-free state to a microstructure with a three-dimensional dislocation cell-like structure. Two typical measures among many that could be used to characterize changes in the microstructure during deformation are

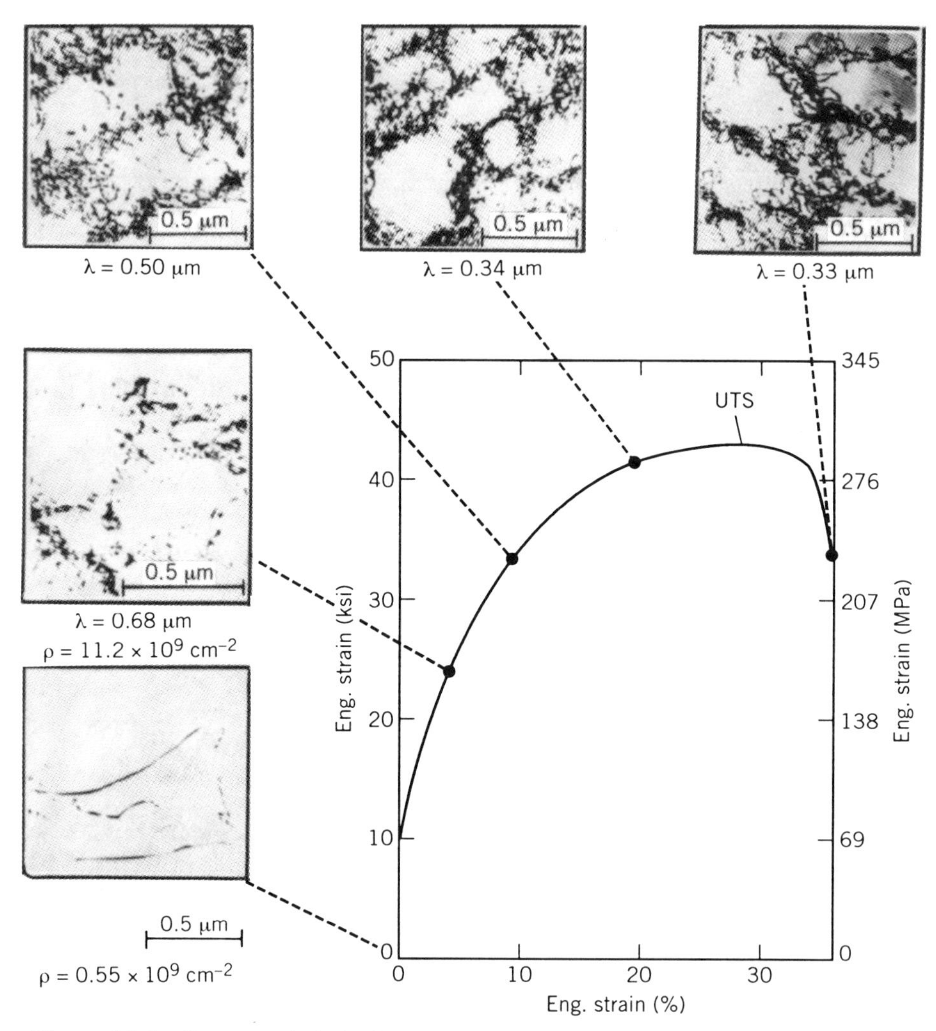

Figure 1.1.1 Stress–strain behavior of type 304 stainless steel at 650°C at a nominal strain rate of 3.17×10^{-4} sec^{-1}, together with the corresponding dislocation substructure at five different strain levels. (After Moteff, 1980.)

the dislocation density and average cell size. The values of these variables can be correlated to a state of stress or strain on the tensile response curve. Thus there is a correlation between the material microstructure and the mechanical response. Our objective is to use this type of information to enhance our understanding of material response and to develop material models.

The first step in understanding this correlation is to consider the microstructural levels (or magnification level) at which different events occur as outlined in Table 1.1.1. The macroscopic scale is where most experiments are

TABLE 1.1.1 Typical Events Observed at Various Levels of Magnification

Lowest Magnification

Macroscopic level (approximately 10^{-3} m and larger)
  Structural size and mechanical response
Micron level (approximately 10^{-6} m)
  Grain and precipitate size and shape, subgrain size, grain boundary properties, and grain orientation
Angstrom level (approximately 10^{-10} m)
  Type of dislocation, dislocation density, arrangement, stacking faults
Electron level (approximately 10^{-14} m)
  Atomic structure, and electron and nucleus properties

Highest Magnification

conducted, and the response is measured with load cells and extensometers. Local sampling regions of a specimen are usually measured in centimeters or millimeters, although the specimen may be much larger. The stresses and strains observed are usually volumetric averages over the sampling regions in the specimen. The first level at which the microstructure is observed is the grain and precipitate level, where distances are typically measured in microns. This is the closest level to the macroscopic level. Characteristics of grains include the grain size and orientation, grain boundary properties, and associated energies. The microstructural characteristics for precipitates or particles in the lattice structure depend on size, lattice type, orientation, and particle or grain boundary conditions. Another class of structures and events occur within the grain, where distances are measured in angstrom units. These structures are typically characterized by the type of dislocation, dislocation density, dislocation arrangement, and stacking fault energy. Finally, there are other events at the atomic level that determine the properties of dislocations, grains, and precipitates. The atomic structure is important for understanding the most fundamental characteristics of the material. However, events on the atomic scale are the hardest to observe and quantify and are not generally included in constitutive modeling.

This chapter is not intended to be a complete study of physical metallurgy. The objective is to understand the physical processes resulting in inelastic deformation and to extract information for constitutive modeling on the macroscopic level. There is no attempt in this book to model individual events on the microscopic scale, or individual dislocations, grains or precipitates. Numerous models exist for the propagation of dislocations in the metallurgical literature. Use of individual dislocation models for structural calculations would require summing or integrating the effect of 10^6 to 10^{12} events per cubic centimeter that occur during deformation. There are too many events in an engineering structure for this approach to be practical.

1.2 DEFORMATION MECHANISMS

Inelastic deformation occurs in metals due to slip, climb, or twinning. When the temperature is less than approximately $0.5T_M$, where T_M is the absolute melting temperature, crystalline metals deform due primarily to the propagation of dislocations through the lattice. This results in the slip of one segment of a crystal relative to another crystal segment. At higher temperatures deformation by dislocation climb, a diffusion-controlled process, becomes more important. Twinning, a rotation of atoms in the lattice structure, occurs in crystals that do not easily permit slip deformation. Twinning is of secondary importance since the resulting strains are very small compared to slip and climb. These three types of deformation mechanisms will be examined and analyzed for their role in controlling material response and their impact on material modeling.

1.2.1 Dislocations and Slip

A perfect, defect-free crystal could support stresses much higher than those commonly observed for yield and plastic flow in metals. In the ideal crystal, plastic flow would result from sliding one plane of atoms over another by simultaneous breaking of all the metallic bonds between the atoms. However, the actual yield stress is much lower than the theoretical shear stress, due to the presence of dislocations in the lattice structure.

Dislocations are disruptions in the crystal lattice structure of the material. As shown in Figure 1.2.1, an edge dislocation is an extra plane of atoms in a crystal lattice. As a result, the crystal is severely deformed along the line where the extra plane terminates. When a shear stress is applied to the crystal lattice, the atoms in a continuous plane next to the edge of the extra plane

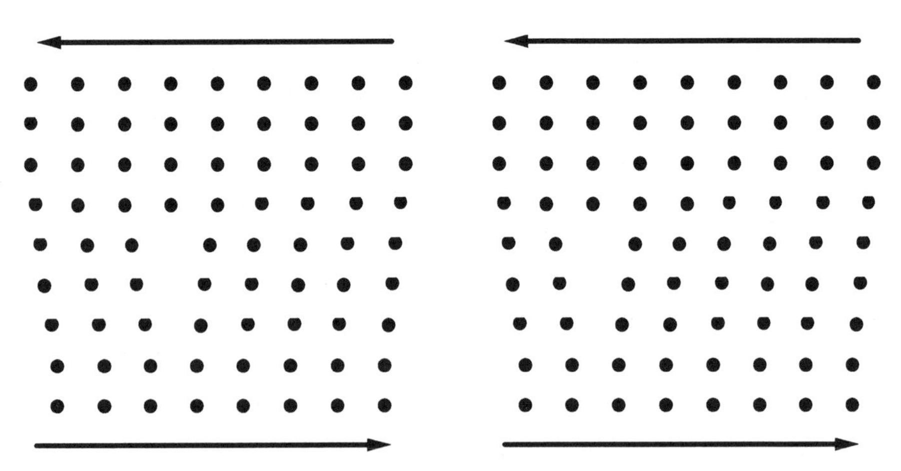

Figure 1.2.1 Propagation of an edge dislocation through a crystal lattice.

may dislodge from their current bonds and form new bonds with the atoms on the edge of the extra plane. As a result, the extra plane has moved one atomic distance along the continuous plane. Application of the stress may actually cause the dislocation to move several steps along the slip plane (plane of motion) in the crystal lattice. Thus the lattice on one side of the slip plane is displaced relative to the lattice on the other side of the slip plane. An important property of edge dislocations is that they only propagate in the direction perpendicular to the plane of the dislocation, and the dislocation line remains in one plane. This statement is valid provided the effect of diffusion is not important, as we discuss in Section 1.2.3.

The screw dislocation shown in Figure 1.2.2 represents another type of disruption that can occur in crystalline metals. In this case the crystal is sheared by the application of a shear stress such that the upper front portion of the crystal moves one atomic lattice distance to the left relative to the lower front portion of the crystal. This type of defect is a "screw dislocation" because the lattice planes of the crystal spiral around the dislocation line BC. The dislocation propagates such that the line BC moves parallel to itself; that is, the line BC can move back into the crystal to a parallel position. Upon continued shearing the entire upper portion of the crystal will move one

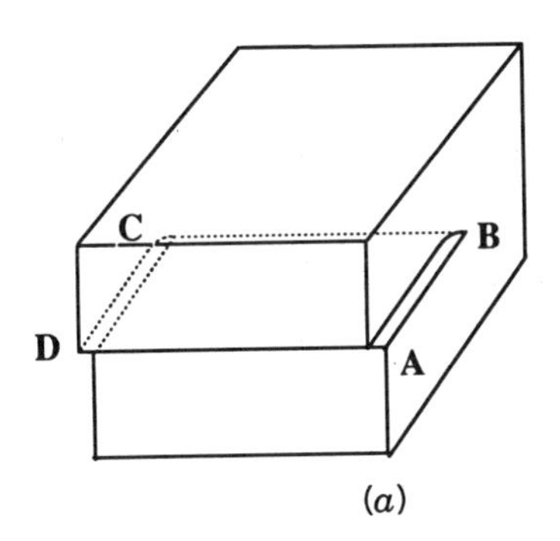

(*a*)

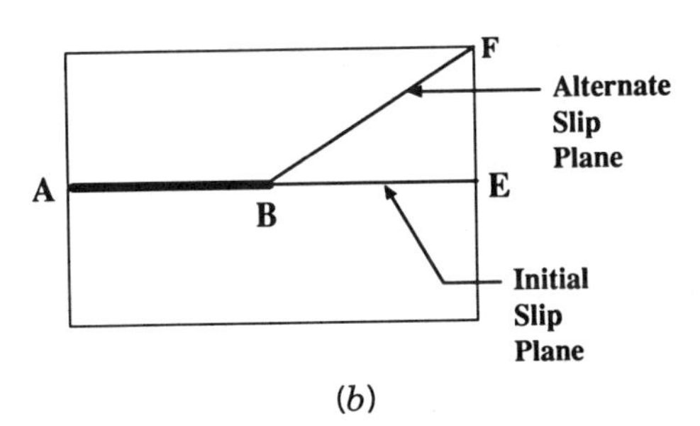

(*b*)

Figure 1.2.2 (*a*) Screw dislocation; (*b*) two slip planes passing through the leading edge of the screw dislocation.

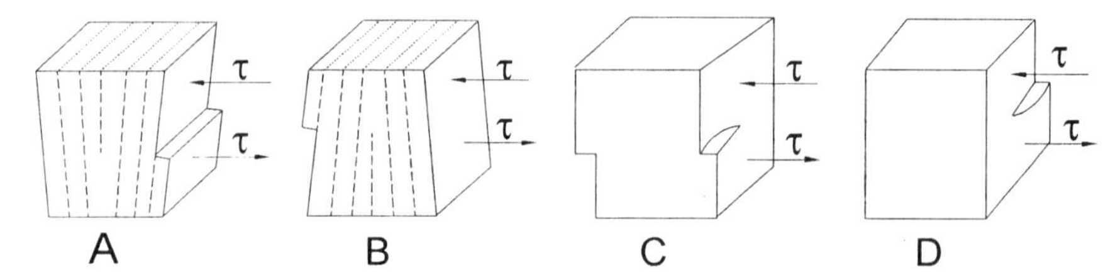

Figure 1.2.3 Definition of four dislocations: (*a*) positive edge dislocation; (*b*) negative edge dislocation; (*c*) left-hand screw dislocation; (*d*) right-hand screw dislocation.

atomic lattice distance to the left. However, a screw dislocation can move along other slip planes passing through *BC* as long as the displaced dislocation line remains parallel to *BC* at point *E* or *F*. The cross slip from one plane to another is enhanced by the presence of shear stress acting on one of the other slip planes passing through *BC*. An important difference between edge and screw dislocations is that the screw dislocations can cross slip onto another slip plane, whereas the edge dislocations always move in the direction perpendicular to the extra plane. A sketch of the propagation sequence for edge and screw dislocations is shown in Figure 1.2.3. This figure also shows the difference between left- and right-hand screw dislocations and negative and positive edge locations. In practice, edge and screw dislocations occur simultaneously in a metal lattice.

The usual method of deformation in metals, shown in Figure 1.2.4, occurs by dislocation propagation, which results in the sliding of one block of a crystal structure over another. Slip occurs in a metal crystal when an applied shear stress exceeds a critical value. The atoms move an integer number of

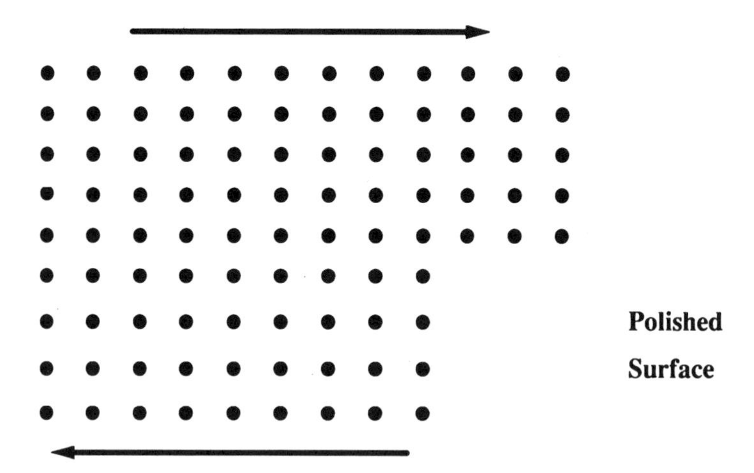

Figure 1.2.4 Planer slip in a crystal resulting slip line on the surface of a polished specimen.

atomic distances along the slip plane and the crystal is perfectly restored if the motion is uniform. As a result of the relative motion between the two crystal blocks a slip line or step is observed experimentally on the surface of a specimen if it is polished. A set of slip lines for copper is shown in Figure 1.2.5. A slip band is a group of closely spaced slip lines that appear as a single slip line at low magnification. In many metals the slip lines are wavy, confirming that screw dislocations are not constrained to move along a single plane.

Since the crystal remains a crystal after slip, there are limitations on the number of ways a crystal can slip. Often, but not always, slip in metal crystals occurs on planes of high atomic density in closely packed directions, so that the distances between atoms is minimum. Under these conditions the shear stress propagating the dislocation corresponds to the minimum stress that can produce an identical lattice structure in the slipped position. Reed-Hill (1973) has shown that slip on high-density planes in the close-pack directions corresponds to the lowest possible energy for slip. The most probable slip planes and slip directions in metals depend on the crystal structure of the metal. The three most common crystallographic structures for metals are face-centered cubic (FCC), hexagonal close pack (HCP), and body-centered cubic (BCC) as shown in Figure 1.2.6. The high-density planes are the basal planes of the hcp crystal, the cube planes of the BCC crystal, and the octahedral planes of the FCC crystal. However, slip has been observed in the FCC crystal in non-close-pack directions on the octahedral plane, and on the cube plane at high

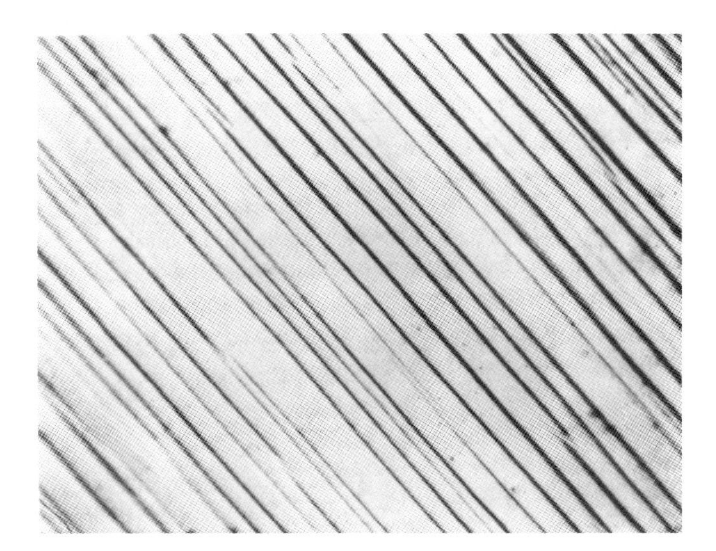

Figure 1.2.5 Straight-up slip lines in copper at 500×. (From G. E. Dieter, *Mechanical Metallurgy,* 3rd ed., McGraw-Hill, Inc., New York, 1986; reproduced with permission of McGraw-Hill, Inc.; micrograph courtesy of W. L. Phillips.)

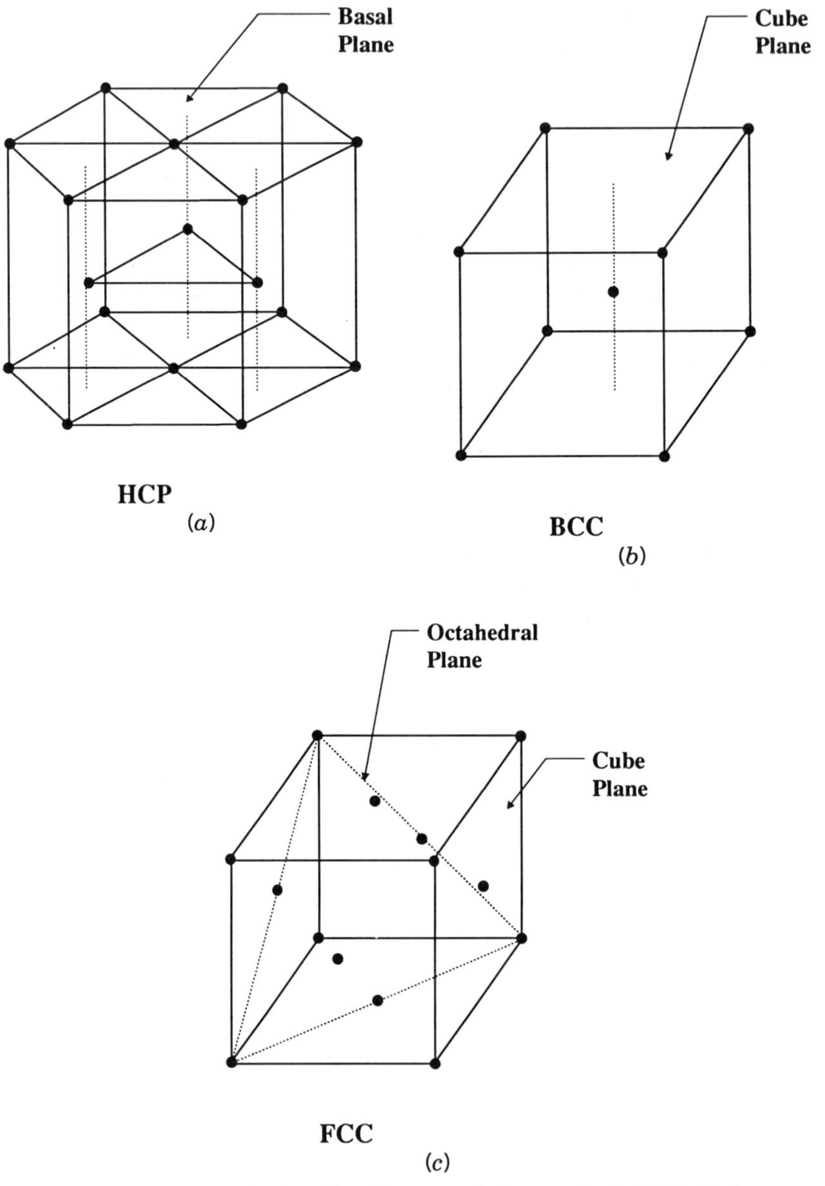

Figure 1.2.6 Examples of unit cells of hexagonal close-packed (HCP), body-centered cubic (BCC), and face-centered cubic (FCC) lattice structures.

temperature. The most probable slip directions for the FCC octahedral, BCC cube, and FCC cube planes are shown in Figure 1.2.7. Notice that the atomic structure of the HCP basal planes is the same as that of the FCC octahedral planes.

Before proceeding with the development of slip mechanisms in the crystal lattice, it is useful to introduce the Miller system of designating slip planes and slip directions. The Miller indices, which are specialized for each type of crystal, will be introduced only for the cubic system, since it will facilitate the development for FCC crystals in Chapter 8. Let $[x, y, z]$ represent the axes of a rectangular coordinate system aligned along the edges of a unit cube. The magnitude of the Miller index is adjusted so that the components are simple integers. Thus [111] is the cube diagonal, [101] is a face diagonal in the x–z plane, and [100] is along the x axis. Crystallographic planes are

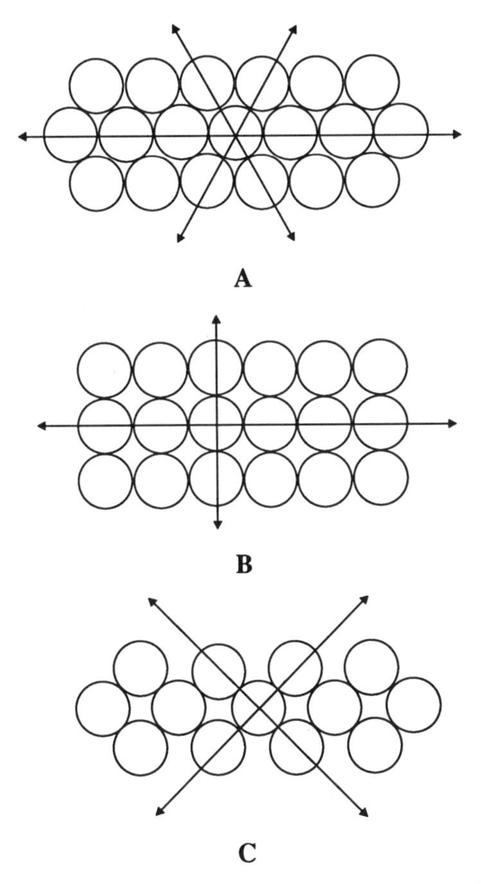

Figure 1.2.7 Slip directions are designated by the arrows on (*A*) close-pack HCP basal plane and close-pack FCC octahedral plane, (*B*) close-pack BCC cube plane, and (*C*) nonclose-pack FCC cube plane.

designated by the Miller index for the normal direction. A typical slip system would be designated as $(100)[10\bar{1}]$, where (100) defines the slip plane, $[10\bar{1}]$ defines the slip direction, and the bar indicates the negative direction. The first index always denotes the slip plane and the second the slip direction.

The three most probable slip systems for the FCC crystal shown in Figures 1.2.8 and 1.2.9 are typical of nickel-based alloys used in gas turbines and Space Shuttle main engine turbopumps. The high-density planes (Figure 1.2.8) are the four octahedral planes. Figure 1.2.8 shows two sets of 12 slip directions on the octahedral planes that are active in different materials at different temperatures. The octahedral planes therefore contain 24 possible slip directions. In addition, there are six possible slip directions in the three cube planes, as shown in Figure 1.2.9. Any one or more of the slip directions could be active, depending on the temperature and magnitude of the shear stress in each slip direction. For example, let us assume that a stress of 1000 MPa is applied to a tensile specimen with the lattice oriented in the [100] direction. The components of shear stress in each of the 30 slip directions in

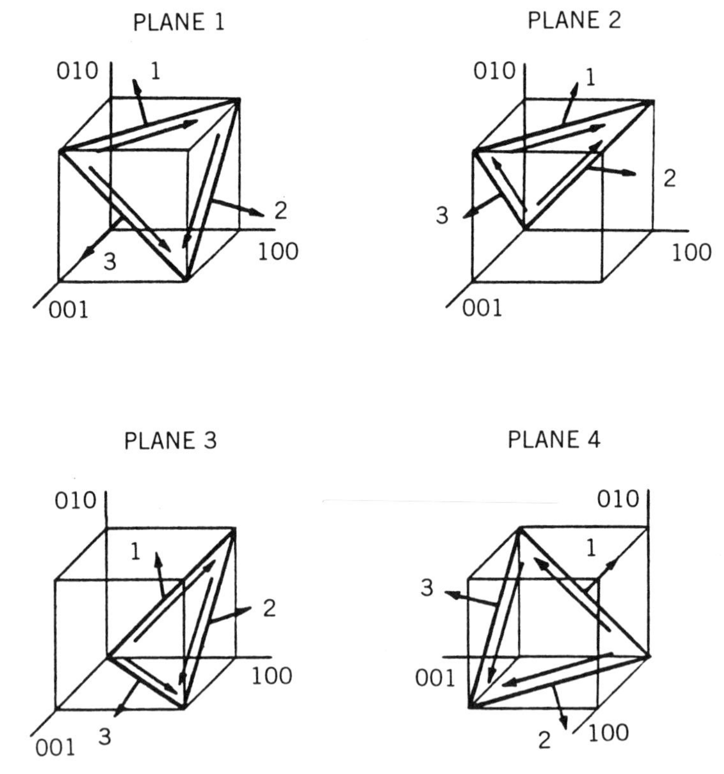

Figure 1.2.8 Four octahedral slip planes for the FCC crystal showing one set of close-pack slip directions (slip numbers 1 to 12 in Table 1.2.2) and one set of non-close-pack slip directions (slip numbers 13 to 24). (After Dame and Stouffer, 1985.)

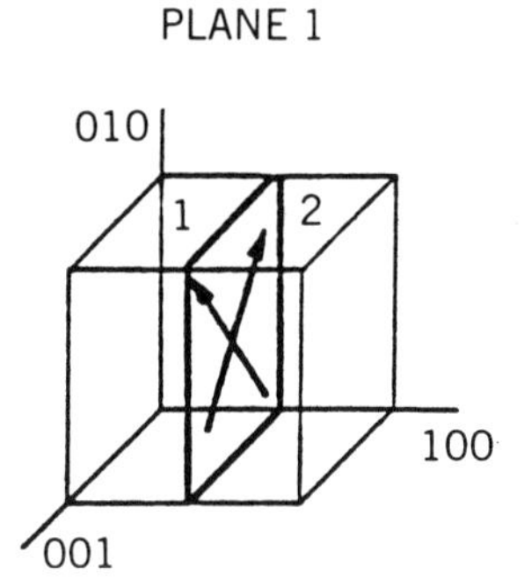

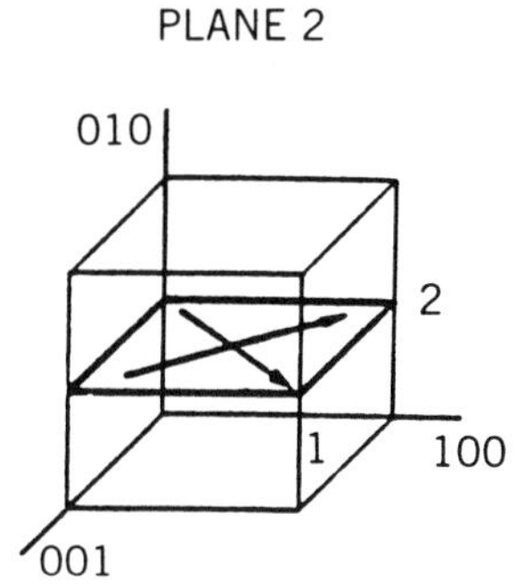

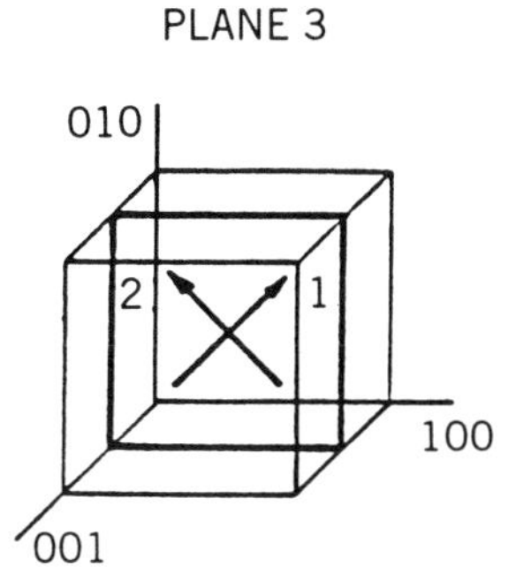

Figure 1.2.9 Thre non-close-pack cube slip planes and the slip directions for the FCC cubic crystal (slip numbers 25 to 30 in Table 1.2.2). (After Dame and Stouffer, 1985.)

Table 1.2.1 shows that the largest stresses (471.4 MPa) are for slip numbers 14, 18, 20, and 22, defined in Table 1.2.2, but slip would not occur in these directions except at rather high temperatures and low strain rates because they are not close-pack directions. Instead, slip would occur in the octahedral slip systems in the eight directions with a stress of 405.25 MPa if the crystal was loaded exactly in the [100] direction. Alternatively, if the load is applied along the [111] direction, slip occurs only in the three cube directions. If the loading is applied in the [321] direction, slip occurs in only one slip system, slip system 11 or 24, depending on temperature and material. Finally, note that if crystallographic rotation occurs during loading, the components of stress in each slip direction would change as the crystal rotates. Thus slip would transfer from one set of planes and directions to others during the loading.

Let us next investigate how slip proceeds in a single slip direction on the octahedral planes of an FCC crystal (or basal planes of an HCP crystal). Slip

TABLE 1.2.1 Distribution of the Resolved Shear Stress in Each Slip System of an FCC Crystal for a Loading of 1000 MPa in the [100], [110], [111], and [321] Directions[a]

	Specimen Orientation			
Slip No.	[100]	[110]	[111]	[321]
1	408.25	408.25	0.00	349.93
2	0.00	−408.25	0.00	−174.96
3	408.25	0.00	0.00	174.96
4	−408.25	0.00	0.00	−116.64
5	−408.25	0.00	−272.17	−291.61
6	0.00	0.00	−272.17	−174.96
7	408.25	0.00	−272.17	0.00
8	0.00	0.00	0.00	0.00
9	408.25	0.00	−272.17	0.00
10	0.00	−408.25	−272.17	−349.93
11	−408.25	−408.25	−272.17	−466.57
12	−408.25	0.00	0.00	−116.64
13	−235.70	235.70	0.00	0.00
14	471.40	235.70	0.00	303.05
15	−235.70	−471.40	0.00	−303.05
16	−235.70	0.00	−314.27	−269.37
17	−235.70	0.00	157.13	33.67
18	471.40	0.00	157.13	235.70
19	−235.70	0.00	157.13	0.00
20	471.40	0.00	−314.27	0.00
21	−235.70	0.00	157.13	0.00
22	471.40	235.70	157.13	336.72
23	−235.70	235.70	157.13	134.69
24	−235.70	−471.40	−314.27	−471.40
25	0.00	353.55	471.40	454.57
26	0.00	353.55	0.00	151.52
27	0.00	353.55	471.40	404.06
28	0.00	353.55	0.00	202.03
29	0.00	0.00	471.40	252.54
30	0.00	0.00	0.00	−50.51

[a]The correspondence between the slip number and slip directions in Figures 1.2.8 and 1.2.9 is given in Table 1.2.2. The slip number was assigned for convenience.

is complicated by the path the atoms actually take when two crystal planes are displaced relative to each other. A portion of the slip plane is shown in Figure 1.2.10*a* with an edge dislocation. The vector describing the total atomic displacement to the next lattice position is **b**. However, an easier path for slip is obtained if the total displacement is divided into two partial displacements **c** and **d** as shown in Figure 1.2.10*b*. The path **c** + **d** is a lower-energy (less work) path than **b** because there is less deformation of the lattice

TABLE 1.2.2 Correspondence between the Slip Number in Table 1.2.1 and Slip Directions in Figures 1.2.8 and 1.2.9[a]

Slip No.	Slip Plane	Slip Direction[b]
	Octahedral Slip a/2⟨110⟩{111}	
1	[1 1 1]	[1 0 −1]
2	[1 1 1]	[0 −1 1]
3	[1 1 1]	[1 −1 0]
4	[−1 1 −1]	[1 0 −1]
5	[−1 1 −1]	[1 1 0]
6	[−1 1 −1]	[0 1 1]
7	[1 −1 −1]	[1 1 0]
8	[1 −1 −1]	[0 −1 1]
9	[1 −1 −1]	[1 0 1]
10	[−1 −1 1]	[0 1 1]
11	[−1 −1 1]	[1 0 1]
12	[−1 −1 1]	[1 −1 0]
	Octahedral Slip a/2⟨112⟩{111}	
13	[1 1 1]	[−1 2 −1]
14	[1 1 1]	[2 −1 −1]
15	[1 1 1]	[−1 −1 2]
16	[−1 1 −1]	[1 2 1]
17	[−1 1 −1]	[1 −1 −2]
18	[−1 1 −1]	[−2 −1 1]
19	[1 −1 −1]	[−1 1 −2]
20	[1 −1 −1]	[2 1 1]
21	[1 −1 −1]	[−1 −2 1]
22	[−1 −1 1]	[−2 1 −1]
23	[−1 −1 1]	[1 −2 −1]
24	[−1 −1 1]	[1 1 2]
	Cube Slip a/2⟨110⟩{100}	
25	[1 0 0]	[0 1 1]
26	[1 0 0]	[0 1 −1]
27	[0 1 0]	[1 0 1]
28	[0 1 0]	[1 0 −1]
29	[0 0 1]	[1 1 0]
30	[0 0 1]	[−1 1 0]

[a]The Miller indices are a crystal-specific coordinate system designating crystallographic planes and directions. In the cubic lattice [100], [010], and [001] correspond to a set of orthogonal axes x, y, and z, respectively, oriented along the edges of the unit cube. The first vector designates the slip plane and second the slip direction.
[b]The direction expressed by the Miller indices does not represent the unit vector in the direction.

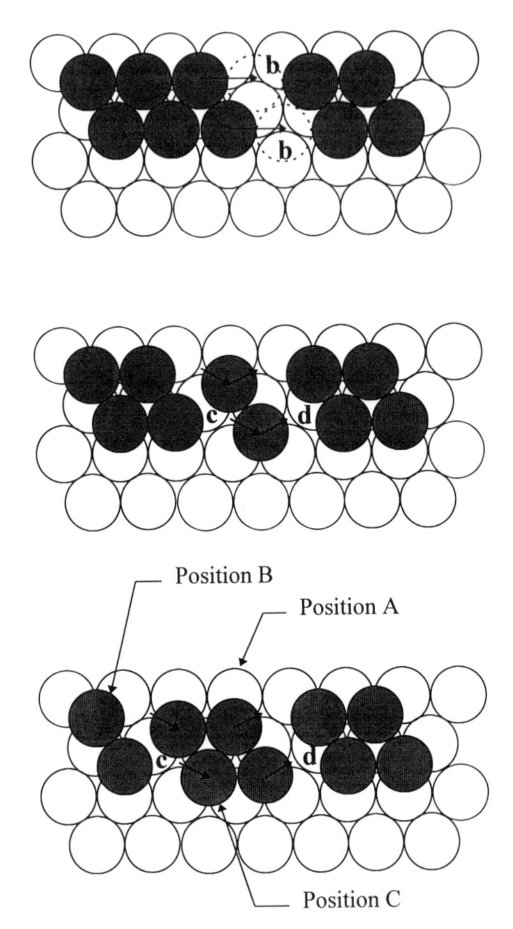

Figure 1.2.10 Top view of an octahedral slip plane in an FCC crystal showing (*a*) an edge dislocation and the total slip vector, (*b*) partial dislocations with the intermediate slip vectors, and (*c*) extended partial dislocations.

directly above and below the dislocation during slip. Since the total energy required for the slip is less with the two partial dislocations, slip occurs most often along path **c** + **d**.

When the total dislocation breaks into two partial dislocations the total strain energy of the lattice is reduced. Since the distances **c** and **d** are approximately equal, the lattice strains on each side of the partial dislocation are approximately equal, and there are two approximately equal repulsive forces keeping the atoms apart. This separation will allow other atoms to be added to the original line of atoms and form an extended dislocation, as shown in Figure 1.2.10*c*. Notice that the atoms in the core (center) of the extended dislocation have a different stacking sequence relative to the underlying plane than the atoms in the initial position. This discontinuity in the stacking sequence is a stacking fault. Since the atoms in the core are not in the position they would occupy in a perfect crystal, a stacking fault possesses an energy that is proportional to the core width. The total stacking fault energy is a function of the distortion energy or bond strength of the lattice itself, and it controls the size or core width of the extended dislocation. The repulsive force between the atoms in positions B and C decreases as the core width increases. Conversely, the stacking fault energy increases as the core width increases. Thus the core width represents a balance between the stacking fault energy and the repulsive energy between the atoms in the extended partial dislocation. Calculations show that the core width is on the order of one row of atoms for high-stacking fault energy materials like aluminum, and 12 atomic spacings for low-energy materials such as copper.

In an ideal setting, slip occurs through the movement of a pair of partial dislocations, separated by a finite distance. The first partial dislocation changes the stacking order as it moves, and the second partial dislocation restores the crystal structure. One portion of the crystal is displaced by the atomic distance **b** relative to the other by the passing of the two partial dislocations. In practice, the actual movement of an extended dislocation is complicated by the presence of other dislocations or alloying elements in the crystal lattice. At high temperature, diffusion can also disrupt the motion.

High-stacking fault energy materials are more prone than low-stacking fault energy materials to cross slip. Recall that cross slip can occur when two or more planes have a common slip direction. In FCC crystals there are common slip directions on the octahedral and cube planes. For example, the first slip direction on plane 1 in Figure 1.2.8 is parallel to the second slip direction on plane 2 in Figure 1.2.9. Figure 1.2.11 shows that in cross slip the slip surface is not a single surface but is made up of segments on the cube and octahedral planes. In an extended dislocation, the atoms in the stacking fault are not in a lattice position consistent for cross slip from the octahedral to the cube plane. Since additional energy is required to create cross slip in an extended dislocation, it should be much easier for cross slip to occur in materials with a large stacking fault energy or small core width.

There are two closing comments. First, the critical resolved shear stress is the shear stress on the slip plane in the slip direction that initiates slip. It corresponds to the value of the yield stress in the ordinary stress–strain curve. It is derived by projecting the yield stress onto the slip plane in the slip direction. In polycrystalline metals with a large number of grains at arbitrary orientation, the slip occurs on the slip planes and slip directions oriented closest to the maximum shear stress that results from the external loading conditions. For example, in a tensile test slip occurs on the slip planes oriented closest to 45° from the tensile axis. Second, not all slip occurs by the disassociation of dislocations into partial dislocations. Figure 1.2.7 shows that total dislocations can slip on the BCC and FCC cube planes in close-pack slip directions.

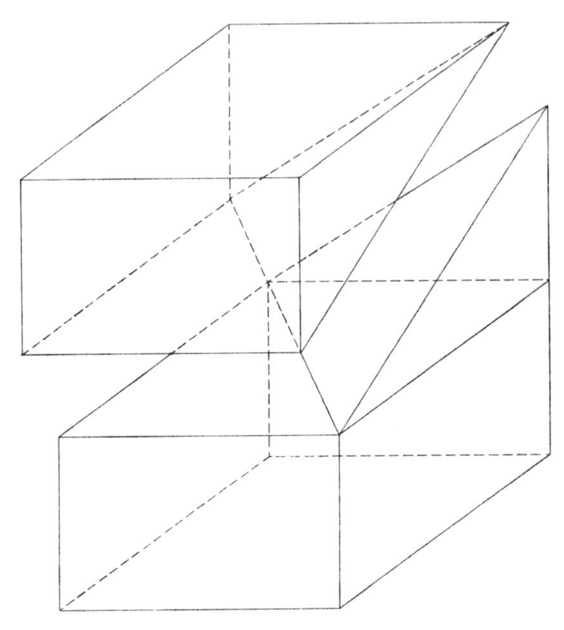

Figure 1.2.11 Cross slip on the octahedral and cube planes in the FCC lattice system.

1.2.2 Deformation Twinning

Twinning is a deformation mechanism that represents reorientation or rotation of a portion of the crystal lattice. Figure 1.2.12 contains a schematic diagram of a crystal structure with a twin after the application of a shear stress. The twin is formed by the rotation of each atom about an axis through the center of the atom. The twin plane is the plane of symmetry. It is perpendicular to the plane of the figure and separates the twinned and undeformed regions. Figure 1.2.13 shows a twin region in a Monel alloy. Twinning occurs very rapidly, as in a "snap-through" mechanism and can produce a loud click. Twinning during a tensile test produces serrations or jumps in the tensile curve.

The individual atoms in the deformed region are shifted very little compared to those in the undeformed region. Thus the inelastic strain resulting from twinning is very small compared to slip, where the lattice can be shifted several atoms. The resulting crystal structures are also much different after deformation by slip and twinning. The orientation of the regions on both sides of the slip plane are the same after slip, but after twinning the orientations are different. Slip involves atoms in a discrete number of slip planes, whereas twinning occurs uniformly over a block of the crystal. Usually, only a small fraction of total volume of a crystal is reoriented by twinning.

Twins are formed in BCC and HCP metals at low temperature with rapid loading. They occur on specific crystallographic planes in each crystal. Twin-

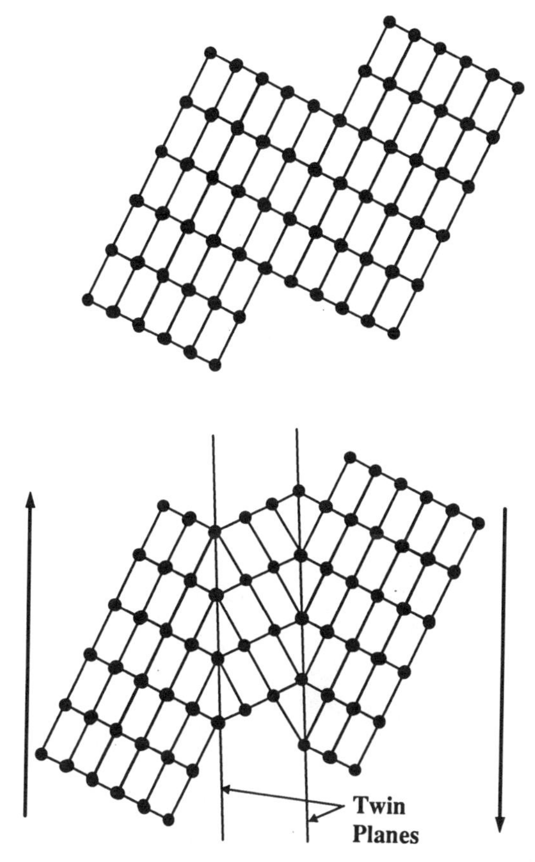

Figure 1.2.12 Representation of a twin.

ning generally occurs when the slip systems are restricted or when the shear stress rises above the critical resolved shear stress required for slip. Twinning is important at low temperature, where the stress to produce slip is relatively high. At high temperature the critical stress required to produce twinning is larger than the critical resolved shear stress for slip. Twinning occurs primarily in BCC and HCP crystals because there are a limited number of slip systems, thereby providing more opportunity for a restricted slip region. Twinning is not common in FCC metals that possess numerous slip systems. Twinning can enhance the opportunity for slip by rotating the crystal structure into a more favorable orientation for slip.

1.2.3 Dislocation Climb

An edge dislocation generally propagates in the slip plane, but under special conditions it can move in a direction perpendicular to the slip plane. Dislo-

Figure 1.2.13 Twin region in Monel fatigued at room temperature (After Strublington and Forsyth, *Journal of the Institute of Metals,* Vol. 56, pp. 90, 1957. Reproduced courtesy of the Institute of Materials.)

cation climb occurs by the diffusion of atoms and vacancies to or away from the site of the dislocation. Positive climb occurs when atoms are removed from the dislocation plane by the diffusion of an atom to a vacant lattice site. This results in the dislocation moving up one lattice distance, as shown in Figure 1.2.14. Simultaneous diffusion of an atom and vacancy can also occur if there is not a vacant site close to the dislocation. It is possible for an atom to break away from a dislocation and move to an interstitial site, but this is

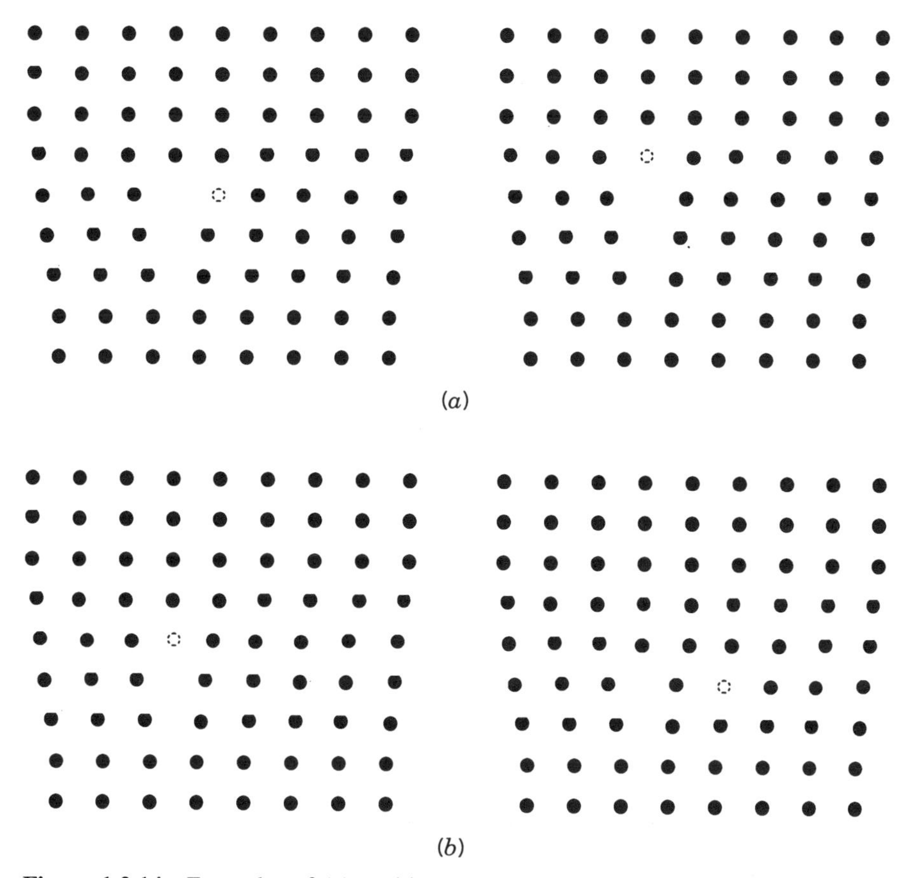

Figure 1.2.14 Examples of (*a*) positive and (*b*) negative dislocation climb resulting from the diffusion of atoms or vacancies to or away from an edge dislocation.

not as energy efficient as diffusion to a vacancy. In negative climb, atoms are added to the dislocation and the dislocation moves down one lattice distance.

Since climb is a diffusion process, the importance of climb as a deformation mechanism increases as the temperature increases. The superposition of stress on a material at high temperature increases the rate of climb. A compressive stress in the slip plane increases the probability that an atom will diffuse from the dislocation to a vacancy at another lattice site. Conversely, a tensile stress in the slip plane will expand the lattice and make it easier for an atom to be added to the dislocation.

The addition or subtraction of atoms to the extra plane of an edge dislocation will result in jogs in the dislocation line since diffusion of atoms is local and not expected to be uniform. Thus the dislocation line may lie in several slip planes simultaneously. This will result in an increase in the shear stress required to produce slip. Thus at high temperature two mechanisms contribute to the observed inelastic strain, slip, and climb. As the temperature increases the potential for climb increases and the potential for slip decreases.

Finally, climb is not possible with a screw dislocation since there is no extra half-plane of atoms. Diffusion of atoms or vacancies to the dislocation line of a screw dislocation does not make the dislocation move. However, an increased temperature will increase the energy at the dislocation and allow slip to occur at a lower critical resolved shear stress.

1.3 DISLOCATION INTERACTIONS

A major factor that controls the movement of dislocations is the intersection of dislocations with each other. As the network of dislocations increases, the complexity of the microstructure increases. Subsequent propagation of a dislocation through a network of dislocations requires a higher shear stress to maintain motion. An increase in the resistance to slip corresponds to strain hardening.

1.3.1 Dislocation Pileups

One of the most fundamental concepts to explain strain hardening is the interaction between individual dislocations in the same plane when they pile up at a barrier in the crystal lattice. The pileups produce a back stress opposite to the applied shear stress on the slip plane.

The concept of back stress is motivated by the interaction of mobile dislocations with a barrier such as a grain boundary, precipitate, or other dislocations. Consider the propagation of dislocations on a slip plane, in the preferred slip direction, under the action of a shear stress in the slip direction. The mobile dislocations will be pinned or stopped at a barrier. A second dislocation, following the same path, will encounter the pinned dislocation and be repelled by forces at the atomic level (Figure 1.3.1). At high homologous temperature (absolute temperature divided by the absolute melting temperature) slip in the reverse direction can result if the applied stress is reduced or removed. Thus the net force producing slip is the difference between the shear stress and repulsion or back stress due to dislocation interaction.

The distribution of edge dislocations in pileups at various barriers can be observed experimentally by transmission electron microscopy (TEM). An ex-

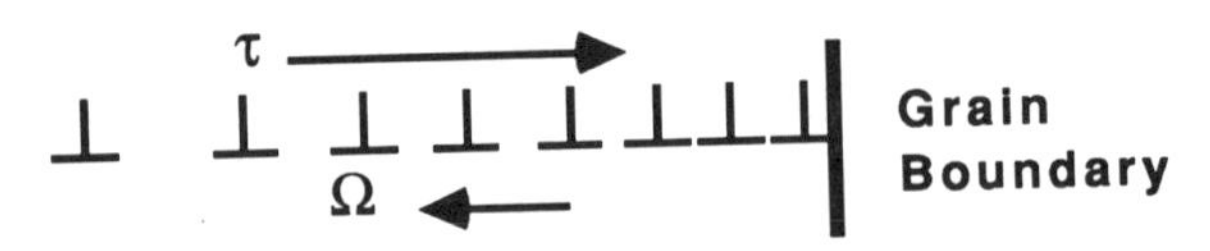

Figure 1.3.1 Edge dislocations piled up at a barrier.

ample of a dislocation pileup is shown in Figure 1.3.2. The dislocations appear as sets of short lines in the micrograph since the slip planes are inclined relative to the plane of the TEM foil and only a small portion of each dislocation is captured in the foil. The magnitude of the distance between dislocations in a slip plane varies inversely to the magnitude of the interatomic forces. Dislocation arrangements in larger pileups are much more complicated due to the presence of internal stress gradients from other dislocations. Hirth and Lothe (1982) reported that in many cases, including pileups containing a large number of dislocations, the approximation for a continuous distribution of pileups is about as good as the approximation for the discrete dislocation pileups. This observation is critical in extending the back stress concept to loading cases with high dislocation density.

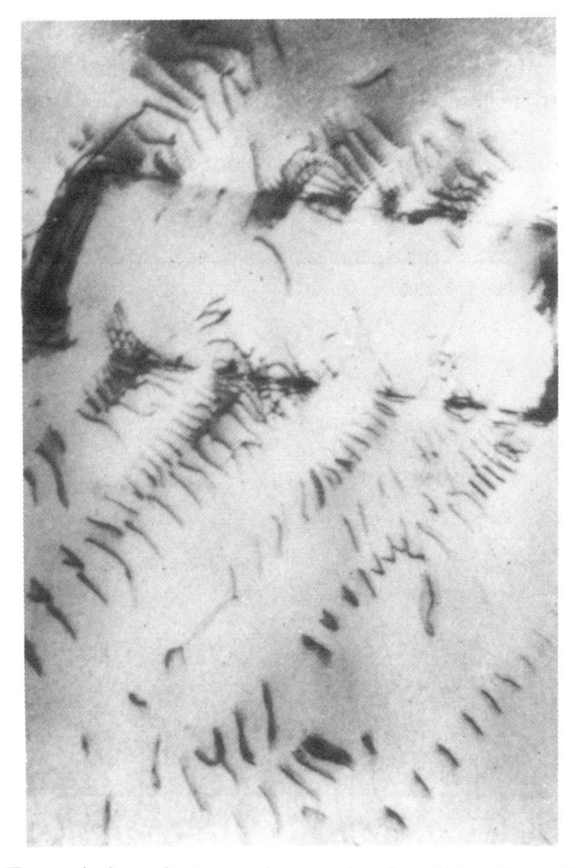

Figure 1.3.2 Transmission electron micrograph of a dislocation pile up at twin boundaries in a Cu–4.5% Al allow deformed 5%. (After Swann and Nutting, *Journal of the Institute of Metals,* Vol. 50, 1961. Reproduced courtesy of the Institute of Materials.)

It is possible to see the effect of back stress in the macroscopic material response of metal at high temperature. As an example, consider a single-crystal superalloy with large precipitates in the lattice structure. In this material the precipitates have a continuous lattice structure with the matrix, but the position of atoms in the lattice is different, creating an antiphase boundary at the precipitate–matrix interface. During tensile loading dislocations pile up at the antiphase boundary. Since the net stress on a slip plane is the difference between the shear stress and back stress, unloading to zero stress after plastic straining could produce negative inelastic flow. Using this hypothesis, two specially designed mechanical tests on nickel-based single-crystal alloy René N4 were conducted to verify the existence of back stress.

Figures 1.3.3 and 1.3.4 show the results of double tensile tests on cubic crystal specimens in the [100] (edge) and [111) (diagonal) orientations, respectively, with a 120-s hold at zero stress. In both tests, samples were first loaded to 1.5% strain at a strain rate of 1×10^{-4} sec^{-1}, unloaded immediately to zero stress within 10 s, and then reloaded at a higher strain rate of 6×10^{-4} sec^{-1} following the 120-s hold period. The higher strain rate was also used to demonstrate the effect of strain rate on the material response that is discussed in Section 1.4.2. Shown in Figure 1.3.5 are inelastic strain-time histories during the 120-s hold period for both samples. A significant amount

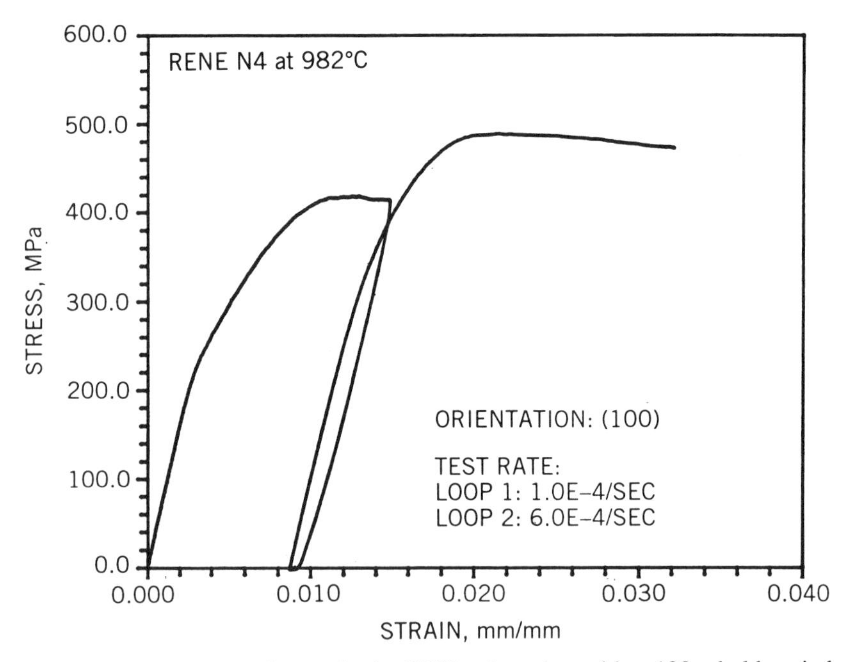

Figure 1.3.3 Double tensile test in the [100] orientation with a 120-s hold period at zero stress. René N4 is an FCC single-crystal alloy with a two-phase structure similar to that shown in Figure 8.1.3. (After Sheh and Stouffer, 1988.)

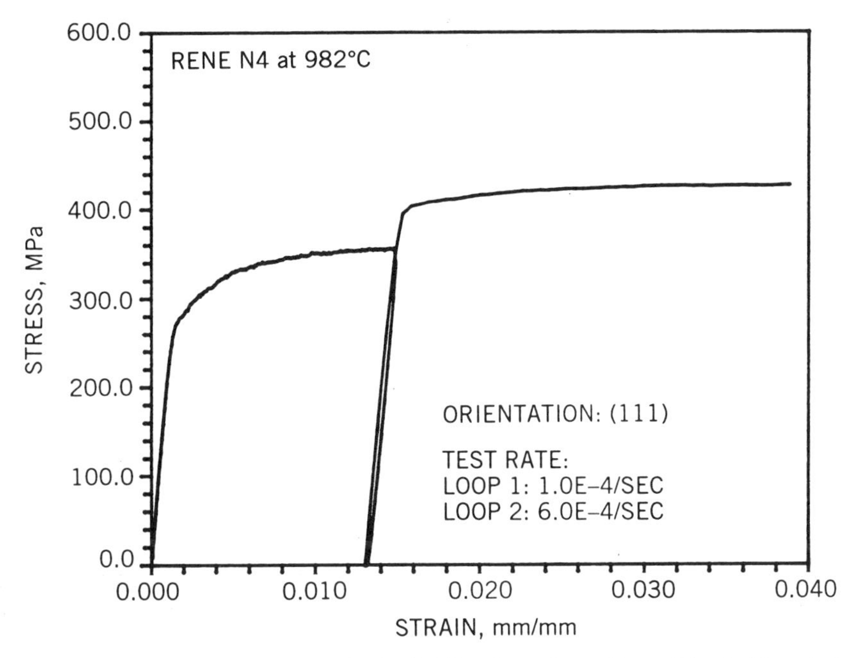

Figure 1.3.4 Double tensile test in the [111] orientation with a 120-s hold period at zero stress. (After Sheh and Stouffer, 1988.)

of reverse flow occurred during the hold period for the [100] sample, whereas the reverse flow is minimal for the [111]-oriented specimen. These results clearly demonstrate the existence of back stress and that it is orientation dependent in single crystals.

1.3.2 Dislocation Intersections

Back stress results from interaction between dislocations in the same plane. But since several slip planes may be active in a crystal at one time, dislocations propagating on different planes may intersect at various angles. Thus a dislocation must cut through other dislocations as it moves through the lattice. Forcing a dislocation through another dislocation requires an element of work and can produce discontinuous dislocations which require more energy to propagate. Thus hardening, or an increased resistance to slip, results from dislocation intersections.

A simple intersection of one dislocation with another can create two types of discontinuities: jogs and kinks. A jog is a sharp discontinuity in a dislocation that is oriented out of the slip plane as shown in Figure 1.3.6*a*. A kink is a sharp discontinuity in a dislocation that lies entirely in the slip plane, as shown in Figure 1.3.6*b*. Jogs occur when dislocation velocity vectors of the two intersecting edge dislocations are not parallel, whereas kinks occur when

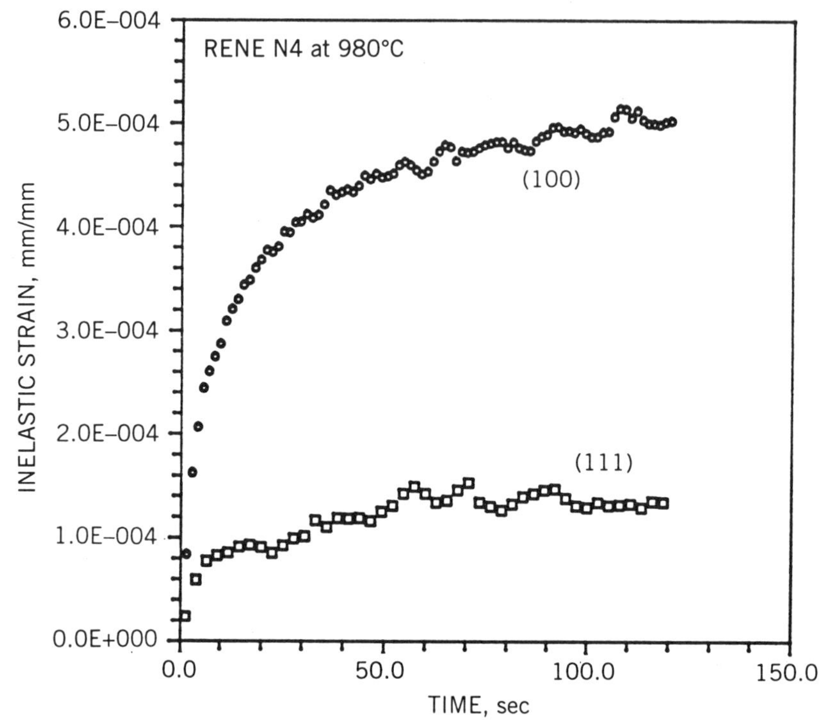

Figure 1.3.5 Inelastic strain-time histories for the [100] and [111] orientations during the 120-s hold period in the double tensile test. (After Sheh and Stouffer, 1988.)

they are parallel. Kinks in edge dislocations are unstable since during glide they can line up and eliminate the offset. The intersection of a screw dislocation and an edge dislocation can produce jogs or kinks depending on orientation, but the intersection of two screw dislocations produces jogs in both screw dislocations. The latter is the most common type of intersection occurring in plastic deformation.

The results of dislocation intersections in a real metal can be very complex. Various intersections at different angles between edge and screw dislocations can produce extensive disruption of the lattice, or in some very special circumstances no disruption at all. The resulting disruption from jogs and kinks can be permanent or temporary. Jogs can also create vacancies in the lattice as they propagate, or they can displace atoms to interstitial positions in the lattice. Intersection of dislocations with jogs can create dislocations with "super jogs" that pin or stop the dislocation at local points in the lattice structure.

1.3.3 Dislocation Cells and Subgrains

Strain hardening due to dislocation interaction is an even more complicated phenomenon when it involves large groups of dislocations. The dislocation

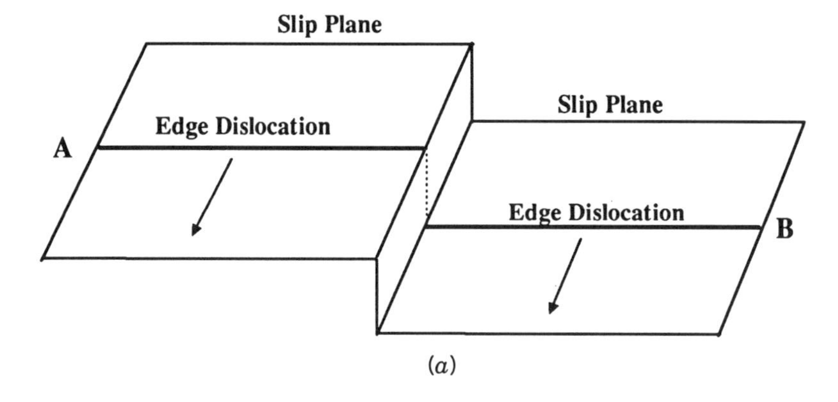

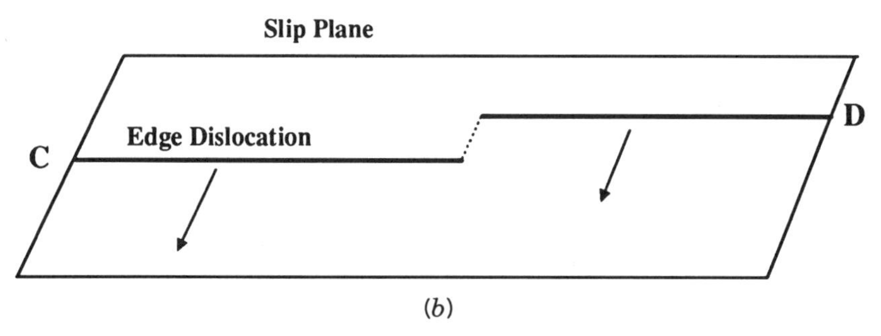

Figure 1.3.6 (*a*) a jog; (*b*) a kink. Lines *AB* and *CD* are edge dislocations normal to the slip planes shown. A jog is a sharp break a few atoms long in the dislocation line out of the slip plane, but a kink is in the slip plane.

density in an undeformed metal can grow from 10^6 cm of dislocation lines per cubic centimeter to 10^{12} cm/cm^3 in a severely deformed metal. This growth in the number of dislocations and the development of the dislocation microstructure in very important to the mechanical properties observed during deformation, as shown in Figure 1.1.1.

One explanation of the source of dislocations in a deforming metal was presented by Frank and Read (1950). In their proposal a dislocation line *a* in Figure 1.3.7 leaves the slip plane at points *P* and *Q*. Points *P* and *Q* are immobilized by the intersection with other dislocations or defects in the lattice structure. A shear stress in the slip plane will cause the dislocation line between points *P* and *Q* to bulge out (lines *c* and *d*) and eventually form a loop. When the loop meets itself, Line *d*, the dislocations at the intersection annihilate each other and form a large loop *e* and another dislocation line *P* and *Q*.

Frank–Read sources have been observed experimentally, but they are not the most common source of dislocations. Other sources which are less understood contribute to the total number of dislocations. For example, a portion

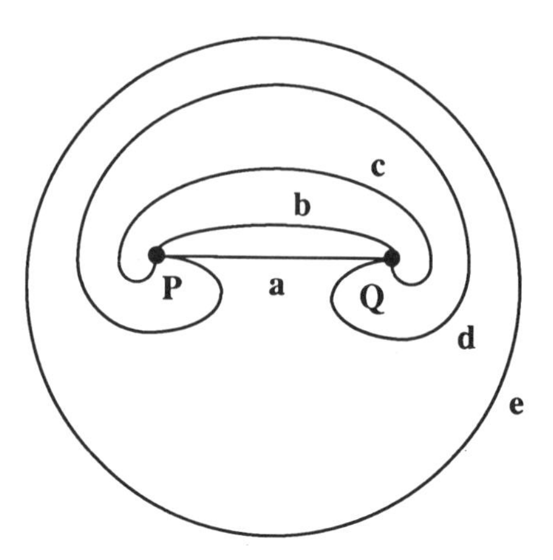

Figure 1.3.7 Frank–Read source showing the sequential development (*a*, *b*, *c*, *d*, *e*) of a dislocation loop from dislocation that is pinned at points *P* and *Q*.

of a screw dislocation in a slip plane could be pinned by the portion not in the slip plane. A shear stress in the slip plane can cause rotation about the pinned segment, producing a spiral slip step. Spiraling around the pinned segment increases the total length of the dislocation line and increases the complexity.

The increase in the number of dislocations clearly affects the response. Taylor proposed in 1931 that one of the most important paramaters affecting hardness and strength is the dislocation density. If dislocations were randomly distributed in plastically deformed metals, no other parameters would be necessary. However, dislocations are not distributed uniformly in a deformed material. Dislocations tend to cluster during plastic deformation and cell structures or subgrains form. The cell or subgrain properties affect the mechanical response.

The development of a cell structure is shown in Figure 1.3.8 for austenitic stainless steels. The relatively dislocation free initial structure evolves into a complex network of dislocations due to cycling. The dislocation density increases up to some critical level and then takes on a three-dimensional modulated dislocation microstructure. The structures appear as small cells that have a size that is inversely related to the shear stress. The maximum state of hardening in the type 304 stainless steel was reached at about 10% of the fatigue life (cycle 80). The cell size also reached the saturation value at cycle 80. The additional energy of the deformation process (remaining 90% life) is attributed to both the propagation of a fatigue crack and supplying the driving force for the cell rotation (increasing misorientation angle). Moteff (1980) estimated the fatigue crack also nucleated at about cycle 80, the point in the test at which the dislocation cells reached an equilibrium size. This appears to be one mechanism that produces crack initiation in low cycle fatigue.

The distinction between cells and subgrains is shown in Figures 1.3.9 and 1.3.10. Cells are bounded by thick fuzzy walls with high dislocation density.

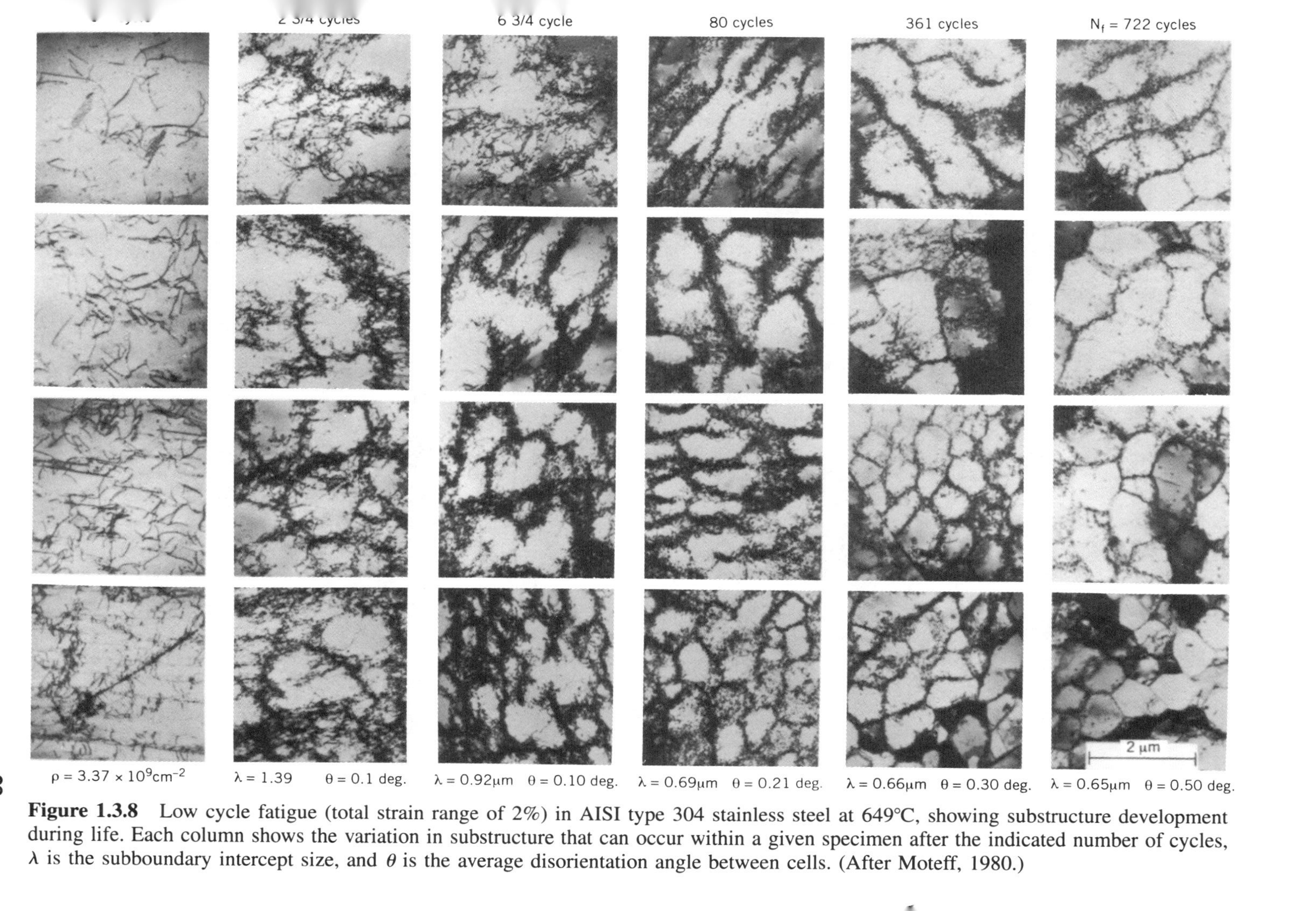

Figure 1.3.8 Low cycle fatigue (total strain range of 2%) in AISI type 304 stainless steel at 649°C, showing substructure development during life. Each column shows the variation in substructure that can occur within a given specimen after the indicated number of cycles, λ is the subboundary intercept size, and θ is the average disorientation angle between cells. (After Moteff, 1980.)

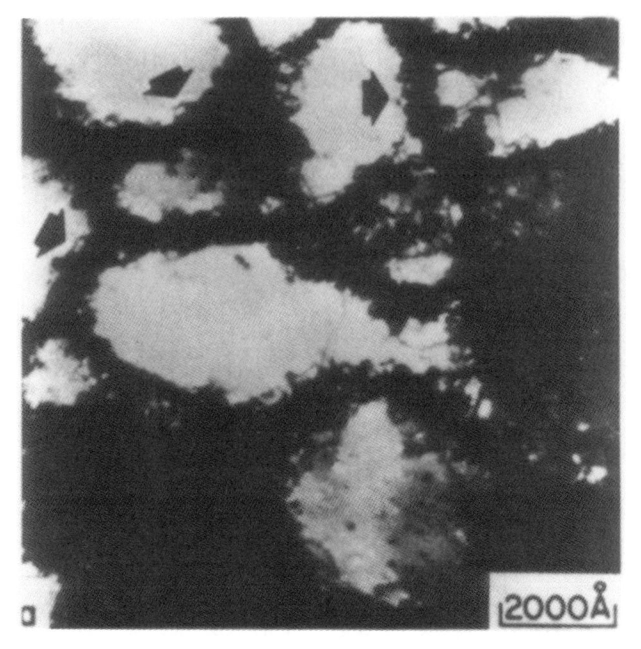

Figure 1.3.9 Cell structure formed in pure nickel after one torsional cycle at room temperature. (From Nix and Gibeling, 1983; photograph courtesy of A. W. Thompson.)

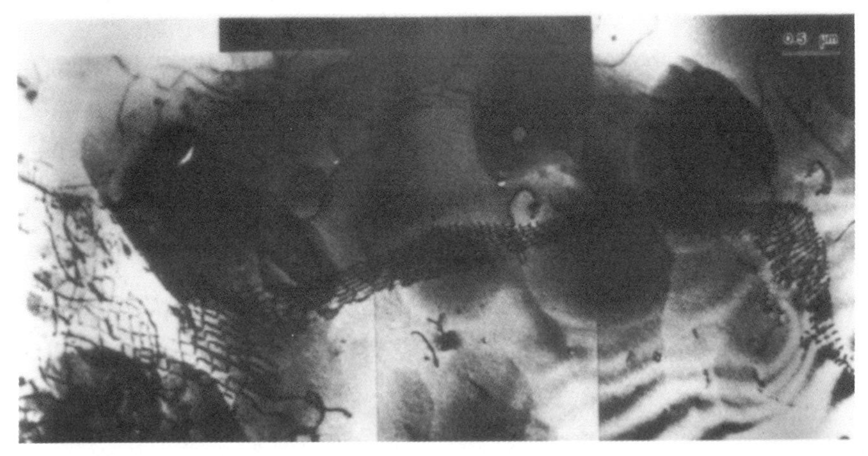

Figure 1.3.10 Subgrains in copper deformed to 9.4% at 1300°F. (From Nix and Gibeling, 1983; photograph courtesy of A. R. Pelton and D. L. Yancy.)

The cell walls contain dislocations with alternating signs such that the misorientation of the lattice is small for the large number of dislocations present. Subgrains are bounded by thin neat walls containing dislocations of mostly one sign. Thus a large misorientation develops across the subgrain walls. Cells are observed at low temperature, whereas subgrains form at high temperature. Both cells and subgrains can be present at intermediate temperatures. The development of subgrains and cells is a recovery mechanism that is discussed next.

1.4 RECOVERY MECHANISMS

Recovery is associated with a reduction in the number of dislocations, a reduction in the stress required to propagate dislocations, and a reduction in the local energy level. The recovery mechanisms in metals are very important, even at low temperature, and can have a profound effect on the observed mechanical response. This section contains a description of static recovery mechanisms occurring when no deformation is present and dynamic recovery that occurs during deformation.

1.4.1 Static Recovery

Static recovery occurs when there are no applied loads, and the movement of dislocations occurs primarily from the interaction stresses between the dislocations themselves. Static recovery is more important at high temperatures when diffusion is present and dislocation mobility is assisted by thermal energy.

The simplest form of recovery occurs in a crystal lattice when it is deformed by slip on a single set of parallel planes. (This mode of deformation does not include activation of multiple slip or bending of the crystal planes, which results in the formation of subgrains or cells.) Recovery by simple slip is achieved by the annihilation of dislocation segments with opposite sign (i.e., positive and negative edge or screw dislocations). For example, two edge dislocations can combine to form a complete continuous plane of atoms. The annihilation of dislocations can be observed experimentally in a single crystal deformed by simple slip. If a specimen is loaded into the plastic range, unloaded, and then reloaded immediately, the material would yield on reload near the stress level achieved just before unloading. But if the specimen is unloaded and held at zero load for a few minutes or longer before reloading, the yield stress on reload will be less than yield in the first case. The reduction of the yield stress on reloading depends strongly on the time and temperature at zero load.

A recovery process in crystals deformed by bending is different. Assume for convenience that a crystal is deformed by pure bending such that the deformation occurs by single slip and that the lattice has an excess of positive (or negative) edge dislocations on the slip planes after deformation. This configuration, shown in Figure 1.4.1*A*, has a relatively high strain energy density. Another configuration containing the same number of dislocations but with a lower state of strain energy is shown in Figure 1.4.1*B*. In this case the dislocations are arranged in a line normal to the slip direction. In the first configuration, when edge dislocations of the same sign are randomly distributed on a slip plane, the strain energy in the lattice due to each dislocation is cumulative. However, since the local stress field in the region around each positive edge dislocation is in compression above the slip plane and tension below the slip plane, the total lattice distortion and strain energy density are much less in Figure 1.4.1*B*. The rearrangement of dislocations from Figure 1.4.1*A* to 1.4.1*B* is polygonization. Since polygonization requires both planar slip and dislocation climb, the rate of recovery is enhanced by increasing the temperature. At low temperature the local forces between the dislocations are present, but polygonization does not occur because the dislocation climb is negligible.

In addition to lowering the strain energy, the dislocations form a low-angle boundary during polygonization. The lattice segments between the boundaries in Figure 1.4.1*B* are nearly flat and stress free. Each crystal segment has a different orientation than its neighbors because of the low-angle boundary. These lattice segments are classified as subgrains that lie inside grains of a

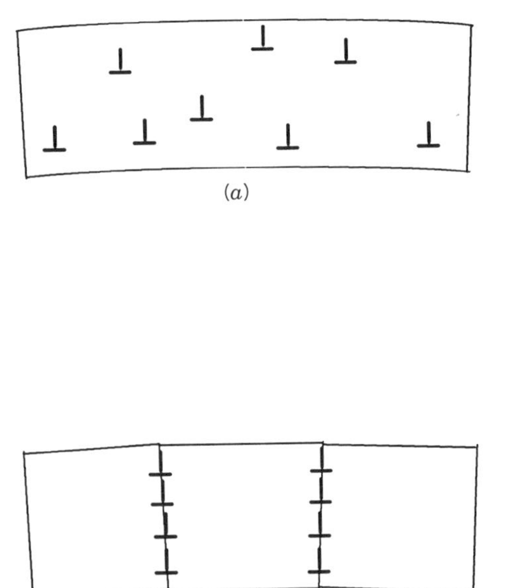

Figure 1.4.1 Position of dislocations (*a*) before and (*b*) after polygonization.

polycrystalline metal. After the initial formation of the low-angle boundaries, there is a coalescence of boundaries to form a smaller number of well-defined boundaries. This process continues as long as the total strain energy is reduced. Larger, well-defined subgrains develop during coalescence of the low-angle boundaries. Polygonization occurs in polycrystalline metals as well as single crystals, but the process is complicated by multiple-slip and complex lattice curvature.

Static recovery is different at high and low temperatures. Polygonization occurs by climb; thus a relatively high temperature is required for rapid polygonization. Therefore, static recovery in deformed polycrystalline metals at high temperature occurs primarily by polygonization and the annihilation of dislocations of opposite sign. At low temperature, static recovery is much less important than at high temperature, and the recovery processes are different. Most theories consider static recovery at low temperature as reducing the number of point defects or vacancies to their equilibrium level. Even though vacancies are mobile at low temperatures, the rate of static recovery is very slow.

1.4.2 Dynamic Recovery

Static recovery occurs after plastic deformation due to the interaction forces between the dislocations. The dislocations tend to form a low-energy dislocation substructure. Dynamic recovery occurs simultaneously with deformation and the resulting substrucutures are similar.

Dynamic recovery results from the formation of two types of substructures: subgrains at high temperature and dislocation cells at low temperature. The size of subgrains formed during deformation is generally similar to those formed in static recovery. The rate of subgrain development is enhanced by the presence of stress, and the substructure is complicated by slip in several slip systems.

The tendency for dislocations to form a cell substructure during deformation is quite strong and can even occur in pure metals deformed in liquid nitrogen (Longo, 1970). At higher temperatures the effect of dynamic recovery is stronger because of the increase in mobility of vacancies and dislocations with temperature. The majority of dislocations are arranged in complex cell walls separated by regions of low dislocation density. The dislocation mechanism for the formation of cell walls is quite complex; however, it is well documented that it occurs more strongly in metals with a high stacking fault energy. The stacking fault energy limits the size of the extended dislocations and promotes cross slip of dislocations from one plane to another. Cell formation can be eliminated in alloys that have a low stacking fault energy. Since the dislocations move from the slip plane to form dislocation walls, the average strain energy stored in the lattice is lower and the presence of cells makes slip easier. The presence of cell walls also serves as a site to nucleate dislocations and enhance the opportunity for slip.

Through the development of cells or subgrains, dynamic recovery acts to lower the dislocation density, distortion, and strain energy density of the lattice structure. Thus the effective rate of strain hardening is reduced by dynamic recovery. Strain hardening and dynamic recovery are competing mechanisms during deformation.

The underlying difference between static recovery and dynamic recovery is important. In static recovery the movement of dislocations occurs from the interaction stresses between the dislocations themselves. In dynamic recovery, the applied stress producing the deformation is added to the interaction stresses between the dislocations. As a result, dynamic recovery can occur at lower temperatures than static recovery.

1.5 GRAINS AND STRENGTHENING MECHANISMS

A physically based understanding of metal behavior should also include the effect of grains and precipitates on the mechanical properties and deformation. In general, the presence of obstacles pin or slow the dislocation propagation and thereby increase the stress required to produce deformation. Thus the presence of grains and precipitates usually strengthens or hardens the metal but in some situations actually promote deformation.

1.5.1 Grains

Grains arise from the nucleation of crystallites at multiple sites in a melt and the growth of these crystallites form a polycrystalline solid. The boundaries between grains are regions of disturbed lattice a few atomic distances wide. The crystallographic orientation can change abruptly across grain boundaries, and a high-angle grain boundary is a region of random misfit between neighboring grains. Dislocations trapped in the misfit are generally not mobile but serve as effective sources of dislocations in the grains. Grain boundaries are preferential sites for diffusion, phase transformations, and precipitation reactions. There are generally thousands of randomly oriented grains in a commercial alloy. Grain sizes commonly range from a few microns to about a millimeter; however, grains sizes on the order of centimeters are found in directionally solidified metals.

Grain boundaries are important in controlling the properties of polycrystalline metals. At low temperatures grain boundaries are usually very strong and do not weaken during deformation. At temperatures above one-half of the melting temperature, deformation can occur by grain boundary sliding. Grain boundaries lose their strength at high temperatures and slow strain rates, leading to large strain and fracture along the grain boundaries. Grain boundary properties are central to the discussion of creep mechanisms presented in Chapter 3.

During deformation the restraint produced by discontinuity at the boundary can cause considerable differences in the strain in adjoining grains. Studies on coarse-grain aluminum by Boas and Hargraves (1948) showed that the strain near the grain boundary was much more severe than in the center of the grain, and the strain was continuous across the grain boundary. As the grain size decreases, the strain throughout the grain becomes more uniform. The constraints of the grain boundary can cause several slip systems to be active, and slip can occur even on non-close-packed planes. Since more slip systems are active at the grain boundaries, there is more hardening near the boundaries. As the grain size is reduced, more hardening occurs in the center of the grain. There is more strain hardening in fine-grain metals than in coarse-grain metals because a larger volume of material is strain hardened in fine-grain metals.

As mentioned above, the grain size has a strong effect on hardness and strength. The smaller the grain size, the greater the hardness and flow stress (stress in tension at a fixed value of strain). Figure 1.5.1 illustrates that the hardness of titanium increases as $d^{-1/2}$, where d is the average grain size. The increase in flow stress with $d^{-1/2}$ is shown in Figure 1.5.2 for three values of strain. The equation of the straight line(s) drawn through the data in Figure 1.5.2 is the well-known Hall–Petch equation,

$$\sigma = \sigma_0 + \frac{k}{\sqrt{d}} \tag{1.5.1}$$

where σ is the flow stress and σ_0 and k are constants.

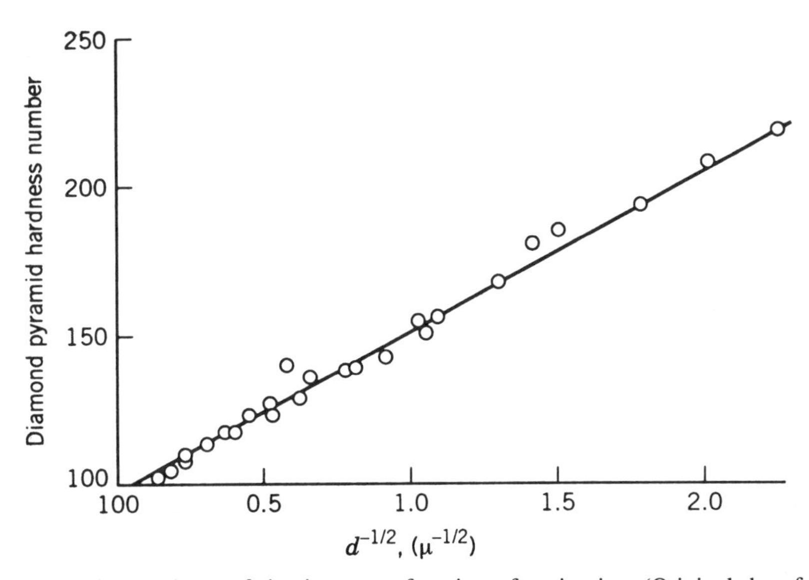

Figure 1.5.1 Hardness of titanium as a function of grain size. (Original data from Hu and Cline, 1968; plot from Armstrong and Jindal, 1968; reproduced with permission of The Minerals, Metals & Materials Society.)

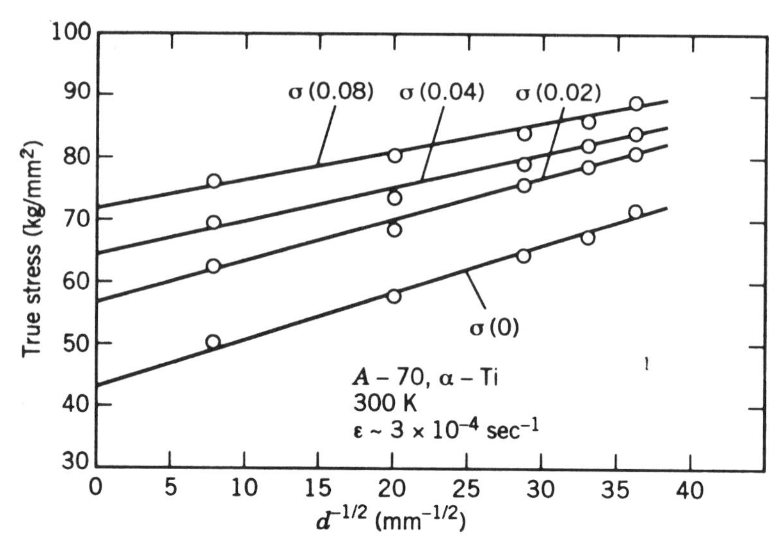

Figure 1.5.2 Flows stress in titanium for three values of strain at room temperature. (After Jones and Conrad, 1969; reproduced with permission of The Minerals, Metals & Materials Society.)

1.5.2 Strengthening Mechanisms

Hardening or strengthening of an alloy can be achieved by slowing the movement of dislocations. This can be accomplished by solid solution strengthening and precipitation hardening. A solid solution is a mixture of elements that do not have a strong chemical affinity for each other. If the solvent and solute atoms are close to the same size, the solute atoms occupy solvent lattice positions as in a substitutional solid solution, but if the solute atoms are much smaller than the solvent atoms, the solute atoms occupy the interstitial spaces in the lattices as shown in Figure 1.5.3. The introduction of solute atoms into a solvent atomic lattice produces an alloy that is stronger than a pure metal. However, only a relatively small number of alloy systems permit extensive solid solubility between two or more elements. Consequently, most commercial alloys are heterogeneous structures of two (or more) distinct phases or types of crystals. The notable exceptions are the carbon steels, which permit a significant amount of solid solution hardening. The response of carbon steels is discussed in Chapter 2.

Generally, precipitates are a second material phase of fine particles dispersed in a grain; however, the particles can be the size of grains dispersed throughout the alloy. The strengthening produced by the second phase is very complex and depends on many factors, such as the size, shape, number, and distribution of second-phase particles, the strength, ductility, and hardening properties of the matrix, and the bonding between the two phases. The precipitates in the microstructure provide obstacles that dislocations must either

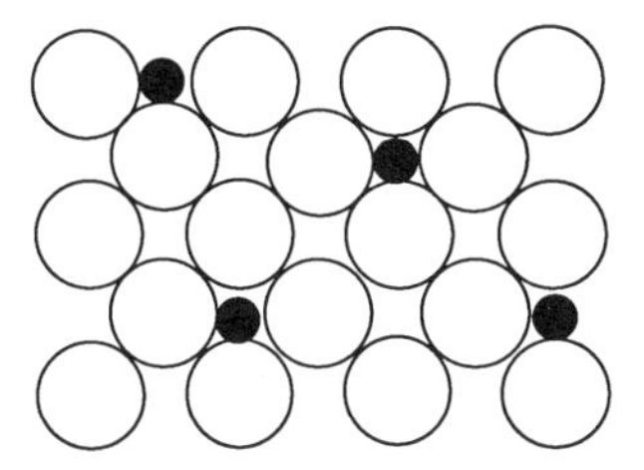

Interstitial Solid Solution

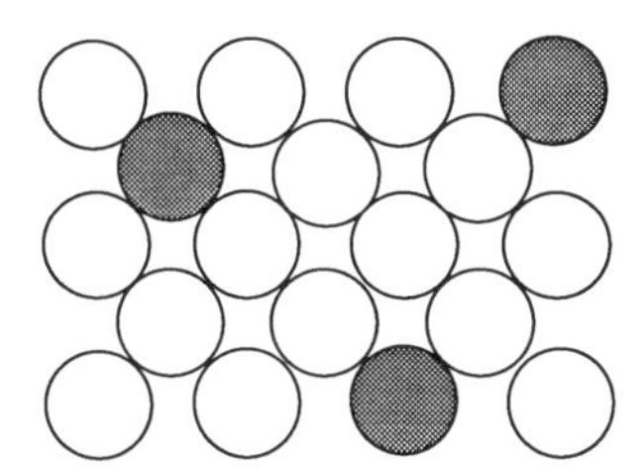

Substitutional Solid Solution

Figure 1.5.3 Interstitial solid solution or mixture of carbon and iron atoms. The large size of the iron atoms in an FCC lattice allows the smaller carbon atoms to fit in the interstitial spaces.

cut through, climb over, or move between. In any case, an increase in stress is necessary to move the dislocations through a lattice containing precipitates. The computer simulation in Figure 1.5.4 shows the effect of precipitates pinning a dislocation until the local stress field is sufficient to produce shear, climb, or looping. The dislocation then moves to the next precipitate and is again pinned for a period of time. The presence of a large number of small precipitates is more effective than a few large precipitates in slowing dislocation movement.

The interaction between precipitates and dislocations can produce hardening or softening. Let us first examine how strain hardening could occur. The looping mechanism proposed by Orowan (1947), as shown in Figure 1.5.5*A*, is based on the assumption that dislocations bend and form loops around the precipitate. The dislocation line intersects on the far side of the precipitate, connects locally at the point of intersection, and continues to move through the lattice. In the process a dislocation ring remains surrounding the particle as shown in Figure 1.5.5. The stress field produced by the loop increases the resistance to motion of the next dislocation. To overcome the increase in resistance to dislocation motion, the stress must be increased. This

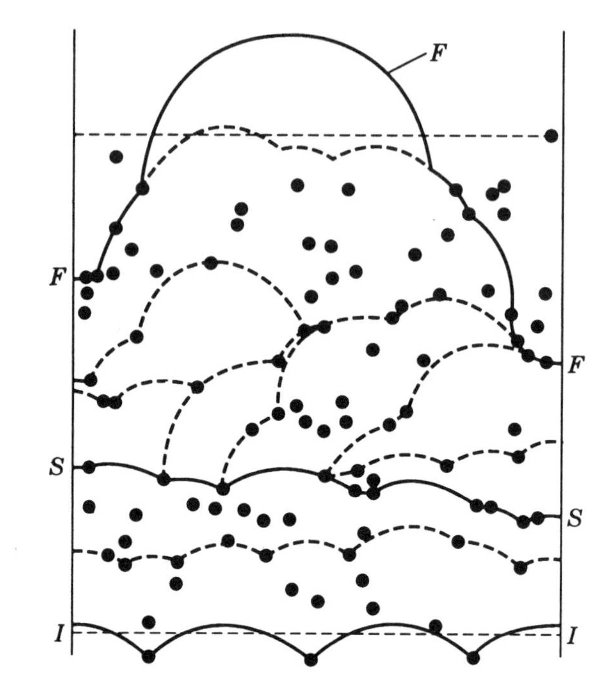

Figure 1.5.4 Computer model of the sequence of positions ($I \rightarrow S \rightarrow F$) of a dislocation through an array of weak obstacles under the action of an upward force. (From P. Haasen, *Physical Metallurgy*, 2nd ed., Cambridge University Press, Cambridge, 1986.)

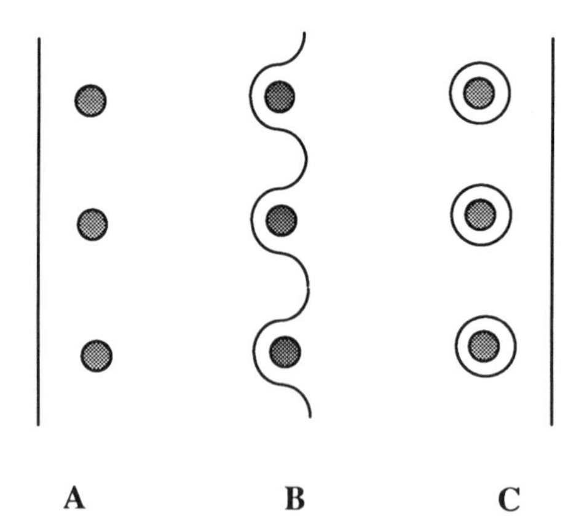

Figure 1.5.5 Orowan process ($A \rightarrow B \rightarrow C$) of dislocations bowing between particles and then bypassing the particles by leaving a dislocation loop surrounding each particle.

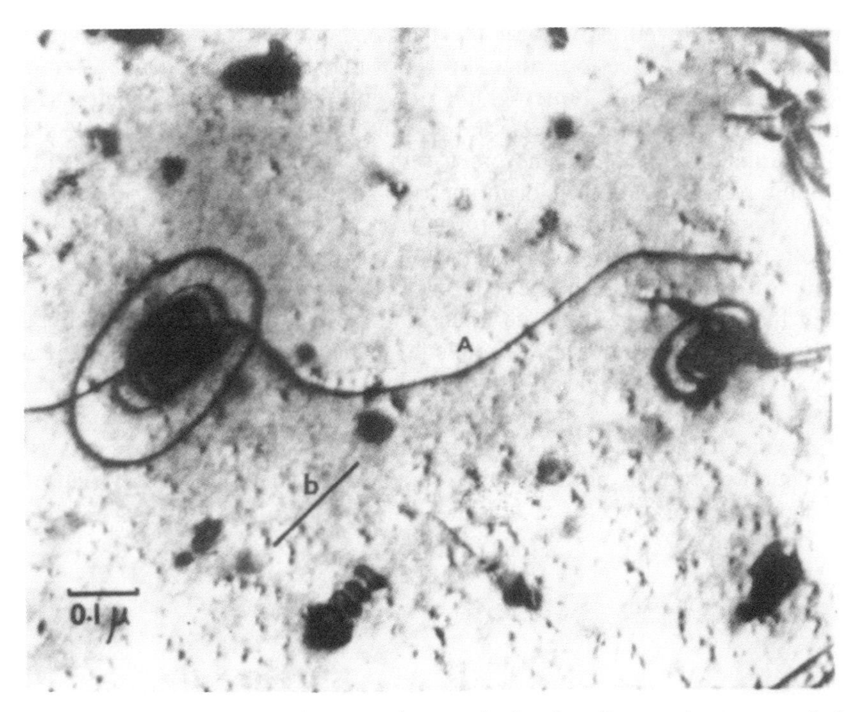

Figure 1.5.6 Transmission electron micrograph showing Orowan loops around alumina particles in a single crystal of Cu–30% wt Zn deformed to 6% at 77 K. (From P. B. Hirsch and F. J. Humphreys, Plastic deformation of two phase alloys containing small deformable particles, in *Physics of Strength and Plasticity,* A. S. Argon, ed., MIT Press, Cambridge, MA, 1969.)

is observed as strain hardening in a stress–strain curve. An example of dislocation looping is shown in the micrograph in Figure 1.5.6.

Softening can result from dislocation shearing precipitates. In the shear mechanism, the dislocation passes through the precipitate, displacing the precipitate lattice on one side of the slip plane relative to the other side of the slip plane as shown in Figure 1.5.7. Several dislocations passing through the

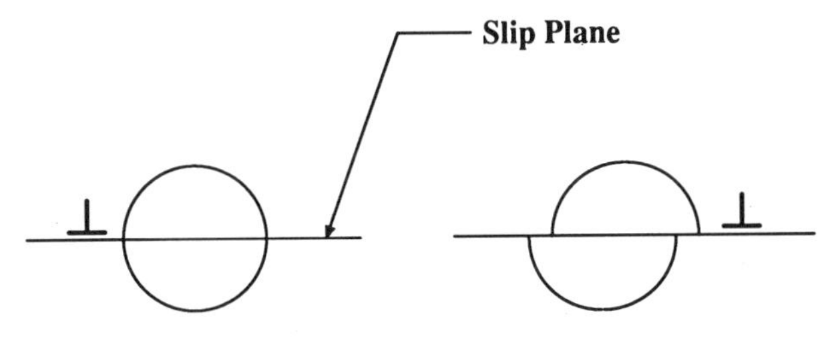

Figure 1.5.7 Dislocation passing through a particle, resulting in the shearing of the particle.

precipitate will eventually shear the particle into two or more subparticles, reducing the resistance to further slip. Thus the shearing mechanism can produce cyclic softening. A micrograph showing the shearing of precipitates in Waspaloy is shown in Figure 1.5.8. In addition to shearing and looping, climb is important in controlling the creep rate, as discussed in Chapter 3. In all cases the propagation of dislocations past particles can produce a back stress in the opposite direction of the propagation.

Both precipitates and solid solution atoms are point defects that slow the motion of dislocations. Their effect on the dislocations are roughly independent of the direction of slip. The effect on slip of grains, cells, and subgrains is also roughly orientation independent. Johnson and Gilman (1959) define drag stress as the resistance to slip (hardening) that is orientation independent or isotropic. Drag stress is like a "frictional" stress that can be represented by a scalar hardening variable. Conversely, back stress, which results from dislocation pileups, is an orientation-dependent hardening property that must be represented be a vector or tensor.

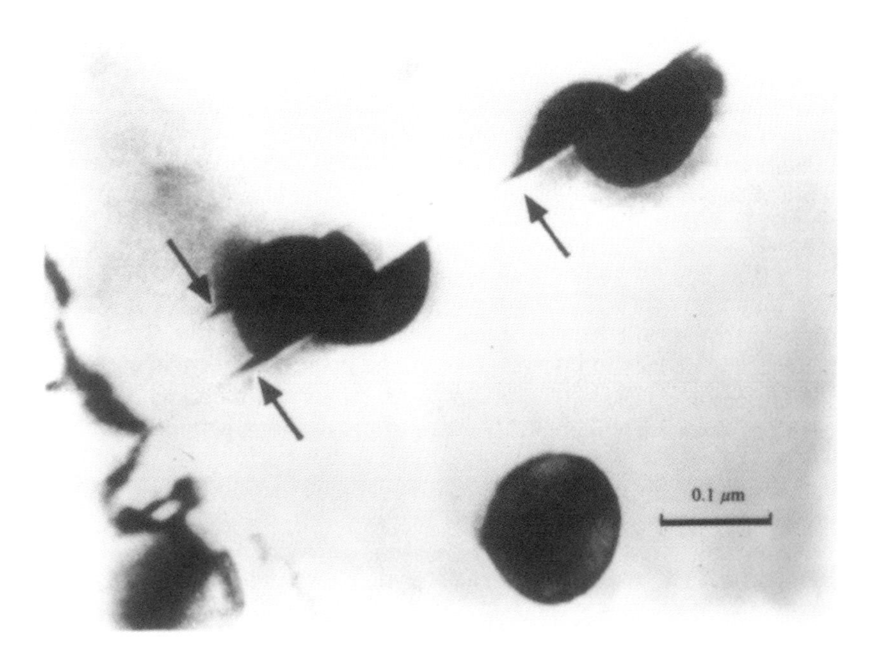

Figure 1.5.8 Transmission electron micrograph showing the effect of dislocations passing through particles and displacing one part of a particle relative to the other (From P. Haasen, *Physical Metallurgy,* 2nd ed., Cambridge University Press, Cambridge, 1986.)

1.6 SUMMARY

There are several concepts from physical metallurgy that are key to understanding the mechanical response of alloys. These concepts are also the basis of constitutive modeling. Inelastic deformation occurs in metals due to the propagation of dislocations on the slip plane (the planes of high atomic density) in the close-pack direction. The critical resolved shear stress, or the stress that produces slip, is the component of the yield stress on the slip plane in the slip direction. In polycrystalline metals slip occurs on the slip planes most closely oriented to the maximum shear stress.

The net stress to produce slip is the difference between the shear stress and the back stress. The back stress results from dislocation pileups at barriers and increases as the number of dislocations in the pileup increases. The back stress gives rise to the Bauschinger effect and can even produce reverse inelastic flow on unloading at high temperature. The increase in the resistance to slip is observed as strain hardening. Back stress is orientation dependent and must be represented by a vector or tensor.

At temperatures above about one-half the melting temperature, deformation occurs by slip and climb. Climb results from the diffusion of atoms and vacancies to or from an edge dislocation and allows the dislocation to move normal to the slip plane. Climb increases the resistance to slip. Climb is more important at higher temperatures.

As inelastic deformation increases the number of dislocations increase and develop a dislocation network. The dislocations are not distributed uniformly throughout the crystal lattice. Dislocation cells form at low temperatures from cross slip, and dislocation subgrains develop from polygonization (climb).

Recovery is associated with a reduction in the number of dislocations and the stored energy. Dynamic recovery occurs simultaneous with deformation. Static recovery occurs due to the forces between the dislocations themselves. The rate of recovery increases with increasing temperatures. Cells and subgrains are recovery mechanisms.

Grains arise from nucleation and growth of crystallites in a melt. Grain boundaries are regions of crystallographic misorientation, immobile dislocations, and severe strain. Grain boundaries maintain strength at low temperatures but serve as diffusion paths at high temperature. Grain boundaries retard slip, so the smaller the grain size, the greater the resistance to slip, and the larger the flow stress.

Precipitates and solid solution atoms in the lattice increase the resistance to slip and are strengthening mechanisms. The interaction of dislocations and precipitates can produce hardening or softening. The interaction of dislocations with precipitates, solid solution atoms, grains, cells, and subgrains are roughly isotropic hardening effects that can be represented by a scalar hardening variable or drag stress.

REFERENCES

Armstrong, R. W., and P. C. Jindal (1968). *Transactions of the Metallurgical Society of AIME,* Vol. 242, p. 2513.

Ashby, M. F. (1973). The microstructure and design of alloys, in *Proceedings of the 3rd International Conference on Strength of Metals and Alloys,* Vol. 2, Cambridge.

Boas, W., and M. E. Hargreaves (1948). *Proceedings of the Royal Society of London, Series A,* Vol. A193, p. 89.

Courtney, T. H. (1990). *Mechanical Behavior of Materials,* McGraw-Hill, New York.

Dame, L. T., and D. C. Stouffer (1985). Anisotropic constitutive model for single crystal alloys: model development and finite element implementation, *NASA CR 175015.*

Dieter, G. E. (1986). *Mechanical Metallurgy,* 3rd ed., McGraw-Hill, New York.

Frank, F. C., and W. T. Read (1950). *Physical Review, Vol. 79.*

Haasen, P. (1986). *Physical Metallurgy, 2nd ed., Cambridge University Press, Cambridge.*

Hall, E. O. (1951). *Proceedings of the Physical Society, London,* Vol. B64, p. 747.

Hirsch, P. B., and F. J. Humphreys (1969). Plastic deformation of two phase alloys containing small deformable particles, in *Physics of Strength and Plasticity, A. S. Argon, ed., MIT Press, Cambridge, MA.*

Hirth, J. P., and J. Lothe (1982). *Theory of Dislocations,* 2nd ed., Wiley-Interscience, New York.

Hu, H., and R. S. Cline (1968). *Transactions of the Metallurgical Society of AIME,* Vol. 242, p. 1013.

Jackson, P. J. (1985). *Progress in Materials Science,* Vol. 29,pp. 139–175.

Jackson, P. J. (1986). *Materials Science and Engineering,* Vol. 81, pp. 169–174.

Johnson, W. G., and J. J. Gilman (1959). *Journal of Applied Physics,* Vol. 30, pp. 129–144.

Jones, R. L., and H. Conrad (1969). *Transactions of the Metallurgical Society of AIME, Vol. 245, p. 779.*

Lindholm, U. S., K. S. Chan, S. R. Bodner, R. M. Weber, K. P. Walker, and B. N. Cassenti (1984). *NASA CR 174718.*

Longo, W. P. (1970). Work softening in polycrystalline metals, unpublished doctoral dissertation, University of Florida.

Millian, W. W., and S. D. Antolovich (1986). *Metallurgical Transactions,* Vol. 18A, p. 491.

Miner, R., R. Voigt, J. Gayda, and T. Gabb (1986). Orientation and temperature dependence of some mechanical properties of the single crystal nickel-base superalloy rené N4, Part I: Tension compression anisotropy, *Metallurgical Transactions,* Vol. 17.A, p. 507.

Moteff, J. (1980). Deformation induced microstructural changes in metals, *Proceedings of a Workshop on a Continuum Mechanics Approach to Damage and Life Prediction,* D. C. Stouffer, ed., Carrollton, KY.

Nix, W. D., and J. C. Gibeling (1983). Mechanisms of time dependent flow and fracture of metals, in *Flow and Fracture at Elevated Temperatures,* R. Raj, ed., ASM Science Seminar, American Society for Metals, Metals Park, Ohio.

Orowan, E. (1947). The creep of metals, *Transactions of the West of Scotland Iron and Steel Institute,* Vol. 54.

Petch, N. J. (1953). *Journal of the Iron and Steel Institute,* Vol. 174, p. 25.

Read, W. T. (1953). *Dislocations in Crystals,* McGraw-Hill, New York.

Reed-Hill, R. E. (1973). *Physical Metallurgy Principles,* 2nd ed., D. Van Nostrand, New York.

Sheh, M. Y., and D. C. Stouffer (1988). A crystallographic model for tensile and fatigue response of René N4 at 982C, *Journal of Applied Mechanics,* Vol. 57, p. 25.

Stouffer, D. C., V. G. Ramaswamy, J. H. Laflen, R. H. Van Stone, and M. R. Williams (1989). A constitutive model for the inelastic multiaxial response of René 80 at 871C and 982C, *Journal of Engineering Materials and Technology,* Vol. 112, p. 241.

Strublington, C. A., and P. J. E. Forsyth (1957). Some metallographic observations on the fatigue behavior of copper, nickel and certain alloys, *Journal of the Institute of Physics,* Vol. 56, p. 90.

Swann, P. R., and J. Nutting (1961). Influence of stacking fault energy on the modes of deformation of polycrystalline copper, *Journal of the Institute of Physics,* Vol. 60, p. 133.

Swanson, G. (1984). Private communication.

Swearengen, J. C., and J. H. Holbrook (1985). Internal variable models for rate dependent plasticity: analysis of theory and experiment, *Research Mechanica,* Vol. 13, p. 93.

Taylor, G. I. (1934). *Proceedings of the Royal Society, Series A,* Vol. 145, p. 362.

Thompson, A. W. (1975). *Acta Metallurgica,* Vol. 23, p. 1337.

Von Mises, R. (1928). *Zeitschrift fur Angewandte Mathematik und Mechanik,* Vol. 8, p. 161.

Weertman, J. (1955). *Applied Physics,* Vol. 26, p. 1213.

PROBLEMS

1.1 Estimate the theoretical shear stress required to produce slip in a perfect crystal. Consider two planes of atoms in a cubic array. Observe that the shear stress is zero when the lattice is perfectly aligned, and it is a maximum when one row of atoms is displaced one-half an atomic distance since the displaced atoms are equally attracted to the undisplaced atoms. Relate the theoretical shear stress to the shear modulus of the material. (*Hint:* What is the shear stress as a function of position during the deformation?)

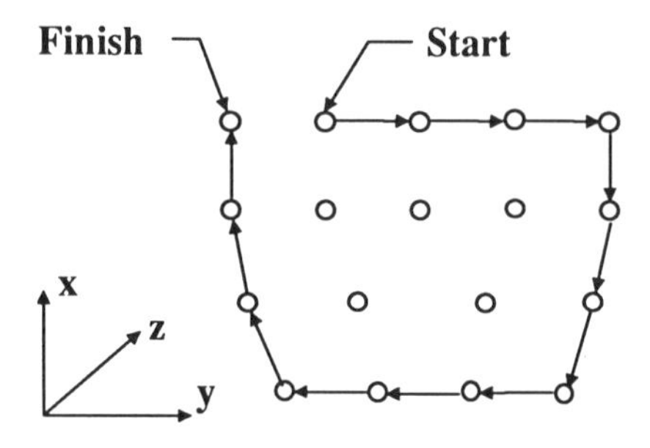

Figure P1.3

1.2 Discuss quantitatively why the presence of edge dislocations allows slip in crystals at a much lower stress value than that of the theoretical shear stress. Is the same argument valid for screw dislocations?

1.3 The Burgers circuit of an edge dislocation is defined by arbitrarily choosing a unit vector **e** tangent to the dislocation line and tracing a right-hand circuit around the dislocation with an equal number of lattice sites in each direction. The Burgers vector **b** is the vector drawn from the final position to the starting position that closes the circuit. The Burgers circuit and vectors are shown in Figure P1.3 for a unit vector **e** into the paper. Show that the same Burgers vector is obtained for any starting position. What is the Burgers vector for a unit vector in the opposite direction? What is the Burgers vector when the extra plane is below the slip plane? What is the Burgers vector for a perfect crystal?

1.4 What are the Burgers vectors for the two screw dislocations shown in Figure 1.2.3*C* and *D*? Which screw dislocation could be described as positive?

1.5 A screw dislocation lies along the axis of the cylinder with **b** parallel to **e** as shown in Figure P1.5. If *AB* is a ledge, is it recessing or

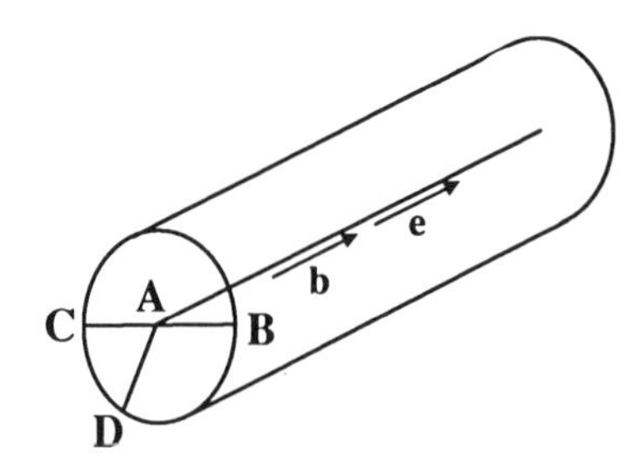

Figure P1.5

overhanging? If *AC* is a ledge, is it recessing or overhanging? Can a ledge occur in any position *AD*?

1.6 The Schmid factor is a parameter between 0 and 1 that transforms a tensile stress to the critical resolved shear stress. Determine the Schmid factor for an angle α between the tensile axis and slip plane normal vector and an angle β between the tensile axis and slip direction.

1.7 Show that the basal planes of the hexagonal close-packed system have the same lattice structure as the octahedral planes of the face-centered cubic system. Are there any additional slip planes in the HCP system? Draw a basal slip plane and add another plane of atoms on top. Show that there is an *ABAB* stacking sequence. Find the pyramidal and prismatic planes in the HCP lattice structure.

1.8 It is known that a certain body-centered cubic lattice system has a critical resolved shear stress of 115 MPa. What is the yield stress for a tensile specimen made of this material with the tensile axis in the [100] direction?

1.9 Determine the total crystal shear strain caused by twinning and the local shear strain between the twin boundaries for the crystal shown in Figure P1.9. Develop a relationship between the local strain in the twin region and the average shear strain for the entire specimen as a function of the ratio of the twin volume to the total volume.

1.10 Briefly discuss the differences between kinks and jogs. Show that jog configurations are formed such that when one dislocation intersects another, each acquires a jog normal to its own slip plane and equal in magnitude to the other dislocation's Burgers vector.

1.11 The yield stress in torsion is usually about one-half the yield stress in tension. Use the observation that polycrystalline metals deform under the action of a shear stress on the slip plane to explain this observation.

1.12 The following data were obtained for the flow stress of carbon steel at 0.02 strain as a function of grain size.

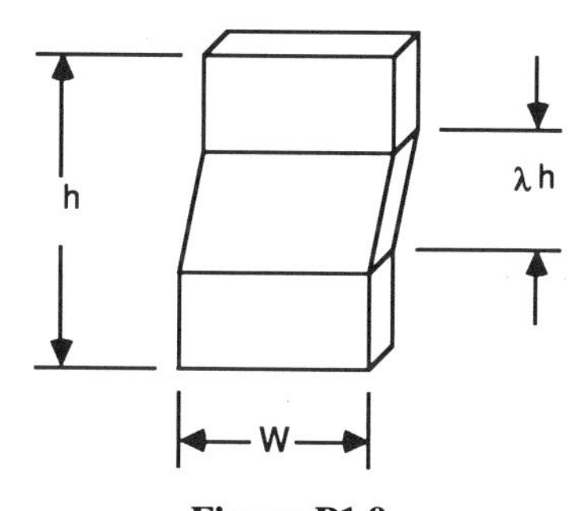

Figure P1.9

d (μm)	σ (MPa)
406	93
106	129
75	145
43	158
30	189
16	233

Show that the data are consistent with the Hall–Petch effect and determine the constants in the Hall–Petch equation. What is the meaning of the parameter σ_0? Suppose that grain size were reduced to 1 μm by a special alloying process. What is the corresponding increase in the flow stress? Discuss your result.

1.13 The strengthening from subgrains and cells is similar to that from grains, and the Hall–Petch equation is a reasonable model for the strengthening. In the case of cells and subgrains the parameter k_y is 0.5 to 0.2 of the value for grains. Plot the Hall–Petch equation for grains for the data in Problem 1.12 and superimpose the effect of cell formation from cold working in a pure crystal of the same material. At approximately what ratio of grain size to cell size does the stress change from grain control to cell control in a material with both grains and cells?

1.14 Consider a two-phase material containing volume fractions of V_1 and $(1 - V_1)$ for phases P_1 and P_2, respectively. Assuming that the flow behavior is not affected by the presence of the other phase and that the two phases undergo uniform strains, show how the tensile response of the two-phase system can be determined from the rule of mixtures. That is, show that

$$\sigma_c(\varepsilon) = V_1\sigma_1(\varepsilon) + (1 - V_1)\sigma_2(\varepsilon)$$

where $\sigma_c(\varepsilon)$ is the flow stress at strain ε of the two-phase system, and σ_1 and σ_2 are the flow stresses for phases P_1 and P_2 at the same strain.

1.15 Calculate the modulus of elasticity for a two-phase alloy of 50 vol % cobalt and 50 vol % tungsten on the basis of

(a) uniform strain throughout the alloy and

(b) uniform stress throughout the alloy.

Assume that $E = 30 \times 10^6$ psi for cobalt and $E = 100 \times 10^6$ psi for tungsten.

CHAPTER 2

Tensile, Compressive and Cyclic Characteristics of Metals

The uniaxial test is probably the most fundamental type of mechanical experiment. An enormous amount of tensile, compressive, and cyclic data is available in the literature for all classes of materials. Much of the current understanding of metals (i.e., how and why they deform) has been deduced from tensile tests. Thus this is an appropriate starting place to develop an understanding of the correlation between mechanical response and the corresponding deformation mechanisms.

2.1 TENSILE RESPONSE AND STRAIN MEASURES

The most fundamental measures of the mechanical properties of a material are determined from a tensile test. The axial deformations resulting from an axial force on a long slender rod are frequently quoted using the engineering stress, σ, and engineering strain, ε. The engineering stress and strain are defined using the undeformed area A_0 and length L_0, respectively, as

$$\sigma = \frac{F}{A_0} \quad \text{and} \quad \varepsilon = \frac{L - L_0}{L_0} \tag{2.1.1}$$

where F is the axial force and L is the current length of the tensile test section at any instant in the loading history. These measures of stress and strain are generally used only for elastic deformations or relatively small total strains, up to two or three times the yield strain. The stretch ratio, λ, is defined as the ratio of the deformed length to the undeformed length,

$$\lambda = \frac{L}{L_0} \tag{2.1.2}$$

and is used for both small and large deformations. The stretch ratios are defined as an alternative normal strain; thus only the principal stretch ratios are used in a three-dimensional deformation.

A typical engineering stress–strain curve is shown in Figure 2.1.1 for a strain-hardening metal. The response of solution-hardened metals such as low-carbon steel is different, as shown in Section 2.2. Yield or onset of significant plastic or inelastic deformation can be defined by the proportional limit or an offset plastic strain, typically 0.2%. A metal generally hardens, and the stress increases at a decreasing rate until the ultimate stress or tensile strength is achieved. These measures are almost always defined for the original cross-sectional area. Necking, the reduction in cross-sectional area due to inelastic straining, is responsible for the reduction in stress at failure. Ductility is a measure of the ability of the material to withstand strain. It is usually defined by the strain or elongation at failure, or the reduction in area at failure. The strain energy density is a measure of the ability of a metal to absorb energy, and it is defined as the area under the stress–strain curve. Note that the ductility and strain energy density correlate closely with each other. Brittle metals exhibit relatively small elongation, reductions in area, and strain energy densities at failure.

Experimentally, the total strain ε can be decomposed into two components, elastic strain ε^{E} and inelastic strain ε^{I}:

$$\varepsilon = \varepsilon^{\mathrm{E}} + \varepsilon^{\mathrm{I}} = \frac{\sigma}{E} + \varepsilon^{\mathrm{I}} \tag{2.1.3}$$

Since the stress and total strain are measured quantities and the elastic modulus is a measure of the resistance to lattice distortion that does not change during inelastic deformation, the inelastic strain can be defined throughout

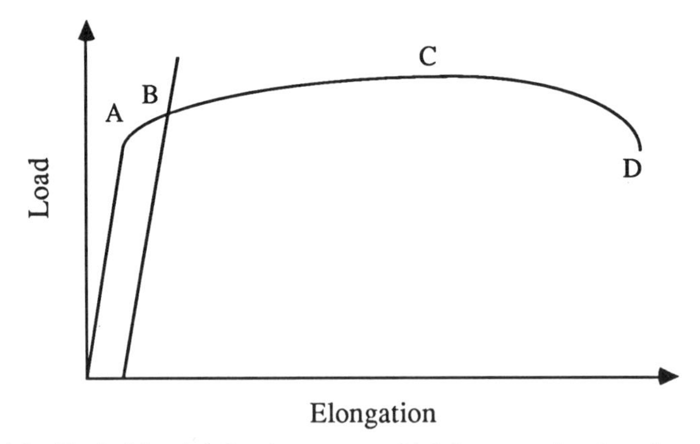

Figure 2.1.1 Typical load-deflection curve, which is proportional to the engineering stress–strain curve, showing (*A*) the proportional limit, (*B*) the offset yield stress, (*C*) the ultimate tensile stress or tensile strength, and (*D*) the engineering stress at failure.

the test as the difference between the total and elastic strains. The inelastic strain is well defined except when the inelastic strain is less than or near the resolution of the strain extensometer. In this manuscript the inelastic strain is defined by Equation 2.1.3 and means the total amount of strain that is not elastic. Plastic strain is limited to inelastic strain that does not depend on time, whereas creep strain is inelastic strain that does depend on time. Plastic and creep strain measures and their interaction are the topics of Chapters 3 and 5. The importance of using a single, unified inelastic strain measure rather than plastic and creep strains has become more evident in the last decade as the field has developed.

The mechanical response of metals in a tensile test depends on the test temperature and the strain rate at which the specimen is loaded. For most metals tested in closed-loop mechanical testing machines where the strain rate is less than 1.0 $\sec^{-1}$, the response will not depend on the strain rate if the temperature is less than about 50% of the melting temperature on the Kelvin scale ($T = 0.5 \times T_h$, homologous temperature). A typical engineering stress–strain curve is shown in Figure 2.1.1 and the stress–strain response for a relatively hard aluminum alloy is shown in Figure 2.1.2. The closed-loop testing machine used to load the aluminum alloy was run about as fast and slow as possible. It is interesting to note that even though the aluminum alloy did not show any strain rate dependence during loading, the stress did relax significantly during the 2-minute hold at constant strain. Stress relaxation, a

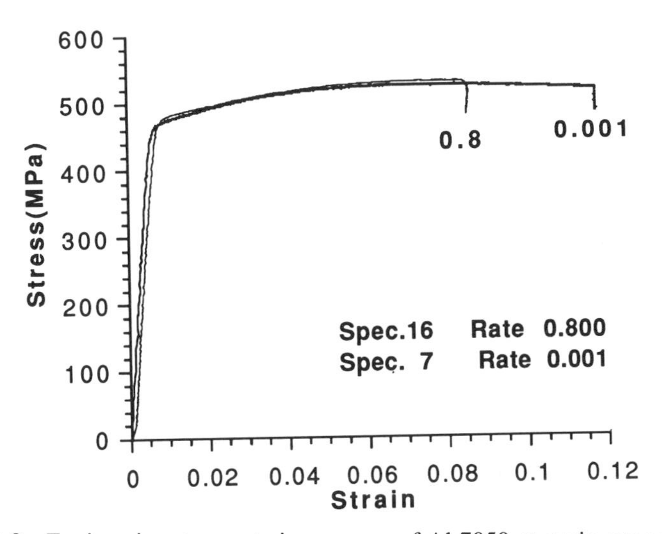

Figure 2.1.2 Engineering stress–strain response of Al 7050 at strain rates of 0.8 and 0.001 $\sec^{-1}$. Even though there is no effect of strain rate, the material did exhibit stress relaxation during 2-min hold periods of constant strain at the end of each test. (From Kuruppu et al., 1992; reprinted by permission of the Council of the Institution of Mechanical Engineers from *Journal of Strain Analysis*.)

reduction of stress while the strain is held constant, can be present even if the material response does not depend on the strain rate.

The influence of temperature on tensile response at one constant strain rate is shown in Figure 2.1.3. It is clear that temperature has a profound effect on the elastic modulus, the maximum stress, and the shape of the curve. The melting temperature of the material is approximately 1100°C. There is probably a change in the deformation mechanism between 538 and 649°C. At lower temperatures the deformation is by planar slip and the flat shape of the curve indicates that the material is initially very hard. The source of the stress overshoot was not reported. At higher temperatures, above 649°C, deformation is by planar slip and climb, and diffusion processes are more important. The shape of the curves at higher temperatures indicate that there is some strain hardening due to the development of a dislocation network.

The shape of the tensile curve also depends on the initial state of the material. In initially soft materials hardening occurs due to the development of a dislocation substructure. As a result, the flow stress (stress at a fixed value of strain) increases until the onset of necking, as shown in Figure 2.1.4. In initially work-hardened materials, the additional hardening from inelastic straining is small compared to the initial hardness created during processing,

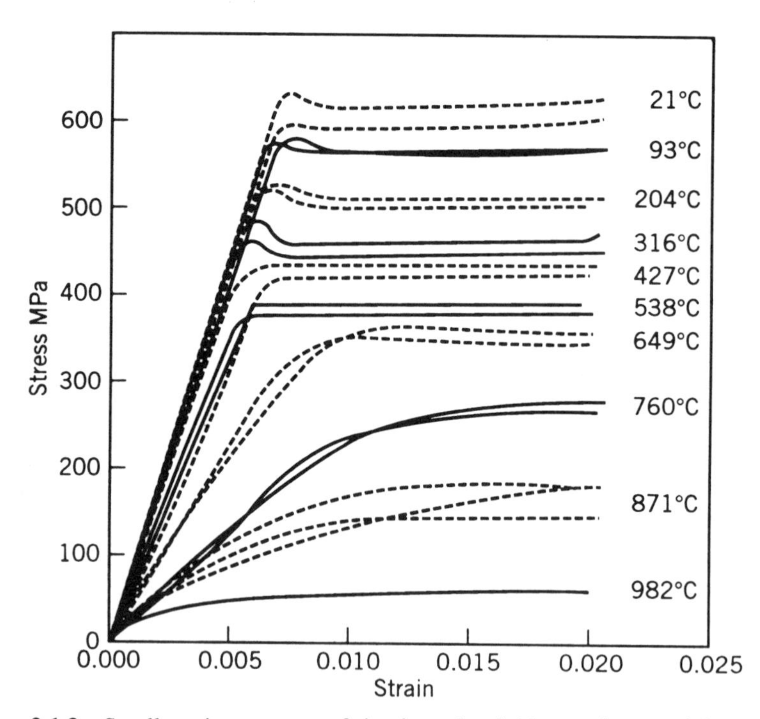

Figure 2.1.3 Small strain response of titanium aluminide matrix material at a strain rate of 0.000833 sec^{-1} showing the effect of temperature. (After Stouffer and Kane, 1988.)

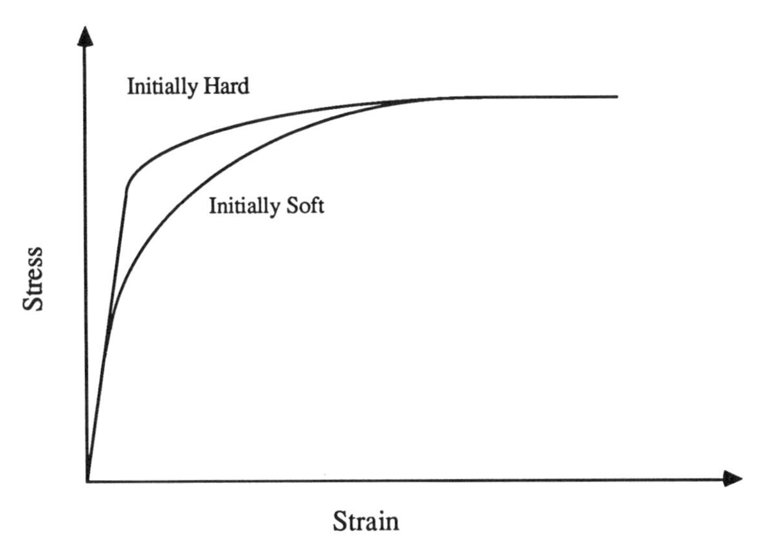

Figure 2.1.4 Tensile characteristics of initially work-hardened and initially soft (fully annealed) metals.

and the stress–strain curve remains nearly flat. Hard materials are generally less ductile than are initially soft materials.

It is also expected that the shape of two stress–strain curves, one in tension and one in compression for the same material at the same temperature ($T <$ .5 T_m), would be approximately symmetric. Since plastic deformation results from slip on a slip plane, only the direction of slip depends on the sign of the applied load. This symmetry is observed experimentally in cyclic loading. For example, the hysteresis loops in Figure 2.5.5 result from symmetric strain range loading, and the stress is symmetric in tension and compression.

There are several types of round specimen designs used for tensile and cyclic testing. Figure 2.1.5 shows a typical specimen design and two types of test sections and ends for three gripping systems. The cylindrical test section gives a uniform stress and strain profile in the test section when the specimen is perfectly aligned and loaded axially in tension. An axial extensometer (Figure 2.1.6), with a gage length of about two test section diameters is mounted axially in the center of the test section.

The hourglass specimen is used for axial loading in both tension and compression. The hourglass shape is better than the cylindrical test section for controlling buckling in compression, but the stresses are not uniform in the test section. Finite-element studies show that in an optimally designed specimen the surface stresses are 4 to 6% larger than the average stress. The strains are measured with a diametrical extensometer and the axial strain is approximated mathematically by assuming that there is no change in the inelastic (plastic) volume. Thus the axial strain is not measured directly. This

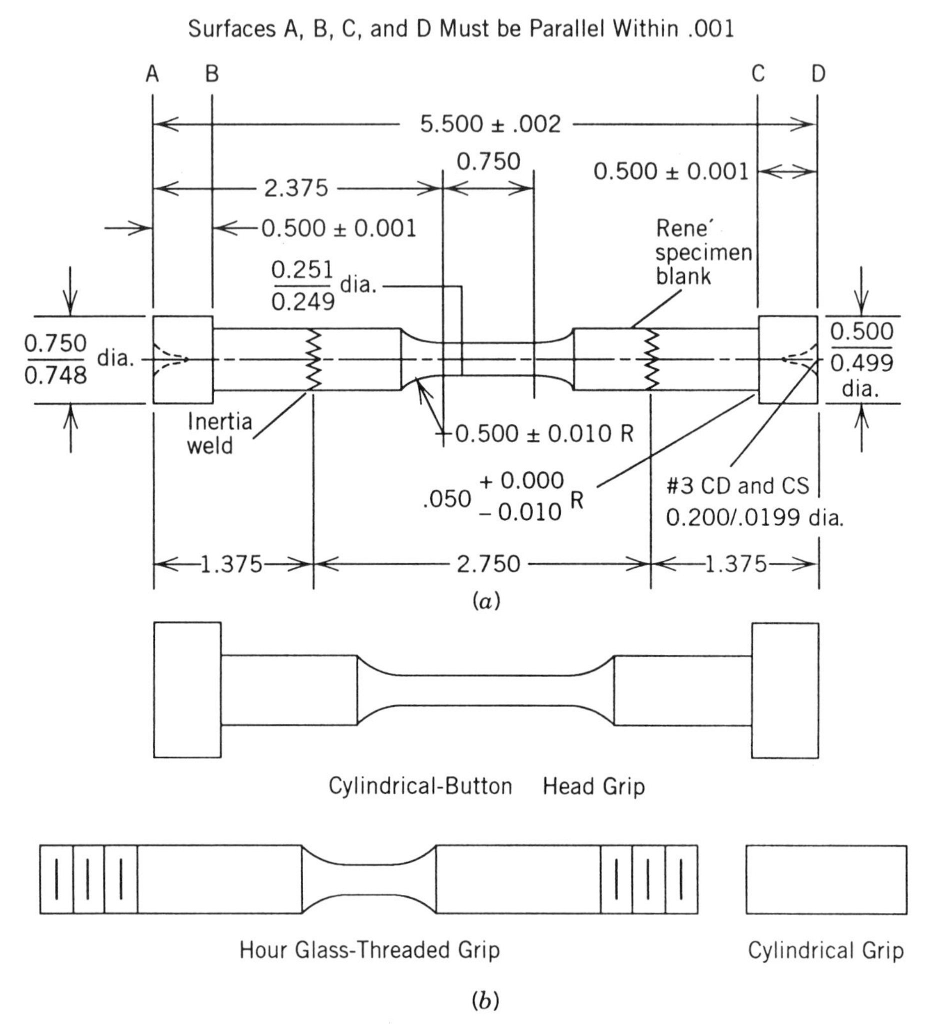

Figure 2.1.5 Geometry of a tensile specimen after inertial welding Inconel 718 cylinders on to a René 95 specimen blank. Diagram of two test specimens with three griping arrangements.

can be a source of experimental error. Another disadvantage of the hourglass specimen is that the volume of material actually being tested is very small compared to the cylindrical specimen. This can lead to more scatter in the test data, due to the lack of volumetric averaging of local material defects. Despite these inherent disadvantages, the hourglass specimen is widely used for low cyclic fatigue testing since it is less likely to buckle in compression.

The specimen grips vary with equipment and specimen costs. The threaded specimen is cost-effective, but axial alignment cannot be guaranteed, due to

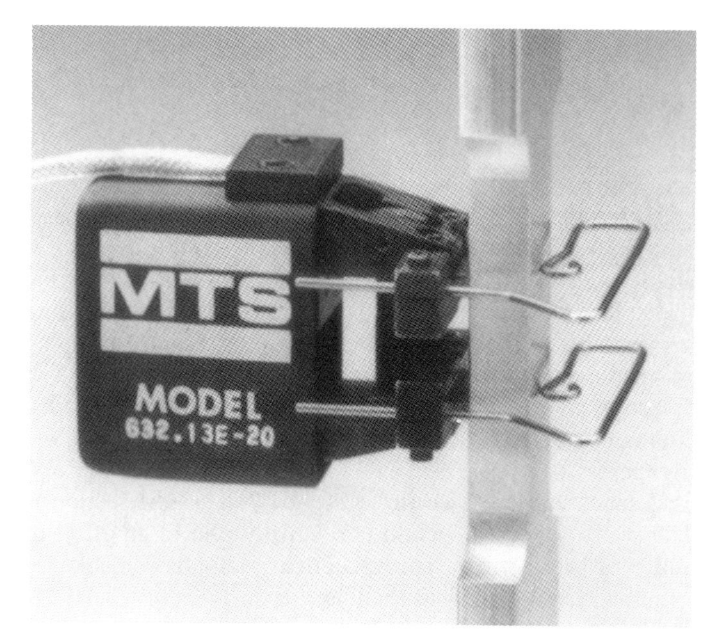

(*a*)

(*b*)

Figure 2.1.6 Axial (*left*) and diametrical (*right*) extensometers used for measuring displacements in many types of materials for tension–compression, low cycle fatigue, creep, and strain-rate testing. (Courtesy of MTS Systems Corporation, Eden Prairie, MN.)

the thread clearance. Further, it may be difficult to separate the specimen from the grip after high-temperature testing of close-fitting threaded specimens due to diffusion between the specimen and grip. The button head specimen requires a simple collet system that fits over the button and grips the specimen diametrically just under the button. The button is used for loading in tension. Alignment can be controlled accurately with this type of gripping system, but the test specimens are relatively expensive to manufacture. The straight cylindrical section can be gripped securely with hydraulically loaded collet grips and alignment can be accurately controlled. However, hydraulic grips are relatively expensive.

A typical torsional specimen is shown in Figure 2.1.7. The test section is tubular in shape. The wall of the test section is relatively thin in an effort to obtain an almost uniform shear stress. The specimen is also acceptable for axial loading or combined axial–torsional loading. Strain is recorded with a torsional extensometer. The specimen can be gripped with a hydraulic collet, and a square or D-shaped cross section is used for torque transfer.

The torsional response of René 80 is shown in Figure 2.1.8 together with the response in tension. It is obvious that the flow stress in torsion is about

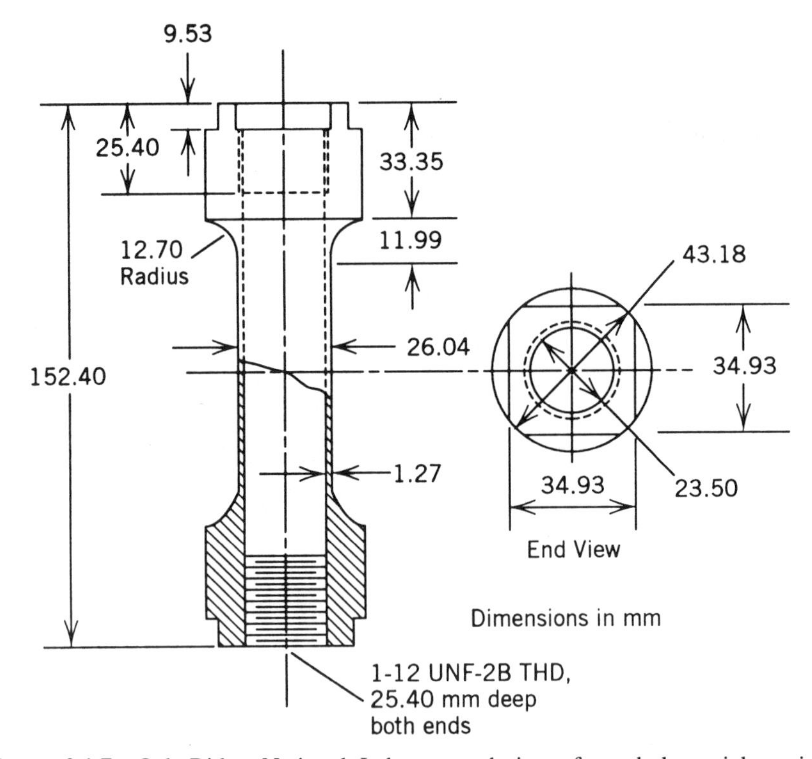

Figure 2.1.7 Oak Ridge National Laboratory design of a tubular axial–torsional stress strain specimen. (After Ellis and Robinson, 1985.)

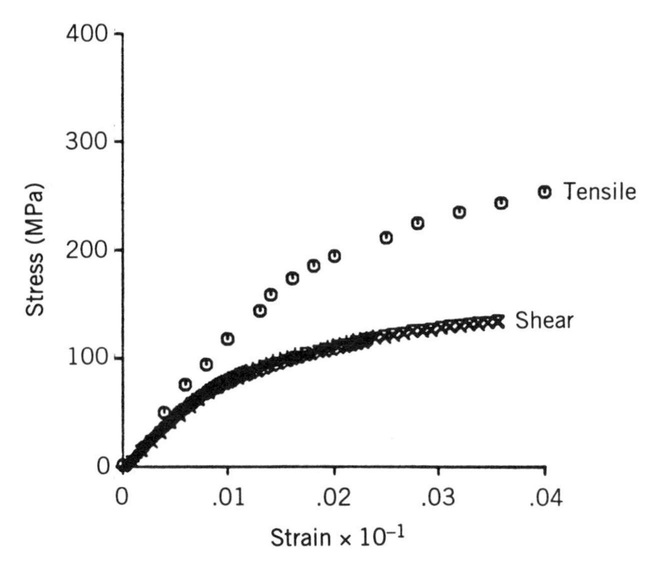

Figure 2.1.8 Comparison of tensile and shear response of René 80 at 982°C and a strain rate of 0.002 min^{-1}. (Data after Ramaswamy, Stouffer, and Laflen, 1986.)

half the corresponding value in tension. This follows since inelastic deformation occurs by lattice slip under the action of shear stress. Since the shear stress on the slip plane required to produce slip is expected to be the same in both tests, the relationship between the applied loads and local shear stress producing slip must be different. In the case of axial loading the slip planes are oriented at 45° to the tensile axis, and the shear stress on these planes in the direction of slip is half the applied axial stress. In the case of torsion, the stress plotted is the shear stress on the slip plane producing slip. This type of experiment confirms the fundamental notion of slip.

2.2 LOW-CARBON STEELS

There are two types of response in carbon steel, depending on the processing. For cold- and hot-rolled steel the material is solution hardened with the addition of carbon. In cold-rolled steel there is a significant amount of permanent strain hardening due to the dislocation structure developed at low temperature. The strain hardening dominates the response. In hot-rolled steel the high temperature allows the dislocation network to recover and there is very little permanent strain hardening. In this case the response is controlled by the carbon in solution in the iron.

The cold-rolled steel in Figure 2.2.1 exhibits a gradual, smooth yielding with a yield stress defined by the strain offset method. Strain hardening starts at a relatively high rate, and the rate decreases with increasing strain. Necking

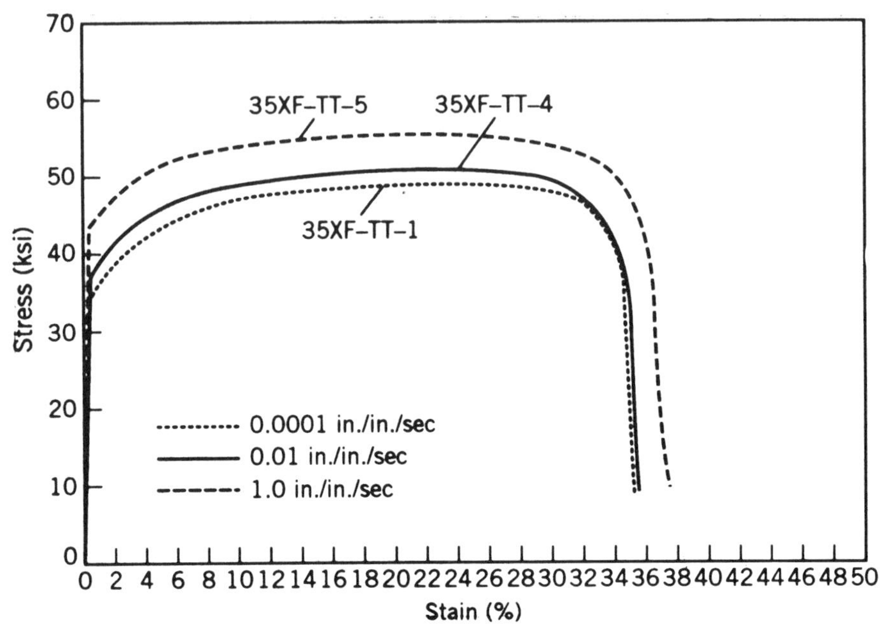

Figure 2.2.1 Room-temperature engineering stress–strain curves at three strain rates for 35XF cold-rolled low-carbon steel. (After Kassar and Yu, 1989a.)

causes the ultimate engineering stress to be lower than the true peak stress, which occurs just before separation. Hot-rolled steel has a sharp upper yield stress and a lower yield stress, as shown in Figure 2.2.2. The lower yield stress is often used in design because the upper yield stress is strain-rate sensitive. The sharp upper yield stress is followed by a period of continuous plastic flow with no strain hardening. At some value of strain, hardening begins and deformation proceeds similar to cold-worked steel. Both the hot-rolled and cold-rolled steels exhibit strain rate sensitivity that is discussed in Section 2.4.

The period of constant stress flow at the lower yield stress in hot-rolled steels is related to an interaction between the material and the design of the test specimen. Slip initiates at a stress concentration, such as a fillet of a tensile specimen as shown in Figure 2.2.3*a*. Once inelastic deformation starts in a local region, a high rate of plastic flow occurs in that region. The plastic deformation spreads to an adjoining region, due to stress concentration at the boundary of the deformed and undeformed regions. The line separating the two regions is a Luder's band (Figure 2.2.3*b*). Luder's bands include many slip planes and occur near the orientation of the maximum shear stress. The bands can form at one or both ends of the specimen and can spread at essentially constant stress. The metal begins to harden once the specimen is covered with Luder's bands. Thus the strain at which hardening begins is a

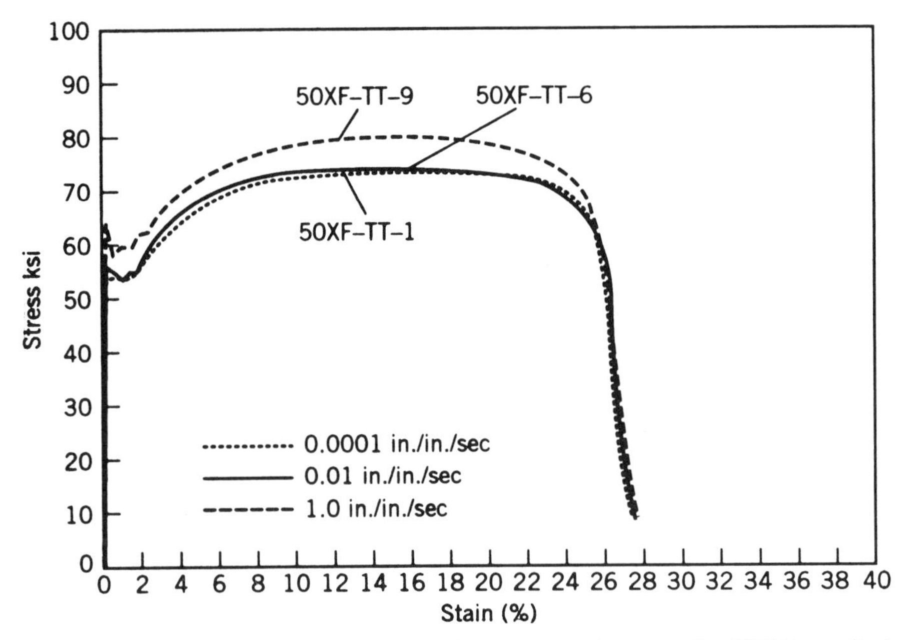

Figure 2.2.2 Room-temperature engineering stress–strain curves for 50XF hot-rolled low-carbon steel at three strain rates. The initial stress overshoot is a result of freeing dislocations that are pinned by an atmosphere of carbon atoms in solid solution. (After Kassar and Yu, 1989a.)

function of the length of the specimen's test section and the rate at which the Luder's bands spread.

Hot-worked steel also exhibits strain aging. Consider the response of a hot-worked carbon steel in two different loading sequences. Begin with case (a) in Figure 2.2.4. The specimen is loaded, then the load is removed and the specimen is reloaded within a short period of time (hours). The specimen responds elastically up to the point of unloading and then deforms plastically following the original response curve. Next consider case (b). After the initial loading identical to case (a), the load is removed and the specimen is reloaded after several months. This time the specimen responds elastically to a point well above the stress at unloading. The specimen then deforms plastically following a path that converges with the original response curve. The stress overshoot on reloading after the extended period with no load is strain aging. The time required for strain aging depends on temperature. Strain aging of carbon steels generally takes several months at room temperature but can occur much faster at high temperature. An example of high-temperature strain aging is shown in Figure 2.2.5.

The initial stress response and the strain aging of hot-worked steel are different from those of precipitate-hardened metals or cold-rolled steel and

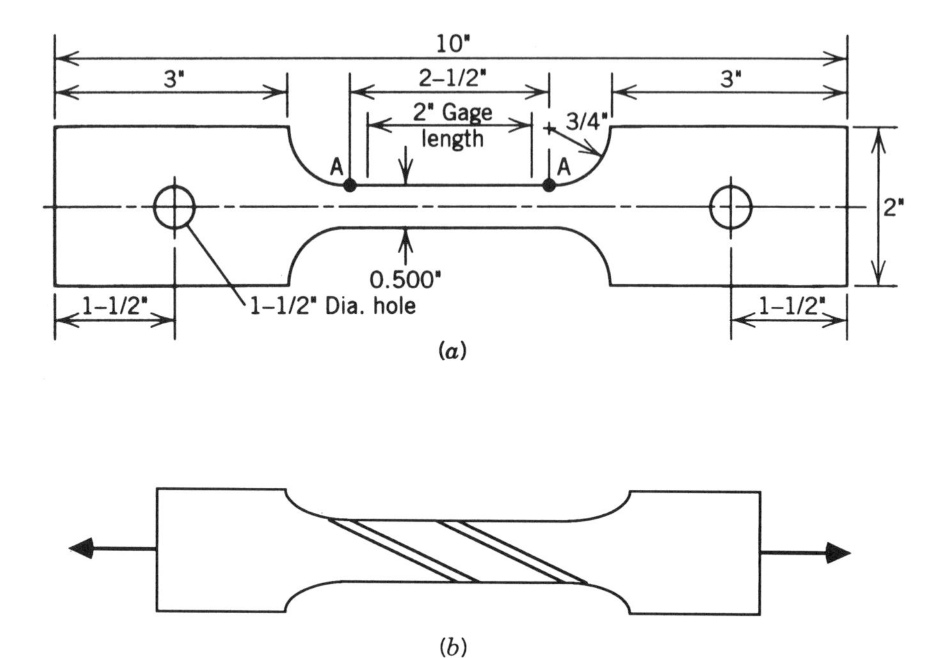

Figure 2.2.3 (*a*) Design of a test specimen used for testing sheet steel; (*b*) development of Luder's band at the stress concentrations in the transition sections of the specimen. (Specimen design after Kassar and Yu, 1989a.)

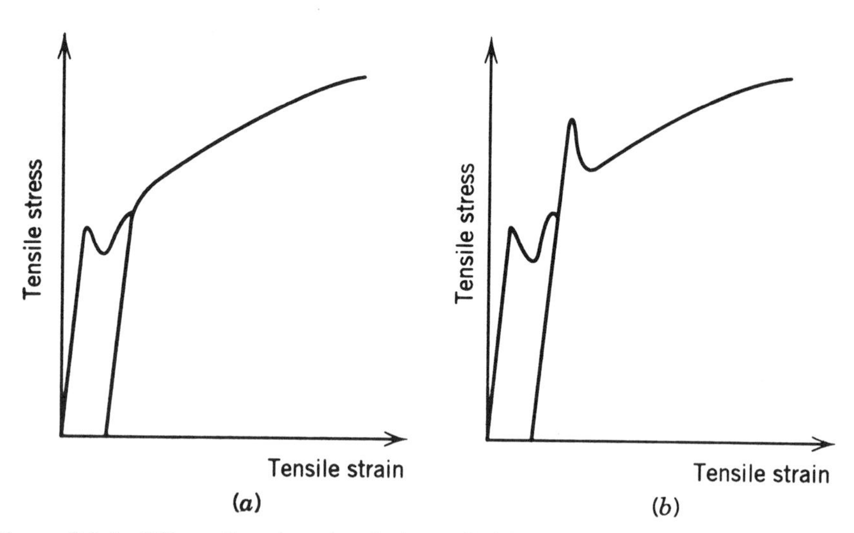

Figure 2.2.4 Effect of strain aging in hot-rolled low-carbon steel. On loading, unloading and reloading in a short period of time, the curve follows the original path; but if the material is loaded, unloaded, and held in the unloaded state for several months (at room temperature), a stress overshoot occurs on reloading.

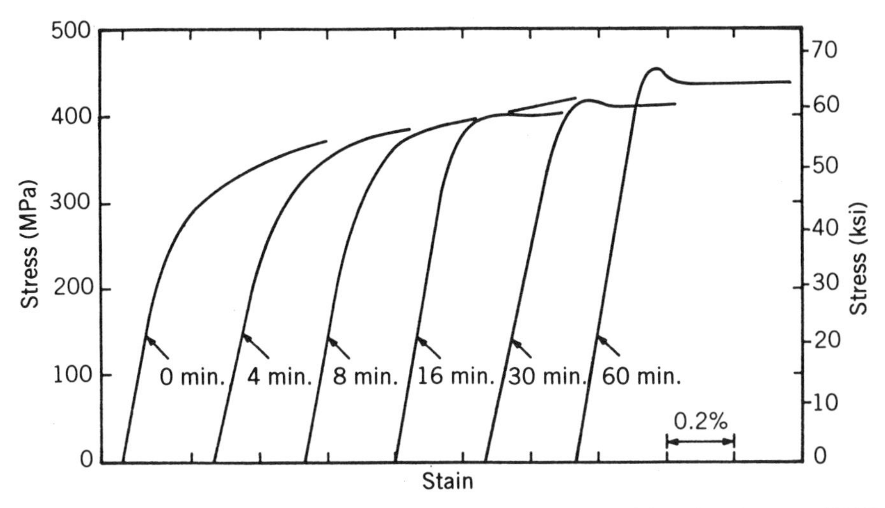

Figure 2.2.5 Change in the shape of stress–strain curves with aging time at 620°C for P1679 steel. (After Wada, Smith, and Lauprecht. Proceedings of the Third National Congress on Pressure Vessels and Piping, 1979, Copyright © The American Society of Mechanical Engineers, 345 East 47th St., New York, NY 10017. Used with permission.)

result from the strengthening mechanism. Recall that steel is an iron-carbon solid solution. The carbon, being relatively small, fits in the interstitial spaces between the larger iron atoms as shown in Figure 1.5.3. There is no chemical bonding between the carbon and iron, and the solubility of carbon in body-centered cubic iron is low (face-centered cubic iron is stable only above 732°C). The presence of an edge dislocation, with an extra plane, introduces local lattice distortion and local stresses, as shown schematically in Figure 2.2.6. The interstitial atoms are repelled from the compressed regions around a dislocation and attracted to the extended regions of the lattice to reduce the local stresses and achieve a lower energy state. The rate at which the interstitial atoms move is controlled by the rate of diffusion of carbon through iron. At high temperatures the diffusion rate is higher and the carbon can move more quickly. Thus the dislocation develops an atmosphere of carbon atoms (a dislocation atmosphere) that affects the propagation of the dislocation.

The sharp initial yield point results from an atmosphere of carbon atoms that collect around dislocations and pin the dislocations. Additional stresses above that normally required for dislocation motion are necessary to free the dislocation from its atmosphere. The increased stress required for initial movement of dislocations corresponds to the upper yield stress. The lower yield stress corresponds to the stress required to move dislocations that are free of their atmospheres. In strain aging the dislocations develop a dislocation atmosphere during periods of no load. Since the atoms must diffuse through

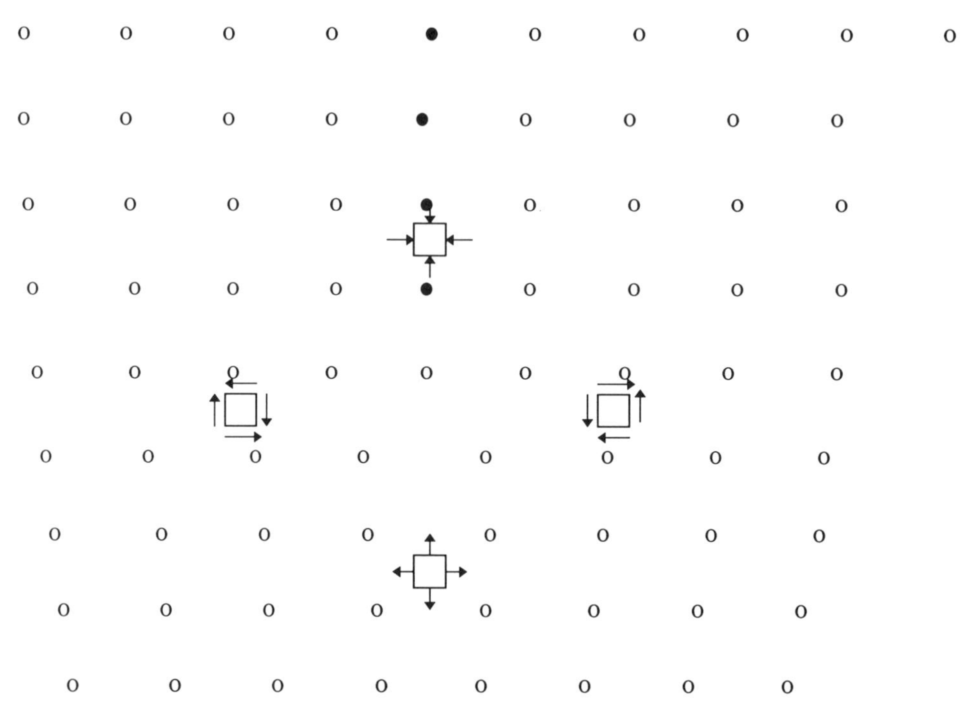

Figure 2.2.6 Local stresses that result from the extra plane of an edge dislocation.

the lattice, the increase in stress observed on reloading is a function of the time and temperature during the no-load period.

2.3 TENSILE RESPONSE MODELS

In this section two relatively simple modeling approaches are presented for the tensile response curve. First is the piecewise linear approximation, followed by the Ramberg–Osgood equation. An exponential model is presented in Problem 2.6.

The most elementary approximation uses one or two straight line segments to represent the tensile response curve, as shown in Figure 2.3.1. The line segments are picked to match the actual response as closely as possible. The two "rigid" models are used for large inelastic strains when the linear elastic response can be neglected. The two models with an elastic modulus are useful for small inelastic strain situations. All four representations may be useful for determining the stress–strain response at inelastic strains away from the knee of the data curve. The parameters in these representations can also be picked to estimate the accumulated strain energy or area under the tensile response

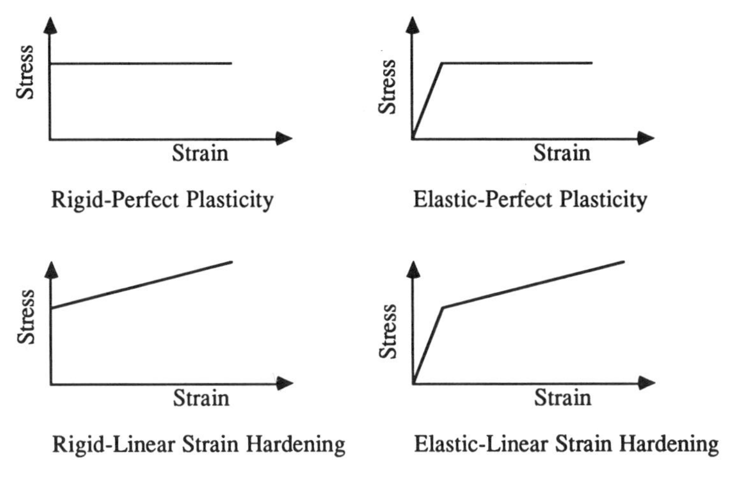

Figure 2.3.1 Four idealizations of stress–strain response curves.

curve. For example, Figure 2.3.2 shows a trilinear representation for Hastelloy X at high temperature for use in a life prediction analysis.

To develop a representation for the entire stress–strain curve to failure it is necessary to introduce the true stress and true strain variables, $\tilde{\sigma}$ and $\tilde{\varepsilon}$, respectively. These variables are defined by taking into account the continuous

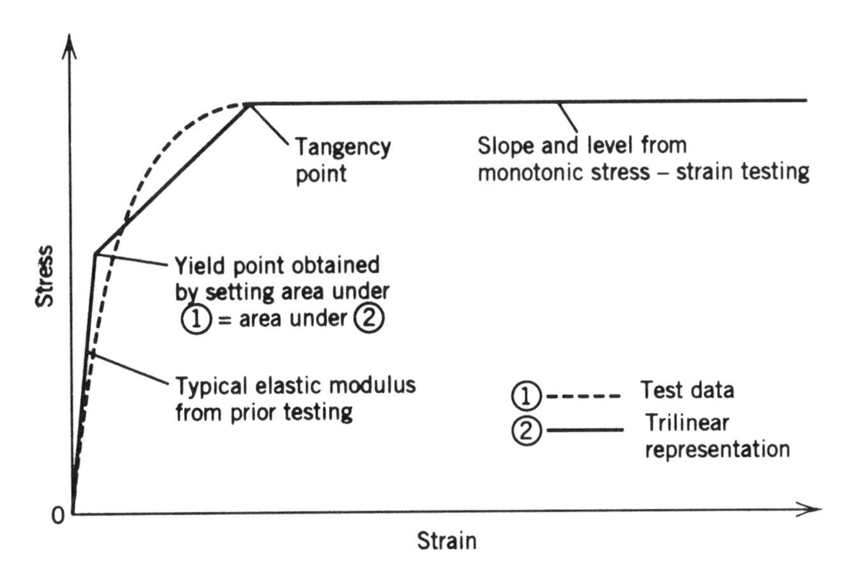

Figure 2.3.2 Construction of a trilinear stress–strain model to approximate the strain energy density of Hastelloy X for use in a MARC finite-element durability analysis. (After Moreno, 1981.)

change in length and area as a test section changes during deformation. The true stress $\tilde{\sigma}$ is defined by

$$\tilde{\sigma} = \frac{F}{A} \tag{2.3.1}$$

where A is the actual area during deformation. The true stress is related to the engineering stress and the initial area A_0 by

$$\tilde{\sigma} = \left(\frac{A_0}{A}\right) \sigma \tag{2.3.2}$$

The true strain increment is defined as an increment of change in the current length L, $d\tilde{\varepsilon} = dL/L$, which can be integrated to obtain

$$\tilde{\varepsilon} = \int_{L_0}^{L} \frac{dL}{L} = \ln \frac{L}{L_0} = \ln \frac{L_0 + \Delta L}{L_0} = \ln(1 + \varepsilon) \tag{2.3.3}$$

where ΔL is the incremental change in length and ε is the engineering strain.

These representations can be simplified by the assumption of a constant plastic volume. Since a metal deforms by planar slip there is almost no permanent change in the volume of the specimen. If a specimen of initial length and area L_0 and A_0, become L and A after loading into the inelastic region and unloading, it is expected that $L_0A_0 = LA$. This assumption is valid up to necking as long as the test section remains cylindrical. The true stress and strain can now be written as

$$\tilde{\sigma} = (1 + \varepsilon)\sigma \tag{2.3.4}$$

and

$$\tilde{\varepsilon} = \ln \frac{A_0}{A} \tag{2.3.5}$$

where σ and ε are the engineering stress and strain, respectively. The true stress and strain are usually assumed equal to engineering stress and strain when the total strain is no greater than two or three times the yield strain.

It is necessary to reflect upon implications of these equations when using them to model a tensile test. The engineering stress and strain are defined and can be used for all values of strain. Equations 2.3.3 and 2.3.4 are valid only up to the onset of necking when an average value of strain is measured with an axial extensometer. Once necking starts the strain and area in the test section are not uniform. If the cross-sectional area is measured during testing, the true stress and true strain up to fracture can be determined from Equations

2.3.1 or 2.3.2 and 2.3.5, respectively. A summary of the range of application for the engineering and true stress and strain is given in Figure 2.3.3.

An example of the difference in engineering and true stress and strain measures is shown in Figure 2.3.4 for hot-rolled AISI 1020 steel. The true stress is always greater than the engineering stress and can be much larger at failure. Before the onset of necking, engineering strain is slightly larger than the true strain according to Equation 2.3.3. However, after necking starts, the true strain as calculated from the measured area in the neck and the value of Equation 2.3.5 is much larger than the engineering strain.

Indentation hardness testing is frequently used to estimate the engineering ultimate stress. There are a number of different standard indentation tests (Brinnell, Vickers, and Rockwell hardness tests) with varying indenter geometries, deformation patterns, and force systems. The resulting mechanical test is highly nonuniform, and the local stresses and inelastic strains are difficult to understand. Time-dependent deformations may occur that affect the experimental procedures and quantitative results. As a result, the measurements are difficult to correlate to the results from a tensile test and are generally compared to calibration curves for the particular machine used in the testing. However, one approximate relationship exists between the Brinell hardness number and the ultimate engineering tensile stress, σ_u, that is,

$$\sigma_u \approx 345\text{BHN}, \quad \text{MPa} \qquad \text{or} \qquad \sigma_u \approx 0.5\text{BHN}, \quad \text{ksi} \tag{2.3.6}$$

Correlation between the Brinell, Vickers, and Rockwell hardness numbers is

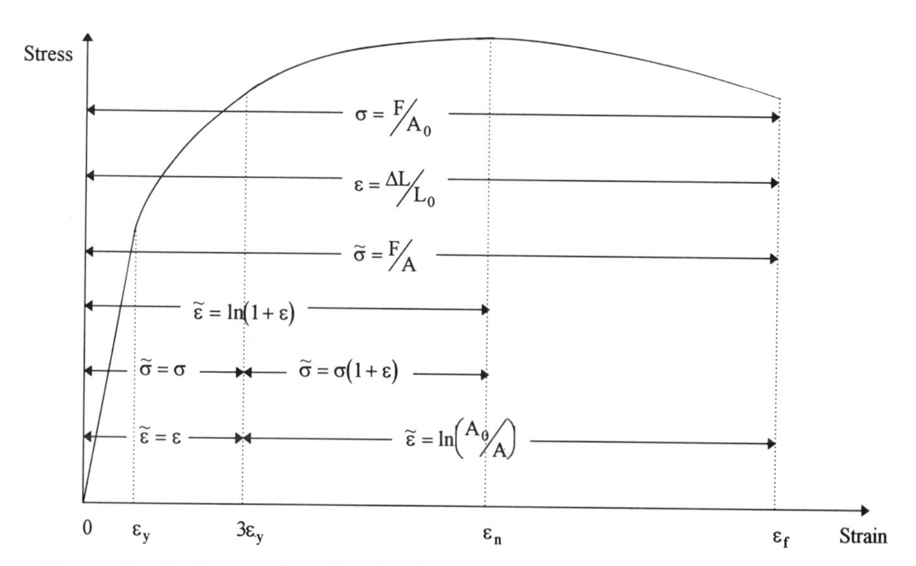

Figure 2.3.3 Range of application for the various measures of stress and strain.

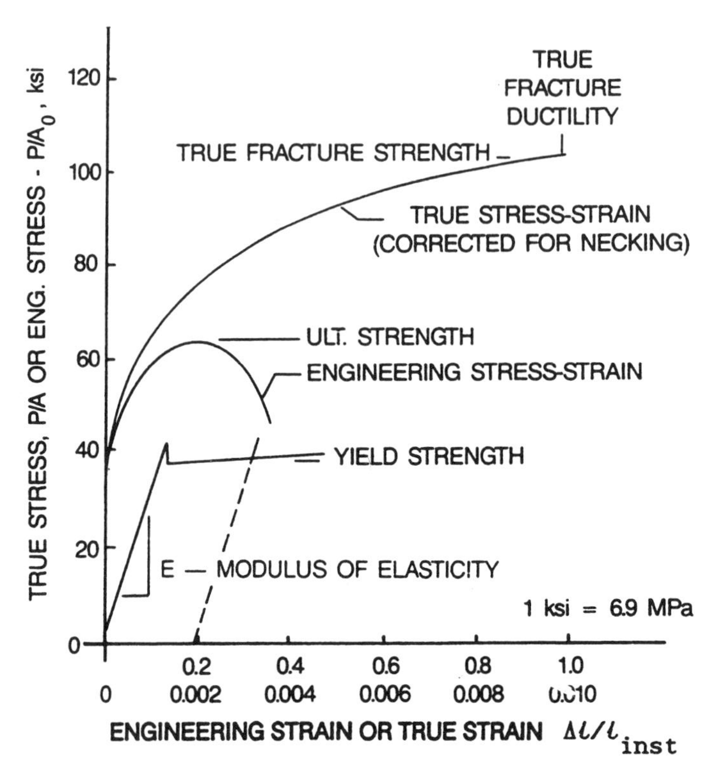

Figure 2.3.4 Comparison of engineering and true stress–strain curves from a tension test on 1020 hot-rolled steel. The small strain response with a stress overshoot is also shown. (After La Pointe, 1982; reprinted with permission from SAE Paper 820679 © 1982 Society of Automotive Engineers, Inc.)

given in Appendix 2. 1. Caution must be exercised when using data of this type.

Recall that the total strain can be decomposed into elastic and inelastic components as defined in Equation 2.1.3. For many metals a power law relationship exists between the true stress and true inelastic strain. Thus a representation can be written as

$$\tilde{\sigma} = K(\tilde{\varepsilon}^{\mathrm{I}})^n \qquad \text{or} \qquad \tilde{\varepsilon}^{\mathrm{I}} = \left(\frac{\sigma}{K}\right)^{1/n} \tag{2.3.7}$$

where K is the strength coefficient and n is the strain-hardening exponent. Substituting Equation 2.3.7 into Equation 2.1.3 gives the Ramberg–Osgood equation,

$$\tilde{\varepsilon} = \frac{\tilde{\sigma}}{E} + \tilde{\varepsilon}^{\mathrm{I}} = \frac{\tilde{\sigma}}{E} + \left(\frac{\tilde{\varepsilon}^{\mathrm{I}}}{K}\right)^{1/n} \tag{2.3.8}$$

The true fracture strength and true fracture ductility are two important quantities. The true fracture strength is the true stress at failure (i.e., $\tilde{\sigma}_f = F_f/A_f$, where F_f and A_f are force and area at fracture). The true fracture ductility is the true strain at fracture and is defined as

$$\tilde{\varepsilon}_f = \ln \frac{A_0}{A_f} = \ln \frac{1}{1 - RA} \tag{2.3.9}$$

where RA is the reduction in area, defined as $RA = (A_0 - A_f)/A_0$.

Typical values for the strain-hardening exponent are $0.01 \leq n \leq 0.5$, and the strength coefficient K can be up to several times greater than the true fracture stress. Since Equation 2.3.8 is valid at fracture, it follows that

$$K = \frac{\tilde{\sigma}_f}{(\tilde{\varepsilon}_f)^n} \tag{2.3.10}$$

which provides a useful relationship between the constants K and n. Finally, it is possible to show that the true strain at the onset of necking (corresponding to the ultimate engineering tensile strength) is equal to the strain-hardening exponent, that is,

$$n = \tilde{\varepsilon}_n \approx \tilde{\varepsilon}_n^{\mathrm{I}} \tag{2.3.11}$$

which is approximately equal to the true inelastic strain at necking for ductile materials since the elastic strains at this point are usually negligible. Parameters for the Ramberg–Osgood equation for several materials are given in Appendix 2.2.

Example Problem 2.1: A cylindrical aluminum specimen tested in tension has an elastic modulus of $E = 10{,}000$ ksi, strain-hardening exponent $n = 0.5$, strength coefficient $K = 70$ ksi, and the reduction of area at failure $RA = 0.86$. A second specimen of the same material is prestrained in tension to 20% engineering strain without producing necking. The prestrained specimen then is considered as a second material. Determine the true fracture ductility, the true fracture strength, and the engineering tensile strength for both materials.

SOLUTION: The loading of the first specimen will follow path o–a–b in Figure 2.3.5, and the second specimen will follow path o–a–c during prestraining and path c–a–b on loading to fracture. The distance o–c is 0.2 prestrain. Let us consider the solution for the first material. The true strain and stress at fracture are given by

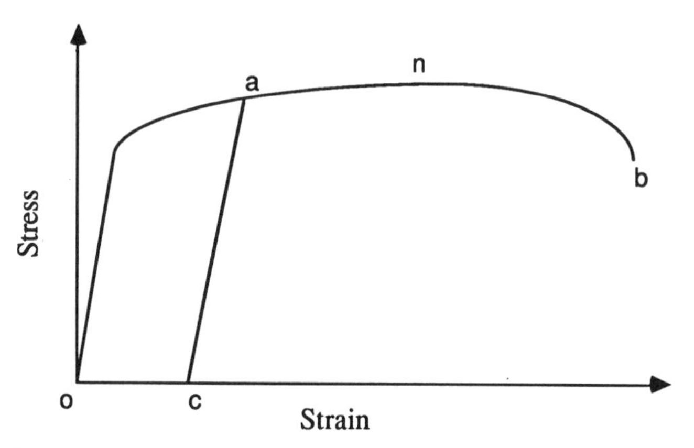

Figure 2.3.5 Engineering stress–strain diagram of the two loading histories in Example Problem 2.1.

$$\tilde{\varepsilon}_{f1} = \ln \frac{1}{1 - \text{RA}} = \ln \frac{1}{1 - 0.86} = 1.966 \approx \tilde{\varepsilon}_{f1}$$

$$\tilde{\sigma}_{f1} = K(\tilde{\varepsilon}_{f1})^n = 70(1.966)^{0.5} = 98.1 \text{ ksi}$$

The inelastic true strain at necking can be approximated as the strain-hardening exponent, $\tilde{\varepsilon}_{n1} = n = 0.5$, using Equation 2.3.11. Thus the true stress at necking can the be found from

$$\tilde{\sigma}_{n1} = K(\tilde{\varepsilon}_{n1})^n = 70(0.5)^{0.5} = 49.5 \text{ ksi}$$

To find the engineering stress at necking, it is necessary first to find the engineering strain at necking.

$$\tilde{\varepsilon}_{n1} = \ln(1 + \varepsilon_{n1}) = 0.5 \qquad \text{or} \qquad \varepsilon_{n1} = 0.649$$

$$\sigma_{n1} = \frac{\tilde{\sigma}_{n1}}{1 + \varepsilon_{n1}} = \frac{49.5}{1 + 0.649} = 30 \text{ ksi}$$

Finally, the area at necking is given by

$$\tilde{\sigma}_n = \sigma_n \frac{A_0}{A_n}, \qquad A_n = \frac{30.0}{49.5} A_0 = 0.606 A_0$$

Consider next the second or prestrained material. To determine the true strain at fracture for the prestrained specimen, it is necessary to subtract the prestrain from the true fracture ductility for the first specimen [i.e., $\tilde{\varepsilon}_{f2} = 1.966 - \ln(1 + 0.200) = 1.78$]. The total true strain at fracture and the RA are the same in both cases. The true stress at fracture is the same in both cases. To find the engineering stress at necking, recall that engineering stress is determined from the initial area, which is different for the two cases. To find the initial area of the prestrained specimen, use conservation of volume,

$$A_0 L_0 = A_{0p} L_{0p} = A_{0p} L_0 (1 + 0.200) \qquad \text{or} \qquad A_{0p} = 0.833 A_0$$

$$\sigma_{n2} = \tilde{\sigma}_n \frac{A_n}{A_{0p}} = 49.5 \left(\frac{0.606 A_0}{0.833 A_0} \right) = 36 \text{ ksi}$$

Thus the prestraining increases the engineering tensile strength and decreases the engineering ductility.

Example Problem 2.2: Estimate the true stress corresponding to a true strain of 0.100 for a material with the same properties as Example Problem 2.1.

SOLUTION: Using the Ramberg–Osgood equation gives

$$\tilde{\varepsilon} = \frac{\tilde{\sigma}}{E} + \left(\frac{\tilde{\sigma}}{K} \right)^{1/n} = \frac{\tilde{\sigma}}{10{,}000} + \left(\frac{\tilde{\sigma}}{70} \right)^{2.0} = 0.100$$

which must be solved for the true stress $\tilde{\sigma}$ using an iterative method. An upper and lower bound for the values of $\tilde{\sigma}$ can be obtained by noting that the elastic contribution to the total strain is usually very small, so an upper bound is obtained by neglecting the elastic strains,

$$\tilde{\sigma}_{\text{UB}} = K(\tilde{\varepsilon})^n = 70(0.1)^{0.5} = 22.136 \text{ ksi}$$

A lower-bound stress can be determined by first finding the inelastic strain associated with the upper-bound stress,

$$\tilde{\varepsilon}^{\text{I}}_{\text{LB}} = \tilde{\varepsilon} - \frac{\tilde{\sigma}_{\text{UB}}}{E} = 0.1 - \frac{22.136}{10{,}000} = 0.097786$$

Then the lower-bound stress is the stress for this inelastic strain,

$$\tilde{\sigma}_{\text{LB}} = 70(0.97786)^{0.5} = 21.889 \text{ ksi}$$

If a solution of $\tilde{\sigma} = 21.889$ ksi is used, the corresponding strain would be

$$\tilde{\varepsilon} = \frac{21.889}{10{,}000} + \left(\frac{21.889}{70} \right)^{2.0} = 0.099970$$

which is very close to the initial value of 0.100. The method would be slower to converge if the inelastic strains were very close to the elastic strains, and the iteration process could be continued if necessary.

2.4 STRAIN-RATE SENSITIVITY

Mechanical tests in hydraulic or screw machines are generally run under strain control using the strain extensometer in a closed-loop control system. In closed-loop testing a load, stroke, or strain control signal is supplied to a controller by an external device such as a computer or function generator. The controller converts the control signal to a motion of the actuator (Figure 2.4.1). Sensors measure the resulting stroke, load, and strain produced in the testing machine. The signals are returned to the controller. The machine response is then compared to the desired value supplied by the computer and the signal to the machine is corrected to provide the desired response. The

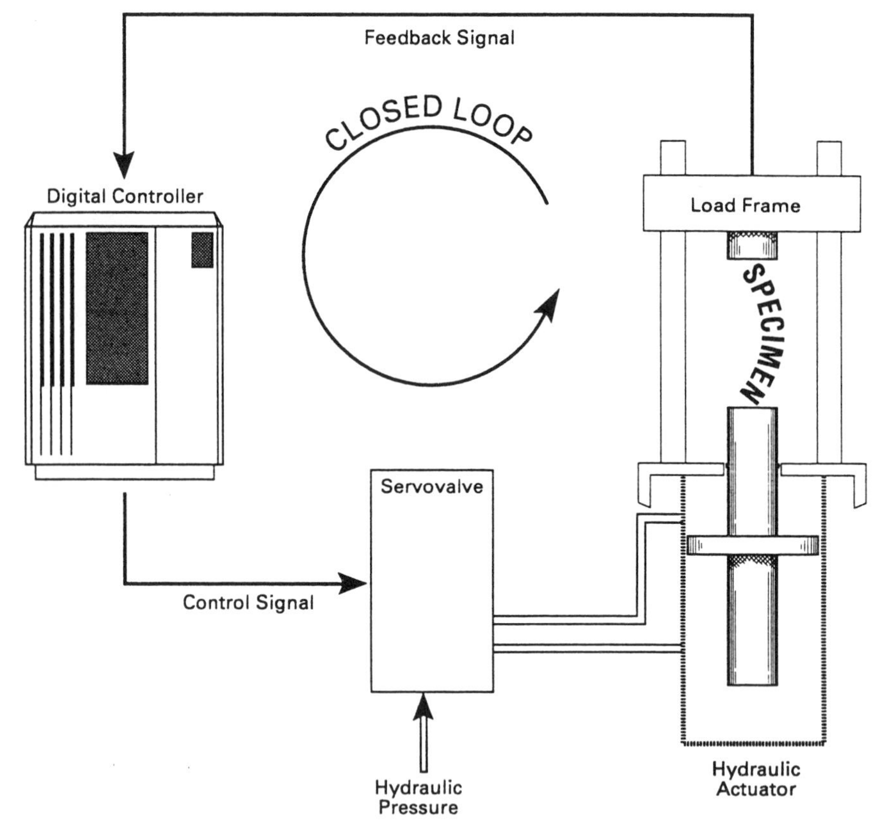

Figure 2.4.1 Closed-loop mechanical testing machine. (Courtesy of MTS Systems Corporation, Eden Prairie, Minnesota.)

feedback loop operates continuously, and the accuracy of electronic control systems is excellent.

The strain rate must be part of the test specification. This is important because the tensile response may be strain-rate sensitive, depending on the material, strain rate, and homologous temperature (ratio of absolute temperature to absolute melting temperature). For example, the stress–strain response curves for René 80 at 535°C and 871°C are shown in Figure 2.4.2. There is significant strain-rate dependence at 871°C but essentially no strain-rate effect at 538°C. The low-carbon steels shown in Figures 2.2.1 and 2.2.2 are strain-rate dependent at room temperature, with the degree of sensitivity increasing with increasing strain rate. Figure 2.4.3 shows the strain-rate-dependent response of Udimet 700 at 927°C for diametrical strain rates from −0.01 to −2.5% per minute. Tensile response curves obtained under constant diametrical strain rate are different from test results from constant axial strain rate. Note that the negative diametrical strain rates correspond to positive axial strain rates. The large drop in stress just before failure is a result of specimen necking.

Strain-rate dependence results from the strong effect that dynamic recovery has on the stress–strain curve. A schematic diagram of a stress-strain curve for a material experiencing dynamic recovery is shown in Figure 2.4.4. The overall level of the stress–strain curve increases with increasing strain rate and decreasing temperature. Since the steady-state dislocation density is determined from the balance of strain hardening and dynamic recovery rates, the steady-state level of stress depends on temperature and strain rate. If the temperature decreases, the rate of recovery decreases and a higher value for the steady-state stress is obtained. Similarly, if the rate of straining increases, a higher value of the dislocation density is obtained at steady state, and a larger saturation stress level is achieved. In this case the rate of recovery also increases, due to the increase in the dislocation density, but the net effect is an increase in the number of dislocations. At low homologous temperatures the strain-rate effect essentially disappears because the rate of dynamic recovery is much slower and becomes negligible compared to the rate of hardening.

The effect of dynamic recovery can also produce stress relaxation. In a stress relaxation test the strain is held constant at a fixed temperature and the reduction in stress is measured as a function of time. Figure 2.1.2 shows the stress relaxation in 2 min, in two tensile tests that were run at 0.800 and 0.001 sec^{-1} An example of the relaxation in stress in the same material at room temperature is shown as a function of time in Figure 2.4.5. It is interesting to observe that this material was not strain-rate sensitive but did exhibit stress relaxation.

The dynamic properties of strain hardening and recovery can be observed in strain-rate jump tests. In these tests the strain rate is changed rapidly by two or three orders of magnitude. For example, the strain-rate jump response of AISI type 304 stainless steel at room temperature is shown in Figure 2.4.6.

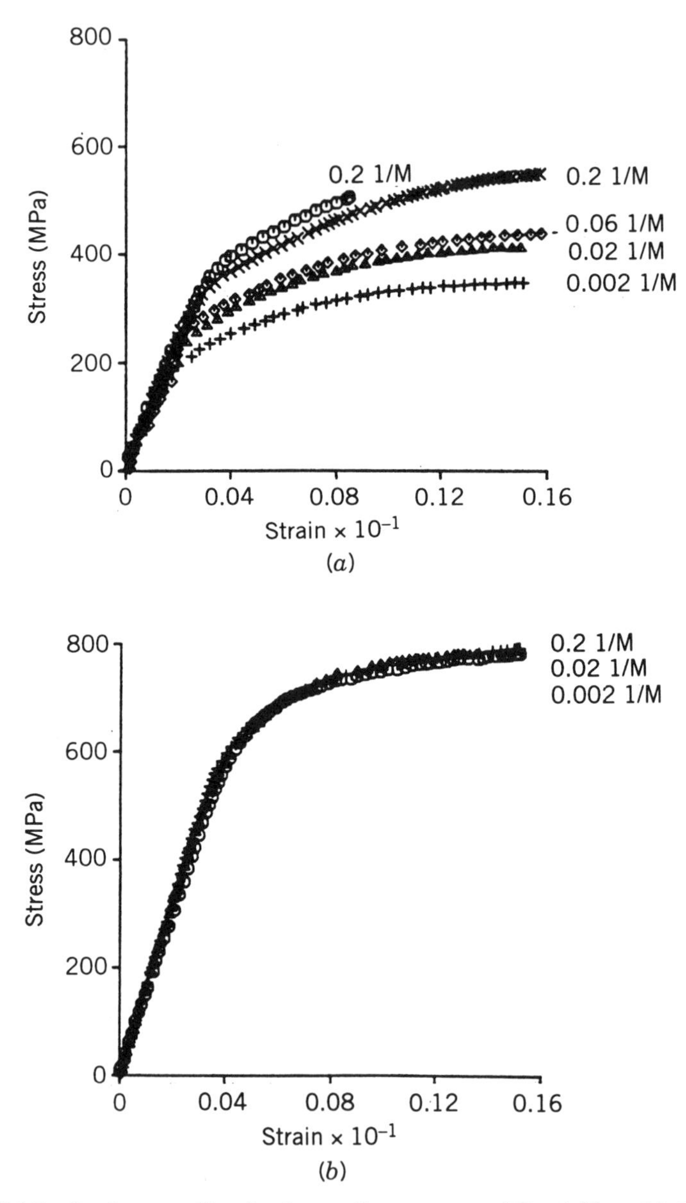

Figure 2.4.2 Strain-rate effect in the tensile response of René 80 at (a) 891°C and (b) 538°C. (After Ramaswamy, Stouffer, and Laflen, 1986.)

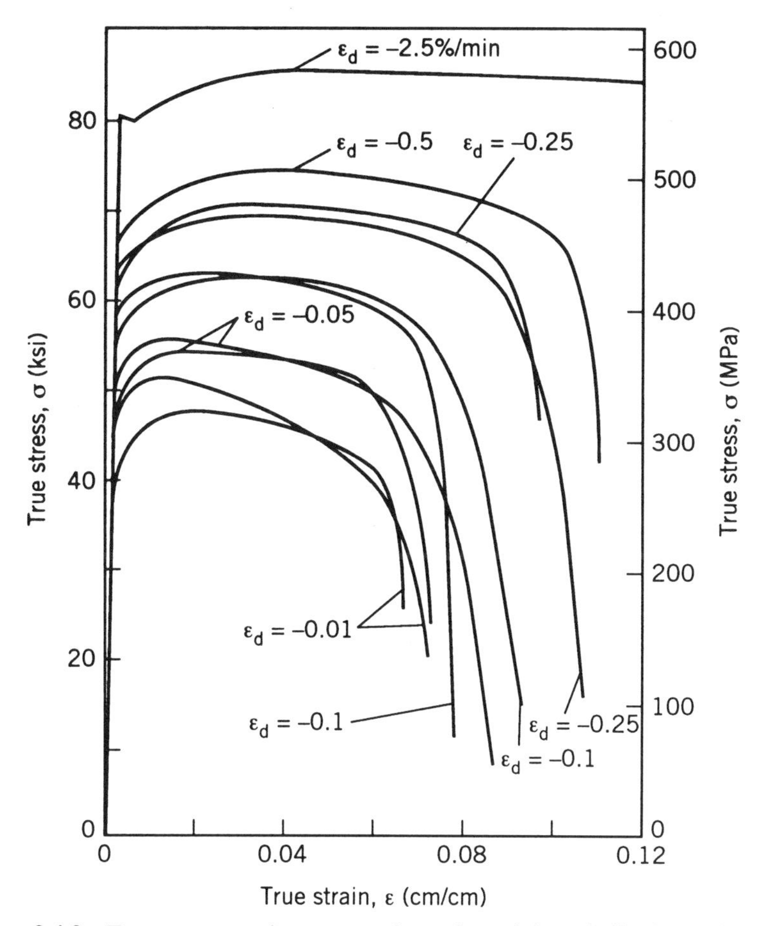

Figure 2.4.3 True stress–strain curves where the axial strain is determined from diametrical strain measurements at several constant diametrical strain rates for wrought Udimet 700 at 927°C. (After Laflen and Stouffer, Journal of Engineering Materials and Technology, V100, 1978. Copyright © The American Society of Mechanical Engineers, 345 East 47th St., New York, NY 10017. Used with permission.)

The response after the jump in strain rate was ordered and near the expected value, and the previous strain rate does not appear to influence the response after the jump. Krempl (1979) reported that there was an elastic slope at the change in strain rate in all tests except for one jump test from 10^{-5} to 10^{-2} sec^{-1}. In this case, the shallow slope was attributed to a slow XY-recorder response time. Thus the change in the balance between strain hardening and recovery during the jump is faster than the characteristic time of events in the experiment. Dynamic recovery really is dynamic.

Strain-rate effects at very high strain rates are observed in split Hopkinson bar testing. The Hopkinson bar assembly consists of a striker, a pressure bar,

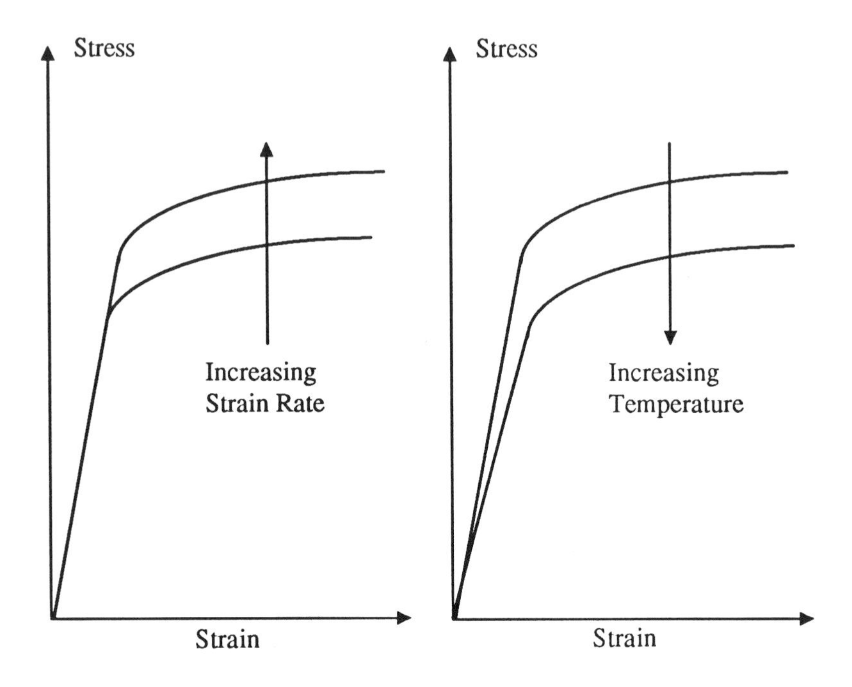

Figure 2.4.4 Diagram showing the effect of temperature and strain rate on the tensile response of metals. Notice that the strain rate generally does not effect the elastic modulus, but the modulus is a function of temperature.

a test specimen, and a transmission bar, as shown in Figure 2.4.7. In operation, the striker is fired and contacts the pressure bar. An elastic–plastic compression wave results that propagates downs the pressure bar, through the specimen and transmission bar, then reflects off the end of the transmission bar and returns as a tensile wave. For tensile testing the specimen can be supported with a split shoulder bar that falls off on return of the tensile wave. The displacement and force supplied to the specimen are determined from the wave characteristics in the pressure and transmission bars that are observed at two locations by strain gages. The data from split Hopkinson bar tests is noisy in comparison to that from hydraulic or screw-driven machines. The strain rate–time and true stress–true strain results from two tests are shown in Figure 2.4.8. The signals observed are compared by fitting a smooth curve through the data, as shown in Figure 2.4.9. The remaining data in this section have been reduced using a smoothing technique and can only be viewed as approximate.

Experimental results from static and dynamic tests are shown in Figure 2.4.10 for C1008 carbon steel. The static tests are from closed-loop hydraulic or screw machine testing and are assumed to be rate independent. The high-rate (1750 sec^{-1}) and low-rate (300 sec^{-1}) results are near the extremes of the method (approximately 10^2 to 10^4 sec^{-1}). There is a range of strain rates

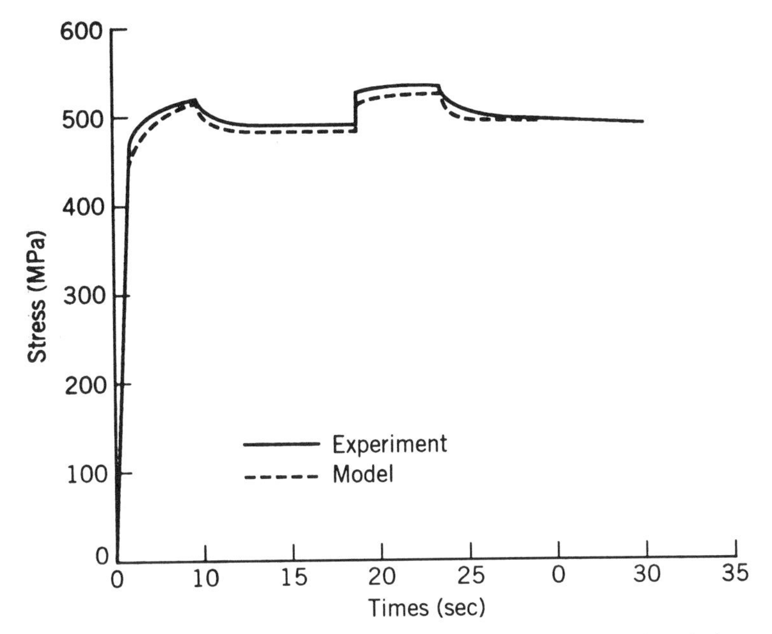

Figure 2.4.5 Stress–strain record with two short, 10-s, strain hold periods at strain levels of 0.045 and 0.085 for Al 7050–T7451 at room temperature. The model is presented in Chapter 6 and the material parameters are given in Appendix 6.3. (From Kuruppu et al., 1992; reprinted by permission of the Council of the Institution of Mechanical Engineers from the *Journal of Strain Analysis*.)

between 1 and 10^2 sec^{-1} where pneumatic or special mechanical machines are required. Extremely high strain rates (10^4 to 10^7 sec^{-1}) can be obtained with shock-wave propagation methods.

A simple empirical model for the tensile stress or ultimate dynamic flow stress in strain-rate-dependent materials was proposed by Cowper and Symonds (Symonds, 1965). In their model the strain-rate-dependent ultimate flow stress σ' is given by

$$\frac{\sigma'}{\sigma_y} = 1 + \left(\frac{\dot{\varepsilon}}{C}\right)^p \tag{2.4.1}$$

where σ_y is the yield stress associated with the static response of the material and $\dot{\varepsilon}$ is the strain rate from a constant-strain-rate tensile test. Typical values for parameters p and C for a few common structural materials are: $C = 40.4$ sec^{-1} and $p = 5$ for mild steel, $C = 6500$ sec^{-1} and $p = 4$ for structural aluminum, and $C = 100$ sec^{-1} and $p = 10$ for type 304 stainless steel.

In conclusion, strain-rate sensitivity depends on the material, strain rate, and temperature. A summary of the effect of strain rate for about six orders of magnitude is given in Figure 2.4.11 for several materials. Stress relaxation

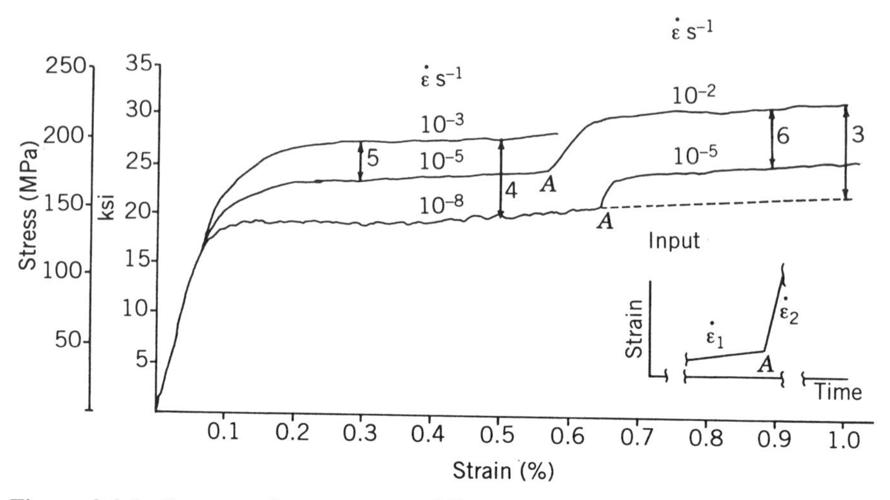

Figure 2.4.6 Stress–strain response at different strain rates of annealed AISI type 304 stainless steel at room temperature. The strain rate was instantaneously changed at point A in the loading history. (Preprinted from *Journal Mechanics and Physics of Solids,* V27, Krempl, "An Experimental Study of Room Temperature Rate Sensitivity Creep and Stress Relaxation of Type AISI 304 Stainless Steel," p. 363, 1979, with kind permission from Elsevier Science Ltd.)

can be present even if there is no strain-rate sensitivity. The models discussed in Section 2.3 are limited to strain-rate-independent response unless different material constants are used for each strain rate.

2.5 UNIAXIAL CYCLIC RESPONSE

Cyclic response characteristics are a very important part of the mechanical response characteristics of metals. It is necessary to know the correct elastic and inelastic strains and the associated stress states to determine the life or durability of structures. In fact, there is frequently much difficulty in predicting life because the cyclic response under the actual loading and environmental conditions is not known. In this section it will be shown that the uniaxial cyclic response of metals is much different from the tensile and compressive response. There can be hardening or softening that is not present in monotonic loading.

To begin the discussion of cyclic response, suppose that two test specimens are loaded, one in tension and one compression, to two points in the inelastic range, points A and A', as shown in Figure 2.5.1, respectively. Before arriving at points A and A', the material yielded in tension and compression, and two yield stresses, σ_{yt} and σ_{yc}, respectively, are recorded. Recalling the discussion

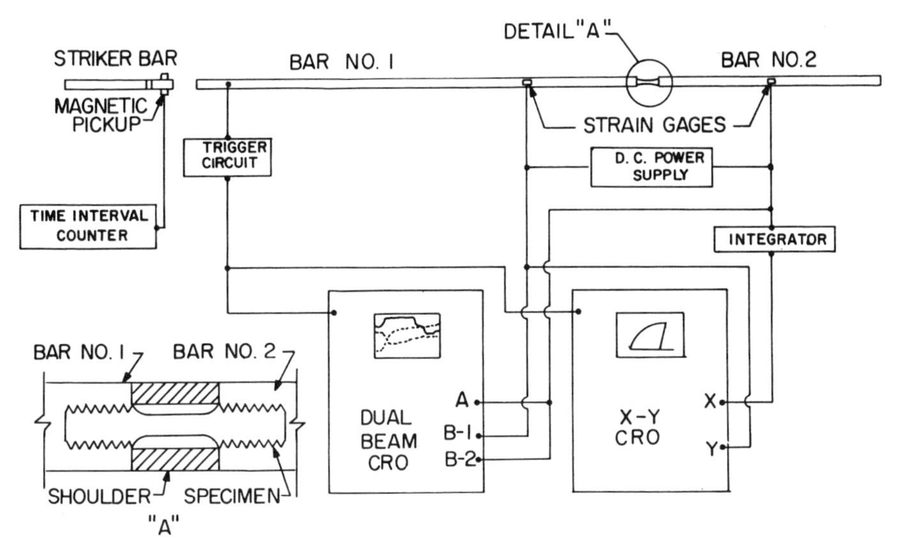

Figure 2.4.7 Major components and instrumentation of a Split Hopkinson Bar. The strain gages are mounted precisely at the quarter point and center of bars 1 and 2, respectively, so the transmitted and reflected pulses are time coincident. The lengths of the striker and transmission bars are chosen so that no spurious pulses interfere with the data pulses. Two cathode-ray oscilloscopes are used to record the data. (After Nicholas, 1980.)

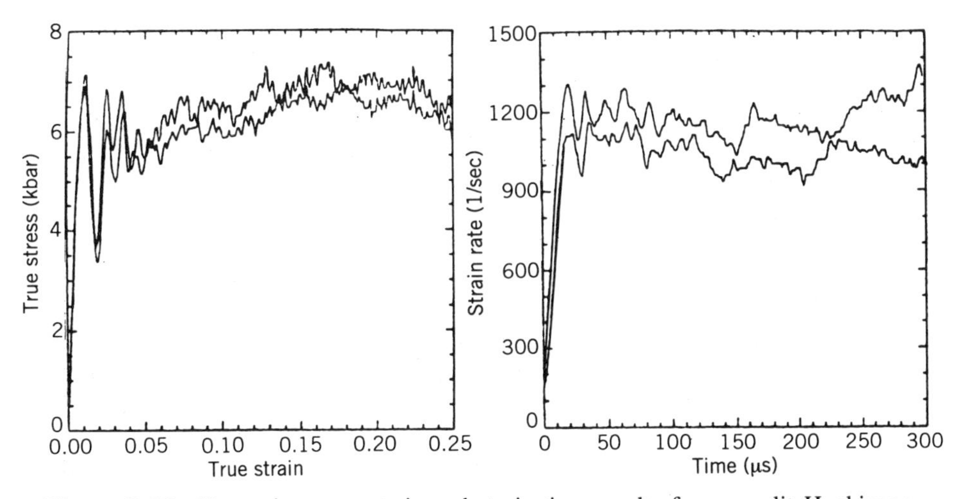

Figure 2.4.8 Dynamic stress–strain and strain-time results from a split Hopkinson bar test on 1020 cold-rolled steel at room temperature. (After Rajendran and Bless, 1985.)

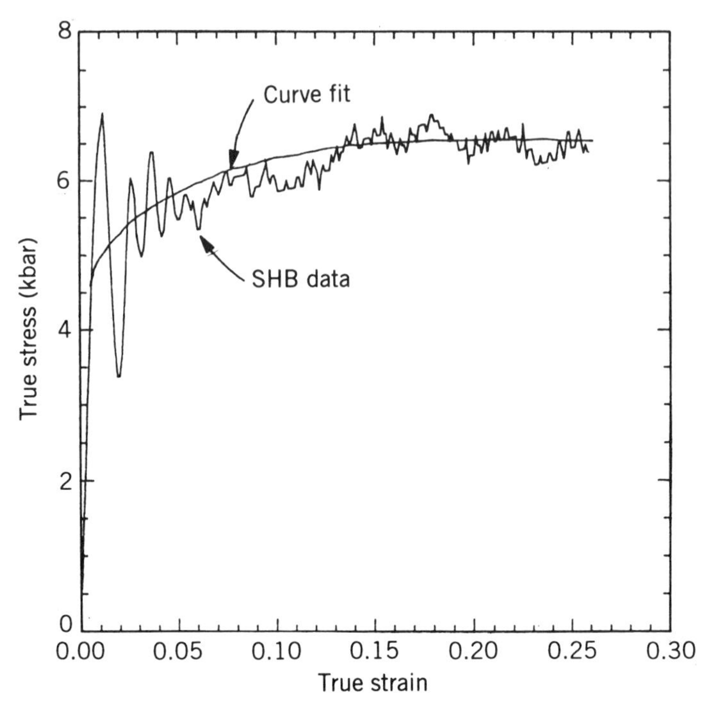

Figure 2.4.9 Correlation between the original data and a constitutive model for split Hopkinson bar data from a test on 1020 cold rolled steel. (After Rajendran and Bless, 1985.)

about slip in tension and compression from Chapter 1, it follows that the two yield stresses in tension and compression are approximately equal and opposite in sign, $\sigma_{yt} = -\sigma_{yc}$. Now consider a third specimen. On loading the specimen in tension to point A and unloading, the response curve will follow a nearly elastic line with slope E, cross into compression, and yield at point B that is less in magnitude than σ_{yc}. In fact, the yield stress in compression is approximately $2\sigma_{yt}$ below point A. Thus the strain hardening in tension reduced the magnitude of the subsequent yield stress in compression. This phenomenon is known as the Bauschinger effect.

The Bauschinger effect is caused by the back stress between dislocations that pile up at barriers such as precipitates, grain boundaries, and other dislocations. Recall that on loading in one direction there are repulsive forces between dislocations, a back stress, that must be overcome for slip to continue. On unloading and reloading in the opposite direction, slip occurs at a lower value of shear stress since the existing back stress from the prior deformation is in the same direction as the shear stress. As loading in the reverse direction proceeds, dislocations will pile up on the opposite side of the barriers and hardening in compression will be developed.

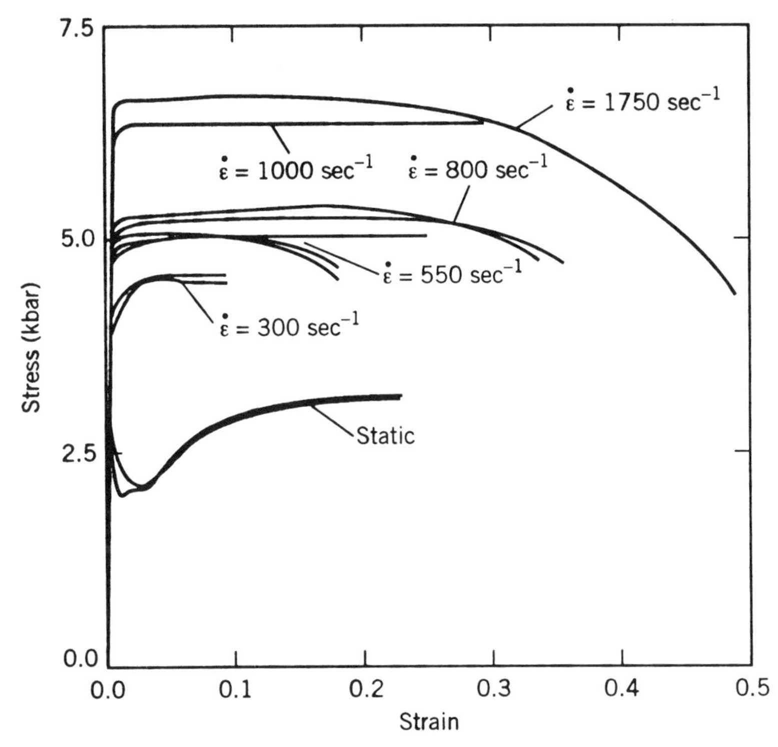

Figure 2.4.10 Results from dynamic and static tests on C1008 low-carbon steel show the strong effect of strain rate in the presence of stress waves. (After Rajendran and Bless, 1985.)

Materials that are strain-rate dependent in tension and compression will also have strain-rate dependence in cyclic loading. For example, consider the experimental results shown in Figure 2.5.2. In Figure 2.5.2*a* two AISI type 304 stainless steel test specimens were loaded at room temperature to the same value of inelastic strain in tension at two different strain rates and then loaded in compression at the same strain rate. In this case the maximum stresses at points *A* and *B* are different, reflecting the strain-rate-dependent characteristics of the material, but the two curves in compression are identical. Thus the strain-rate effect in tension has no effect on the subsequent yield response in compression. This confirms that the balance between the rate of strain hardening and dynamic recovery is dynamic. In the second set of experiments two specimens are loaded in tension at the same rate, then loaded in compression at different rates. In all four experiments the strain rate effects are ordered as expected, higher strain rates correspond to higher flow stresses, and the subsequent yield in compression shows the Bauschinger effect. Finally, note that defining the subsequent yield in compression as $-2\sigma_{yt}$, as

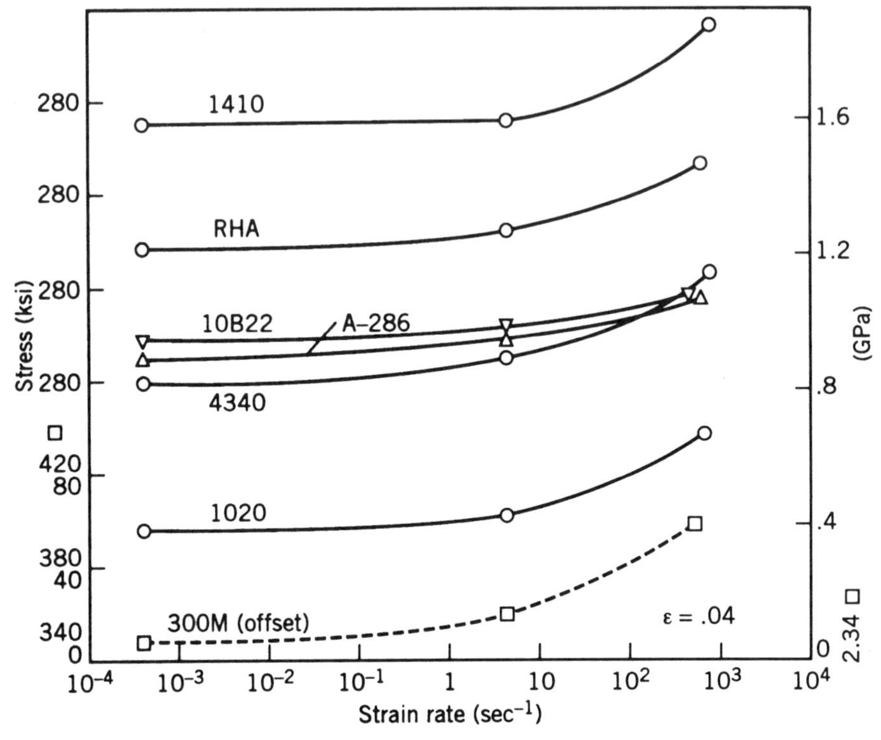

Figure 2.4.11 Effect of strain rate on stress at 4% strain for several structural steels. (After Nicholas, 1980.)

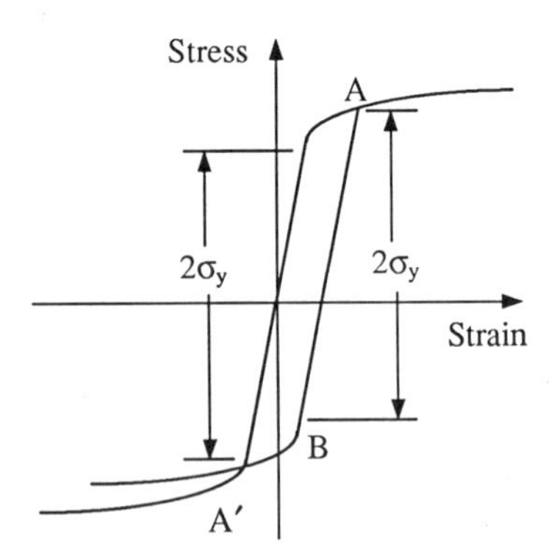

Figure 2.5.1 Bauschinger effect correlating the yield in compression and strain hardening in tension.

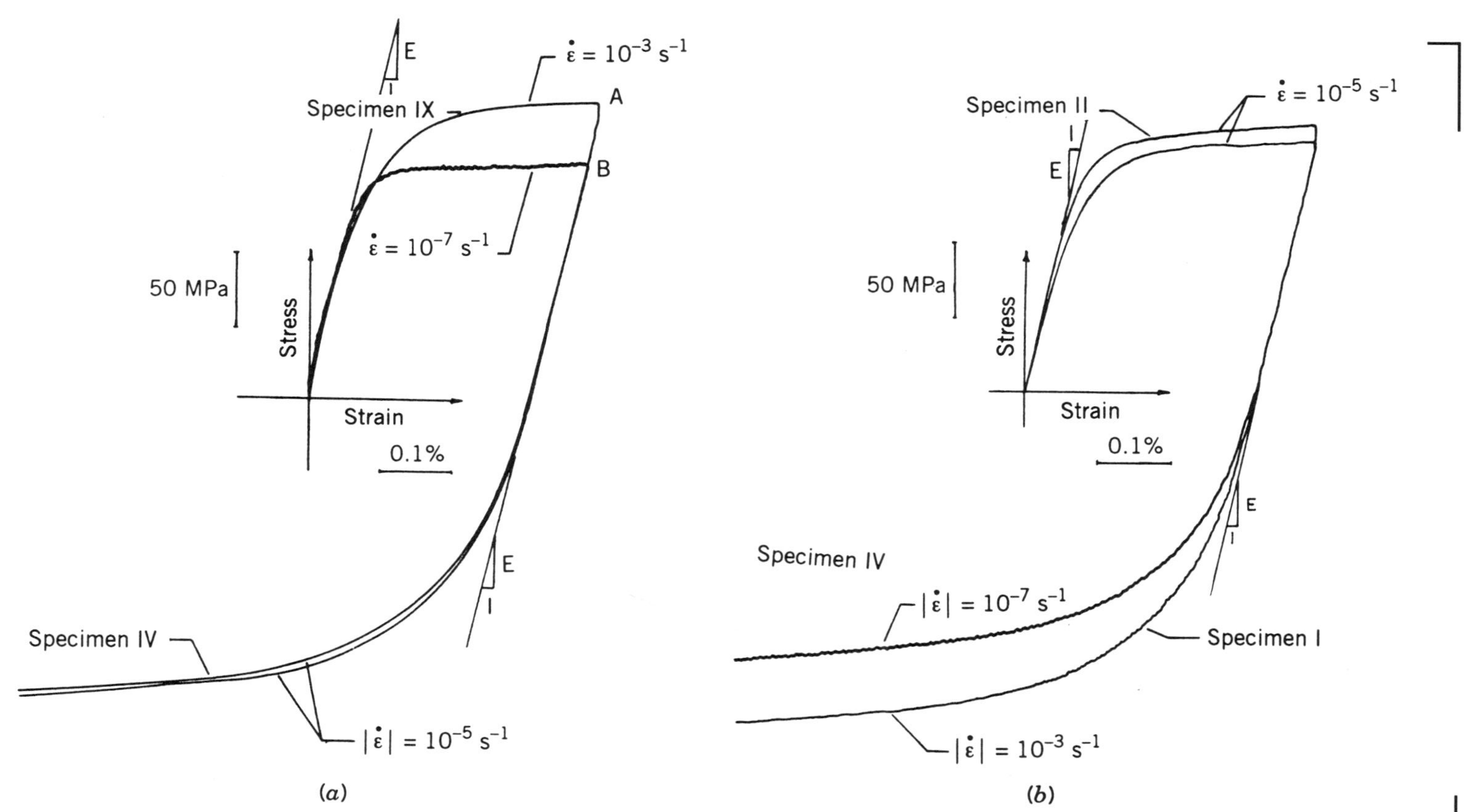

Figure 2.5.2 Effect of strain rate on the Bauschinger effect in AISI type 304 stainless steel at room temperature. (Reprinted from the *Journal of Mechanics and Physics of Solids,* V32, Krempl and Kallianpur, "Some Critical Uniaxial Experiments for Viscoplasticity at Room Temperature," p. 301, 1984, with kind permission from Elsevier Science Ltd.)

shown in Figure 2.5.1, is only a rule of thumb, and there is even some scatter in the initial yield stress in tension in the four tests.

Experiments in cyclic fatigue must be designed to use test specimens that are stable in compression. The hourglass design or the cylindrical design with a short test section (see Figure 2.1.5) is frequently used to limit buckling. Another issue in cyclic experiments is the collection of data. A cyclic test with 100,000 cycles, for example, can be overwhelming in data. There are two issues in data collection, the number of data points per cycle and the number of cycles per test. Data collection guidelines depend on the material under study and the purpose of the test. For work in low-cycle fatigue (life less than 100,000 cycles when inelastic strains are present) it is necessary to obtain enough data to characterize the shape of the fatigue loop. Data collection exercises with fatigue loops containing 100, 150, 200, 400, and 600 data points were conducted to find parameters such as the inelastic strain rate from the stress, strain, and time test records. It was found that data sets with fewer than 200 data points were not sufficient for accurate definition of slope with a sliding spline, and there was no increase in accuracy with more the 200 data points per loop. Thus we generally collect 200 data points per loop to define the shape of a loop for later inelastic analysis. Next, it is necessary to decide how many fatigue loops of data are to be collected for each specific test and when to collect most of the data. Frequently, it is useful to characterize transients in the response at the beginning of the test, near fracture, and at changes in the load cycle. Thus these events are considered in the data collection scheme.

The parameters used in cyclic loading are defined to characterize the dimensions and position of the fatigue loop in the stress and strain coordinates. Letting $\sigma_{\max}$ and $\sigma_{\min}$ designate the maximum and minimum engineering stress in a cyclic experiment, the following definitions can be established:

$$\text{Stress range:} \quad \Delta\sigma = \sigma_{\max} - \sigma_{\min}$$

$$\text{Stress amplitude:} \quad \sigma_a = \frac{\sigma_{\max} - \sigma_{\min}}{2}$$

$$\text{Mean stress:} \quad \sigma_m = \frac{\sigma_{\max} + \sigma_{\min}}{2}$$

$$\text{Stress ratio } R\text{:} \quad R_\sigma = \frac{\sigma_{\min}}{\sigma_{\max}}$$

$$\text{Amplitude ratio } A\text{:} \quad A_\sigma = \frac{\sigma_a}{\sigma_m}$$

The definitions of these parameters are shown graphically in Figure 2.5.3*a* for constant-amplitude engineering stress. Similar definitions can be applied

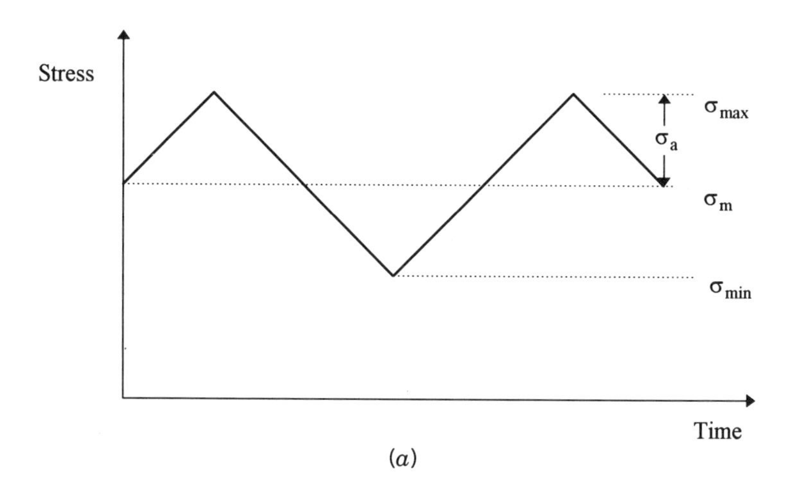

(a)

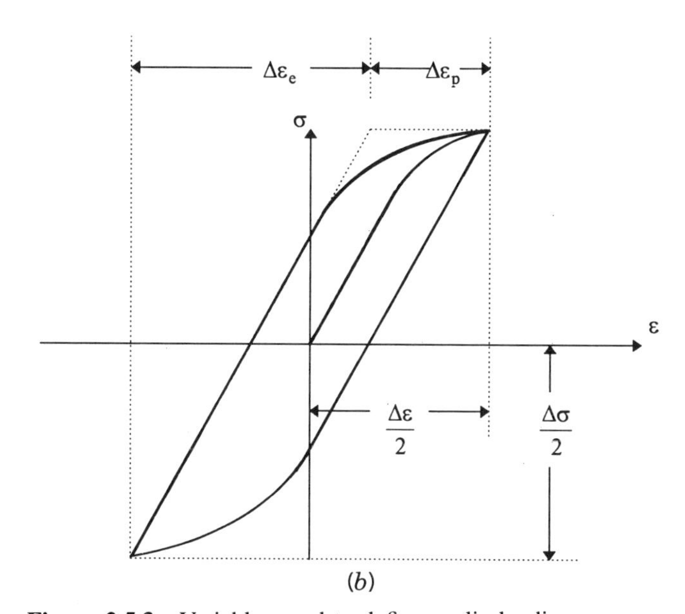

(b)

Figure 2.5.3 Variables used to define cyclic loading.

to engineering strain ε, true stress $\tilde{\sigma}$, and true strain $\tilde{\varepsilon}$. The definitions applied to the elastic and inelastic strain ranges and the stress range of a hysteresis loop are also shown graphically in Figure 2.5.3*b*.

In cyclic loading there are three general types of effects—hardening, softening, and cyclically stability—as shown in Figure 2.5.4. In cyclic hardening materials the stress range increases and the inelastic strain range decreases

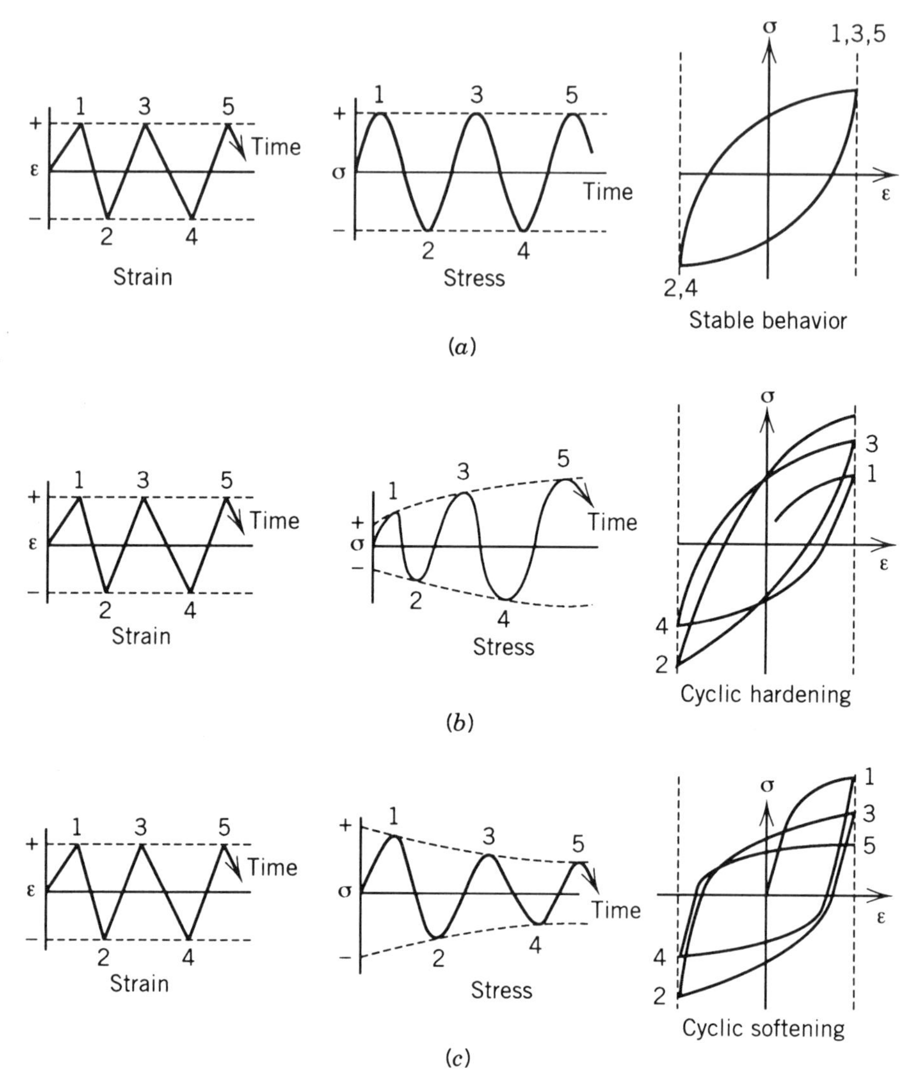

Figure 2.5.4 Cyclic hardening, cyclic softening, and cyclic stability for constant-strain-range controlled test. (From Barsom, Klippstein, and Shoemaker, 1980.)

for a constant total strain range experiment. Conversely, in cyclic softening materials, the stress range decreases and the inelastic strain range increases. Cyclically stable metals do not produce much hardening or softening during cycling and the size and shape of the hysteresis loop are nearly constant.

The three different types of hardening effects are present in cyclic loading of metals (hardening, softening, relatively stable behavior), depending on the initial microstructure of the material. An example of the effect of initial hardness on the cyclic response of oxygen-free high-conductivity (OFHC) copper was determined by Morrow (1965) and is shown in Figure 2.5.5. The material that was fully annealed demonstrated the most hardening, and the material that was fully hardened by cold working exhibited the most softening. The partially hardened material hardened initially, then softened as cycling continued, and ultimately stabilized relatively close to the initial state. An example of the inelastic strain history in a fully reversed stress-controlled experiment is shown in Figure 2.5.6 for René 95 at 649°C. Initial specimen hardening is recognized by the reduction in the inelastic strain range. The response passed through a quasi-stable state relatively early in life and then slowly softened until fracture. Note that the changes in the inelastic strain range are relatively slow after the initial hardening. The lack of symmetry in the response resulted from the stress-controlled experiment and the Bauschinger effect.

Generally, materials that are initially soft will cyclic harden and materials that are initially hard will cyclic soften. For initially soft materials, the initial dislocation density is relatively low and cyclic loading increases the dislocation density, hence cyclic hardening results. Initially hard materials were processed to produce a strengthening mechanism such as cold working, addition of a solute atoms, or precipitate hardening. The softening mechanism during cycling depends on the initial hardening mechanism. For example, in cold-worked materials the cyclic deformation frees initially pinned dislocations and recovery results since the dislocation density is not in an equilibrium state. Softening of precipitate-hardened metals can occur from dislocations shearing the precipitates. In most cases the cycling disrupts the initial hardening mechanism, and softening results.

A method to characterize the cyclic properties of a metal for a variety of strain ranges is the cyclic stress–strain curve. The basic construction of a cyclic stress–strain curve is shown schematically in Figure 2.5.7. A material is cycled at a constant fully reversed strain range, $R_\varepsilon = -1$, until a stabilized hysteresis loop is obtained. The strain range is then increased a number of times and the process is repeated each time. The stabilized values of the stress and strain amplitudes are then plotted to form the cyclic stress–strain curve. The stress–strain curve corresponds to the mechanical response of the material with a fully developed microstructure. Depending on the strain rate and temperature, the microstructure will be a dislocation network of cells, subgrains, or both, as discussed in Chapter 1. The material state corresponding to a fully developed microstructure is generally used for predicting the low-cycle fatigue life of the material when inelastic strains are present.

Martin (1973) presented an efficient method to develop a cyclic stress–strain curve experimentally, following the definition outlined in Figure 2.5.8. In this case a test specimen is cycled while the strain range is slowly increased and decreased. The specimen generally fails in 20 to 30 blocks. A cyclic stress–strain curve is constructed from the stress range and strain range data after three to four loading blocks, near stabilization, or at midlife. Ban-

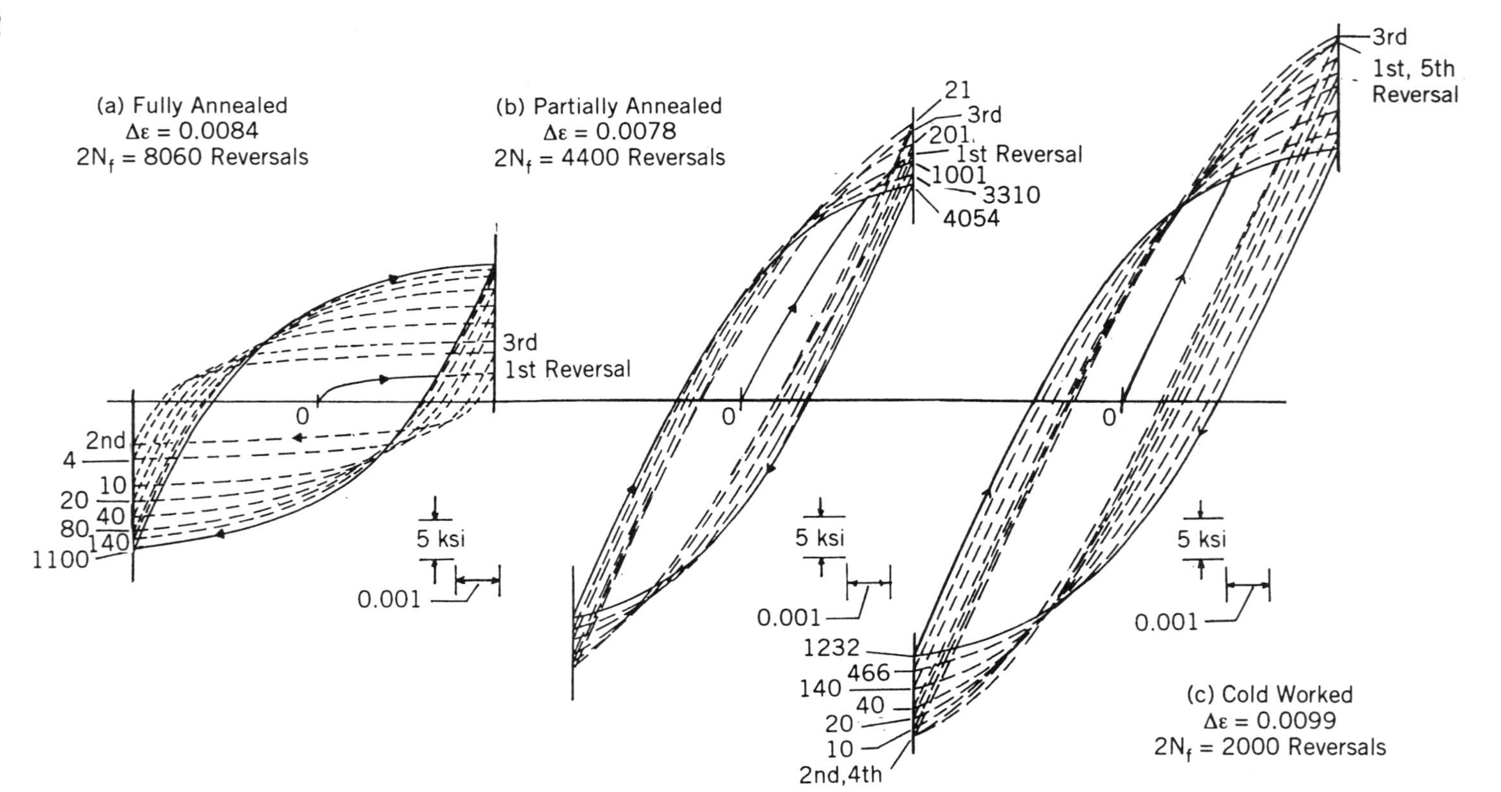

Figure 2.5.5 Cyclic response of OFHC copper in three different initial conditions. (After Morrow, 1965; copyright ASTM, reprinted with permission.)

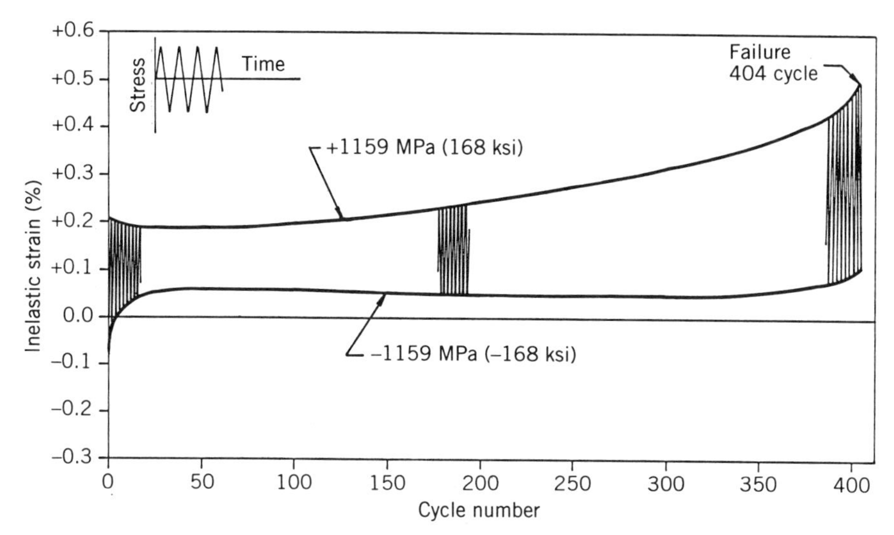

Figure 2.5.6 Envelope of the inelastic strain for the entire life of René 95 at 649°C. The test was started in compression and fully reversed cycled at 168 ksi. (From Stouffer, Papernik, and Bernstein, 1980.)

nantine, Comer, and Handrock (1990) reported that if the specimen is pulled to failure after three or four loading blocks, the resulting stress–strain curve is very close to the cyclic stress–strain curve. The response to block or uniform cyclic loading never does truly stabilize. Cyclic softening materials tend to soften continuously at a decreasing rate until the initiation of a crack; then the "apparent softening" accelerates. Cyclic hardening materials harden at a decreasing rate and approach a "pseudo stable" period when the rate of hard-

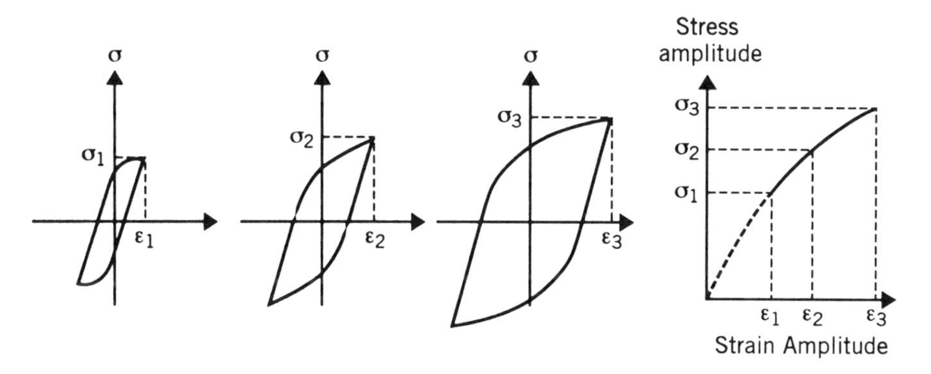

Figure 2.5.7 Construction of a cyclic stress–strain curve, which passes through the peak of stabilized hysteresis curves obtained at various strain ranges.

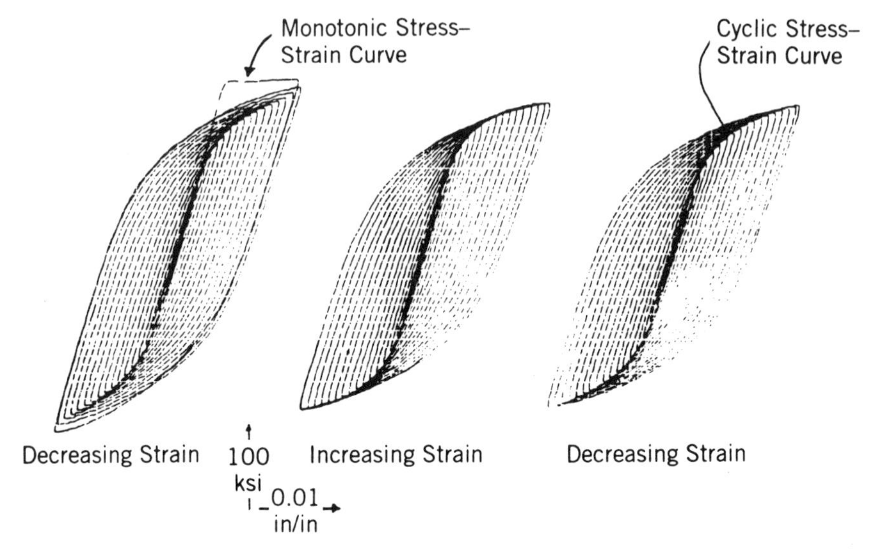

Figure 2.5.8 Typical strain history in an incremental step test and cyclic stress–strain response of quenched and tempered SAE 4142, 380 BHN at room temperature. (Landgraf, Morrow, and Endo, 1969; copyright ASTM, reprinted with permission.)

ening is offset by microcracking. The material looses strength when a macroscopic crack forms and grows. There can be a reasonable amount of cyclic life between crack initiation and failure.

A comparison between the cyclic and monotonic stress strain curves for the same materials is shown in Figure 2.5.9. This figure shows that the cyclic and monotonic stress–strain curves can be relatively close or significantly different. In all cases the cyclic stress–strain curve is smooth and monotonically increasing at a decreasing rate, and the initial monotonic strain-hardening effects, such as the stress overshoot in SAE4340, are not present. The observation that soft materials harden and hard materials soften was used by Manson and Hirshberg (1964) to develop the following rules to predict the cyclic response from the true ultimate stress and yield stress in a monotonic test:

If $\tilde{\sigma}_{\text{ult}} > 1.4\sigma_{\text{yield}}$	material will cyclically harden
If $\tilde{\sigma}_{\text{ult}} < 1.2\sigma_{\text{yield}}$	material will cyclically soften
If $1.2\sigma_{\text{yield}} \leq \tilde{\sigma}_{\text{ult}} \leq 1.4\sigma_{\text{yield}}$	material is nearly stable and will not cyclically harden or soften very much

Manson's rule is only a general trend and predictions should be used with great care. For example, Waspaloy A (see Figure 2.5.9) does not follow Manson's rule.

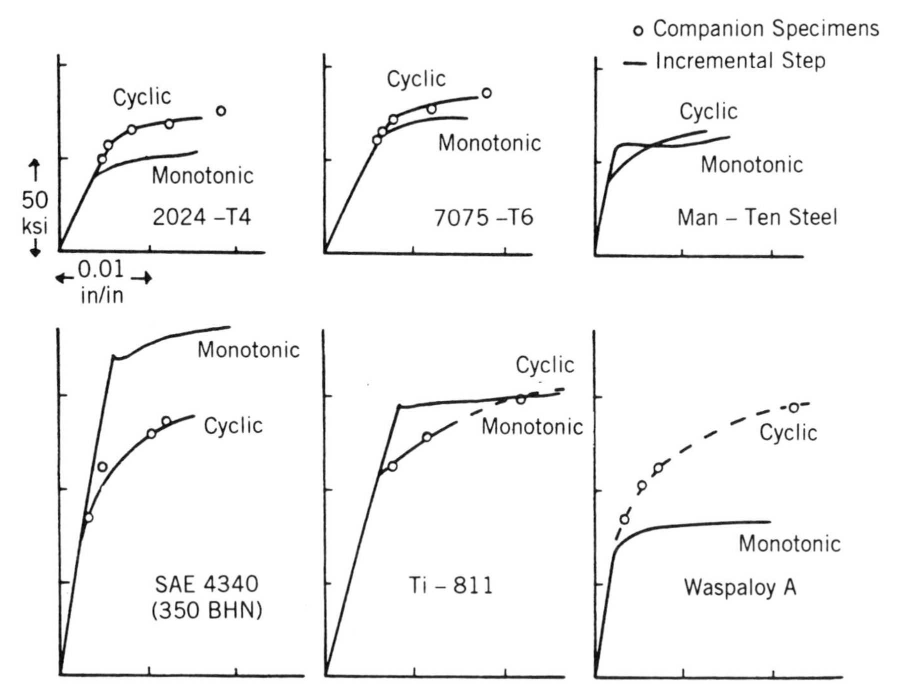

Figure 2.5.9 Several cyclic and monotonic stress–strain curve for engineering materials. (From Landgraf, Morrow, and Endo, 1969; copyright ASTM, reprinted with permission.)

Cyclic loading does not always produce low cyclic fatigue with inelastic strain in every hysteresis loop. The presence of mean strain can change the shape and size of the hysteresis loops observed in a test. Under certain conditions elastic shakedown results. In fact, this condition is highly desirable from a life management perspective.

Consider, for example, the hysteresis loops in Figure 2.5.10 that were predicted by a state-variable model that will be presented in Chapter 6. The model and experimental results were conducted with a tensile mean strain, $A_\varepsilon = +1$. Notice that the peak stresses decrease, the mean stress also decreases, and the response along paths *AA* and *CC* are fully elastic. The loops shown in the figure represent the entire test of 183 cycles. The reversed inelastic flow in cycles *BB* produce mean stress relaxation. The test data in Figure 2.5.10 have the same strain range but a very different stress range, due to the inelastic strain history. Also notice the strain range from the origin to *A* produces inelasticity, but path *AA* with the same strain range is fully elastic.

The cyclic stress–strain curve can generally be modeled better than the monotonic stress–strain curve using the Ramberg–Osgood equation. Cyclic stress–strain data are easier to model than monotonic stress–strain data be-

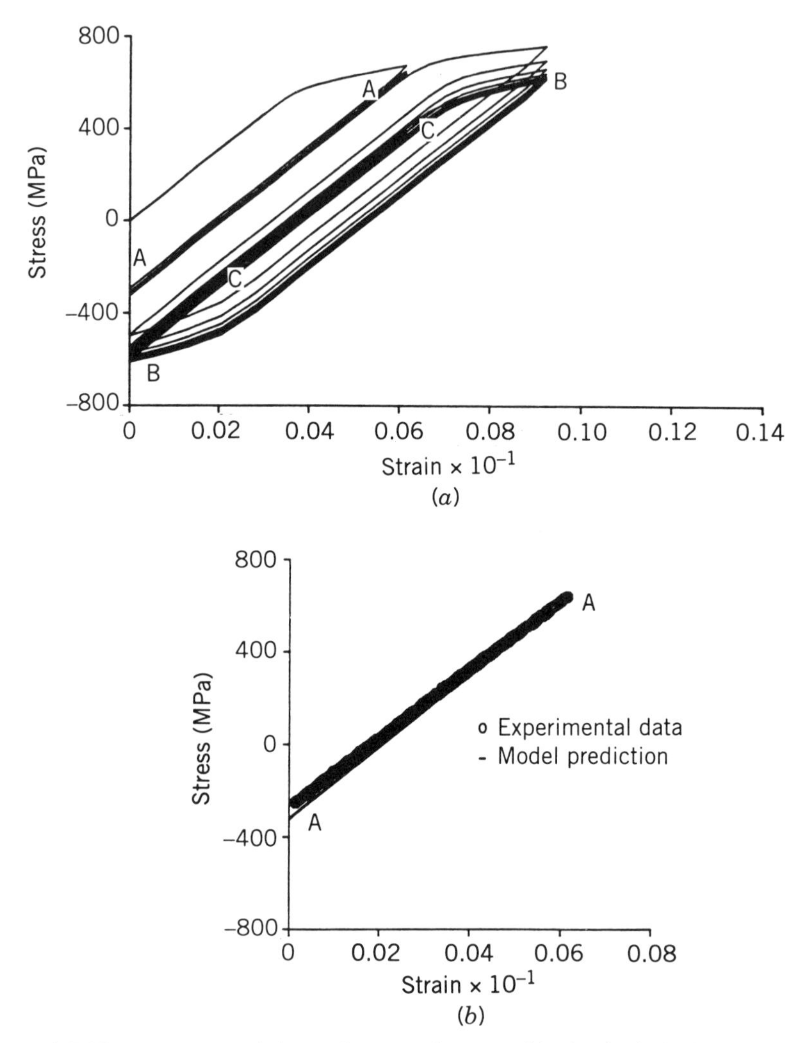

Figure 2.5.10 (*a*) Hysteresis loops for an entire test with elastic shakedown and mean stress shifts; (*b*) experimental and model results (the model is presented in Chapter 6) for the same test showing the elimination of the mean stress. (After Ramaswamy, 1986.)

cause the slope of the cyclic stress–strain curve is almost always smooth, continuous, and monotonically increasing (recall Figure 2.5.9). Relating the alternating true stress amplitude to the true inelastic strain amplitude by a logarithmic linear relationship gives

$$\tilde{\sigma}_a = K'(\tilde{\varepsilon}_a^{\mathrm{I}})^{n'} \tag{2.5.1}$$

The cyclic strength coefficient and cyclic hardening exponent are denoted by

K' and n', respectively. Solving for the true inelastic strain amplitude and using the left-hand side of equation 2.1.3 yields a representation for the cyclic stress–strain curve:

$$\tilde{\varepsilon}_a = \frac{\tilde{\sigma}_a}{E} + \left(\frac{\tilde{\sigma}_a}{K'}\right)^{1/n} \tag{2.5.2}$$

The engineering stress and strain are frequently used because the strain amplitude is usually less than two or three times the yield strain. The hysteresis curve is a plot of the cyclic stress and strain range, and the model for the hysteresis curve can be written as

$$\Delta\tilde{\varepsilon} = \frac{\Delta\tilde{\sigma}}{E} + 2\left(\frac{\Delta\tilde{\sigma}}{2K'}\right)^{1/n} \tag{2.5.3}$$

The magnitudes of the hysteresis curve variables are double those of the cyclic stress–strain curve (i.e., $\Delta\tilde{\varepsilon} = 2\tilde{\varepsilon}_a$ and $\Delta\tilde{\sigma} = 2\tilde{\sigma}_a$). This hysteresis equation is used for peak-to-peak calculations, whereas equation 2.5.2 is used for loading to the first peak.

Example Problem 2.3: Determine the cyclic strain-hardening exponent and strength coefficient for 2024-T351 aluminum alloy, and plot the cyclic stress–strain curve. The modulus of elasticity is assumed to be 10,000 ksi, and the following engineering stress range and strain range data were obtained experimentally:

$\Delta\sigma$ (ksi)	$\Delta\varepsilon$ (in./in.)
84	0.0091
100	0.0325
116	0.0976
126	0.1756

The yield stress in tension is known to be about 42 ksi.

SOLUTION: True stress and strain are used in this example because there is almost a 20-fold increase in the strain-range data. The constants for the data set can be determined by taking the natural logarithm of equation 2.5.1,

$$\ln \tilde{\sigma}_j = n' \ln \tilde{\varepsilon}_j^{\mathrm{I}} + \ln K' \tag{a}$$

and recalling that

$$\tilde{\sigma}_j = \frac{\Delta\sigma_j}{2}\left(1 + \frac{\Delta\varepsilon_j}{2}\right) \qquad \text{and} \qquad \tilde{\varepsilon}_j^{\mathrm{I}} = \ln\left(1 + \frac{\Delta\varepsilon_j}{2}\right) - \frac{\tilde{\sigma}_j}{E}$$

The subscript j in the equations above corresponds to each data pair ($\tilde{\sigma}_j$, $\tilde{\varepsilon}_j^{\mathrm{I}}$). Thus it is necessary to solve a system of linear equations. The variables in equation (a) for each data pair in the data set are

σ	ε	$\tilde{\sigma}$	$\tilde{\varepsilon}^{\mathrm{I}}$	$\ln \tilde{\sigma}$	$\ln \tilde{\varepsilon}^{\mathrm{I}}$
42	0.00455	42.2	0.00032	3.7422	−8.0454
50	0.01625	50.8	0.01104	3.9281	−4.5064
58	0.04880	60.8	0.04156	4.1081	−3.1805
63	0.08780	68.5	0.07730	4.2273	−2.5600

Using a linear regression analysis in a hand-held calculator, the values of the material parameters are $n' = 0.0830$ and $K' = 79.92$ ksi, and the correlation coefficient for the log linear fit to the data is 0.9599. The plot of equation 2.5.2 using the foregoing constants is shown in Figure 2.5.11 together with the original data. The difference between the cyclic stress–strain curve and the data results because the data are not exactly logarithmic linear, as shown in Figure 2.5.12. This result suggests that a semilogarithmic linear representation might be better (see Problem 2.6). The exercise problem is also a good example of the level of error that might be expected using the Ramberg–Osgood model.

2.6 RESIDUAL STRESSES

On loading a simple structure such as a plate with a hole in the center, the distribution of stress and strain throughout the body are not uniform. At low

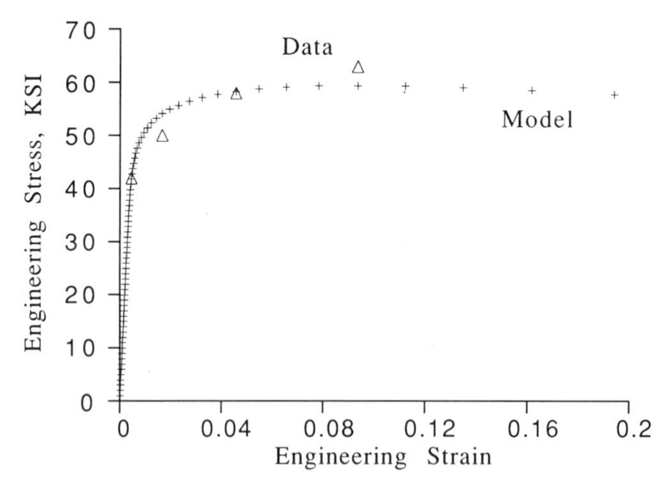

Figure 2.5.11 Comparison of model and experimental results for cyclic stress–strain curve from Example Problem 2.3.

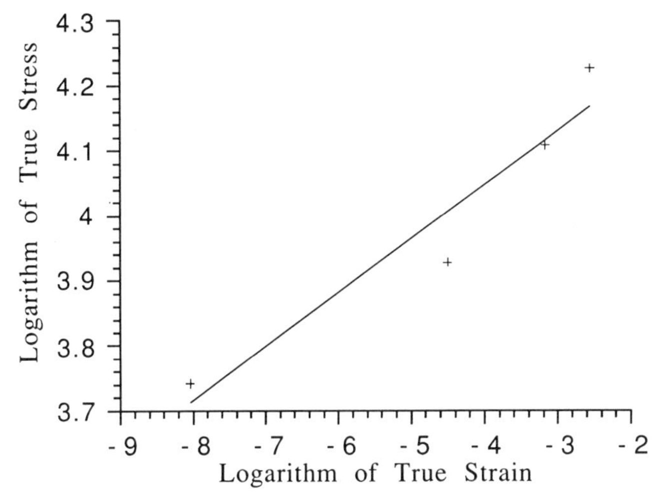

Figure 2.5.12 Logarithmic plot of data from Example Problem 2.3.

levels of external load, the strains in the body will be elastic everywhere, but the magnitude of the strains and stresses will be much larger near the hole, as shown in Figure 2.6.1. As the load is increased, the stress at the hole will exceed the yield stress and inelastic strain will occur. The specimen will not fail because much of the cross section at the hole is still elastic and the amount of inelastic straining at the hole is constrained by the surrounding elastic structure. On unloading to zero external load, the stresses in the body will

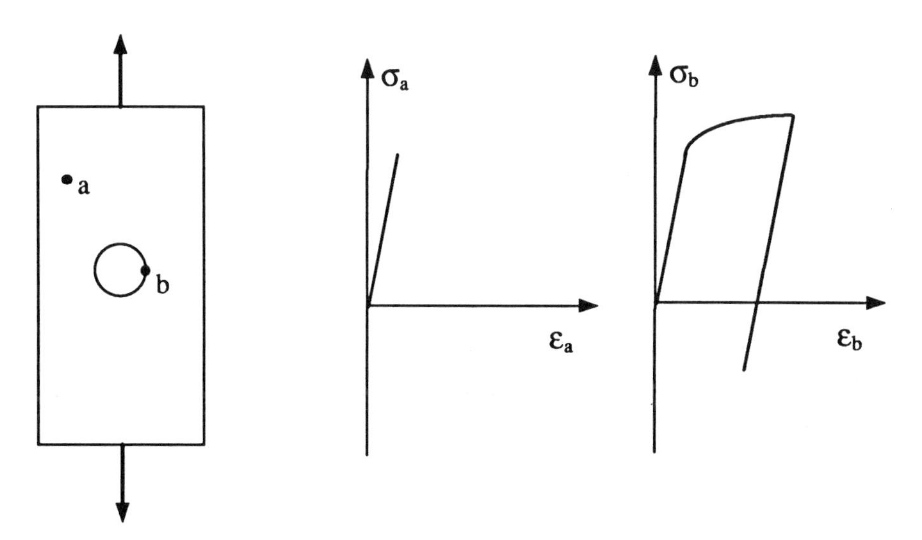

Figure 2.6.1 Residual stress at point *b* near a hole and at a remote location after loading and unloading in tension.

not return to zero. The region near the hole that experienced the inelastic strain will be in compression and the elastic region surrounding the inelastic region will be in tension, so the net section forces are in equilibrium. Thus residual stresses are developed because of local yielding when the stress distribution thoughout a body is not uniform.

A mechanical testing machine with a specimen in place can be considered as a structural system. The deformation history developed in the specimen under strain control can be used to generate and measure residual stresses. During loading the structure will remain elastic everywhere except in the test section of the specimen. Cycling the specimen under strain control is equivalent to developing inelastic deformation in a structure with a stress concentration. Returning the actuator to the initial position after testing will result in residual stresses.

Example Problem 2.4: Determine the cyclic response for the three strain sequences shown in Figure 2.6.2. The material is steel with the following properties: $E = 30000$ ksi, $n' = 0.202$, and $K' = 174.6$ ksi.

SOLUTION: The stress response for the first quarter-cycle of each strain history can be determined for the strain amplitude equation. Since the differences between engineering stress and strain and true stress and strain are small for the deformations under consideration, the Ramberg–Osgood equation can be written as

$$\varepsilon = \frac{\sigma}{E} + \left(\frac{\sigma}{K'}\right)^{1/n'} = \frac{\sigma}{30{,}000} + \left(\frac{\sigma}{174.6}\right)^{4.9505}$$

To determine the response between peaks, the hysteresis equation must be used:

$$\Delta\varepsilon = \frac{\Delta\sigma}{E} + 2\left(\frac{\Delta\sigma}{2K'}\right)^{1/n'} = \frac{\Delta\sigma}{30{,}000} + \left(\frac{\Delta\sigma}{349.2}\right)^{4.9505}$$

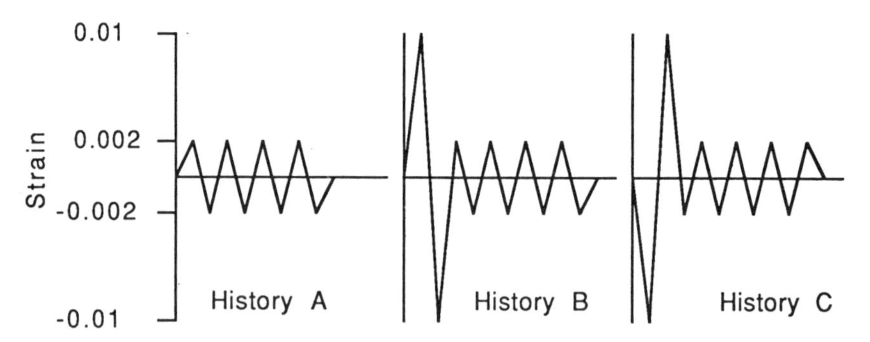

Figure 2.6.2 Strain histories for Example Problem 2.4.

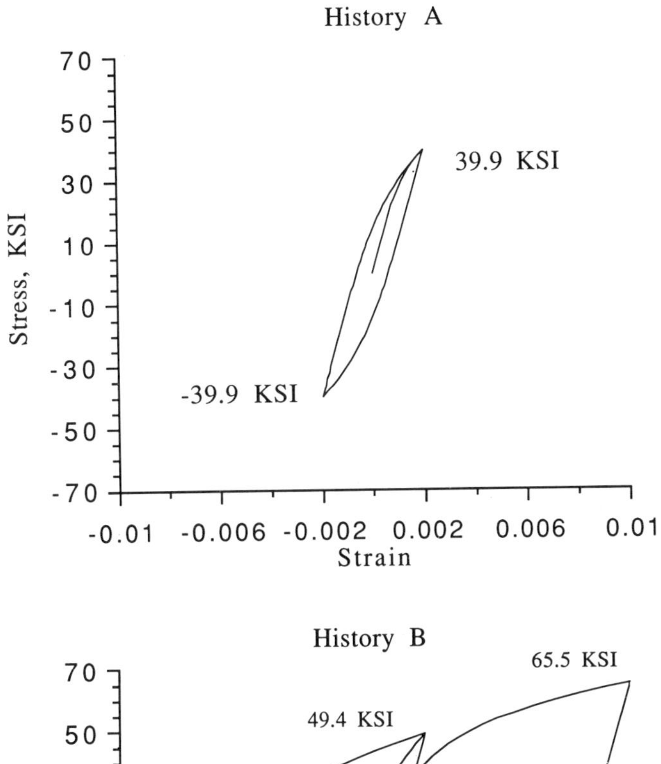

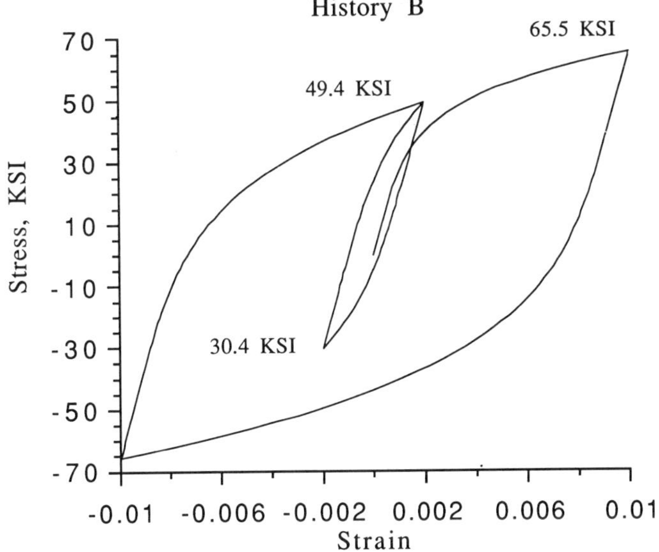

Figure 2.6.3 Calculated response for the three strain histories in Example Problem 2.4.

The equations must be solved numerically for the stress as discussed in Example Problem 2.2. The results of a spreadsheet calculation for the three loading histories is shown in Figure 2.6.3. The resultant hysteresis loops from history A are symmetric (no mean stress) and the stress range is 79.8 ksi. The results from history B have the same stress range, but there is a tensile mean stress of 9.5 ksi, whereas there is a compressive mean stress of −9.5 ksi from history C. Thus the loading sequence can change the response significantly by the creation of residual stresses.

2.7 SUMMARY

The tensile response of metals is independent of the strain rate at temperatures below about one-half of the melting temperature. At higher temperatures the combined effect of strain hardening and dynamic recovery produces strain-rate sensitivity. Thus the tensile response of a metal is a function of strain rate and temperature. If inelastic deformation occurs primarily by slip, deformation in tension, compression, and torsion are essentially the same. However, the observed torsional stress is about half the tensile stress since the shear stress on the slip plane that produces slip in a tensile specimen is half of the applied tensile stress. Strain hardening resulting from unidirectional loading occurs, at least in part, due to dislocation pileups at barriers. The response of hot-worked solid solution (carbon) steels is different from the

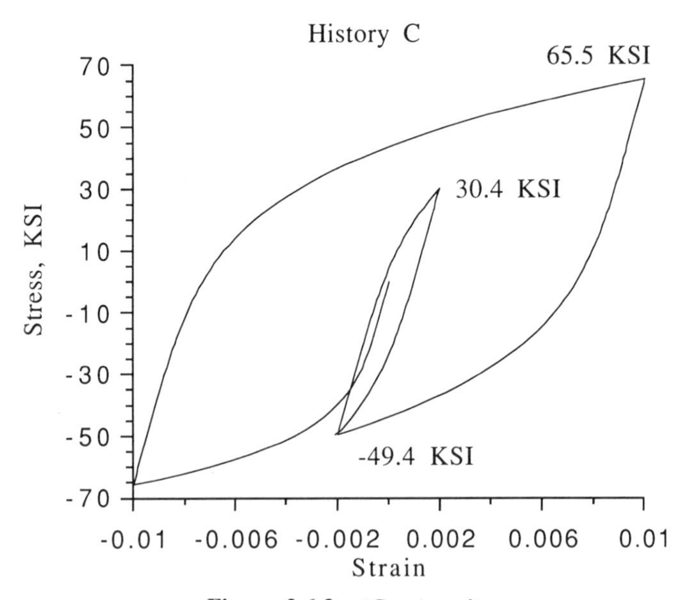

Figure 2.6.3 *(Continued)*

response of cold-worked or precipitate-hardened steels because the carbon atoms trapped in the lattice structure are free to diffuse and reduce the local stress around dislocations.

Inelastic strain is defined as the difference between the total and elastic strains. Plastic strain is usually assumed to mean strain-rate-independent inelastic strain response. The engineering stress and strain variables are valid up to about three times the yield strain and the true stress and strain variables are valid up to necking. The ultimate (tensile) stress in ksi is equal to about half the Brinell hardness number.

Continuous cyclic loading produces an extensive dislocation cell and subgrain microstructure generally not found in monotonic loading. These microstructures are isotropic. In general, initially soft materials will harden during cyclic loading, initially hard materials will soften, and half-hard metals do not exhibit much cyclic hardening or softening.

The Ramberg–Osgood equation is valid only for time-independent (plastic) response. The constants K and n are easy to determine from experimental data, but they are different for monotonic and cyclic loading. The model is easy to implement numerically even though it is necessary to iterate for the stress. The model is reasonably accurate for tensile response, but it cannot model some of the mean strain effects observed in metals.

REFERENCES

Bannantine, J. A., J. J. Comer, and J. L. Handrock (1990). *Fundamentals of Metal Fatigue Analysis,* Prentice Hall, Englewood Cliffs, NJ.

Barsom, J. M., J. H. Klippstein, and A. K. Shoemaker (1980). "State of the art report on the fatigue behavior of sheet steel for automotive applications, *Report to the American Iron and Steel Institute, AISI Project 1201-409D,* United States Steel Corp.

Boyer, H. E. and T. L. Gall, eds. (1985). *Metals Handbook,* American Society for Metals, Metals Park, OH.

Chambers, B. (1959). *Physical Metallurgy,* Wiley, New York.

Dieter, G. E. (1986). *Mechanical Metallurgy,* 3rd ed., McGraw-Hill, New York.

Dowling, N. E. (1993). *Mechanical Behavior of Materials,* Prentice Hall, Englewood Cliffs, NJ.

Ellis, J. R. and D. N. Robinson (1985). Some advances in experimentation supporting development of viscoplastic constitutive models, *NASA CR 174555.*

Kassar, M., and W. W. Yu (1989a). The effect of strain rate on the mechanical properties of sheet steel, *11th Progress Report, Design of Automotive Structural Components Using Sheet Steels,* Department of Civil Engineering, University of Missouri–Rolla, January.

Kassar, M., and W. W. Yu (1989b). The effect of strain rate on the mechanical properties of sheet steel, *12th Progress Report, Design of Automotive Structural Com-*

ponents Using Sheet Steels, Department of Civil Engineering, University of Missouri–Rolla, August.

Krempl, E. (1979). An experimental study of room temperature rate sensitivity, creep, and relaxation of AISI type 304 stainless steel *Journal of the Mechanics and Physics of Solids,* Vol. 27, pp. 363–375.

Krempl, E., and V. V. Kallianpur (1984). Some critical uniaxial experiments for viscoplasticity at room temperature, *Journal of the Mechanics and Physics of Solids,* Vol. 32, p. 301.

Kuruppu, M. D., J. F. Williams, N. Bridgeford, R. Jones, and D. C. Stouffer (1992). Constitutive modeling of the elastic–plastic behavior of 7075-T7451 aluminum alloy, *Journal of Strain Analysis,* Vol. 27, No. 2, pp. 85–92.

Laflen, J. H., and D. C. Stouffer (1978). An analysis of high temperature metal creep, Part 1: Experimental definition of an alloy, *Journal of Engineering Materials and Technology,* Vol. 100, pp. 363–370.

Landgraf, R. W., J. Morrow, and T. Endo (1969). Determination of the cyclic stress strain curve, *Journal of Materials,* Vol. 4, No. 1, pp. 176–188.

La Pointe, N. R. (1982). Monotonic and fatigue characterization of metals, *Proceedings of the SAE Fatigue Conference P-109,* Paper 820679.

Manson, S. S., and M. H. Hirshberg (1964). *Fatigue: An Interdisciplinay Approach,* Syracuse University Press, Syracuse, NY, p. 133.

Martin, J. F. (1973). Cyclic stress strain behavior and fatigue resistance of two structural steels, *Fracture Control Program Report 9,* University of Illinois at Urbana–Champaign.

Moreno, V. (1981). Combuster liner durability analysis, *NASA CR 165250,* NASA Lewis Research Center, Cleveland, OH.

Morrow, J. (1965). Cyclic plastic strain energy and fatigue resistance of metals, *ASTM STP 378,* p. 35.

Nicholas, T. (1980). Dynamic tensile testing of structural materials using a split Hopkinson bar apparatus, *AFWAL-TR-80-4053,* Materials Laboratory, Wright-Patterson AFB, OH, October.

Rajendran, A. M., and S. J. Bless (1985). High strain rate material behavior, *AFWAL-TR-85-4009,* Materials Laboratory, Wright-Patterson AFB, OH.

Ramaswamy, V. G., D. C. Stouffer, and J. H. Laflen (1990). A unified constitutive model for the inelastic uniaxial response of René 80 at temperatures between 538C and 982C, *Journal of Engineering Materials and Technology,* Vol. 112.

Ramaswamy, V. G. (1986). A constitutive model for the inelastic multiaxial cyclic response of a nickel base superalloy René 80, NASA CR 3998.

Reed-Hill, R. E. (1973). *Physical Metallurgy Principles,* 2nd ed., D. Van Nostrand, New York.

Richards, F. D., and R. M. Wetzel (1970). Mechanical testing of materials using an analog computer, *Materials Research and Standards,* Vol. 11, No. 2, pp. 19–21.

Stouffer, D. C., and B. Kane (1988). Inelastic deformation modeling of titanium aluminide matrix material, *Final Report from the University of Cincinnati to General Electric Aircraft Engines on Contract 201-LS-L1P15321.*

Stouffer, D. C., C. L. Papernik, and H. L. Bernstein (1980). An experimental evaluation of the mechanical response of René 95, *AFWAL-TR-80-4136,* Materials Laboratory, Wright-Patterson AFB, OH.

Stouffer, D. C., V. G. Ramaswamy, J. H. Laflen, R. H. Van Stone, and R. Williams (1990). A constitutive model for the inelastic multiaxial response of René 80 at 871C and 982C, *Journal of Engineering Materials and Technology,* Vol. 112.

Symonds, P. S. (1965). Viscoplastic behavior in response of structures to dynamic loading, in *Behavior of Materials Under Dynamic Load,*, N. J. Huffington, ed., ASME, New York, p. 106.

Wada, T., Y. T. Smith, and W. E. Lauprecht (1979). Mn–Mo–V–Cb steel for pressure vessels, ASME 3rd National Congress on Pressure Vessels and Piping, San Francisco.

PROBLEMS

2.1 Volumetric strain is defined as the ratio of the change in volume, ΔV, to the initial volume, V. For small strains show that the volumetric strain for a cube of material initially of volume $V = dx\, dy\, dz$ is given by

$$\frac{\Delta V}{V} = \varepsilon_x + \varepsilon_y + \varepsilon_z$$

where ε_x, ε_y, and ε_z are the engineering axial strains in each of the three coordinate directions. If the inelastic (plastic) volumetric strain is zero, show that the inelastic value for Poisson's ratio is 0.5.

2.2 The engineering diametrical strain, ε_d, is defined as the ratio of the change in diameter to initial diameter. Show that the corresponding axial strain, ε_{ax}, is given by

$$\varepsilon_{ax} = \left(1 - \frac{\nu^e}{\nu^p}\right)\frac{\sigma_{ax}}{E} - \frac{1}{\nu^p}\varepsilon_d$$

where ν^e and ν^p are the elastic and plastic values of Poisson's ratio, E the elastic modulus, and σ_{ax} the axial stress.

2.3 Knowing that the engineering strain and the engineering stress are 0.004 and 40,000 psi, respectively, determine the true stress and true strain. What are the engineering stress and strain values corresponding to true stress and strain values of 70,000 psi and 0.035, respectively?

2.4 Use the power law relation between true plastic strain and stress to show that

$$\tilde{\varepsilon}^p = \tilde{\varepsilon}_f \left(\frac{\tilde{\sigma}}{\tilde{\sigma}_f}\right)^{1/n}$$

2.5 Use the data in Figure 2.4.2*b* to determine the parameters in the Ramberg–Osgood model and compare the model to experimental results. Discuss the capabilities of the model. How could you use it to model the data in Figure 2.4.2*a*?

2.6 Another representation for tensile response is the exponential hardening model, where

$$\varepsilon = \frac{\sigma}{E} \quad \text{for } \sigma \le \sigma_y \qquad \text{and} \qquad \varepsilon = \varepsilon_y \left(\frac{\sigma}{\sigma_y}\right)^{1/n} \quad \text{for } \sigma \ge \sigma_y$$

The quantities σ and ε are engineering stress and strain, and the subscript y corresponds to the value at yield. Determine the parameters for René 80 given in Figure 2.4.2*b* and compare the model to experimental results. Compare the results to Problem 2.5 and discuss the differences between the two models.

2.7 Show that the strain-hardening exponent, n, in the power law relationship

$$\tilde{\sigma} = K(\tilde{\varepsilon}^{1})^{n}$$

is equal to the true strain at the point where the engineering stress is equal to the ultimate stress; that is, verify that $n = \tilde{\varepsilon}_u$.

2.8 A hot-worked low-carbon-steel test specimen within a cylindrical test section (50 mm diameter by 250 mm long) is loaded in tension. The following loads and deflections were recorded:

Load (MN)	Deflection (mm)	Observation
0	0	Begin test
0.40	0.22	Elastic response
0.37	0.24	Sudden drop in load
0.37	1.10	Continuous deformation at constant load
0.60	45.00	Maximum load, necking begins
0.30	60.0	Failure

Determine the elastic modulus, upper and lower yield stresses, ultimate (tensile) strength, true stress and strain at necking, true fracture strength, and ductility if the measured diameter at failure is 28 mm, and the strain hardening exponent and strength coefficient.

2.9 Determine the strain-hardening exponents and strength coefficients for each of the three strain rates from the true stress–strain curves for the cold-worked low-carbon steel shown in Figure 2.2.1. Plot an estimate of the stress–strain response for a strain rate of 0.1 sec^{-1}.

2.10 Determine the true stress–strain curve in an 2024 T4 aluminum alloy loaded in tension to a true strain of 0.300. The material parameters for 2024 T4 are E = 10,600 ksi, n = 0.200, and K = 117 ksi.

2.11 The true stress and strain of a metal at failure are 200 ksi and 0.8, respectively. It is also known that the strain-hardening exponent is 0.4 and E = 30,000 ksi. Determine the strength coefficient and plot the true stress–strain curve. Find the 0.2% offset yield stress, ultimate strength, and the fractional reduction in area at failure.

2.12 Two identical test samples are loaded in tension differently to examine the effect of cold working on material properties. One specimen is loaded to failure and the other is prestrained to 15% before loading to failure. Compare the 0.2% yield stress, ultimate tensile strength, true fracture ductility, and the reduction in area at fracture of the two samples by treating the prestrained sample as a different material. The properties of the undeformed material are E = 30,000 ksi, n = 0.25, $\tilde{\varepsilon}_f$ = 1.10, and $\tilde{\sigma}_f$ = 250 ksi.

2.13 To evaluate the accuracy of the approximation relating the BHN to the ultimate stress, determine the coefficient k in $\sigma_u = k$ BHN for each material in Appendix 2.1, and then calculate mean value and standard deviation of k for the entire data set.

2.14 What are the values of the strain ratio, R_ε, and amplitude ratio, A_ε, for fully reversed, zero to ε_{max} and zero to ε_{min} strain-controlled cycling?

2.15 Use the cyclic stress–strain model, equation 2.5.2, and derive the hysteresis curve model, equation 2.5.3. Plot the cyclic stress–strain and hysteresis curves on the same set of axes for a material with the following material properties: E = 30,000 ksi, K' = 175 ksi, and n' = 0.200.

2.16 Apply the Ramberg–Osgood model to the cyclic data given in Appendix 2.3. Develop a plot of the model parameters as a function of strain rate for the tensile data. Use these results to predict the cyclic response and compare the model result to the experimental data. What are the advantages and/or disadvantages of this approach?

2.17 A metal has the following monotonic tensile properties: $E = 193$ GPa, $\sigma_y = 325$ MPa, $\sigma_u = 650$ MPa, $\tilde{\sigma}_f = 1400$ MPa, $\tilde{\varepsilon}_f = 1.1731$, and $n = 0.193$.

(a) Will the material harden of soften under cyclic loading?

(b) Determine the engineering strain at the end of each loading step for the stress history: 0 to 400 MPa to -375 MPa to 0.

CHAPTER 3

Creep Of Metals

Creep of metals is very different from the tensile and fatigue properties discussed in Chapter 2. Tensile and fatigue deformations are controlled by planar slip and climb. Hardening results from the interaction of dislocations with various features in the microstructure as well as other dislocations. Creep is a high-temperature deformation and is controlled by the diffusion of atoms and vacancies through the crystal lattice and along grain boundaries. The creep rate is affected by the presence of dislocations, but the primary driving force is temperature, not shear stress.

In this chapter the observed characteristics of creep and the creep-related effects of stress relaxation and recovery are presented. The underlying deformation mechanisms, creep testing, and the prediction of long-time creep response are discussed. A number of widely used physical and empirical creep models are included. The chapter concludes with a discussion of the interaction between creep and cyclic loading.

3.1 PHENOMENOLOGICAL ASPECTS OF CREEP, RECOVERY AND STRESS RELAXATION

When a constant load is applied to a tensile specimen at a temperature above about $0.5T_h$, the specimen will elongate over a period of time even though the stress is below the yield stress. The initial creep rate can be relatively high, as shown schematically in Figure 3.1.1. The creep strain rate usually decreases to a relatively steady (minimum) value for a prolonged period of time. Eventually, the creep rate increases from the steady-state value and unrestrained flow occurs, leading to rupture.

In standard uniaxial creep tests the creep stress is constant and is defined as the applied load divided by the initial test specimen area. Planar slip is not the dominant deformation mechanism, and there is no ongoing "plasticity," although there may be some plasticity during initial loading if the creep stress is greater than the yield stress. Temperature is usually greater than one-half

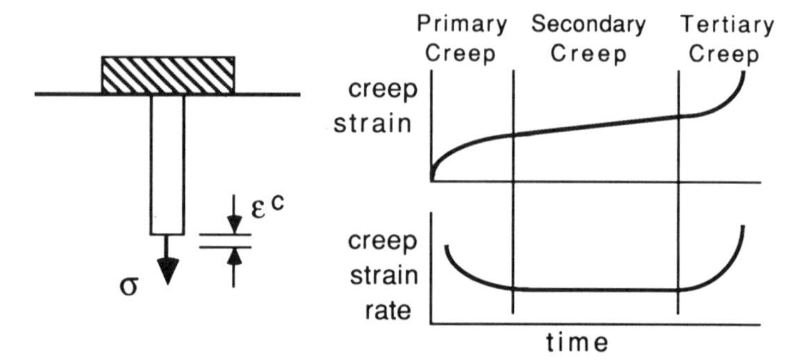

Figure 3.1.1 Creep strain and creep strain rate.

the absolute melting temperature, and the creep time is generally measured in hundreds or thousands of hours.

The primary creep regime shown in Figure 3.1.1 is defined as the time from application of the load to the onset of the minimum or steady-state creep rate. The secondary creep regime is the time during which the creep rate is at its minimum value. Tertiary creep is defined as an increase in strain rate leading to rupture. Depending on the material, the balance among primary, secondary, and tertiary creep can be skewed. The primary, secondary, and tertiary creep rates and the time to creep rupture are strong functions of the creep stress and temperature. Increasing stress and temperature increases the creep rate and reduces the time to rupture.

Creep strain is an inelastic strain that occurs over time due to the action of temperature and stress. Mathematically and experimentally, the creep strain is defined as the difference between the total measured strain and the calculated elastic strain,

$$\varepsilon^T = \varepsilon^E + \varepsilon^I = \frac{\sigma}{E} + \varepsilon^C \tag{3.1.1}$$

In the classical theory of plasticity and creep, the creep strain is considered additive to the rate-independent plastic strain, which is usually associated with low-temperature inelastic straining:

$$\varepsilon^T = \varepsilon^E + \varepsilon^P + \varepsilon^C \tag{3.1.2}$$

If a material exhibits creep, it will probably exhibit stress relaxation and strain recovery. As discussed earlier, these characteristics are defined by the response to special loading conditions. Strain recovery (also called creep recovery, anelasticity, or elastic after effect) results when the stress is removed after an inelastic straining event. For example, a creep and recovery test is shown schematically in Figure 3.1.2. A creep stress is applied for a period of time t' and is then removed. Up to time t' the specimen will exhibit creep, and at time t' there will be a reduction in strain corresponding to instantaneous elastic recovery. However, as time advances from t', the strain will continue to recover from a rearrangement of dislocations and diffusion of atoms and vacancies. The driving force for recovery comes from the energy stored in the microstructure and temperature (recall the definition of static recovery). The amount of time associated with static recovery is similar to the time for creep. Static recovery is much slower than dynamic recovery, although it can be rather fast at extremely high temperatures.

Stress relaxation is the decay in stress over time while the total strain is held constant, that is,

$$\dot{\varepsilon} = 0 = \frac{\dot{\sigma}}{E} + \dot{\varepsilon}^C \qquad \text{or} \qquad \dot{\sigma} = -E\dot{\varepsilon}^C \tag{3.1.3}$$

Stress relaxation results from the redistribution of elastic and inelastic strain. Thus the mechanism driving the long-time stress relaxation is the same as creep. These mechanisms are different from the rapid relaxation rate observed in Figure 2.1.2 when there was a continuous deformation process. This point is discussed is Section 3.3.

3.2 CREEP MECHANISMS

Creep is generally viewed as a thermally driven process that is influenced by the presence of stress. Temperature rather than stress is the primary driving force. Depending on the temperature and stress level, there are several pos-

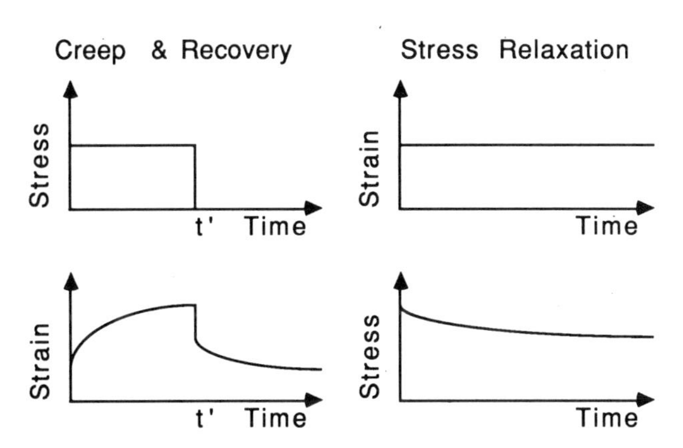

Figure 3.1.2 Definitions of creep and recovery, and stress relaxation.

sible creep mechanisms: dislocation glide, bulk diffusion, grain boundary diffusion, and dislocation creep. More than one deformation mechanism can be active at once, and their interactions are not well understood.

3.2.1 Dislocation Glide

Lattice vibrations resulting from thermal energy affect the magnitude of the shear stress required for dislocation mobility. For example, consider the schematic diagram in Figure 3.2.1*a*, which shows the force between a dislocation. and an obstacle as a function of position. The variation of force with distance is from the outer limit of influence to the position when the dislocation is directly above or below the obstacle. The work or energy required for a dislocation to pass the obstacle is the area under the force–position curve. At 0 K all the work must be supplied by external stress; however, at a finite temperature thermal vibrations provide some of the energy, and the mechanical work required for dislocation mobility is reduced. At higher temperatures even less external work is required, and if the temperature is high enough, no external work is necessary for the dislocation to pass the obstacle.

To illustrate the roles of stress and temperature in dislocation glide, let us define U_0, the area under the curve in Figure 3.2.1*a*, as the activation energy required to move a dislocation past a barrier. A diagram of the variation in energy with position for a dislocation–obstacle system is shown in Figure 3.2.1*b*. The diagram shows that the energy distribution is symmetric in position about the obstacle when there is no stress, and there is equal potential for a dislocation to pass the obstacle in either direction due to thermal energy alone. However, when stress is applied, the thermal energy required to pass the barrier is reduced on one side of the barrier, and a lower-energy state could be obtained by the dislocation passing the barrier. The presence of the stress produces a bias for dislocation flows in the direction of the applied stress, and macroscopic creep results.

Dislocation glide involves dislocations moving on slip planes and overcoming barriers by the action of thermal energy and assisted by stress. Dis-

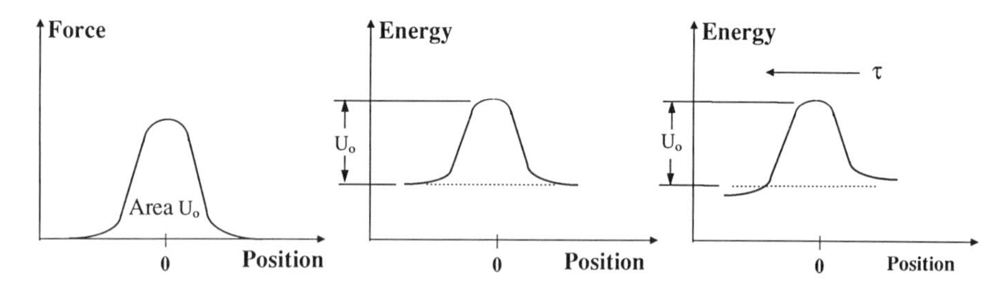

Figure 3.2.1 (*a*) Force as a function of position of a dislocation with respect to an obstacle; (*b*) energy as a function of position for free and loaded dislocation–barrier system, and (*c*) a loaded dislocation barrier system.

location glide is different from diffusional flow and climb. The creep rate in dislocation glide is controlled by the intrinsic activation energy, U_0, for the specific dislocation–barrier system and the magnitudes of external and thermal energies.

3.2.2 Nabarro–Herring Creep

Nabarro–Herring creep is accomplished by the diffusion of atoms and vacancies. It is the primary deformation mechanism at high temperature and low stress, and occurs at much lower stresses than dislocation glide and dislocation creep. Nabarro–Herring creep does not depend on the movement of dislocations and is therefore very different from the dislocation glide mechanism.

The Nabarro–Herring creep mechanism is shown symbolically for a single grain of a polycrystalline metal in Figure 3.2.2. A low-level load applied to a material will produce nonuniform stresses in a grain with regions of tension and compression that will expand and compress the lattice volume at different sites in the grain. The local change in lattice volume will reduce the mass concentration below the equilibrium level in the expanded regions, reduce the vacancy concentration below the equilibrium level in the compressed regions, and establish mass and vacancy concentration gradients. The combination of stress and temperature will provide the energy for diffusion. Mass will flow toward the low mass concentration or tensile regions, with vacancies moving toward the compressed regions. The mass flow will allow the grain to grow

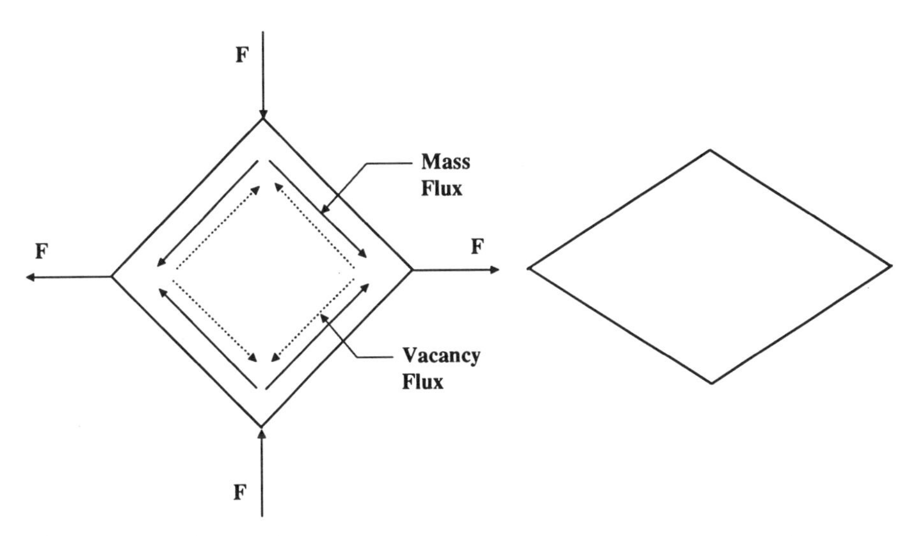

Figure 3.2.2 Nabarro–Herring creep showing the mass flux toward the tensile regions and the vacancy flux toward the compressive regions, and a hypothetical change in shape.

locally in tensile regions, whereas vacancy diffusion allows the grain to shrink at sites of compression. The shape of the grain changes due to the mass and vacancy flows, as indicated in Figure 3.2.2*b*. The creep rate is controlled by the diffusion properties of the material and the mechanical and thermal loads.

3.2.3 Coble Creep

Coble creep is driven by concentration gradients similar to Nabarro–Herring creep; however, it occurs by mass and vacancy diffusion along grain boundaries. The disruptions in the lattice structure at the grain boundaries provide favorable paths for the mass and vacancy fluxes.

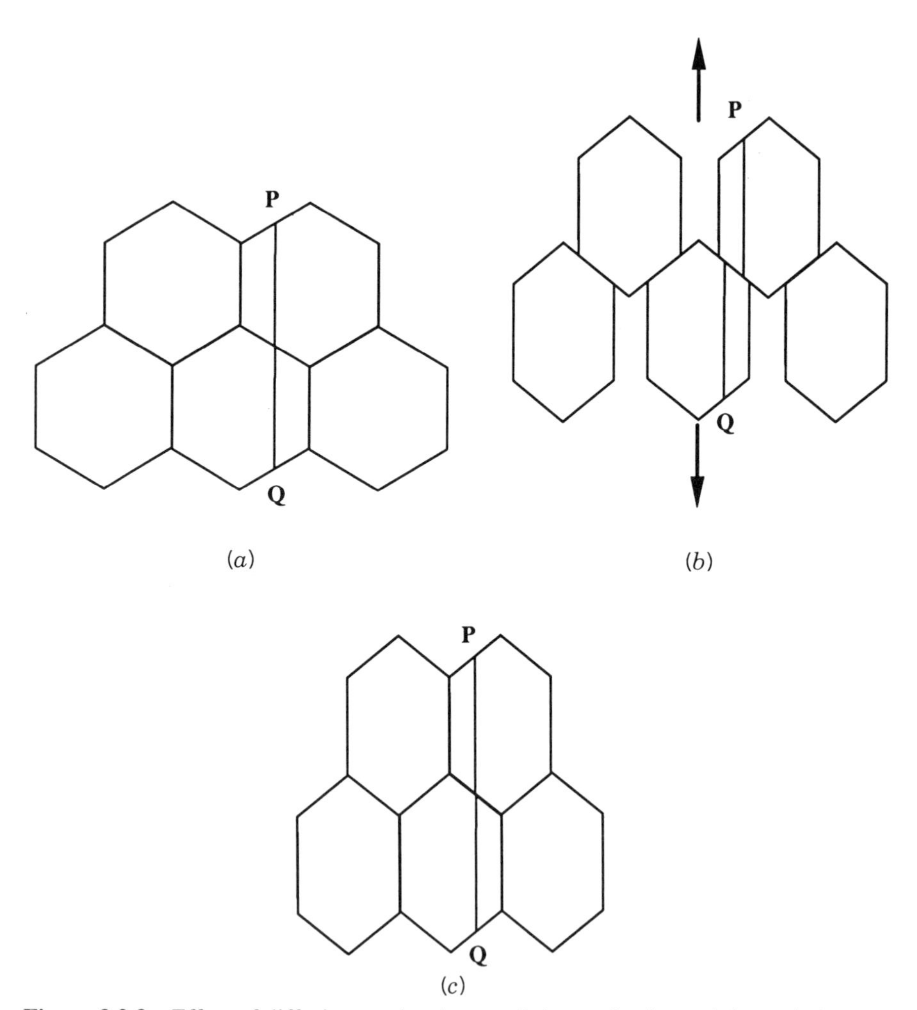

Figure 3.2.3 Effect of diffusion on the change of shape of gains and the grain boundary slip required to maintain closed boundaries.

Coble creep may occur simultaneously with Nabarro–Herring creep, and their creep rates are generally considered as additive. Coble creep dominates at lower temperatures since diffusion along grain boundaries can occur at lower energy levels than are favorable for bulk diffusion. At higher temperatures bulk diffusion processes dominate since there are a limited number of paths for grain boundary diffusion. Thus the Coble creep rates are less than the Nabarro–Herring creep rates at high temperature. Nabarro–Herring creep is present only at the higher temperatures, whereas Coble creep is also present at lower temperatures. Coble creep is also more important in fine-grain materials than in coarse-grain materials, where there are more paths for grain boundary diffusion.

Both grain boundary and bulk diffusion promote the movement of mass and shape changes of grains that would tend to produce voids or cracks at grain boundaries. Grain boundary movements such as sliding occur simultaneously with diffusion to prevent the development of an internal free surface and higher-energy state. As shown schematically in Figure 3.2.3, diffusional creep could produce grain boundary separation that must be accommodated by sliding. Grain boundaries are rarely uniform, so three-dimensional exchanges of mass and vacancies produce complex grain boundary movements.

3.2.4 Dislocation Creep

Dislocation creep occurs through a combination of dislocation glide and vacancy diffusion. It occurs at lower stress than dislocation glide, and at higher temperatures, where diffusion is important. Dislocation creep is a sequential process. For example, consider a glide plane with a number of obstacles of different activation energies. A creep stress will produce dislocation glide until the dislocation encounters an obstacle with a larger activation energy than available. The dislocation will be pinned while diffusion occurs, allowing the dislocation to pass the barrier by climb. Glide then continues until another large energy barrier is encountered.

The rate of creep is determined by the lesser of the climb and glide rates. In most coupled glide–climb situations the creep rate is controlled by the climb velocity (i.e., the climb velocity is less than the glide velocity). For example, if precipitates are present, the climb mechanism controls the creep rate. Weertman (1968) proposed that creep resistance in multiphase alloys is controlled by dislocation climb around particles. During glide, dislocations loop around particles and loops will build up around the bypassed particle. The presence of back stress will prevent further glide until the first dislocation loop can climb around the particle. When this happens the dislocation loop collapses and is annihilated, as shown in Figure 3.2.4. The pinned dislocation then forms a loop and continues to propagate by dislocation glide. This is another recovery mechanism associated with precipitates. Creep in multiphase materials is controlled by dislocation climb at particles.

In some cases, dislocation glide controls the creep rate. For example, observe that dislocation glide is a deformation mechanism that does not involve

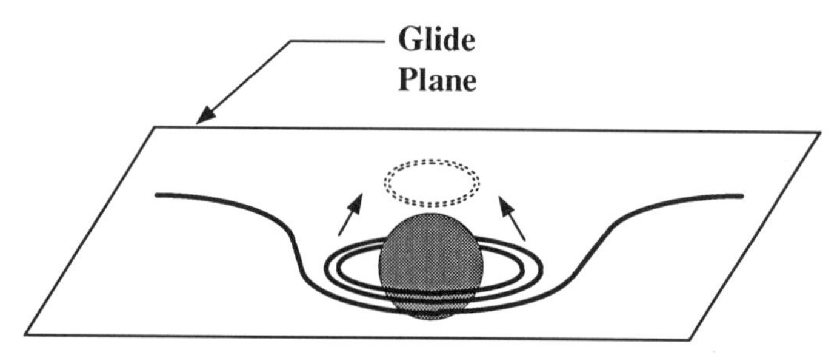

Figure 3.2.4 Residual dislocation loops around a precipitate preventing dislocation glide until climb occurs.

transport of mass. However, if solute atoms are present, they may slow the glide rate. In the case of a stationary positive edge dislocation there are regions of compression and tension above and below the dislocation. These regions will establish concentration gradients of solute atoms and vacancies in the neighborhood of the dislocation until equilibrium occurs. Then the chemical kinetic forces required to produce glide must also move the solute atoms and vacancies or "dislocation atmosphere" with the dislocation. Thus the dislocation atmosphere increases the energy required to produce motion and slows the glide velocity. In this case the glide velocity is less than the climb velocity, and the creep rate is controlled by glide.

The glide and climb mechanisms working together correspond to the competition between hardening and recovery. The hardening corresponds to resistance to glide that results from dislocation pileups at precipitates. Recovery occurs due to the collapse of dislocation loops. A steady-state creep rate is obtained when hardening and recovery rates are in equilibrium.

The combination of glide–climb mechanisms can also lead to the development of subgrains and cells during creep. This provides a recovery mechanism as discussed in Section 1.4. Nabarro–Herring creep can occur in subgrains just as in grains. The subgrain walls serve as a source and sink for vacancies, and mass is transferred by volume diffusion. The subgrain walls must also migrate to accommodate the mass diffusion.

3.2.5 Deformation Maps

A convenient summary of the various active creep mechanisms as a function of stress and temperature has been developed by Ashby and Frost (1975) in the form of a deformation map. Deformation maps have been constructed for creep for a large number of engineering materials. The maps are useful for explaining and understanding the inelastic response of metals that is observed.

The deformation map axes are shear-stress normalized by the shear modulus, G, and temperature normalized by the melting temperature. The theoretical shear stress is an estimate of the stress required to produce lattice slip in the absence of dislocations and serves as an upper limit for stress. Typical values for the theoretical shear stress are $0.01G$ to $0.1G$. A schematic diagram of a deformation map is shown in Figure 3.2.5. The map shows only the dominate mechanism in each region, even though more than one deformation mechanism may be active. The boundaries correspond to the points where two or more mechanisms produce the same slip rate. Three mechanisms are equally active at the intersection of three lines on the map. The maps are constructed from a combination of experimental data and mathematical models for the specific mechanisms shown and only indirectly represent the actual mechanisms. Thus the maps are an approximation of the active deformation mechanism and the boundaries are only as good as the models.

The deformation maps for pure nickel with two different grain sizes are shown in Figure 3.2.6. Figure 3.2.6*a* shows the dominate deformation mechanism in each stress and temperature regime with the contours of constant strain rate for a grain size of 10 μm. Figure 3.2.6*b* is for nickel with a grain size of 1 mm instead of 10 μm. Notice how the increase in grain size reduces the amount of grain boundary diffusion and the Coble rates in the low stress high temperature region of the map. The strain rate contour allows the correlation of three variables: stress, temperature, and strain rate.

Construction of these maps requires a considerable amount of data or modeling information; thus it may be possible to construct only a small portion of a map for a research program on any one material. Further, changes in the microstructure can affect the map significantly. It is hard to correlate data from similar alloys, as seen from the difference in the maps for fine and coarse nickel. However, these maps clearly show why high temperature alloys,

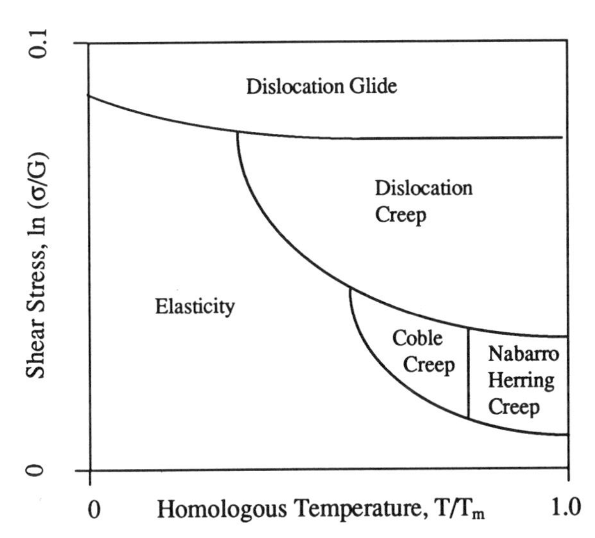

Figure 3.2.5 Primary deformation mechanisms at approximate stress and temperature regimes.

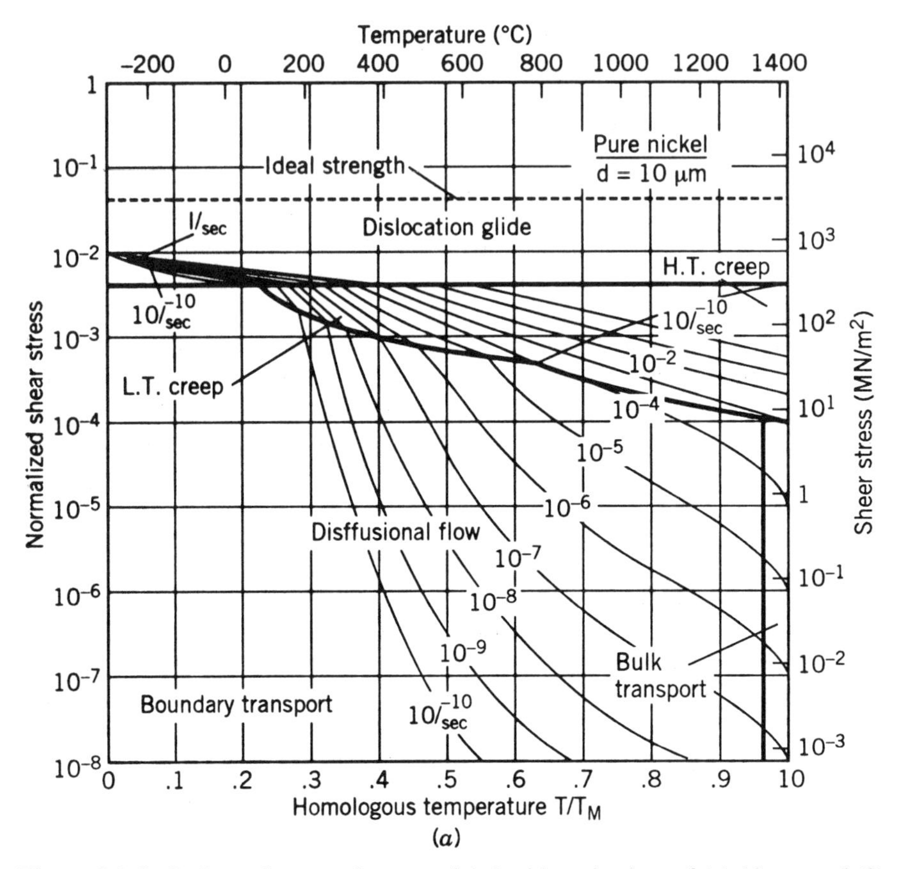

Figure 3.2.6 Deformation map for pure nickel with grain sizes of (*a*) 10 μm and (*b*) 1 mm. Note that power law creep is equivalent to dislocation creep. The maps show that low-temperature creep is eliminated by increasing the grain size. This is why many high-temperature alloys have large grains. (After Ashby and Frost. Constitutive Equations in Plasticity, ed. Argon, MIT Press, Cambridge, Copyright © 1975 by The Massachusetts Institute of Technology.)

like those used in gas turbines, have large grains. Despite the difficulties, the maps do give an excellent overview of the dominate deformation mechanisms as a function of stress, temperature, and strain rate.

3.3 CREEP TESTING AND PRESENTATION OF TEST RESULTS

Most long-time creep testing is done in creep frames that produce a constant load in the test specimen. As shown in Figure 3.3.1, these machines are simple levers with the load applied through loading weights. The specimen is heated and strains are measured by an extensometer that attaches to the specimen

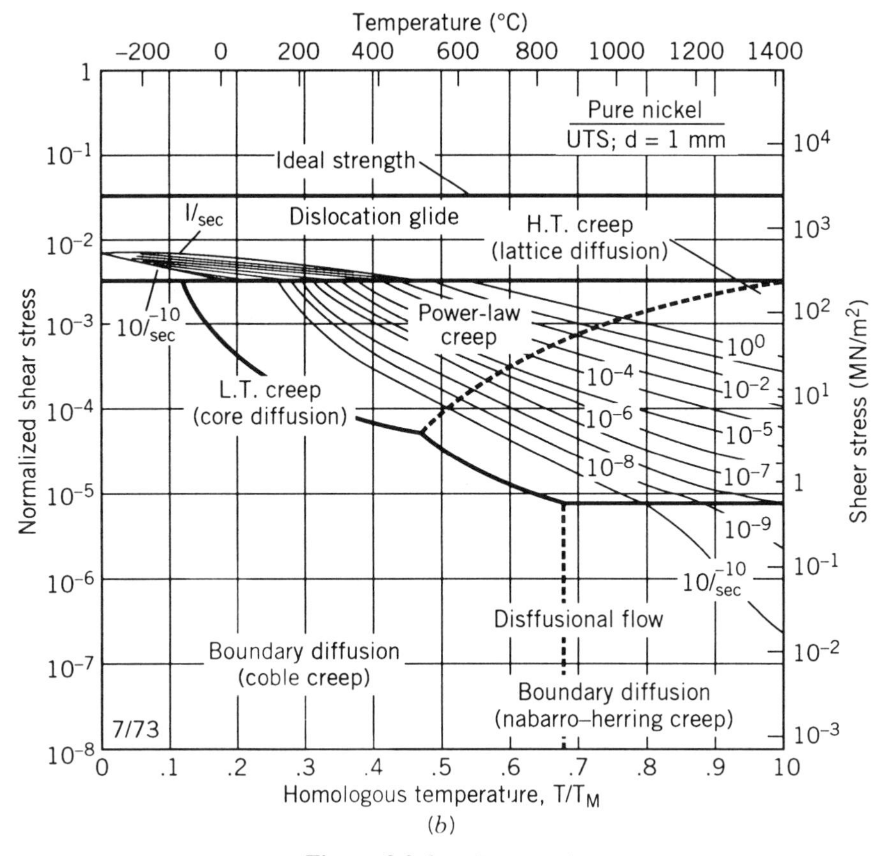

(*b*)

Figure 3.2.6 (*Continued*)

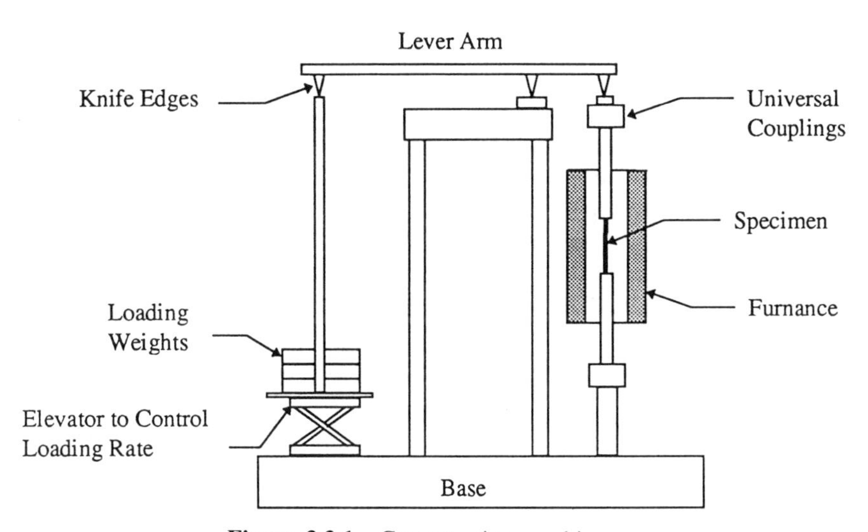

Figure 3.3.1 Creep testing machine.

with long arms so that the sensing element is outside the heated region. The extensometer arms are generally ceramic to limit thermal distortion and provide reliable strain data. If alignment of the specimen and load chain is not perfect, specimen bending can give unreliable primary creep strain data, although long time data are acceptable as the specimen straightens during deformation. Screw threads may produce some misalignment, as discussed in Section 2.1.

Experimental creep results obtained in creep frames are frequently not consistent. Experimental errors can arise from poor specimen alignment, thermal distortion of the strain extensometer arms, nonuniform heating of the test specimen, variation in the initial loading rate, and specimen and material variability. Thus engineering design is frequently based on multiple tests at each temperature and creep stress level. The data are then analyzed and mean strain curves developed for use in design. These curves represent, by definition, mathematical representations of the actual data.

The most common representation of creep data for metals is a plot of creep strain (not total strain) as a function of time, as illustrated in Figures 3.3.2 and 3.3.3. In this presentation, each curve represents the creep strain response for constant values of temperature and stress or load. The creep strain is obtained by dividing creep elongation (total elongation minus the elastic part of the deformation) by the initial gage length. Sometimes the total strain is plotted as a function of time.

Several alternative representations of the creep data may be convenient for special purposes or for use in a simplified analysis. One common method of plotting creep data is the isochronous stress–strain curve shown schematically in Figure 3.3.4. A single isochronous stress–strain curve represents stress as a function of strain at a particular time. The curve represents data from many tests. The data plotted from a single test specimen produce points along a horizontal line at the stress level of the test. Creep data are sometimes plotted as time to a particular strain level or time to rupture. An example of a time-to-rupture curve is shown in Figure 3.3.5.

The creep strength and rupture strength are frequently reported engineering data. Creep strength is the stress that produces the minimum creep rate at a specified temperature. It is frequently quoted for creep rates of 0.00001 and/or 0.001 hr^{-1}. The creep strength can also be defined as the stress to produce a specified creep strain (i.e., 1%) for a specified time and temperature. The rupture strength is the stress that produces rupture at a fixed temperature in a reference period of time, typically 1000, 10,000, or 100,000 hr.

Engineering design data for the short-time creep response of René 80 (a face-centered cubic precipitate-hardened nickel-base superalloy) is shown in Figures 3.3.2 and 3.3.3. The variation of creep strain with stress is given in Figure 3.3.3. The increase in the primary creep strain with temperature for seven temperatures at 140 MPa is given in Figure 3.3.2. The apparent decrease in sensitivity with increasing temperature is a function of the logarithmic strain scale. The variation in the secondary, minimum creep rate can be

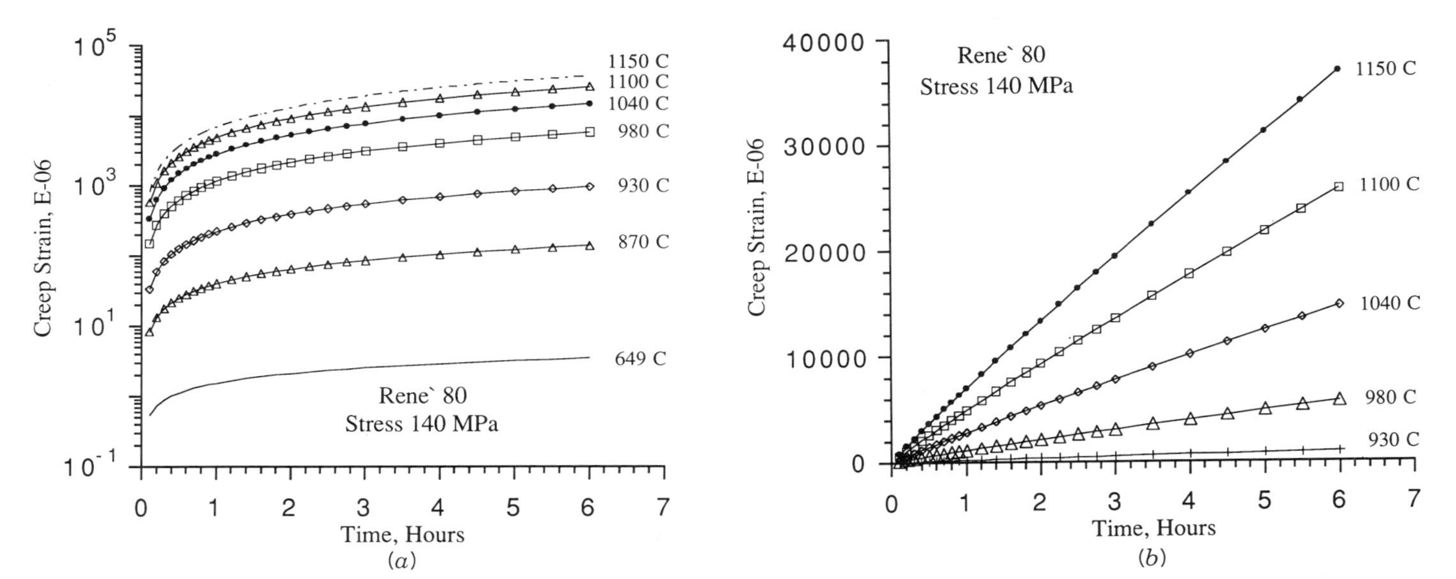

Figure 3.3.2 Short-time nominal creep response of René 80 at 140 MPa (20 ksi) for seven values of temperature as plotted on (*a*) logarithmic and (*b*) linear strain scales. (Curves obtained from data given by McKnight, Laflen, and Spamer, 1982.)

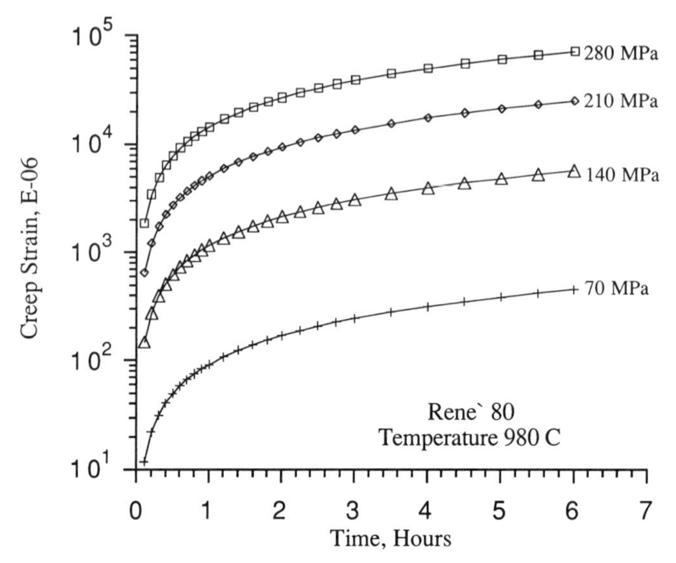

Figure 3.3.3 Effect of stress on the short-time nominal creep response of René 80 at 982°C. (Curves obtained from data given by McKnight, Laflen, and Spamer, 1982.)

seen in Figure 3.3.2*b* for the five highest temperatures using a linear strain scale. In this representation of the data the amount of primary creep is small, and the creep rates at 870 and 649°C were essentially zero on the 7 hour time scale.

The variation in creep strain with time for four values of stress is given on a logarithmic scale in Figure 3.3.3. Primary and minimum creep rate both increase with increasing stress. The yield stress of René 80 at 980°C is a function of strain rate and vanes between approximately 180 and 350 MPa. Thus the two highest stress levels are near or above the yield stress. Deformation can occur by action of several simultaneous mechanisms, which include dislocation climb, dislocation glide, dislocation creep, and bulk and grain boundary diffusion. ln all cases, increasing the tensile stress will increase the rate of deformation.

Another method to evaluate the constant stress and temperature creep response of metals is to plot the data in the form of a hodograph. This is a plot of the logarithm creep rate as a function of the logarithm of the accumulated creep strain. A hodograph representation of the true creep strain and true creep strain rate of Udimet 700W at 1700°F for three values of stress is given in Figure 3.3.6. The plot provides a method to identify the initial or primary creep rate and the minimum or secondary creep rate.

Creep rates are different in tension and compression. The short-time creep response for René 95 at 649°C is shown in Figure 3.3.7 for several values of stress. At this temperature the yield stress in tension is about 175 ksi, and the tensile stress–strain response is almost strain-rate independent. The creep

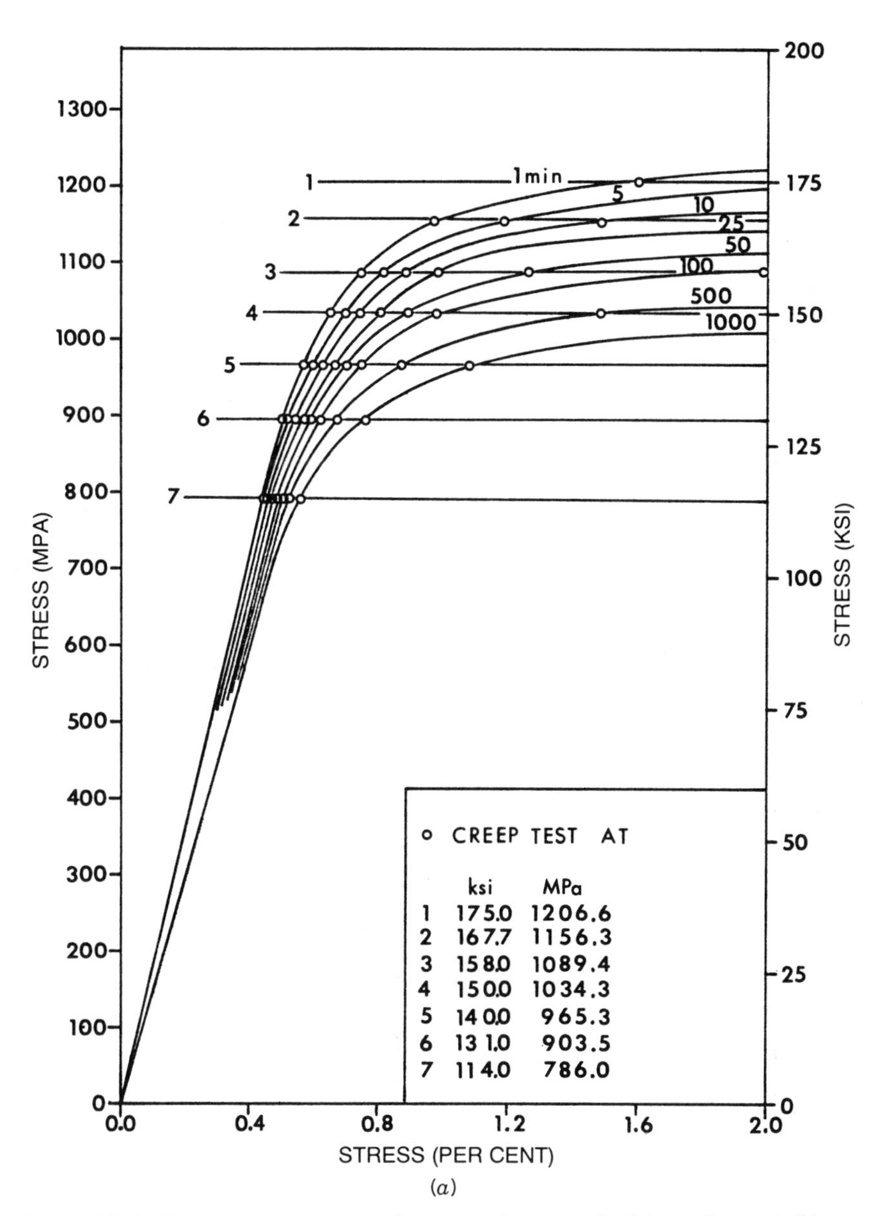

(*a*)

Figure 3.3.4 Isochronous stress–strain curves for creep in (*a*) tension and (*b*) compression of René 95 at 649°C.

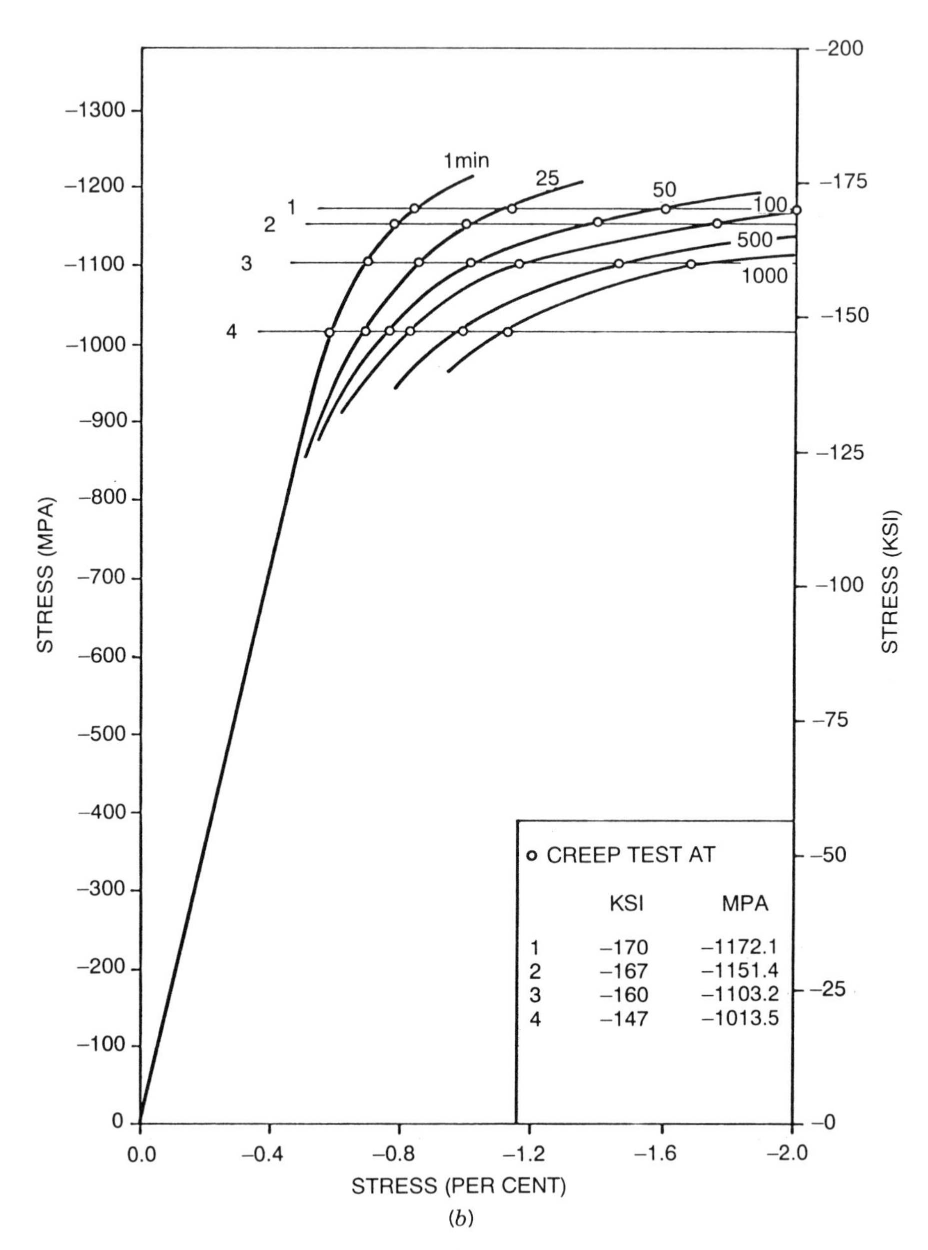

(*b*)

Figure 3.3.4 (*Continued*)

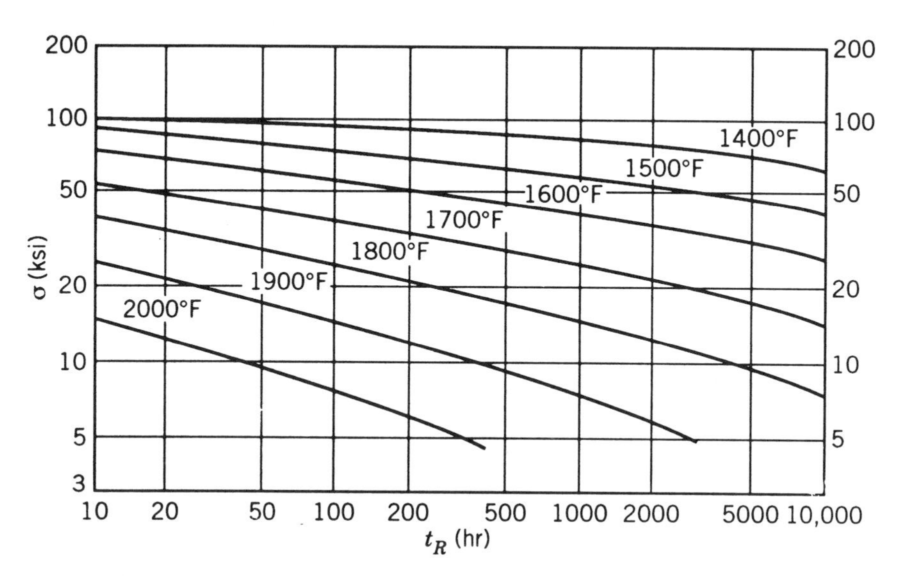

Figure 3.3.5 Typical stress versus time to rupture as a function of temperature for an engineering alloy. (From H. Kraus, *Creep Analysis,* John Wiley & Sons, Inc., New York, 1980; reprinted by permission of John Wiley & Sons, Inc.)

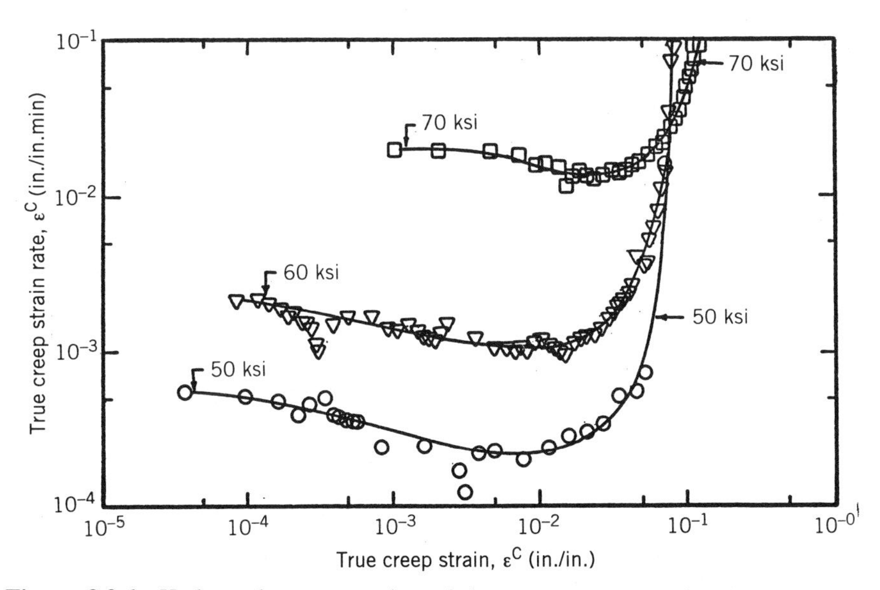

Figure 3.3.6 Hodograph representation of the creep response of Udimet 700W at 1700°F. The initial load rate in all three tests was 120 ksi/min. (After Laflen and Stouffer, Journal of Engineering Materials and Technology, V100, p366, 1978, Copyright © The American Society of Mechanical Engineers, 345 East 47th St., New York, NY 10017. Used with permission.)

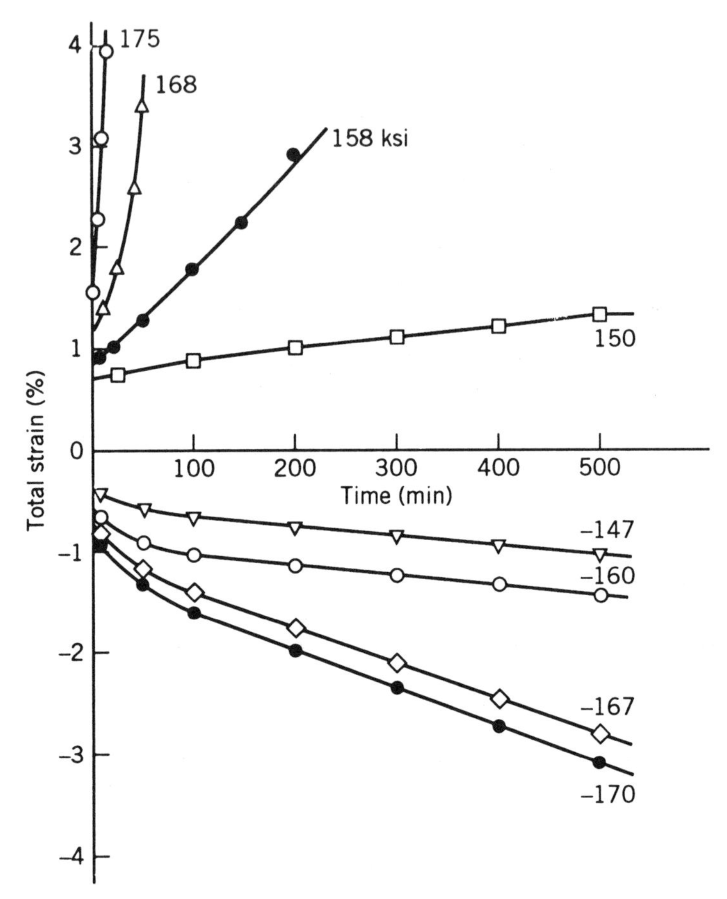

Figure 3.3.7 Creep in tension and compression for René 95 at 649°C. Creep data obtained from experiments conducted in a closed-loop testing machine using a button head specimen and special extensometer to reduce experimental errors arising for misalignment.

rates in both tension and compression are rather high, and the creep tests in compression were terminated at the onset of buckling. The results show that for nearly the same magnitude of stress in tension and compression, the magnitude of the creep rates in tension are much greater than the corresponding rates in compression. Unlike plastic stress–strain response, creep response is not skew symmetric. This difference is due partially to the fact that the creep tests are constant-load, not constant-stress tests. The creep stress is defined as the constant load divided by the initial specimen area. Therefore, in a tensile test the true stress increases as the specimen area decreases, and in compression the true stress decreases as the specimen area increases. The most important reason for different responses in tension and compression is that diffusion is the dominate deformation mechanism. Tensile stresses expand the lattice and reduce the resistance to diffusion, whereas compressive stresses reduce the lattice dimensions and increase the resistance for diffusion. Design engineers should note that many of the models in the literature are based on the assumption that creep in tension and compression is skew symmetric, but this assumption is usually not valid.

When considering a test program for a particular material and application, best results will be achieved if the creep data are obtained for temperatures at or near the operating temperature of the component of interest. Because creep is thermally activated and the creep deformation mechanism can change dramatically with temperature, it is advisable to use creep data that bracket the operating range of the component to be analyzed (especially on the high-temperature end). Creep strain rate is also a strong function of stress and it is best if the stress levels in the creep tests span the operational stress range of interest.

Finally, it is possible to show how creep is different from the time effects observed in constant-strain-rate tensile testing. Recall that the aluminum alloy 7050 at room temperature was strain-rate independent in tensile testing (Figure 2.1.2) but exhibited about 50 MPa of stress relaxation during a strain hold of 10 sec (Figure 2.4.5). Figure 3.3.8 shows that there is almost no creep in 400 min at a creep stress of 449 MPa, which is near the yield stress. These results clearly show that the mechanism producing stress relaxation in the tensile test is different from that required to produce creep. Aluminum 7050

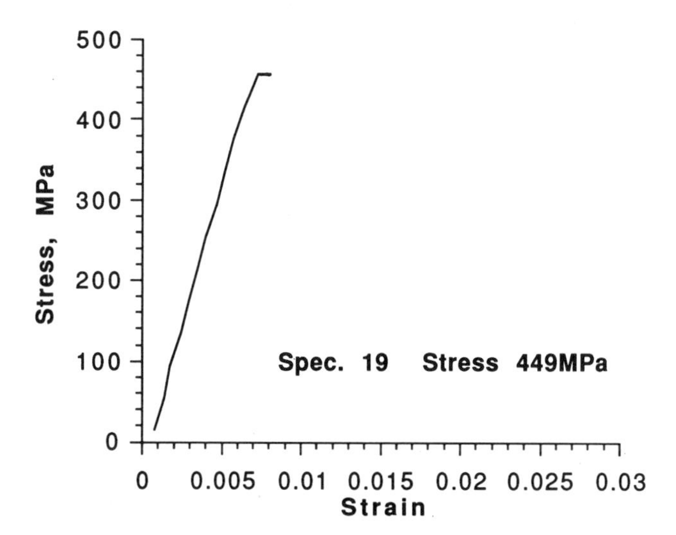

Figure 3.3.8 Stress–strain plot showing the accumulated creep strain response only 0.001 for Al 7050 after 400 min at room temperature. The creep stress was 449 MPa. (From Kuruppu et al., 1992; reprinted by permission of the Council of the Institution of Mechanical Engineers from the *Journal of Strain Analysis*.)

exhibits dynamic recovery at room temperature, but static thermal recovery was not observed. Thus all time effects in metals are not the same; they depend on the temperature and loading history.

3.4 CHARACTERIZATION OF STEADY-STATE CREEP DATA

Figure 3.3.2*b* suggests a typical creep response for many metals; that is, the creep time curve is almost linear. Thus engineering properties of metal creep are frequently characterized by the minimum creep rate, $\dot{\varepsilon}^C_{\min}$, and the time to rupture, t_r. If secondary creep constitutes a significant portion of the total creep curve, the minimum creep rate and time to failure are close to inversely related. This observation is the basis of the Larson–Miller parameter and a method to predict the long-time creep behavior from short-time creep data.

The temperature and stress dependence of steady state creep strain rate is often modeled in the form of an Arrhenius equation. Functions of this type are used to model thermally activated processes. The Arrhenius equation has the form

$$\dot{\varepsilon}^C_{\min} = f(\sigma) \exp\left(-\frac{Q}{RT}\right) \tag{3.4.1}$$

where $f(\sigma)$ is a function of only stress, Q the activation energy of the rate-controlling process, R the universal gas constant, and T the absolute temperature. The gas constant is given by $R = kN$, where k is the Boltzmann constant and N is Avogadro's number. The function $f(\sigma)$ is sometimes called the Zener parameter.

The activation energy is an important quantity in thermally driven deformation processes. For example, Q may represent the energy barrier that an atom or vacancy must overcome to jump to another position. The jump can occur only if the energy of the atom or vacancy is greater than Q. Conversely, if the energy of the atom or vacancy is less than Q, a jump cannot occur. Equation 3.4.1 gives the creep rate as a function of the thermal and mechanical energy present. The activation energy is a material property that is a function of the material, temperature, stress, and the specific deformation process. That is, there are a number of activation energies for a material, one for each deformation mechanism.

The activation energy for a particular mechanism, say Coble creep, can be determined for the creep response of a specimen deforming by Coble creep. A logarithmic plot of equation 3.4.1 with the creep rate as a function of inverse time, $1/T$, will give a set of parallel straight lines, one for each value of stress. The slope of the lines is Q/R. The activation energy Q can be determined from the slope of the lines. If two temperatures T_1 and T_2 are

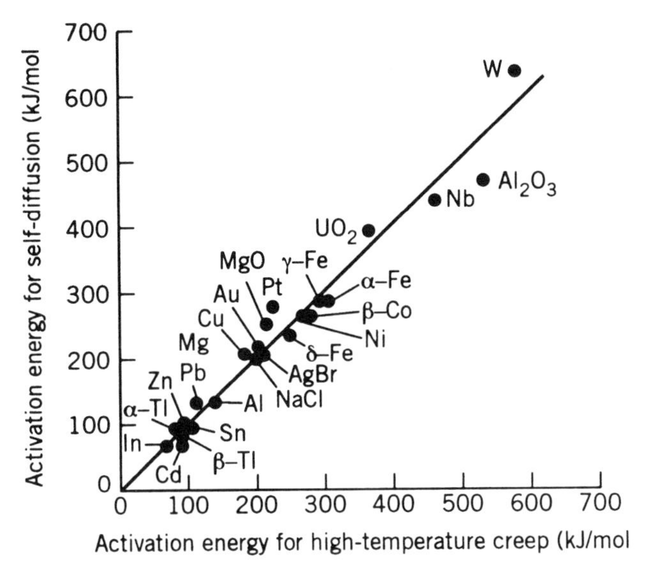

Figure 3.4.1 Correlation between creep activation energy and the energy for self-diffusion for several materials. (From Sherby and Burke, reprinted from Progress in Material Science, V13, p325, 1967; with kind permission from Elsevier Science Ltd.

picked close to each other so that the deformation mechanism does not change, then for the same value of stress,

$$f(\sigma) = \dot{\varepsilon}_1^C \exp\left(\frac{Q}{RT_1}\right) = \dot{\varepsilon}_2^C \exp\left(\frac{Q}{RT_2}\right)$$

The activation energy, Q, can then be written as

$$Q = \frac{R \ln(\dot{\varepsilon}_1^C / \dot{\varepsilon}_2^C)}{1/T_1 - 1/T_2} \tag{3.4.2}$$

Recall that the activation energy is, in general, a function of stress and temperature. Sherby and Burke (1967) have shown in Figure 3.4.1 that the activation energy for high-temperature creep is equal to the energy for self-diffusion, which is independent of stress and temperature for $T > 0.5T_m$, approximately. Thus the activation energy for high-temperature diffusional creep is independent of stress and temperature.

The creep rate in equation 3.4.1 depends on the stress function, $f(\sigma)$, as well as temperature. However, referring to the deformation map for creep (Figure 3.2.5), there are three distinct regions where different deformation mechanisms are active (recall that Coble creep and Nabarro–Herring creep are both diffusion driven). Thus there is a different representation for each type of creep mechanism. Freed and Walker (1993) completed a study of physically based creep models and found that the models for the dislocation creep and dislocation glide mechanisms are closely related; that is, the function $f(\sigma)$ in equation 3.4.1 is given as follows:

$$\text{Dislocation glide:} \qquad f(\sigma) = A' \exp\left(\frac{|\sigma|}{D'}\right) \tag{3.4.3}$$

$$\text{Dislocation creep:} \qquad f(\sigma) = A\left(\frac{|\sigma|}{D}\right)^n \tag{3.4.4}$$

$$\text{Power law breakdown:} \quad f(\sigma) = A\left[\sinh\left(\frac{|\sigma|}{D}\right)\right]^n \tag{3.4.5}$$

where equation 3.4.5 is used at the transition between dislocation creep and dislocation glide. The parameters A, D, and n are the same in equations 3.4.4 and 3.4.5, and $A' = A/2^n$ and $D' = D/n$ in equation 3.4.3. The ability of this representation to correlate the steady-state creep of NARloy Z is shown in Figure 3.4.2. The constants in Appendix 3.1 were determined from the data in Figure 3.4.2*a* (equation 3.4.3) and used to predict the data in Figure 3.4.2*b* with the dislocation creep (power law) model. These three representations are the starting point for most state-variable models.

It is often necessary to extrapolate creep data for long times. This may be required because of the exceedingly long service life of a particular component or the expense of obtaining long time data for a new or different alloy. The basic idea is to approximate long-time data by testing at a higher temperature for the same stress. Recall the deformation maps show the dominant creep mechanisms do not change with changes in temperature. Caution should be exercised not to stretch this idea too far since changes in the secondary deformation mechanism with temperature or accumulated creep strain may make the results invalid.

The most common procedure for extrapolating to long creep times uses the Larson–Miller parameter. To derive the Larson–Miller parameter, it is assumed that the temperature and stress dependence of the minimum creep strain rate are given by the Arrhenius function (equation 3.4.1). The appropriate stress function must be selected, and for this example equation 3.4.4 will be used in the form $f(\sigma) = B|\sigma|^m$, where $B = AD^{-n}$. It is also assumed that the secondary or minimum creep rate, $\dot{\varepsilon}^C_{\min}$, multiplied by the time to rupture, t_r, or the time to some value of creep strain is a constant, C'':

$$(\dot{\varepsilon}^C_{\min})t_r = C'' \tag{3.4.6}$$

Combining equations 3.4.1 and 3.4.6 produces the relationship

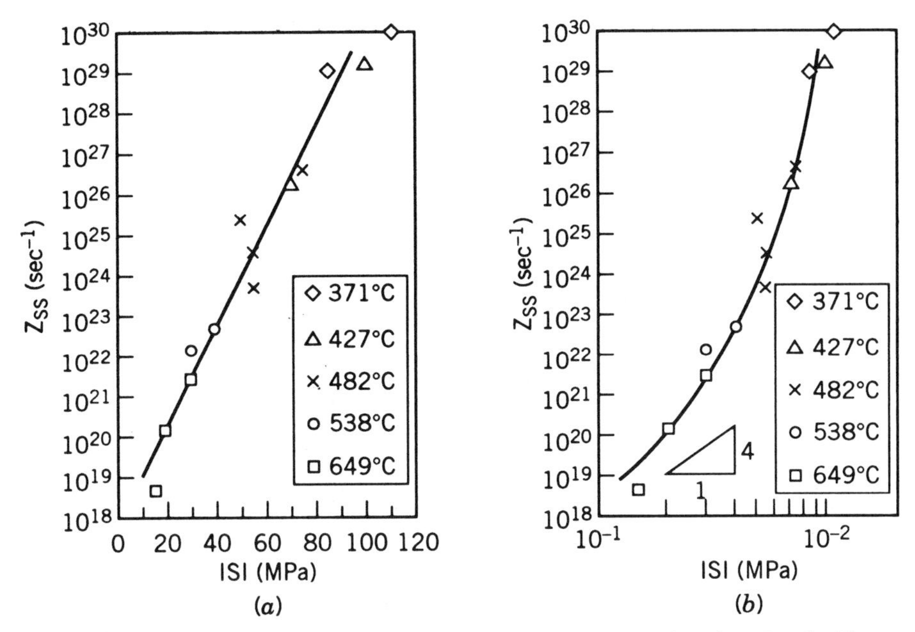

Figure 3.4.2 Steady-state creep behavior of NARloy Z. The function Z_{ss}, the Zener parameter, is the same as $f(\sigma)$. Observe the scale of Z_{ss}. (After Freed and Walker, Materials Parameter Estimation for Modern Constitutive Equations, MD V43, AMD V168, pg 74, 1993, Copyright © The American Society of Mechanical Engineers, 345 East 47th St., New York, NY 10017. Used with permission. Data after Lewis, 1970, with the reported material composition: Cu–2.89% Ag–0.22% Zr.)

$$B|\sigma|^m \exp\left(-\frac{Q}{RT}\right) t_r = C'' \tag{3.4.7}$$

For a fixed value of stress the parameters Q and $B|\sigma|^m$ are constants (Q is constant for $T > 0.5T_m$); then taking the log (base 10) of both sides of equation 3.4.7 gives

$$\log t_r - \frac{MQ}{RT} = \log \frac{C''}{B|\sigma|^n} \equiv -C \tag{3.4.8}$$

where $M = \log e = 0.4343$. Equation 3.4.8 is linear in $1/T$ and $\log t_r$, as shown in Figure 3.4.3. The slope of the line is a function of stress, and $\log t_r = -C$ is the intercept at $1/T = 0$. Equation 3.4.8 is generally rewritten in the form

$$T(C + \log t_r) = P \tag{3.4.9}$$

where P is the Larson–Miller parameter, and the Larson–Miller constant C

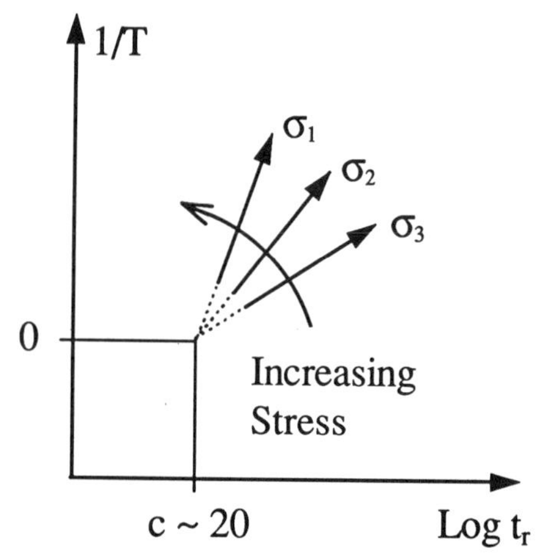

Figure 3.4.3 Plot of the Larson–Miller equation as a function of stress and temperature.

varies between 15 and 25 log(hr) but is usually assumed to be 20 log(hr) for most materials. The absolute temperature, T, is in degrees Rankine or Kelvin, and t_r is the time to rupture, in hours, or the time to some other value of strain.

The master curve for P in Figure 3.4.4 is constructed from creep rupture data at different stresses and temperatures. To construct a point on the master curve, the Larson–Miller parameter is computed for the time to rupture and test temperature and plotted against the stress. It is also simple to find the time to rupture at a particular stress and temperature. The first step is to find the Larson–Miller parameter for the applied stress from the master curve of

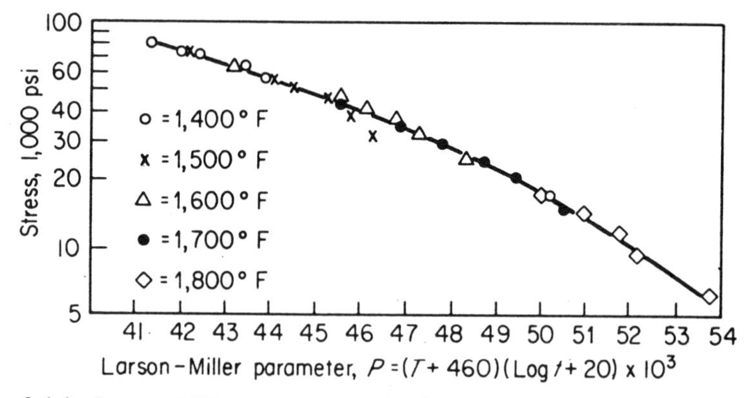

Figure 3.4.4 Larson–Miller master curve for Astroloy. (From G. E. Dieter, *Mechanical Metallurgy,* 3rd ed., McGraw-Hill, Inc., New York, 1986; reproduced with permission of McGraw-Hill, Inc.)

the material. Then substitute the Larson–Miller parameter and temperature into equation 3.4.9 and solve for rupture time or the time required to reach a specified value of strain.

Since it is very time consuming (and therefore expensive) to obtain long-term creep data, the Larson–Miller parameter is frequently used to predict low-stress long-time creep response to time to rupture. Although it may be necessary to make such approximations when data are not available, these methods should not be regarded as a substitute for data. It is very questionable to predict the creep response of a particular alloy by extrapolating in time, and particularly for different stresses. In fact, over 30 parameters have been suggested in the literature. Three such parameters are given by Orr and Sherby (1954), Manson and Herferd (1953), and White and LeMay (1978)

3.5 CREEP CURVE MODELS

Most of the models for the uniaxial creep of metals are empirically based. They arose after years of experimental observations and experience by many workers in the field. The results are widely used and are implemented as options in several commercial finite-element codes. In this section models for uniaxial creep are presented and one model is applied to creep data as an example. Extension to three-dimensional creep is discussed in Chapter 5.

The uniaxial creep strain or creep strain rate is expressed as a function of stress, time, and temperature, and some models also use the accumulated creep strain to model hardening, that is,

$$\varepsilon^C = \mathfrak{C}(\sigma, t, T) \tag{3.5.1}$$

or

$$\dot{\varepsilon}^C = \dot{\mathfrak{C}}(\sigma, t, \varepsilon^C, T) \tag{3.5.2}$$

where $\mathfrak{C}$ and $\dot{\mathfrak{C}}$ are creep and creep rate functions, respectively. Temperature is often not explicit in the creep equation, and the material parameters in the equation are considered to be functions of temperature. When temperature dependence is explicit in the creep equation, it is often in the form of the Arrhenius equation, 3.4.1. Establishing the constants for various creep equations ordinarily requires use of experimental creep strain data. In all the models summarized below, the creep strain is defined as the inelastic component of the total strain, t is time, and stress is defined as the constant creep stress.

$$\text{Bailey–Norton law:} \quad \varepsilon^C = A\left(\frac{|\sigma|}{\sigma_0}\right)^m \left(\frac{t}{t_0}\right)^n \operatorname{sgn} \sigma \tag{3.5.3}$$

The coefficient, A, and exponents, $m > 1$ and $n < 1$, are functions of temperature. The reference stress and time, σ_0 and t_0, are included for dimensional homogeneity and are frequently taken as one unit of stress and one unit of time. The form of the Bailey–Norton equation above can be used for creep in both tension and compression, but it is reasonable to expect that the parameters would be different in tension and compression. Since the exponent $n < 1$ the creep rate at $t = 0$ is not defined. This is not consistent with the hodograph for Udimet 700W and may not be valid for other materials. This also introduces numerical difficulties when integrating the creep rate for the accumulated creep strain (see Section 5.9). Parameters for René 80 at several temperatures are given in Appendix 3.2 for the Bailey–Norton creep law.

$$\text{Hyperbolic sine law:} \quad \varepsilon^C = A + \text{B} \sinh\left[C\left(\frac{t}{t_0}\right)^{1/3}\right] \tag{3.5.4}$$

The parameters A, B, and C are usually taken as functions of stress and temperature. The creep strain has an offset, A, at $t = 0$ to approximate the total amount of accumulated primary creep.

$$\text{Marin–Pao equation:} \quad \varepsilon^C = A(1 - e^{-kt}) + Bt \tag{3.5.5}$$

The first term is used to model the primary creep and the second parameter, B, represents the minimum or secondary creep rate. The coefficients, A and B, and exponent k are functions of stress and temperature. The initial creep rate of the Marin–Pao equation is defined, and it is a criterion to determine some of the properties of A and k. An example of the stress dependence of A, B, and k is given in Appendix 3.3 for the Udiment 700W at 1700°F shown in Figure 3.3.6.

$$\text{Logarithmic equation:} \quad \varepsilon^C = A + B\log(1 + Ct) \tag{3.5.6}$$

This equation gives a logarithmic linear transition between the primary and secondary creep regions. The coefficients are functions of stress and temperature.

$$\text{Andrade's } \tfrac{1}{3} \text{ Law:} \quad \varepsilon^C = A\left[1 + B\left(\frac{t}{t_0}\right)^{1/3}\right]e^{kt} \tag{3.5.7}$$

This equation models primary and secondary creep. The parameters are functions of stress and temperature.

$$\text{Lacombe equation:} \quad \varepsilon^C = A + B\left(\frac{t}{t_0}\right)^m + C\left(\frac{t}{t_0}\right)^n \tag{3.5.8}$$

The Lacombe equation is used to model all three creep regions. Primary creep is modeled by the strain offset, A. The parameters A, B, C, m, and n are all functions of stress and temperature.

Example Problem 3.1: Use the data given below to determine the constants in the Marin–Pao equation and verify the result by comparing a plot of the experimental data and model results.

Time (hr)	Strain at: 57 MPa	86 MPa	117 MPa
0.0	0.0000	0.0000	0.0000
0.5	0.0011	0.0019	0.0028
1.0	0.0015	0.0027	0.0041
2.0	0.0019	0.0037	0.0075
4.0	0.0029	0.0057	0.0120
8.0	0.0044	0.0091	0.0190
12.0	0.0060	0.0119	0.0250
16.0	0.0074	0.0149	0.0310
24.0	0.0103	0.0202	0.0440
32.0	0.0131	0.0258	0.0570
40.0	0.0158	0.0334	0.0730

SOLUTION: Since the data are given at only one temperature, the parameters A, B, and k are taken to be functions only of stress for the response at the test temperature. The schematic plot of the Marin–Pao equation shown in Figure 3.5.1 shows that B is the minimum creep rate and the coefficient A is the total amount of primary creep. The exponent k controls the total time period of primary creep.

A plot of the data as a function of time is given in Figure 3.5.2. Lines through the secondary creep data intercept the strain axes on the left (0 hr) and right (40 hr) at the values shown below, and the minimum creep rate, B, is found by dividing their difference by the elapsed time, 40 hr.

Stress (MPa)	Strain Intercept: Left	Strain Intercept: Right	Strain Rate (hr^{-1})
57	0.0018	0.0160	3.55×10^{-4}
86	0.0035	0.0315	7.00×10^{-4}
117	0.0080	0.0650	1.43×10^{-3}

Further, it is observed the all three curves reach secondary creep in about 8 hr.

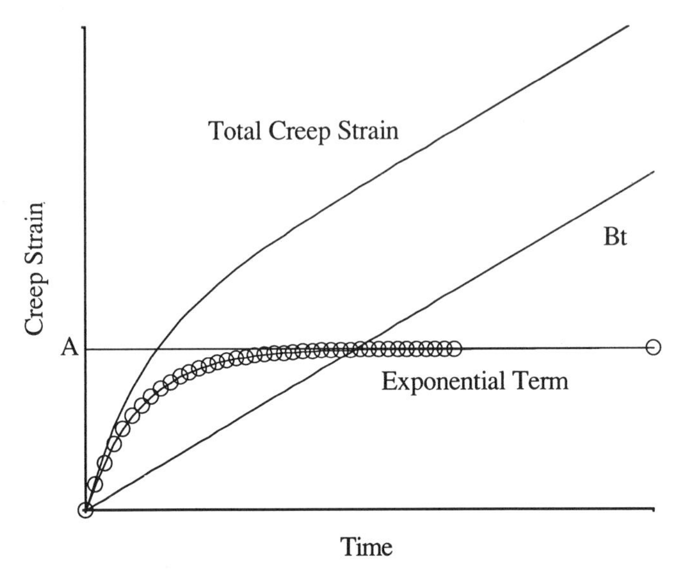

Figure 3.5.1 Schematic plot of the Marin–Pao equation showing the relative contribution of the primary and secondary creep terms.

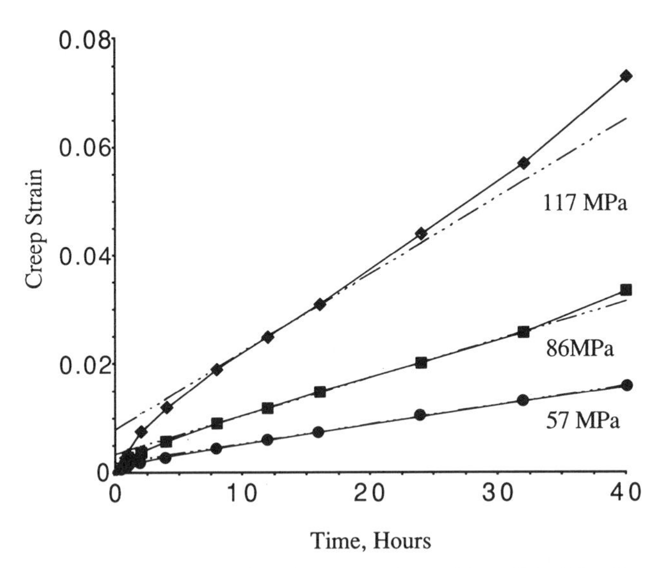

Figure 3.5.2 Plot of the creep data in Example Problem 3.1. The lines are drawn through the curves to estimate the minimum creep rate from the strain levels at 0 and 40 hr.

The next step is to determine representations for A, B, and k as a function of stress. The material parameters in the creep models are usually logarithmic or exponential functions of stress (and temperature). Using graphics software it was found that the coefficient A, the left intercept, is a semilogarithmic linear function of stress as shown in Figure 3.5.3. A mathematical representation was found in the form

$$A = c_1 \exp(c_2 \sigma) \tag{a}$$

were the values of the parameters are as given in Figure 3.5.3. The correlation coefficient of the curve fit is 0.99906. Similarly, the minimum creep rate, B, was also found to be an exponential function of the stress, that is,

$$B = c_3 \exp(c_4 \sigma) \tag{b}$$

This relationship and the associated constants are shown in Figure 3.5.4, and the correlation coefficient is 0.999989. Since all three curves reach the minimum creep domain in about 8 hr, the exponent k is a constant. Assuming that the exponential is within 1% of zero at 8 hr, that is,

$$\exp(-8k) = 0.01 \tag{c}$$

the parameter $k = 0.575$.

The representations from equations (a), (b), and (c) were used in the Marin–Pao equation to determine the creep response. A comparison of the model

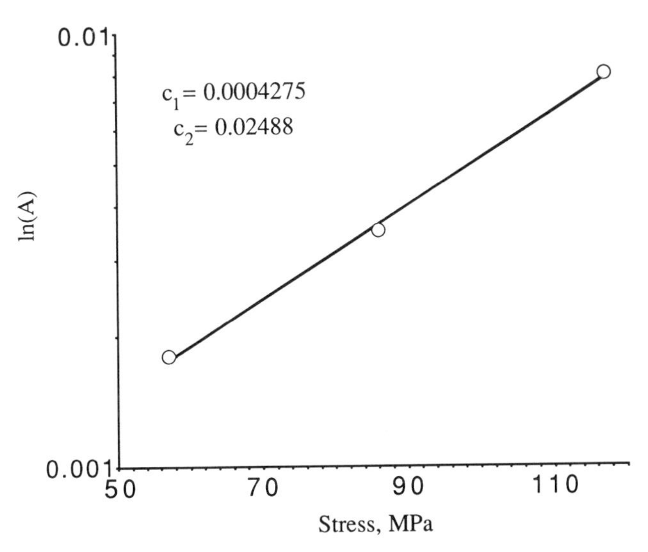

Figure 3.5.3 Evaluation of parameter A in Example Problem 3.1.

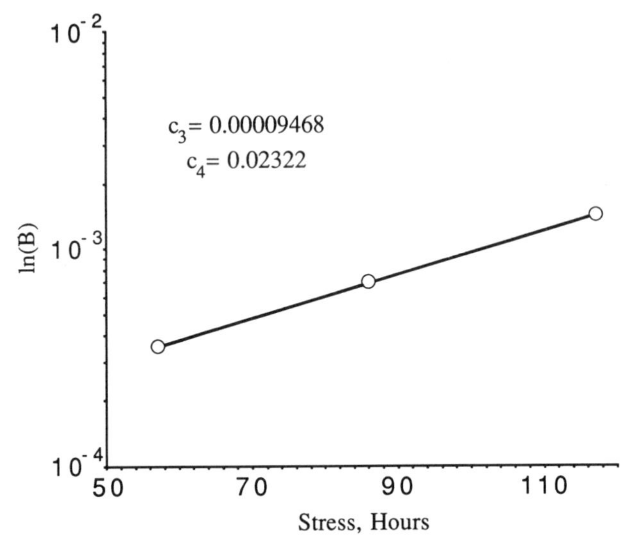

Figure 3.5.4 Evaluation of parameter B in Example Problem 3.1.

and data is given in Figure 3.5.5. The representation deviates from the data at the longer times and higher stresses since tertiary creep is not included in the model.

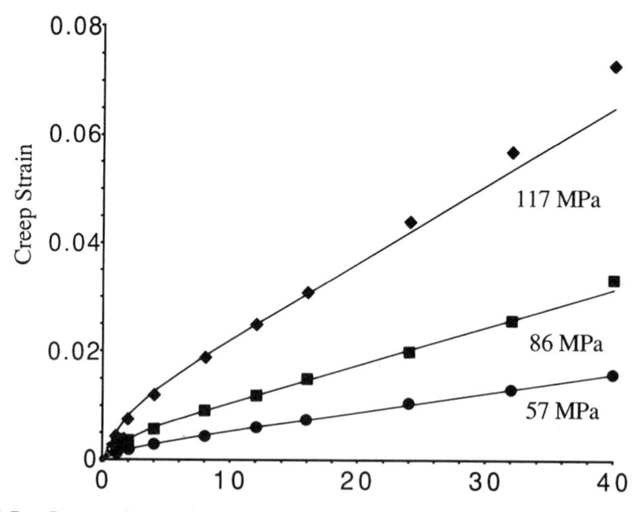

Figure 3.5.5 Comparison of the data (points) and model (lines) for Example Problem 3.1.

3.6 INTERACTION BETWEEN CREEP AND FATIGUE

The response characteristics involving creep coupled with tensile or cyclic loading is more complicated than creep alone. The interaction between the two types of loading is really an interaction between the active deformation mechanisms.

Consider first the conditions of tensile and creep testing. In tensile testing of strain-rate-sensitive materials at a constant total strain rate, the stress and inelastic strain rate are constant in the flat portion of the tensile curve (Figure 3.6.1), that is,

$$\dot{\varepsilon} = \frac{\dot{\sigma}}{E} + \dot{\varepsilon}^I = \frac{0}{E} + \dot{\varepsilon}^I = \text{constant} \tag{3.6.1}$$

The rate of hardening and recovery are balanced and the microstructure is stable. Similarly, in creep testing during the minimum creep portion of the test, the stress and inelastic strain rate are constant. Once again, the microstructure is in equilibrium, but the active deformation mechanisms are different. At high temperature several mechanisms can be active at once (recall: planar slip, dislocation climb, dislocation glide, dislocation creep, and bulk and grain boundary diffusion), and the dominate mechanism for a given temperature and material depends on the stress level. Thus the equilibrium inelastic strain rate at different stress levels would be expected to change gradually as the stress changes. Therefore, the transition in equilibrium inelastic strain rate between stress-controlled creep testing and strain-rate-controlled tensile testing is expected to be smooth and continuous.

This hypothesis has been tested for several materials. For example, the results from creep and tensile tests on René 95 at 650°C are given in Figure

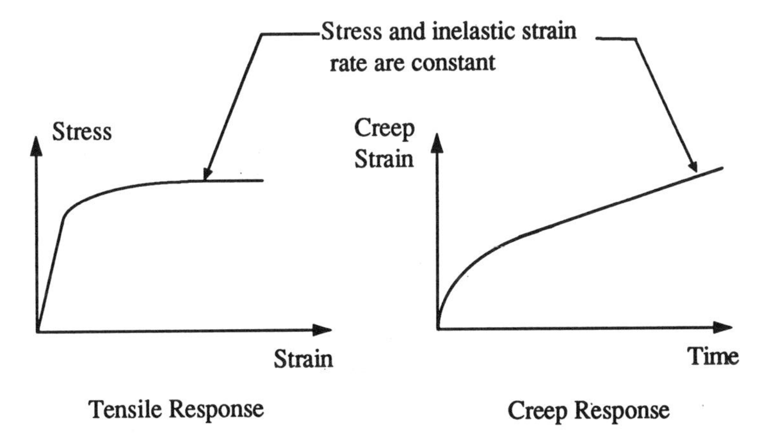

Figure 3.6.1 Conditions corresponding to a stable microstructure in tensile and creep testing.

3.6.2. The stresses plotted in Figure 3.6.2 are the stable tensile stresses and the creep stresses. The strain rates are the minimum creep rates and the strain rates at constant stress from the tensile tests. The results show a continuous change in inelastic strain rate with stress over a very large range of strain rates. This result also suggests that if a material exhibits creep or strain rate sensitivity, both effects are present. Further, the minimum creep rates can be estimated from the tensile tests, or the amount of strain-rate sensitivity can

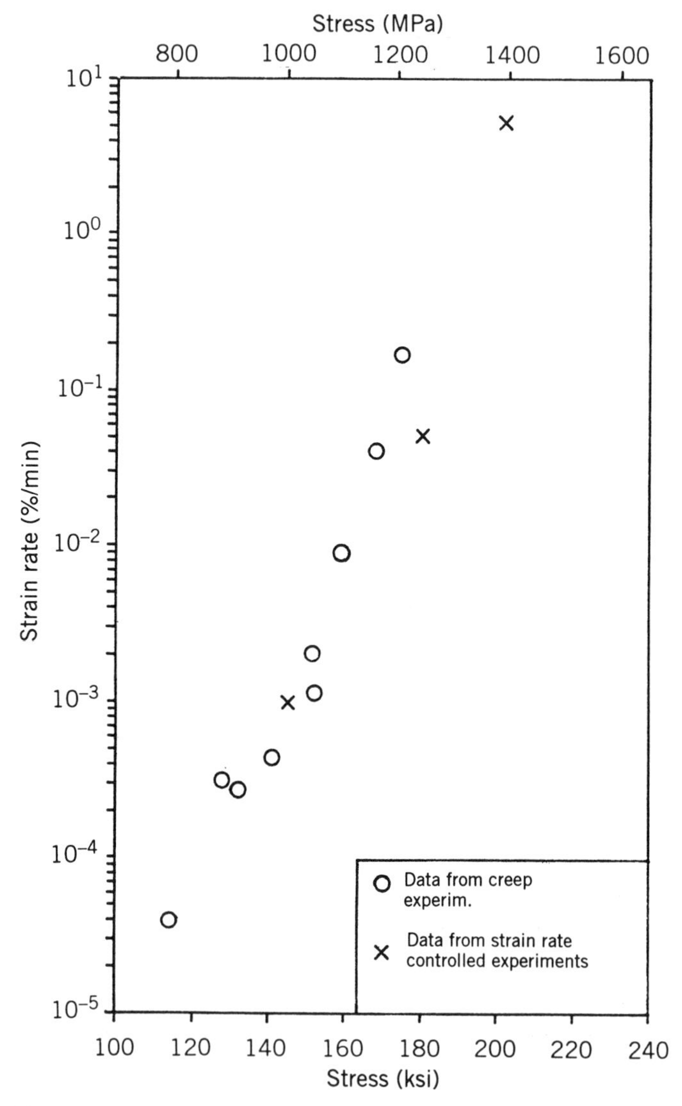

Figure 3.6.2 Stable values of stress and strain rate from tensile and creep experiments on René 95 at 650°C.

be estimated from the creep results if a representation for the curve in Figure 3.6.2 is known. Further, creep and strain-rate sensitivity are both generally accepted to become important above $0.5T_h$, so it is reasonable to expect both if one is present.

The details of how a creep test is performed can also influence the response considerably. Normally, only the value of the creep stress is specified in a test matrix. However, the experiment cannot be executed without raising the stress from zero to the specified creep stress level to start the test (Figure 3.6.3). This initial stress rate has an impact on the creep response observed. Creep curves of three specimens of AISI type 304 stainless steel at room temperature are shown in Figure 3.6.4. All three tests are at the same creep stress level. Two of the tests were loaded at an initial stress rate of 68.8 kPa s^{-1} and the third (specimen I) at 6890 kPa s^{-1}. There is a considerable effect of the initial stress rate on the subsequent amount of primary creep. The minimum creep rate does not appear to be influenced by the initial stress rate for 304 stainless steel, but this is not always the case (Abuelfoutouh, 1983). Thus it is necessary to specify and control both the creep stress and initial stress rate in creep experiments. This is another source of scatter in creep results from creep frame testing (Figure 3.3.1) because the loading weight platform is generally not controlled.

The source of the extra amount of primary creep resulting from a higher stress rate depends on the active creep mechanism. The higher stress rate produces a higher initial dislocation density. Consider first the dislocation glide mechanism. At a given level of activation energy for dislocations to move past barriers, an increase in the number of dislocations will increase the number of dislocations passing barriers per a unit of time and increase the creep rate observed. In diffusion-controlled creep, an increase in the number of dislocations in the lattice will enhance the opportunity for diffusion and increase the creep rate. In situations where the minimum creep rate is not affected, the initial dislocation structure does not appear to be permanent, and the microstructure evolves into an equilibrium state corresponding to the stress and temperature levels.

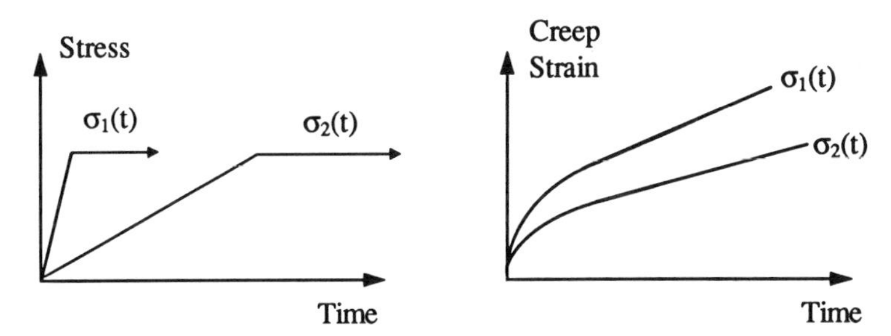

Figure 3.6.3 Effect of the initial stress rate on subsequent creep response. The minimum creep rate and initial creep rate can both be influenced by the initial load rate.

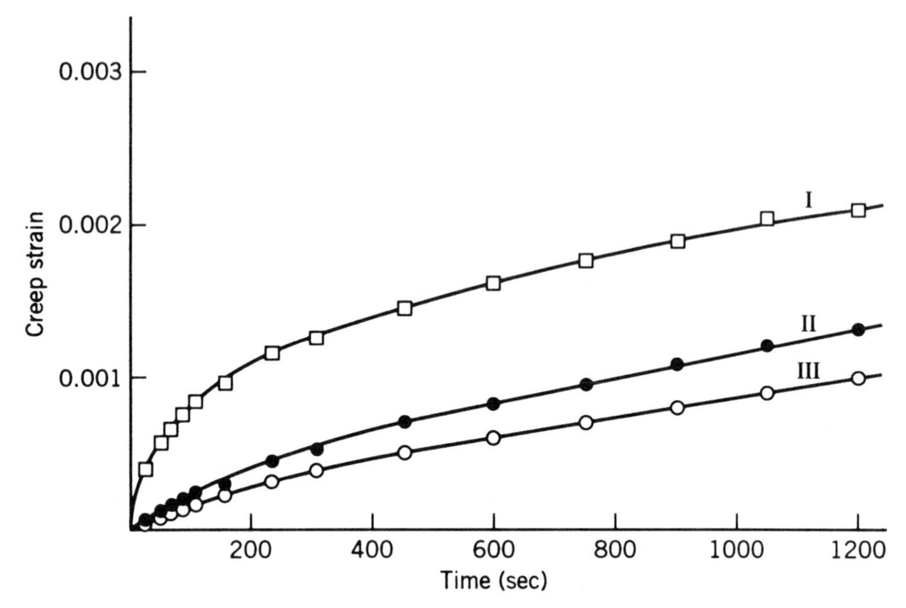

Figure 3.6.4 Influence of the initial stress rate on the creep response of AISI type 304 stainless steel at room temperature. All three specimens were loaded to 186 MPa. The initial load rate was 6890 kPa/sec for test I, and 68.9 kPa/sec for tests II and III. (After Krempl, Journal of Engineering Materials and Technology, V101, pg 385, 1979. Copyright © The American Society of Mechanical Engineers, 345 East 47th St., New York, NY 10017. Used with permission.)

The mechanical response under a combination of cyclic and creep loading can produce creep ratcheting. The loading cycle used in a test on Udimet 700W at 1700°F is shown in Figure 3.6.5. The 12-min cycle was defined to include stresses at 70 ksi for 2 min and 50 ksi for 10 min. The true creep strain response as a function of time is also given in Figure 3.6.5. The primary creep increased in each cycle. In fact, the accumulated amount of primary creep in the third cycle is almost double that found in the first cycle. It can also be seen that the secondary creep rate is higher in the second cycle than in the first cycle. This type of effect is particularly dangerous in structures because it can produce failure much sooner than predicted from constant-load creep testing.

A number of analytical and conceptual approaches have been proposed to explain the effect of fatigue in creep. One possibility suggested by Wigmore and Smith (1971) is the acceleration of void cavity growth along grain boundaries. Grain boundary diffusion promotes the growth of voids along the grain boundaries. A dislocation passing through a void displaces part of the void boundary relative to the rest of the void cavity. This creates a diffusion flux to restore the equilibrium configuration of the void boundary, resulting in void cavity growth. The process is repeated each time a dislocation passes

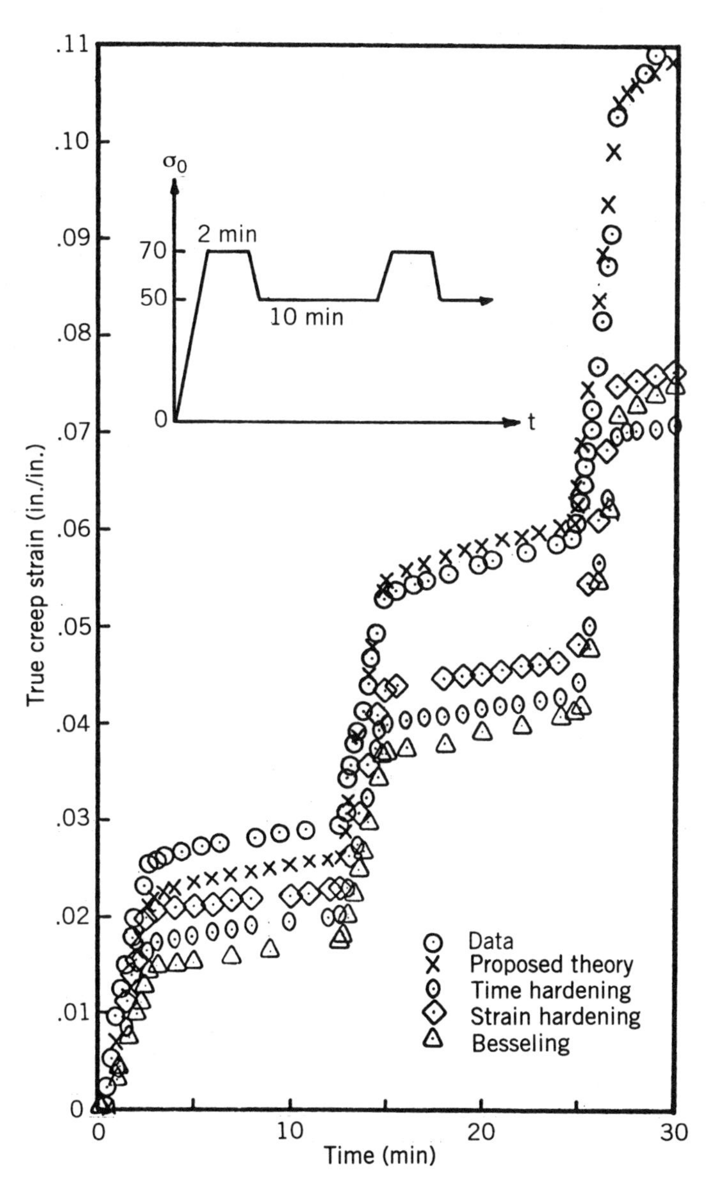

Figure 3.6.5 Growth of inelastic strain in each cycle, typical of creep ratcheting, in Udimet 700W at 1700°F. The time- and strain-hardening rules are discussed in Chapter 6. The other models are given by Laflen and Stouffer, Journal of Engineering Materials and Technology, V100, pg 367, 1978. Copyright © The American Society of Mechanical Engineers, 345 East 47th St., New York, NY 10017. Used with permission.)

through a void cavity. It is also possible that the increase in the dislocation density from cycling could promote deformation as discussed above.

3.7 SUMMARY

The mechanisms of creep are much different from those of slip. Creep occurs at stresses below the critical resolved shear stress. Temperature is the driving force. At low values of stress, the diffusion of atoms and vacancies is influenced by local regions of volumetric tension and compression. At high values of stress, dislocations glide on the slip plane, overcoming barriers by the action of thermal energy and assisted by stress. Dislocation creep occurs at intermediate values of stress by dislocation glide between barriers and climb at barriers.

The creep models presented in Sections 3.4 and 3.5 are limited to constant stress. The steady-state creep representations and whole-curve creep models are simple and reasonably accurate from many metals. The constants can generally be found by exponential or power law plots of the data, although fitting the data may require creativity in some cases. Since the models are empirical, significant inaccuracies may result when using them for conditions other than those used to determine the parameters. Since the creep models are for constant stress applications, extensions need to be developed to model time-varying stress histories (Chapters 5 and 6). Effects of the initial loading rate and creep ratcheting have not been included in many models.

Numerical implementation is simple for constant-stress conditions. The creep models are used in finite-element simulations, and they are callable options in several commercial codes.

REFERENCES

Abuelfoutouh, N. M. (1983). A thermodynamically consistent constitutive model for inelastic flow of materials, Ph.D. dissertation, University of Cincinnati.

Andrade, E. N. da C. (1914). Creep and recovery, *Proc. Royal Society, London,* Vol. 90A, pp. 329–342.

Ashby, M. F. (1973). The microstructure and design of alloys, *Proceedings of the 3rd International Conference on Strength of Metals and Alloys,* Vol. 2, Cambridge.

Ashby, M. F., and H. J. Frost (1975). Kinematics of inelastic deformation above 0 K, in *Constitutive Equations in Plasticity,* A. Argon, ed., MIT Press, Cambridge, MA.

Conway, S. B. (1967). *Numerical Methods for Creep and Rupture,* Gordon and Breach, New York.

Courtney, T. H. (1990). *Mechanical Behavior of Materials,* McGraw-Hill, New York.

Dieter, G. (1986). *Mechanical Metallurgy,* 3rd ed., McGraw-Hill, New York.

Evans, R. W., and B. Wilshire (1985). *Creep of Metals and Alloys,* Institute of Metals, London.

Freed, A. D., and K. P. Walker (1993). Viscoplastic model development with an eye towards characterization, in *Material Parameter Estimation for Modern Constitutive Equations,* MD Vol. 43, AMD Vol 168, ASME, New York, pp. 71–88.

Grant, N. J., and A. G. Buchlin (1950). *Transactions of ASM,* Vol. 42, p. 720.

Kraus, H. (1980). *Creep Analysis,* Wiley, New York.

Krempl, E. (1979). Viscoplasticity based on total strain: the modeling of creep with special considerations of initial strain and aging, *Journal of Engineering Materials and Technology,* Vol. 101, pp. 38O–386.

Kuruppu, M. D., J. F. Williams, N. Bridgeford, R. Jones, and D. C. Stouffer (1992). Constitutive modeling of the elastic-plastic behavior of 7075-T7451 aluminum alloy, *Journal of Strain Analysis,* Vol. 27, No. 2, pp. 85–92.

Lacombe, P., and L. Beaujard (1947). *Journal of the Institute of Metals,* Vol. 74, p. 1.

Laflen, J. H., and D. C. Stouffer (1978a). An analysis of high temperature metal creep, Part 1: Experimental definition of an alloy, *Journal of Engineering Materials and Technology,* Vol. 100, pp. 363–370.

Laflen, J. H., and D. C. Stouffer (1978b). An analysis of high temperature metal creep, Part 2: A constitutive formulation and verification, *Journal of Engineering Materials and Technology,* Vol. 100, pp. 71–380.

Lewis, J. R. (1970). Creep behavior of NARloy Z, *Report MA-SSE-70-902,* Rockwell International, Rockerdyne Division, Canoga Park, CA, July.

Manson, S. S., and A. M. Herferd (1953). *NACA TN 2890.*

McKnight, R. L., J. H. Laflen, and G. T. Spamer (1982). Turbine blade tip durability analysis, *NASA Lewis Research Center Contractor Report 165268.*

Orr, R. I., and O. D. Sherby (1954). *Transactions of SAM,* Vol. 46, p. 113.

Pao, Y. H., and J. Marin (1957). Analytical theory of creep deformation of materials, *Journal of Applied Mechanics,* Vol. 20, No. 2, pp. 245–252.

Reed-Hill, R. E. (1973). *Physical Metallurgy Principles,* 2nd ed., D. Van Nostrand, New York.

Sherby, 0. D. (1962). *Acta Metallurgica,* Vol. 10, pp. 135–147.

Sherby, 0. D., and P. M. Durke (1967). *Progress in Materials Science,* Vol. 13, p. 325.

Sherby, O. D., R. L. Orr, and J. E. Dorn (1954). *Transactions of AIME,* Vol. 200, pp. 71–80.

Stouffer, D. C., L. Papernik, and H. L. Bernstein (1980). An experimental evaluation of the mechanical response characteristics of René 95, *AFWAL-TR-80-4136,* Materials Laboratory, Wright Patterson AFB, OH.

Weertman, J. (1965). *Transactions of ASM,* Vol. 61, p. 681.

White, W. E., and I. LeMay (1978). *Journal of Engineering Materials and Technology,* Vol. 100, p. 319.

Wigmore, C., and G. C. Smith (1971). *Metals Science Journal,* Vol. 5, p. 58.

PROBLEMS

3.1 Compare the Nabarro–Herring and Coble creep mechanisms. Which mechanism is more temperature sensitive? Which mechanism will dominate at high and low temperatures?

3.2 Steady-state creep can be phenomenologically considered as the balance between work hardening and recovery. Let stress be written as a function of strain and time, $\sigma = \sigma(\varepsilon, t)$, and expand $d\sigma$ as a function of the independent variables. What is the physical meaning of the two terms? What happens during steady-state creep? Steady-state creep can also be considered as a constant-dislocation-density deformation process. Let the dislocation density be a function of strain and time, $\rho = \rho(\varepsilon, t)$ and find $d\rho$. What happens now during steady-state creep? How does this result compare to the expansion in stress?

3.3 Which creep models in Section 3.5 give an undefined creep rate at $t = 0$? How can this be avoided numerically? What effect does infinite creep rate at $t = 0$ have on the subsequent predictions?

3.4 Sketch the curve corresponding to the Bailey–Norton equation (3.5.3) and identify the effect of the various of parameters on the shape of the curve. Determine the parameters for the creep data given in Example Problem 3.1, and make a plot comparing the model and data. What are the attributes and deficiencies of the model?

3.5 Repeat Problem 3.4 but for the **(a)** hyperbolic sine law (equation 3.5.4); **(b)** logarithmic model (equation 3.5.6); **(c)** Andrade $\frac{1}{3}$ law (equation 3.5.7); **(d)** Lacombe model (equation 3.5.8).

3.6 A steel rod supporting a stress of 8000 psi at 1000°F is not to exceed 5% creep strain. Knowing that the steady-state creep rate can be expressed by an equation of the form

$$\dot{\varepsilon}_s^C = B|\sigma|^n \exp\left(\frac{-Q}{kT}\right)$$

where Q is the creep activation energy, determine the constants from the data for the steel in Figure P3.6 and estimate the life of the rod. (°R = °F + 460.)

3.7 The following data describing the secondary creep rate of austenitic stainless steel at 1500°F were determined by Grant and Buchlin (1950). Plot the data and determine an appropriate mathematical representation.

Stress (psi)	10,000	15,000	20,000	30,000	40,000	50,000
Minimum creep Rate (% min)	0.00008	0.0026	0.025	2.00	30.0	320.0

3.8 An open-ended, thin-walled pressure vessel of austenitic stainless steel has a 0.75-in. wall thickness and a 20-in. diameter and operates at

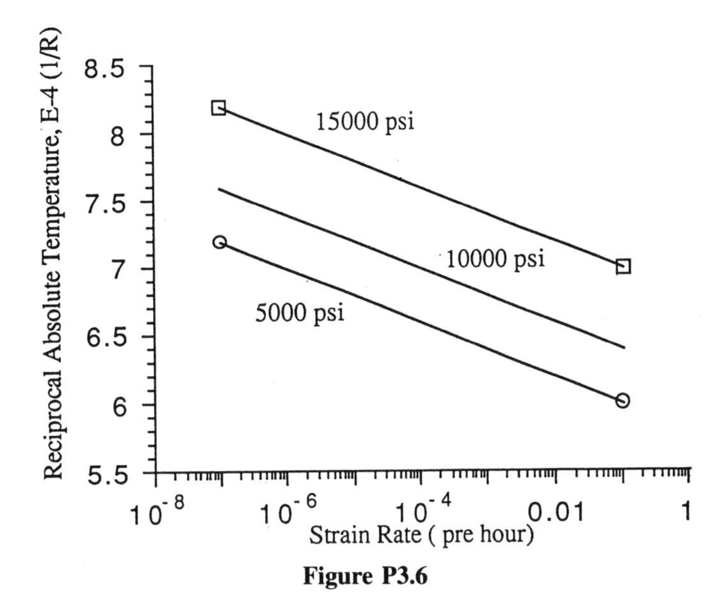

Figure P3.6

1500°F. Determine the maximum allowable pressure if the diameter is not to increase more than 0.2 in. in three years. Assume steady-state creep conditions and use the results from Problem 3.7 to estimate the creep rate.

3.9 Conceptually develop a method to draw isochronous stress–strain curves from creep data. Use the data in Figure 3.3.2 to develop three isochronous stress–strain curves for René 80 at 982°C at 2, 4, and 6 hr.

3.10 The steady-state creep rate is often considered to scale inversely with the time to rupture. Under what conditions is the approximation closely followed? (*Hint:* Consider the shape of the creep response curve.)

3.11 McKnight, Laflen, and Spamer (1982) determined the constants for the high-temperature creep of René 80 at several temperatures that are given in Appendix 3.2. Determine the activation energy for the high-temperature creep of René 80, and represent the data by Equation 3.4.1. Compare the creep strain and creep strain-rate response of both models and discuss their advantages and disadvantages.

PART TWO

MULTIAXIAL PLASTICITY AND CREEP

Part One contains a description of the physical metallurgy and physical properties of metals and the uniaxial mechanical response characteristics that can be observed in metals, and many simple uniaxial models for tensile, cyclic, and creep response. However, the models presented in Part One have two major shortcomings. First, they are limited to uniaxial loading. Most real structures experience three-dimensional loads and deformation, so it is necessary to develop methods for three-dimensional plasticity and creep. This is the subject of Part Two. The second shortcoming is that many of the uniaxial models are not very good when the response depends on time or strain rate. The uniaxial creep models are for constant stress creep. Most of the plasticity models are for applications when strain-rate effects are not present. They are not very acceptable for combined creep and strain-rate-dependent applications or situations when stress relaxation or strain recovery are present. The models are also empirical and do not contain any of the results from physical metallurgy. Part Three brings together concepts from metallurgy and mechanics for modeling when rate and time effects are present.

CHAPTER 4

Principles of Mechanics

In this chapter the principles of continuum mechanics that will be used in the development of three-dimensional constitutive models are reviewed. This is not a general discussion of mechanics but is focused on the relevant topics for constitutive modeling. The material begins with a discussion of Cartesian tensors and a summary of results from elasticity, including the properties of stress, strain, and the elasticity equations for linear isotropic materials and single crystal alloys with cubic symmetery. Deviatoric stress and strain are introduced for use in plasticity and creep models. The principle of objectivity is presented as a basic requirement in constitutive model development.

4.1 NOTATION AND PROPERTIES OF CARTESIAN TENSORS

All vector and tensor components are defined with respect to a particular coordinate system. The tensors and vectors used in this book are defined with respect to orthogonal or Cartesian coordinate systems. When orthogonal coordinate systems are used, there is only one set of base vectors that are in the direction of the coordinate axes, and the rules for transforming stress or strain from one coordinate system to another nonparallel coordinate system are relatively simple. In this section we describe the notation, transformation rules, and basic properties for Cartesian vectors and tensors in a three-dimensional Euclidean space (a vector space on which a real-valued inner product is defined).

4.1.1 Notation

Most of the current literature in solid mechanics makes use of index (subscript) notation and the summation convention. This notation allows the use of compact expressions for long, tedious equations. This feature makes the index notation useful for theoretical developments but is not very convenient for numerical calculations. The matrix method of representation of vectors and tensors is also widely used, especially for performing computations. Thus to do both theoretical developments and numerical calculations it is necessary to use both types of notation and to be able to pass from one to the other at any step in a development.

The index notation is defined for vectors and tensors in a three-dimensional space with three right-handed orthogonal unit vectors parallel to the coordinate axes (orthonormal basis). The coordinate axes x, y and z may alternatively be labeled x_1, x_2, and x_3. Using the index notation, x_i represents a point or position vector in the space of the coordinate axes (x_1, x_2, x_3), and the index i is defined to take on the values 1, 2, and 3. The base vectors parallel to the three coordinate axes are designated as $\mathbf{e}_1$, $\mathbf{e}_2$, $\mathbf{e}_3$, or $\mathbf{e}_i$. Ordinarily, when index notation is used the summation convention is also adopted. For example, the vector $\mathbf{v}$ may be written as

$$\mathbf{v} = v_1\mathbf{e}_1 + v_2\mathbf{e}_2 + v_3\mathbf{e}_3 = \sum_{i=1}^{3} v_i\mathbf{e}_i \equiv v_i\mathbf{e}_i \tag{4.1.1}$$

Whenever an index is repeated, summation with respect to that index is implied. The range of summation is equal to the number of dimensions of the space, three in this case. The summation convention implies that summation occurs over the repeated index, but the summation sign is not written. The letter used for the repeated index is not unique and has no special significance; therefore, $v_i\mathbf{e}_i$ is identical to $v_k\mathbf{e}_k$. A repeated index is a summing or dummy index. Furthermore, if more than one pair of dummy indices occurs in a term, the choice of the index letter must be distinct and summation is implied for each index separately; for example:

$$\begin{aligned} u_{ij}v_{ij} = u_{1j}v_{1j} + u_{2j}v_{2j} + u_{3j}v_{3j} &= u_{11}v_{11} + u_{12}v_{12} + u_{13}v_{13} \\ &+ u_{21}v_{21} + u_{22}v_{22} + u_{23}v_{23} + u_{31}v_{31} + u_{32}v_{32} + u_{33}v_{33} \end{aligned} \tag{4.1.2}$$

An unrepeated index such as i or j that occurs only once in every term of an expression is called a free index. Free indices are used to designate the number of equations in a system of equations; for example, the expression $v_i = a_{im}y_m$ is a system of three equations given by

$$\begin{aligned} v_1 &= a_{1m}y_m = a_{11}y_1 + a_{12}y_2 + a_{13}y_3 \\ v_2 &= a_{2m}y_m = a_{21}y_1 + a_{22}y_2 + a_{23}y_3 \\ v_3 &= a_{3m}y_m = a_{31}y_1 + a_{32}y_2 + a_{33}y_3 \end{aligned} \tag{4.1.3}$$

An index appearing more than twice in a term has no meaning and is an error. Therefore, a letter used as a free index is never used as a dummy index in the same expression.

4.1.2 Kronecker Delta and the Permutation Symbol

The Kronecker delta and permutation symbol are used to define the vector dot and cross products, respectively. The Kronecker delta, δ_{ij}, has the definition

$$\mathbf{e}_i \cdot \mathbf{e}_j = \delta_{ij} = \begin{cases} 1 & \text{if } i = j \\ 0 & \text{if } i \neq j \end{cases} \tag{4.1.4}$$

where $i = j$ means that the numerical value of i is equal to the numerical value of j. The summation convention implies that

$$\delta_{tt} = \delta_{11} + \delta_{22} + \delta_{33} = 1 + 1 + 1 = 3$$

Products involving a summation convention with a Kronecker delta and another term, such as $\delta_{pq}v_q$, become

$$\delta_{pq}v_q = \delta_{p1}v_1 + \delta_{p2}v_2 + \delta_{p3}v_3 = v_p$$

since, for example, the term δ_{p1} will be zero unless $p = 1$. Thus the Kronecker delta in the expression $\delta_{pq}v_q$ changes the index of the vector $\mathbf{v}$ from q to p. The dot product of two vectors $\mathbf{a}$ and $\mathbf{b}$ can be written as

$$\mathbf{a} \cdot \mathbf{b} = (a_i\mathbf{e}_i) \cdot (b_j\mathbf{e}_j) = a_ib_j(\mathbf{e}_i \cdot \mathbf{e}_j) = a_ib_j\delta_{ij} = a_ib_i$$

Note that the double sum, $a_ib_j\delta_{ij}$, has nine terms, but only three are nonzero. The six terms with $j \neq i$ are zero and the three non zero terms with $j = i$ become a_ib_i.

The permutation symbol, e_{ijk}, is used to define the vector cross product of the unit vectors, $\mathbf{e}_i$, that is

$$\mathbf{e}_i \times \mathbf{e}_j = e_{ijk}\mathbf{e}_k \tag{4.1.5}$$

where

$$e_{ijk} = \begin{cases} +1 & \text{if } ijk \text{ is an even permutation of 123} \\ -1 & \text{if } ijk \text{ is an odd permutation of 123} \\ 0 & \text{if } ijk \text{ has a repeated number} \end{cases}$$

The nonzero terms of e_{ijk} are $e_{123} = e_{312} = e_{231} = 1$ and $e_{321} = e_{132} = e_{213} = -1$. The cross product of two vectors $\mathbf{a}$ and $\mathbf{b}$ can be expressed in index notation as

$$\mathbf{a} \times \mathbf{b} = (a_m\mathbf{e}_m) \times (b_n\mathbf{e}_n) = a_m b_n(\mathbf{e}_m \times \mathbf{e}_n) = a_m b_n e_{mnk}\mathbf{e}_k$$

The triple scalar product is an important quantity that is used frequently and is defined as

$$\mathbf{u} \cdot (\mathbf{v} \times \mathbf{w}) = u_i\mathbf{e}_i \cdot (v_j\mathbf{e}_j \times w_k\mathbf{e}_k) = u_i\mathbf{e}_i \cdot (v_j w_k e_{jkl}\mathbf{e}_l)$$

$$= u_i v_j w_k e_{jkl}\delta_{il} = v_j w_k u_i e_{jki}$$

Recalling the properties of determinants, it follows that

$$\mathbf{u} \cdot (\mathbf{v} \times \mathbf{w}) = \det\begin{bmatrix} u_1 & u_2 & u_3 \\ v_1 & v_2 & v_3 \\ w_1 & w_2 & w_3 \end{bmatrix} = \det(\mathbf{uvw})$$

$$= u_1(v_2w_3 - v_3w_2) + u_2(v_3w_1 - v_1w_3) + u_3(v_1w_2 - v_2w_1)$$

Then using the definition of the permutation and the Kroneker delta, it can be proven that

$$\det(\mathbf{uvw}) = [\mathbf{uvw}] = [\mathbf{wuv}] = [\mathbf{vwu}] = -[\mathbf{vuw}] = -[\mathbf{wuv}] = -[\mathbf{vwu}]$$

which gives the correct values for permutating or exchanging rows or columns of the determinant.

The vector triple product identity can be written in the form

$$\mathbf{u} \times (\mathbf{v} \times \mathbf{w}) = (\mathbf{u} \cdot \mathbf{w})\mathbf{v} - (\mathbf{u} \cdot \mathbf{v})\mathbf{w}$$

This result can be used to show that the permutation symbol and the Kronecker delta are related by

$$e_{ijk}e_{ilm} = \delta_{jl}\delta_{km} - \delta_{jm}\delta_{kl} \tag{4.1.6}$$

4.1.3 Matrix Properties of Cartesian Tensors

A scalar quantity such as temperature is characterized by a single number that is independent of coordinate system but may be a function of the spatial position. A vector in three-dimensional space is specified by three numbers or components which depend on the base vectors of the coordinate system. Both scalars and vectors are special cases of a class of objects called tensors. The number of components required to characterize a tensor is determined by the order of the tensor and the dimension of the space. A scalar is a zero-order tensor requiring $3^0 = 1$ component, and a vector is a first-order tensor requiring $3^1 = 3$ components. Other physical quantities, such as stress and strain, are second-order tensors requiring $3^2 = 9$ components. The set of constants characterizing the elastic properties of a material is a fourth-order tensor with $3^4 = 81$ components. Tensors are not simply a set of numbers. They possess well-defined mathematical properties. An important subset of tensors is the Cartesian tensors, which are defined with respect to orthogonal or Cartesian coordinate systems and undergo only orthogonal transformations as defined later.

Two tensors with the same base vectors are equal if the corresponding components of the tensors are equal. For example, tensors a_{ijk} and b_{ijk} are equal if the components $a_{ijk} = b_{ijk}$.

The sum (or difference) of two tensors is defined only for tensors of the same order and results in another tensor of that order. The components of the resulting tensor are computed by taking the sum (or difference) of the corresponding components in the original tensors term by term; for example, $c_{ij} = a_{ij} + b_{ij}$ implies that

$$c_{11} = a_{11} + b_{11}$$

$$c_{12} = a_{12} + b_{12}$$

$$\vdots$$

Multiplication of a scalar (zero-order tensor) with a tensor produces a tensor of the same order with its components equal to the product of the scalar and the corresponding component in the original tensor. In this case $c_{ij} = ab_{ij}$ means that

$$c_{11} = ab_{11}$$

$$c_{12} = ab_{12}$$

$$\vdots$$

Multiplication of two tensors produces a tensor whose order is the sum of the orders of the original tensors. The tensor $c_{ijk} = a_i b_{jk}$ has 27 components defined by the product of the three values of a_i with each of the nine values b_{jk}.

When performing numerical calculations, it is often more convenient to use matrix notation than index notation. In the following expressions a square 3×3 matrix is represented by a bold capital Latin letter and a vector by lower case bold Latin letters. The correlation between matrix and indicial notation comes from the definitions of matrix notation. The matrix **A** has components a_{ij}, where the first index designates the row of the component in the matrix and the second index designates the column of the component in the matrix; that is,

$$\mathbf{A} = [a_{ij}] = \begin{bmatrix} a_{11} & a_{12} & a_{13} \\ a_{21} & a_{22} & a_{23} \\ a_{31} & a_{32} & a_{33} \end{bmatrix} \tag{4.1.7}$$

Similarly, a vector of three scalar components v_1, v_2, and v_3 in each of the coordinate directions is represented by a row or column matrix. The choice of row or column matrix for a vector is arbitrary. In this manuscript both row and column matrices will be used. The row matrix representation of a vector is

$$\mathbf{v} = (v_1, v_2, v_3) \tag{4.1.8}$$

The transpose of a second-order tensor $(a_{ij})^T$ is a_{ji}, which is equivalent to interchanging the rows and columns of the matrix a_{ij}. A tensor is symmetric if $a_{ij} = a_{ji}$ and it is skew symmetric if $a_{ij} = -a_{ji}$. The transpose of a row matrix is a column matrix.

The multiplication of two matrices is defined such that

$$\mathbf{A} = \mathbf{BC} \quad \text{implies that} \quad a_{rs} = b_{rt}c_{ts} \tag{4.1.9}$$

where the summation is between the columns of **B** and the rows of **C** as shown. This is the standard form that must be used when converting between matrix and indicial notation. For example, the indicial equation

$$h_{tr} = k_{tl}b_{rl} = k_{tl}b_{lr}^T \quad \text{becomes} \quad \mathbf{H} = \mathbf{KB}^T$$

The product of two vectors (represented as row matrices) can give a scalar or a second-order tensor:

$$\begin{aligned} w &= \mathbf{ab}^T \quad \text{implies that} \quad w = a_k b_k \\ \mathbf{W} &= \mathbf{a}^T\mathbf{b} \quad \text{implies that} \quad w_{pq} = a_p b_q \end{aligned} \tag{4.1.10}$$

A number of equivalent algebraic operations in indicial and matrix notation with vectors written as row matrices are:

$$
\begin{aligned}
\mathbf{A} + \mathbf{B} &= \mathbf{C} & a_{ij} + b_{ij} &= c_{ij} \\
\mathbf{w} &= \mathbf{u} + \mathbf{v} & w_i &= u_i + v_i \\
\mathbf{W} &= \mathbf{u}^T\mathbf{v} & w_{ij} &= u_i v_j \\
\mathbf{v} &= \mathbf{u}\mathbf{A} & v_j &= a_{pj}u_p \\
\mathbf{v}^T &= \mathbf{A}^T\mathbf{u}^T & v_j &= a_{pj}u_p \\
\mathbf{v}^T &= \mathbf{A}\mathbf{u}^T & v_j &= a_{jp}u_p \\
\mathbf{v} &= \mathbf{u}\mathbf{A}^T & v_j &= a_{jp}u_p \\
\mathbf{u}\mathbf{v}^T &= \lambda & u_i v_i &= \lambda \\
\mathbf{A}\mathbf{B} &= \mathbf{C} & a_{ip}b_{pj} &= c_{ij} \\
\mathbf{A}^T\mathbf{B} &= \mathbf{C} & a_{pi}b_{pj} &= c_{ij} \\
\mathbf{A}\mathbf{B}^T &= \mathbf{C} & a_{ip}b_{jp} &= c_{ij} \\
\mathbf{A}^T\mathbf{B}^T &= \mathbf{C} & a_{pi}b_{jp} &= c_{ij}
\end{aligned}
\tag{4.1.11}
$$

where λ is a scalar.

4.1.4 Rotation of Coordinates

Vectors (first-order tensors) and tensors are physical quantities with components that must be expressed relative to a coordinate system. It is frequently necessary to express the components of a vector or tensor in terms of coordinate systems that are oriented differently in space. Consider two Cartesian reference systems x_i and x_i' with common origin. Unit base vectors $\mathbf{e}_i$ and $\mathbf{e}_i'$ are parallel to the x_i and x_i' coordinate directions, respectively, as shown in Figure 4.1.1. The vectors $\mathbf{e}_i'$ can be expressed as functions of components in the $\mathbf{e}_i$ directions in matrix notation. Let $\hat{\mathbf{e}}_1' = a_{11}\hat{\mathbf{e}}_1 + a_{12}\hat{\mathbf{e}}_2 + a_{13}\hat{\mathbf{e}}_3$ where a_{1i} are the components of $\hat{\mathbf{e}}_i'$ in the $\hat{\mathbf{e}}_1$ basis. Similar expansions for $\hat{\mathbf{e}}_2'$ and $\hat{\mathbf{e}}_3'$ give

$$
\begin{Bmatrix} \mathbf{e}_1' \\ \mathbf{e}_2' \\ \mathbf{e}_3' \end{Bmatrix} = \begin{bmatrix} a_{11} & a_{12} & a_{13} \\ a_{21} & a_{22} & a_{23} \\ a_{31} & a_{32} & a_{33} \end{bmatrix} \begin{Bmatrix} \mathbf{e}_1 \\ \mathbf{e}_2 \\ \mathbf{e}_3 \end{Bmatrix} \tag{4.1.12}
$$

in matrix notation, or in indicial notation

$$
\mathbf{e}_i' = a_{im}\mathbf{e}_m
$$

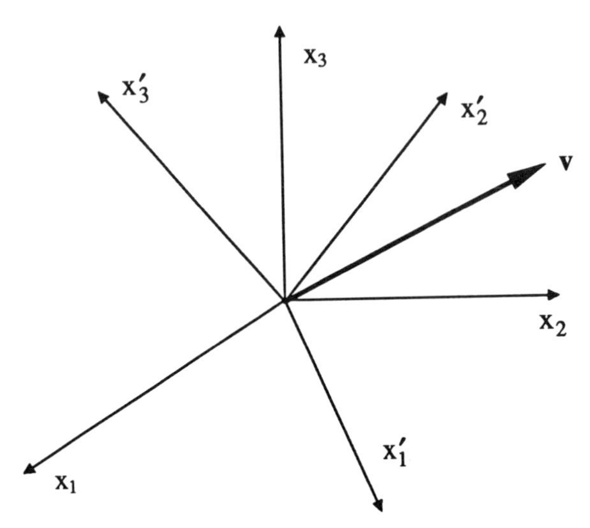

Figure 4.1.1 The vector **v** will have different components in two different coordinate systems.

The direction cosines between the coordinate axes x'_i and the coordinate axes x_j are the components of the matrix a_{ij}. Using properties of the dot product, it follows that

$$\begin{aligned}\mathbf{e}'_i \cdot \mathbf{e}_j &= |\mathbf{e}'_i||\mathbf{e}_j|\cos(x'_i, x_j) = \cos(x'_i, x_j)\\ &= a_{im}\mathbf{e}_m \cdot \mathbf{e}_j = a_{im}\delta_{mj} = a_{ij}\end{aligned}$$

It should be noted that in general $a_{ij} \neq a_{ji}$. Since the rows in the matrix represent the components of the $\mathbf{e}'_i$ base vectors in the unprimed coordinate system and the columns are the components of the $\mathbf{e}_i$ base vectors in the primed system, the sum of the square of the components in each row and column must equal unity. Similarly, the inner product (dot product) of any two rows or columns must be zero.

If matrix a_{ij} transforms a right-hand orthonormal basis $\mathbf{e}_i$ into another right-hand orthonormal basis $\mathbf{e}'_i$, then a_{ij} is a proper orthogonal transform. Since the rows of a_{ij} are components of the $\mathbf{e}'_i$ base vectors, the transformation will maintain the right-hand orientation if $\det(a_{ij}) > 0$. Since the determinant corresponds to the volume of the parallelpiped defined by the base vectors, the transformation will preserve unit volume if $\det(a_{ij}) = \pm 1$. The angles between the $\mathbf{e}'_i$ base vectors will be 90° if

$$\delta_{ij} = e'_i \cdot e'_j = a_{im}e_m \cdot a_{jn}e_n = a_{im}a_{jn}\delta_{mn} = a_{im}a_{jm}$$

Thus a_{ij} is a proper orthogonal transformation if

$$a_{ij}a_{ik} = a_{ji}a_{ki} = \delta_{jk} \qquad \text{and} \qquad \det(a_{ij}) = +1 \tag{4.1.13}$$

Equation 4.1.13 is necessary and sufficient to guarantee that the basis vectors in both the unprimed and primed coordinate systems form right-hand orthonormal bases.

The vector $\mathbf{v}$ can be written in either the primed or unprimed bases so that $\mathbf{v} = v_i\mathbf{e}_i = v_i'\mathbf{e}_i'$, as shown in Figure 4.1.1. It follows that

$$\begin{aligned} v_i' &= \mathbf{v} \bullet \mathbf{e}_i' = v_j\mathbf{e}_j \bullet \mathbf{e}_i' = v_j\mathbf{e}_j \bullet a_{ik}\mathbf{e}_k = a_{ij}v_j \\ v_i &= v \bullet \mathbf{e}_i = v_j'\mathbf{e}_j' \bullet \mathbf{e}_i = v_j\mathbf{e}_j' \bullet a_{ki}\mathbf{e}_k' = a_{ji}v_j' \end{aligned} \tag{4.1.14}$$

Observe the order of the subscripts in the two equations and note that one transformation is the inverse of the other. In a similar manner, a second-order Cartesian tensor with t_{ij} components in the $\mathbf{e}_i$ base vectors may be expressed in terms of the $\mathbf{e}_i'$ base vectors,

$$t_{ij}' = a_{im}a_{jn}t_{mn} \qquad \text{or} \qquad \mathbf{T}' = \mathbf{ATA}^T \tag{4.1.15a}$$

and the components t_{ij}' in the $\mathbf{e}_i'$ base vectors may be transformed to the $\mathbf{e}_i$ base vectors.

$$t_{ij} = a_{mi}a_{nj}t_{mn}' \qquad \text{or} \qquad \mathbf{T} = \mathbf{A}^T\mathbf{T}'\mathbf{A} \tag{4.1.15b}$$

Once again compare the order of the subscripts in equations 4.1.15, and note that they follow the same sequence as equations 4.1.14. The rules for rotating Cartesian tensor components from one coordinate system to another can easily be extended to higher-order tensors. For example, a fourth-order tensor typical of the type used to model anisotropic elastic response is transformed as follows:

$$\begin{aligned} C_{ijkl} &= a_{mi}a_{nj}a_{ok}a_{pl}C_{mnop}' \\ C_{ijkl}' &= a_{im}a_{jn}a_{ko}a_{lp}C_{mnop} \end{aligned} \tag{4.1.16}$$

Example Problem 4.1: Find the transformation matrix a_{ij} that carries $\mathbf{e}_j$ into $\mathbf{e}_i'$ for a counterclockwise rotation of the x_1 and x_2 axes about the x_3 axis as shown Figure 4.1.2. Show that a_{ij} is a proper orthogonal transformation.

SOLUTION: The components of the $\mathbf{e}_i'$ base vectors can be written as

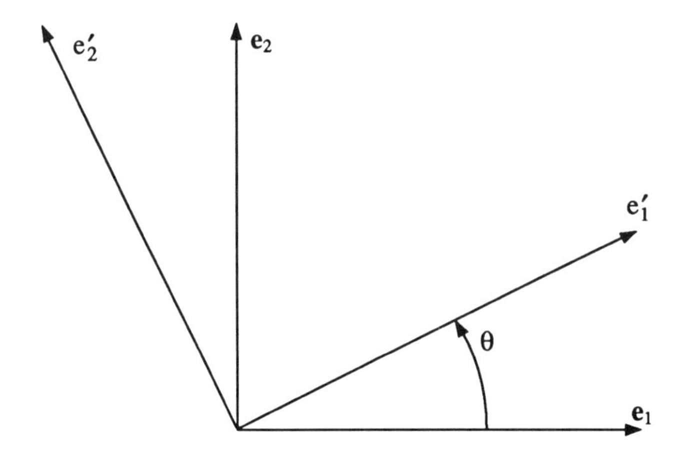

Figure 4.1.2 The x_i' coordinates are obtained by a counter clockwise rotation of the x_i coordinates about the x_3 axis as described in Example Problem 4.1.

$$\begin{aligned}\mathbf{e}_1' &= \cos\theta\,\mathbf{e}_1 + \sin\theta\,\mathbf{e}_2\\ \mathbf{e}_2' &= -\sin\theta\,\mathbf{e}_1 + \cos\theta\,\mathbf{e}_2\\ \mathbf{e}_3' &= \mathbf{e}_3\end{aligned} \qquad \text{or} \qquad \begin{bmatrix}\mathbf{e}_1'\\ \mathbf{e}_2'\\ \mathbf{e}_3'\end{bmatrix} = \begin{bmatrix}\cos\theta & \sin\theta & 0\\ -\sin\theta & \cos\theta & 0\\ 0 & 0 & 1\end{bmatrix} \times \begin{bmatrix}\mathbf{e}_1\\ \mathbf{e}_2\\ \mathbf{e}_3\end{bmatrix}$$

The transformation matrix is a proper orthogonal transformation if equation 4.1.13 is satisfied. It follows that

$$\det\begin{bmatrix}\cos\theta & \sin\theta & 0\\ -\sin\theta & \cos\theta & 0\\ 0 & 0 & 1\end{bmatrix} = +1$$

The second part of the definition can be satisfied by observing that $a_{ij}a_{ik} = a_{ji}^T a_{ik} = \delta_{jk}$, so that

$$\begin{bmatrix}\cos\theta & -\sin\theta & 0\\ \sin\theta & \cos\theta & 0\\ 0 & 0 & 1\end{bmatrix}\begin{bmatrix}\cos\theta & \sin\theta & 0\\ -\sin\theta & \cos\theta & 0\\ 0 & 0 & 1\end{bmatrix} = \begin{bmatrix}1 & 0 & 0\\ 0 & 1 & 0\\ 0 & 0 & 1\end{bmatrix}$$

Therefore, a_{ij} is a proper orthogonal transform.

4.1.5 Tensor Fields and Tensor Calculus

A tensor field assigns to every position **x** in some space S for every time t a tensor $T_{ijk\ldots p}(\mathbf{x}, t)$. The tensor may be of any order, and it is generally assumed to be continuous in **x** and t to whatever order is necessary for a particular calculation. The functions $p(\mathbf{x}, t)$, $v_i(\mathbf{x}, t)$, and $T_{ij}(\mathbf{x}, t)$ are scalar, vector, and second-order tensor fields, respectively. Partial differentiation with respect to the temporal and spatial variables is denoted by ∂t and ∂x_i, respectively, and their values are determined using the rules of partial differentiation. A comma denotes differentiation with respect to spatial variables when using the index notation $\partial \phi / \partial x_i = \phi_{,i}$, and a dot above a variable represents differentiation with respect to time $\partial \phi / \partial t = \dot{\phi}$. The differentiation of the independent spatial variables with respect to the dependent spatial variables gives the following result:

$$\frac{\partial x_i}{\partial x_j} = x_{i,j} = \delta_{ij}$$

The del or gradient operator of vector calculus is defined as

$$\nabla = \frac{\partial}{\partial x_1}\mathbf{e}_1 + \frac{\partial}{\partial x_2}\mathbf{e}_2 + \frac{\partial}{\partial x_3}\mathbf{e}_3$$

The gradient of a scalar is the vector

$$\nabla \phi = \frac{\partial \phi}{\partial x_1}\mathbf{e}_1 + \frac{\partial \phi}{\partial x_2}\mathbf{e}_2 + \frac{\partial \phi}{\partial x_3}\mathbf{e}_3 = \phi_{,i}\mathbf{e}_i$$

The divergence and curl of the vector $\mathbf{v}$ with components defined relative to coordinate axes x_i are written as

$$\text{div } \mathbf{v} = \nabla \bullet \mathbf{v} = v_{i,i} = v_{1,1} + v_{2,2} + v_{3,3} = \frac{\partial v_1}{\partial x_1} + \frac{\partial v_2}{\partial x_2} + \frac{\partial v_3}{\partial x_3}$$

$$\text{curl } \mathbf{v} = \nabla \times \mathbf{v} = (v_{3,2} - v_{2,3})\mathbf{e}_1 + (v_{1,3} - v_{3,1})\mathbf{e}_2 + (v_{2,1} - v_{1,2})\mathbf{e}_3$$

The gradient of the vector $\mathbf{v}$ has nine components $v_{i,j}$ and is usually written as

$$\text{grad } \mathbf{v} = \nabla \mathbf{v} = v_{i,j} = \begin{bmatrix} v_{1,1} & v_{1,2} & v_{1,3} \\ v_{2,1} & v_{2,2} & v_{2,3} \\ v_{3,1} & v_{3,2} & v_{3,3} \end{bmatrix} \tag{4.1.17}$$

although the transpose of equation 4.1.17 appears in some works.

4.2 STRESS

This section contains a brief review of the definition of stress, the stress vector and tensor, and the important properties of stress. The objective is not to derive the results from first principles but only to present the underlying ideas and to write down the important equations that will be used in constitutive modeling. More detail is available from any of the elasticity or continuum mechanics texts listed in the references for this chapter.

4.2.1 Definition of Stress

Stress at a point in a body depends on the loads applied to the body, the shape of the body, and the position of the point in the body. If we pass an imaginary plane with unit normal vector **n** through a body that is loaded by external forces, there will be a set of distributed forces on the internal surface that must satisfy the equation of motion (Figure 4.2.1). Consider the distributed internal forces, **F**, acting on a small portion of the area, ΔA, of the imaginary cutting plane as shown. The force **F** will have a component normal to the plane, F_n, and a component tangent to the plane, F_t. The distributed force per unit area at some point on the plane in the normal direction is $F_n/\Delta A$ and the force per unit area in the tangential direction is $F_t/\Delta A$. The stress vector, $\overset{n}{\mathbf{T}}$, is defined as the force per unit area as $\Delta A \to 0$. The component of force per unit area in the normal direction is the normal stress σ_{nn} and the component of force per area in the tangential direction is the shear stress, σ_{ns}. Mathematically, the stress vector, normal stress, and shear stress components on the plane at some position are defined by

$$\overset{n}{\mathbf{T}} = \lim_{\Delta A \to 0} \frac{\mathbf{F}}{\Delta A}, \qquad \sigma_{nn} = \lim_{\Delta A \to 0} \frac{F_n}{\Delta A}, \qquad \sigma_{ns} = \lim_{\Delta A \to 0} \frac{F_t}{\Delta A} \tag{4.2.1}$$

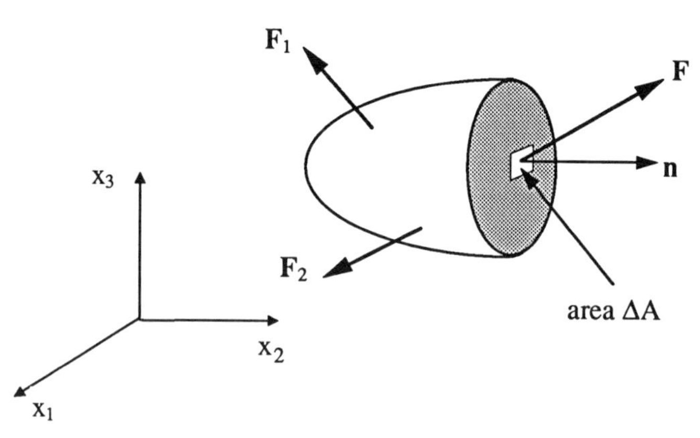

Figure 4.2.1 Increment of force at position A on the surface defined by normal **n** in a body with external loads.

The shear stress σ_{ns} is the maximum value of shear stress at that position, and the total definition of shear stress requires that the direction of the tangential force be known.

The stress vector on the opposite side of the infinitesimal area ΔA with unit normal $-\mathbf{n}$ is designated $\overset{-n}{\mathbf{T}}$. By considering Newton's second law of motion it follows that $\overset{n}{\mathbf{T}}$ and $\overset{-n}{\mathbf{T}}$ are of equal magnitude and opposite in direction; that is,

$$\overset{-n}{\mathbf{T}} = -\overset{n}{\mathbf{T}} \tag{4.2.2}$$

The normal and shear stress components at a point in the body will depend on the orientation of the cutting plane and the position of the point. The state of stress at a point is defined completely when the components of stress are known on three mutually perpendicular planes. The stress vectors on each of the mutually perpendicular coordinate planes as shown in Figure 4.2.2 may be written as

$$\overset{1}{\mathbf{T}} = \sigma_{11}\mathbf{e}_1 + \sigma_{12}\mathbf{e}_2 + \sigma_{13}\mathbf{e}_3$$

$$\overset{2}{\mathbf{T}} = \sigma_{21}\mathbf{e}_1 + \sigma_{22}\mathbf{e}_2 + \sigma_{23}\mathbf{e}_3$$

$$\overset{3}{\mathbf{T}} = \sigma_{31}\mathbf{e}_1 + \sigma_{32}\mathbf{e}_2 + \sigma_{33}\mathbf{e}_3$$

or

$$\overset{i}{\mathbf{T}} = \sigma_{ij}\mathbf{e}_j \tag{4.2.3}$$

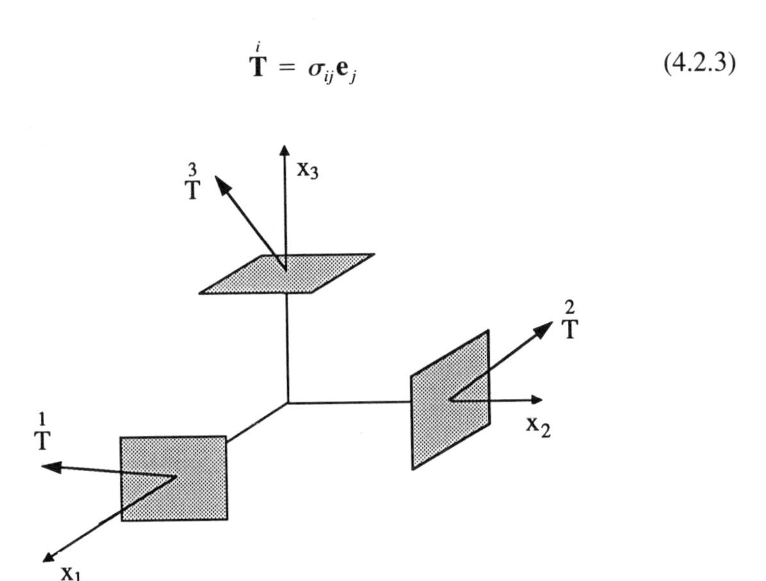

Figure 4.2.2 Stress vectors in three mutually perpendicular planes.

where σ_{1i} are the components of the $\overset{1}{\mathbf{T}}$ stress vector. The standard notation for stress defines the first subscript as the direction of the normal to the plane, and the second subscript is the direction of the stress component on that plane. A stress is positive if the direction of the unit normal vector and direction of the stress component are both in the positive direction or both in the negative direction of the coordinate system. Using this definition, tensile stress is defined as positive and compressive stress is negative.

It can be proven from the equations of motion (or equilibrium) that:

1. The stress at a point is a second-order tensor.
2. The stress tensor is symmetric.
3. The stress vector and stress tensor are related.

Since the stress at a point is a second-order tensor, the stress tensor components in one orthogonal coordinate system at a point can be determined from the stress tensor components in another orthogonal coordinate system at the same point using equations 4.1.15*a* and 4.1.15*b*. Given the stress tensor in the x_i coordinate system, the stress tensor in the x_i' coordinate system is determined from

$$\sigma_{ij}' = a_{im} a_{jn} \sigma_{mn} \tag{4.2.4}$$

Consider next the relationship between the stress vector and stress tensor. The nine components of the stress tensor at a point are the components of three stress vectors on three mutually orthogonal planes at that point. For example, the stress tensor, σ_{ij}, and the stress vectors, $\overset{i}{\mathbf{T}}$, in the $\mathbf{e}_i$ basis are

$$[\sigma_{ij}] = \begin{Bmatrix} \overset{1}{\mathbf{T}} \\ \overset{2}{\mathbf{T}} \\ \overset{3}{\mathbf{T}} \end{Bmatrix} = \begin{bmatrix} \sigma_{11} & \sigma_{12} & \sigma_{13} \\ \sigma_{21} & \sigma_{22} & \sigma_{23} \\ \sigma_{31} & \sigma_{32} & \sigma_{33} \end{bmatrix} \tag{4.2.5}$$

Since the stress tensor is symmetric, it follows that $\sigma_{ij} = \sigma_{ji}$ in any coordinate system (i.e., $\sigma_{12}' = \sigma_{21}'$, $\sigma_{13}' = \sigma_{31}'$, $\sigma_{32}' = \sigma_{23}'$). The relationship between the stress vector and the stress tensor is Cauchy's formula for stress. The magnitude and direction of the stress vector is determined from the stress tensor and the unit outward normal vector, $\mathbf{n}$, using

$$\overset{n}{T}_i = \sigma_{ij} n_j \quad \text{or in matrix notation} \quad \overset{n}{\mathbf{T}}^T = [\sigma]\mathbf{n}^T \tag{4.2.6}$$

Cauchy's formula can be used to determine the normal and shear stresses defined in Equation 4.2.1. The magnitude of the normal stress on any plane with outward unit normal **n** is the dot product of the stress vector $\overset{n}{\mathbf{T}}$ on the plane with the unit normal vector **n**, that is,

$$\sigma_{nn} = \overset{n}{\mathbf{T}} \cdot \mathbf{n} = \overset{n}{T}_i n_i \tag{4.2.7}$$

Using the definition of the stress vector gives the normal component of the stress on any plane with outward normal **n** as

$$\sigma_{nn} = \sigma_{ij} n_i n_j \tag{4.2.8}$$

The magnitude of the square of the stress vector is given by

$$\left(\overset{n}{T}\right)^2 = \overset{n}{\mathbf{T}} \cdot \overset{n}{\mathbf{T}} = \sigma_{ij}\sigma_{ik} n_j n_k \tag{4.2.9}$$

and the magnitude of the shear stress squared can be computed from

$$\sigma_{ns}^2 = \left(\overset{n}{T}\right)^2 - \sigma_{nn}^2 \tag{4.2.10}$$

Two other notations for the stress tensor that are frequently encountered in engineering are

$$\sigma_{ij} = \begin{bmatrix} \sigma_{11} & \sigma_{12} & \sigma_{13} \\ \sigma_{21} & \sigma_{22} & \sigma_{23} \\ \sigma_{31} & \sigma_{32} & \sigma_{33} \end{bmatrix} = \begin{bmatrix} \sigma_{xx} & \sigma_{xy} & \sigma_{xz} \\ \sigma_{yx} & \sigma_{yy} & \sigma_{yz} \\ \sigma_{zx} & \sigma_{zy} & \sigma_{zz} \end{bmatrix} = \begin{bmatrix} \sigma_{x} & \tau_{xy} & \tau_{xz} \\ \tau_{yx} & \sigma_{y} & \tau_{yz} \\ \tau_{zx} & \tau_{zy} & \sigma_{z} \end{bmatrix}$$

The stress tensor components are also frequently written in as a column vector, especially when dealing with computational techniques such as the finite-element method. The component order is not unique, but two of the common definitions are

$$\{\sigma\} = \begin{Bmatrix} \sigma_{11} \\ \sigma_{22} \\ \sigma_{33} \\ \sigma_{12} \\ \sigma_{13} \\ \sigma_{23} \end{Bmatrix} \quad \text{and} \quad \{\sigma\} = \begin{Bmatrix} \sigma_{11} \\ \sigma_{22} \\ \sigma_{33} \\ \sigma_{23} \\ \sigma_{31} \\ \sigma_{12} \end{Bmatrix} \tag{4.2.11}$$

Finally, it is necessary to have an equation of motion that is valid throughout the body. These equations place restrictions on the stress distribution at each instant during the motion of the body. Newton's second law applied to an infinitesimal cube of material with density P and an applied body force **b** per unit volume results in three linear coupled partial differential equations of motion

$$\frac{\partial \sigma_{11}}{\partial x_1} + \frac{\partial \sigma_{12}}{\partial x_2} + \frac{\partial \sigma_{13}}{\partial x_3} + b_1 = \rho \ddot{x}_1$$

$$\frac{\partial \sigma_{21}}{\partial x_1} + \frac{\partial \sigma_{22}}{\partial x_2} + \frac{\partial \sigma_{23}}{\partial x_3} + b_2 = \rho \ddot{x}_2$$

$$\frac{\partial \sigma_{31}}{\partial x_1} + \frac{\partial \sigma_{32}}{\partial x_2} + \frac{\partial \sigma_{33}}{\partial x_3} + b_3 = \rho \ddot{x}_3$$

which can be written in indicial notation as

$$\sigma_{ij,j} + b_i = \rho \ddot{x}_i \tag{4.2.12}$$

Example Problem 4.2: A thin plate is subjected to a uniform plane (two-dimensional) stress as shown in Figure 4.2.3. Use the relationship between the stress vector and stress tensor to find the stress tensor in the orthogonal x and y coordinates.

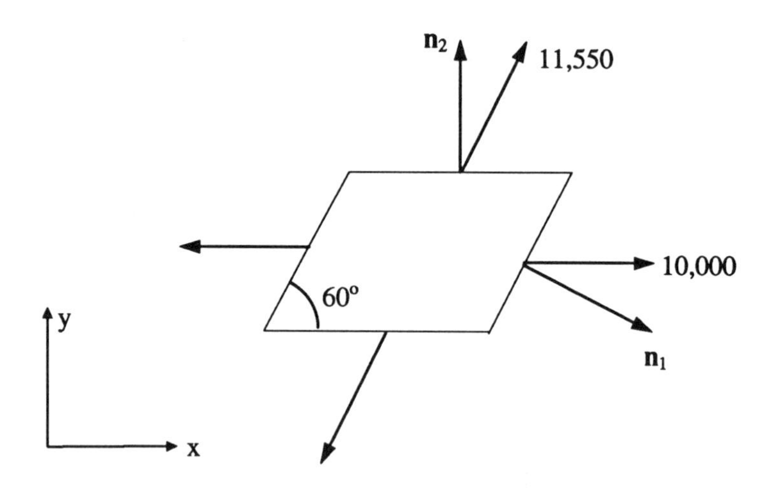

Figure 4.2.3 Definition of the stresses on an inclined element for Example Problem 4.2.

SOLUTION: Cauchy's formula for the stress vector, equation 4.2.6*b*, can be written in two-dimensional matrix notation as

$$\begin{bmatrix} \overset{1}{\mathbf{T}} \\ \overset{2}{\mathbf{T}} \end{bmatrix} = \begin{bmatrix} \sigma_{11} & \sigma_{12} \\ \sigma_{12} & \sigma_{22} \end{bmatrix} \begin{bmatrix} \mathbf{n}_1 \\ \mathbf{n}_2 \end{bmatrix}$$

Since the stress vectors and directions of the normal vectors are known on the side and top surfaces of the element, it is possible to apply Cauchy's equation for each surface and determine the components of the stress tensor. On the right side the stress and unit normal vectors are

$$\overset{1}{\mathbf{T}} = [10{,}000 \quad 0] \qquad \text{and} \qquad \mathbf{n}_1 = [0.866 \quad -0.5]$$

then

$$\begin{bmatrix} 10{,}000 \\ 0 \end{bmatrix} = \begin{bmatrix} \sigma_{11} & \sigma_{12} \\ \sigma_{12} & \sigma_{22} \end{bmatrix} \begin{bmatrix} 0.866 \\ -0.5 \end{bmatrix} \quad \text{or} \quad \begin{aligned} 10{,}000 &= 0.866\sigma_{11} - 0.5\sigma_{12} \quad \text{(a)} \\ 0 &= 0.866\sigma_{12} - 0.5\sigma_{22} \quad \text{(b)} \end{aligned}$$

Following a similar procedure for the top surface, it follows that

$$\begin{bmatrix} 5{,}775 \\ 10{,}000 \end{bmatrix} = \begin{bmatrix} \sigma_{11} & \sigma_{12} \\ \sigma_{12} & \sigma_{22} \end{bmatrix} \begin{bmatrix} 0 \\ 1 \end{bmatrix} \quad \text{or} \quad \begin{aligned} \sigma_{12} &= 5775 \\ \sigma_{22} &= 10{,}000 \end{aligned}$$

Substituting these values into equation (a) gives $\sigma_{11} = 14{,}881$ and equation (b) is satisfied. The system of equations is overspecified because the stress tensor is symmetric; however, the results are consistent and the extra shear equation can be used as a check.

4.2.2 Principal Stresses and Invariants of the Stress Tensor

In a general state of stress at a point there are three mutually perpendicular planes for which the shear stresses are zero, and the only nonzero stresses present are normal stresses. The normal stresses on these planes are the principal stresses and the directions of the corresponding normal vectors are the principal directions. The principal stresses and principal stress directions are determined by requiring that the stress vector on a plane with normal vector **n** be in the direction of the normal **n**. Mathematically, this can be stated as

$$\overset{n}{\mathbf{T}} = \sigma\mathbf{n} \tag{4.2.13}$$

where σ is the unknown value of the normal stress on the plane with normal **n**. Substituting equation 4.2.6 into equation 4.2.13 and rearranging gives

$$\sigma_{ij}n_j - \sigma n_i = (\sigma_{ij} - \sigma\delta_{ij})n_j = 0$$

There are four unknowns in the solution, normal stress, σ, and three components of the normal vector, n_j. The vector equation represents three scalar equations. A fourth equation is obtained by requiring that **n** be a unit vector, $|\mathbf{n}| = 1$. For the set of equations to have a nontrivial solution the inverse of the coefficient matrix $(\sigma_{ij} - \sigma\delta_{ij})$ must not exist. This condition will be satisfied if the determinant of the coefficient matrix is zero, $\det(\sigma_{ij} - \sigma\delta_{ij}) = 0$. Expansion of the determinant results in the characteristic equation

$$\sigma^3 - I_1\sigma^2 + I_2\sigma - I_3 = 0 \tag{4.2.14}$$

which can be solved for three principal (normal) stresses since equation 4.2.14 is a cubic equation. There are three real roots σ_1, σ_2, σ_3 to the cubic equation since stress is a real symmetric tensor. The roots may be distinct or repeated. The coefficients I_1, I_2, and I_3 are the invariants of the stress tensor and are given by

$$\begin{aligned}
I_1 &= \sigma_{11} + \sigma_{22} + \sigma_{33} \\
I_2 &= \begin{vmatrix} \sigma_{11} & \sigma_{12} \\ \sigma_{12} & \sigma_{22} \end{vmatrix} + \begin{vmatrix} \sigma_{22} & \sigma_{23} \\ \sigma_{23} & \sigma_{33} \end{vmatrix} + \begin{vmatrix} \sigma_{11} & \sigma_{13} \\ \sigma_{13} & \sigma_{33} \end{vmatrix} \\
I_3 &= \begin{vmatrix} \sigma_{11} & \sigma_{12} & \sigma_{13} \\ \sigma_{22} & \sigma_{22} & \sigma_{23} \\ \sigma_{31} & \sigma_{32} & \sigma_{33} \end{vmatrix}
\end{aligned} \tag{4.2.15}$$

The quantities I_1, I_2, and I_3 are invariant quantities because their values are identical for all stress tensors at a point, regardless of the orientation of the base vectors. That is, the stress tensor with components σ_{ij} or σ'_{ij} related by equations 4.1.15*a* and 4.1.15*b* will have the same invariants for any choice of transformation matrix, a_{ij} The invariants I_1, I_2, and I_3 may also be written as a function of the principal stresses:

$$\begin{aligned}
I_1 &= \sigma_1 + \sigma_2 + \sigma_3 \\
I_2 &= \sigma_1\sigma_2 + \sigma_1\sigma_3 + \sigma_2\sigma_3 \\
I_3 &= \sigma_1\sigma_2\sigma_3
\end{aligned} \tag{4.2.16}$$

Once the principal stresses are found, each of the three principal directions can be computed by substituting each of the principal stress values into

$$(\sigma_{ij} - \sigma\delta_{ij})n_j = 0$$

The three normal vectors can be arranged into a right-hand orthonormal basis by letting $\mathbf{n}_3 = \mathbf{n}_1 \times \mathbf{n}_2$. If two roots are repeated, the associated normal vectors are not unique and any two vectors that are mutually perpendicular to the unique normal vector may be used to form the right-hand orthonormal basis. If all three roots are repeated, any set of right-hand orthonormal base vectors may be used for the principal stress directions. The matrix of the components of these base vectors is a proper orthogonal transform that relates the original and principal stress tensors.

Example Problem 4.3: Find the principal stresses and principal stress directions for a stress at a point that is given by

$$[\sigma_{ij}] = \begin{bmatrix} 5 & -10 & 8 \\ -10 & 2 & 2 \\ 8 & 2 & 11 \end{bmatrix} \text{ksi}$$

SOLUTION: The first step is to construct the characteristic equation; thus it is necessary to determine the invariants using equation 4.2.15.

$$I_1 = 5 + 2 + 11 = 18$$

$$I_2 = \begin{vmatrix} 5 & -10 \\ -10 & 2 \end{vmatrix} + \begin{vmatrix} 2 & 2 \\ 2 & 11 \end{vmatrix} + \begin{vmatrix} 5 & 8 \\ 8 & 11 \end{vmatrix} = -90 + 18 - 9 = -81$$

$$I_3 = \begin{vmatrix} 5 & -10 & 8 \\ -10 & 2 & 2 \\ 8 & 2 & 11 \end{vmatrix} = 5(18) + 10(-126) + 8(-36) = -1458$$

Substituting these results into equation 4.2.14 gives

$$\sigma^3 - 18\sigma^2 - 81\sigma + 1458 = 0$$

This equation can be solved for the three principal stresses using a standard solution technique for a cubic equation with real roots (for example, see S. M. Selby, 1972). The roots thus obtained are

$$\sigma_1 = -9 \text{ ksi}, \qquad \sigma_2 = 9 \text{ ksi}, \qquad \sigma_3 = 18 \text{ ksi}$$

To find the principal stress directions, $\mathbf{n}$, it is necessary to solve

$$\begin{bmatrix} \sigma_{11} - \sigma & \sigma_{12} & \sigma_{13} \\ \sigma_{21} & \sigma_{22} - \sigma & \sigma_{23} \\ \sigma_{31} & \sigma_{32} & \sigma_{33} - \sigma \end{bmatrix} \begin{bmatrix} n_1 \\ n_2 \\ n_3 \end{bmatrix} = \begin{bmatrix} 5 - \sigma & -10 & 8 \\ -10 & 2 - \sigma & 2 \\ 8 & 2 & 11 - \sigma \end{bmatrix} \begin{bmatrix} n_1 \\ n_2 \\ n_3 \end{bmatrix} = 0$$

for each of the three values of principal stress. Substituting $\sigma_1 = -9$ ksi, the first solution gives three equations,

$$14n_{11} - 10n_{12} + 8n_{13} = 0 \quad \text{(a)}$$

$$-10n_{11} + 11n_{12} + 2n_{13} = 0 \quad \text{(b)}$$

$$8n_{11} + 2n_{12} + 20n_{13} = 0 \quad \text{(c)}$$

Taking equation (a) minus four times equation (b) produces $n_{11} = n_{12}$, and equation (a) plus five times equation (c) gives $n_{11} = -2n_{13}$. This defines a line in the direction of the first principal stress but a unique unit vector is not totally defined. A unit vector is found by normalizing a vector of arbitrary length parallel to the line. For example, letting $n_{11} = 2$ will define a vector $[2, 2, -1]$ that is parallel to the first principal direction. Two acceptable solutions for the unit vector are

$$\mathbf{n}_1 = [-\tfrac{2}{3}, -\tfrac{2}{3}, \tfrac{1}{3}] \quad \text{(d)}$$

$$\mathbf{n}_1 = [\tfrac{2}{3}, \tfrac{2}{3}, -\tfrac{1}{3}] \quad \text{(e)}$$

Let us arbitrarily choose equation (d) as the solution. Repeating the same process for the second principal stress, $\sigma_2 = 9$ ksi, gives

$$\mathbf{n}_2 = [-\tfrac{1}{3}, \tfrac{2}{3}, \tfrac{2}{3}]$$

This result can be checked by observing that $\mathbf{n}_1 \cdot \mathbf{n}_2 = 0$. To construct a right-hand orthonormal set of base vectors, the third principal stress direction can be determined from

$$\mathbf{n}_3 = \mathbf{n}_1 \times \mathbf{n}_2 = [-\tfrac{2}{3}, \tfrac{1}{3}, -\tfrac{2}{3}]$$

This particular set of normal vectors is not unique, but the three mutually normal lines are unique. There are several choices of coordinate systems that lie along the three lines with different positive directions. Any one of these coordinate systems is acceptable. A final check can be made by realizing that the normal vectors, $\mathbf{n}_1$, $\mathbf{n}_2$, $\mathbf{n}_3$, define an orthonormal basis relative to the basis of the original stress tensor, $\mathbf{e}_1$, $\mathbf{e}_2$, $\mathbf{e}_3$. Thus the transformation matrix that carries the original coordinates into the principal directions is

$$\begin{bmatrix} \mathbf{n}_1 \\ \mathbf{n}_2 \\ \mathbf{n}_3 \end{bmatrix} = \frac{1}{3}\begin{bmatrix} -2 & -2 & 1 \\ -1 & 2 & 2 \\ -2 & 1 & -2 \end{bmatrix}\begin{bmatrix} \mathbf{e}_1 \\ \mathbf{e}_2 \\ \mathbf{e}_3 \end{bmatrix} \quad \text{or} \quad A = \frac{1}{3}\begin{bmatrix} -2 & -2 & 1 \\ -1 & 2 & 2 \\ -2 & 1 & -2 \end{bmatrix}$$

Since $\sigma' = \mathbf{A}\sigma\mathbf{A}^T$, where σ is the original stress and σ' is the principal stress tensor, it follows that

$$\sigma' = \frac{1}{9}\begin{bmatrix} -2 & -2 & 1 \\ -1 & 2 & 2 \\ -2 & 1 & -2 \end{bmatrix}\begin{bmatrix} 5 & -10 & 8 \\ -10 & 2 & 2 \\ -8 & 2 & 11 \end{bmatrix}\begin{bmatrix} -2 & -1 & -2 \\ -2 & 2 & 1 \\ 1 & 2 & -2 \end{bmatrix}$$

$$= \begin{bmatrix} -9 & 0 & 0 \\ 0 & 9 & 0 \\ 0 & 0 & 18 \end{bmatrix}$$

4.3 DEFORMATION AND STRAIN

The total motion of a body from one configuration to another can be thought of as a rigid-body displacement plus deformation of the body. In this section we review the key concepts of the analysis of deformation and strain.

4.3.1 Analysis of Deformation

Assume that a particle in a body which initially occupies a region R in the undeformed configuration experiences a deformation and subsequently occupies region R', as shown in Figure 4.3.1. For every point in the initial region R there is a corresponding point in region R'. A point P in the body initially at coordinates (x_1, x_2, x_3) in a fixed Cartesian reference frame moves with the deformation of the particle to position P' with coordinates (a_1, a_2, a_3). The displacement vector $\mathbf{u}$ is defined as the difference between positions P and P', and is written as

$$u_1 = a_1 - x_1, \qquad u_2 = a_2 - x_2, \qquad u_3 = a_3 - x_3$$

or

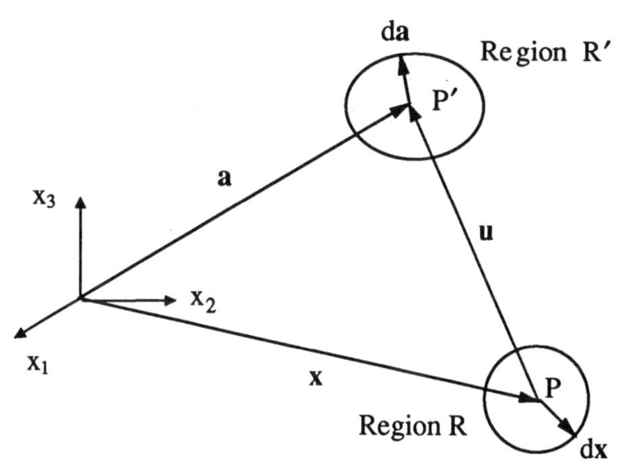

Figure 4.3.1 Deformation and rigid-body motion of a particle in a deformable body.

$$\mathbf{u} = \mathbf{a} - \mathbf{x} \tag{4.3.1}$$

The position of a particle at point P in the deformed configuration can be written as a function of its position in the undeformed configuration. Similarly, the initial position of a particle may be considered to be a function of its position in the deformed configuration. These conditions are defined by the mapping functions

$$a_i = a_i(x_1, x_2, x_3) \qquad \text{and} \qquad x_i = x_i(a_1, a_2, a_3) \tag{4.3.2}$$

It is assumed that the functions in equation 4.3.2 are continuous and produce a one-to-one mapping of points between the deformed and undeformed configurations. Therefore, the displacement of a point can be defined as a function of the coordinates in the original configuration or the deformed configuration,

$$u_i(x_1, x_2, x_3) = a_i(x_1, x_2, x_3) - x_i$$

or

$$u_i(a_1, a_2, a_3) = a_i - x_i(a_1, a_2, a_3) \tag{4.3.3}$$

A description of the deformation in region R can be determined by noting that a material line in the neighborhood of $\mathbf{x}$ defined by the vector $d\mathbf{x}$ becomes the material line $d\mathbf{a}$ at position $\mathbf{a}$. The relation between $d\mathbf{x}$ and $d\mathbf{a}$ can be established from the mapping function in the form

$$da_i = \frac{\partial a_i}{\partial x_j} dx_j$$

The partial derivatives $\partial a_i/\partial x_j$ contain all the information about how the material region R at $\mathbf{x}$ deforms into region R' at $\mathbf{a}$ since the gradient $\partial a_i/\partial x_j$ maps any line $d\mathbf{x}$ into the line $d\mathbf{a}$. The quantity $\partial a_i/\partial x_j$ is the Jacobian of the deformation. A necessary and sufficient condition for a physically possible continuous deformation is that the determinant of $\partial a_i/\partial x_j$ (Jacobian) must be positive,

$$J = \begin{vmatrix} \dfrac{\partial a_1}{\partial x_1} & \dfrac{\partial a_1}{\partial x_2} & \dfrac{\partial a_1}{\partial x_3} \\ \dfrac{\partial a_2}{\partial x_1} & \dfrac{\partial a_2}{\partial x_2} & \dfrac{\partial a_2}{\partial x_3} \\ \dfrac{\partial a_3}{\partial x_1} & \dfrac{\partial a_3}{\partial x_2} & \dfrac{\partial a_3}{\partial x_3} \end{vmatrix} = \begin{vmatrix} a_{1,1} & a_{1,2} & a_{1,3} \\ a_{2,1} & a_{2,2} & a_{2,3} \\ a_{3,1} & a_{3,2} & a_{3,3} \end{vmatrix} = \left| \frac{\partial a_1}{\partial x_j} \right| > 0 \tag{4.3.4}$$

The definition of strain is based on the difference in the square of length of the differential line segments $d\mathbf{x}$ and $d\mathbf{a}$ in the deformed and undeformed configurations, that is,

$$ds^2 = d\mathbf{a} \cdot d\mathbf{a} - d\mathbf{x} \cdot d\mathbf{x} = da_i \, da_i - dx_j \, dx_j \tag{4.3.5}$$

Since ds^2 is the difference of the magnitude squared of the vectors $d\mathbf{x}$ and $d\mathbf{a}$, rigid-body motions are eliminated from the strain measure. The differentials dx_i may be expressed in terms of da_i by using the components of the deformation gradient. Introducing the displacements, the differentials dx_i become

$$dx_i = \frac{\partial x_i}{\partial a_j} da_j = \left(\delta_{ij} - \frac{\partial u_i}{\partial a_j}\right) da_j \tag{4.3.6}$$

Introducing equation 4.3.6 into equation 4.3.5 gives $ds^2 = 2e_{ij} \, da_i \, da_j$, where

$$e_{ij} = \frac{1}{2}\left(\frac{\partial u_i}{\partial a_j} + \frac{\partial u_j}{\partial a_i} - \frac{\partial u_k}{\partial a_i}\frac{\partial u_k}{\partial a_j}\right) \tag{4.3.7}$$

Repeating the process by expanding da_i produces

$$da_i = \frac{\partial a_i}{\partial x_j} dx_j = \left(\frac{\partial u_i}{\partial x_j} + \delta_{ij}\right) dx_j$$

Equation 4.3.5 may be written as $ds^2 = 2E_{ij} \, da_i \, da_j$, where

$$E_{ij} = \frac{1}{2}\left(\frac{\partial u_i}{\partial x_j} + \frac{\partial u_j}{\partial x_i} + \frac{\partial u_k}{\partial x_i}\frac{\partial u_k}{\partial x_j}\right) \tag{4.3.8}$$

The strain tensor E_{ij}, which has displacement derivatives written with respect to the initial configuration, is the Green strain tensor. It is also known as the Green–Lagrange strain or Lagrangian strain. The strain tensor e_{ij}, which has displacement derivatives written with respect to the deformed configuration, is the Almansi strain or Eulerian strain. Both of these strain tensors are symmetric and are valid for finite deformation. A linear strain displacement relationship results if both the displacements and displacement gradients are small. In this case the higher-order terms in equations 4.3.7 and 4.3.8 are negligible and the undeformed and deformed configurations are so close that either $\mathbf{x}$ or $\mathbf{a}$ can be used to determine the strain. Thus the two strain measures are equivalent and the infinitesimal or small strain tensor is

$$E_{ij} \approx e_{ij} \approx \tfrac{1}{2}(u_{i,j} + u_{j,i}) \tag{4.3.9}$$

The gradient of the displacement vector can also be used to establish the linear (small) strain tensor. The displacement gradient may be written as

$$\text{grad } \mathbf{u} = \begin{bmatrix} \dfrac{\partial u_1}{\partial x_1} & \dfrac{\partial u_1}{\partial x_2} & \dfrac{\partial u_1}{\partial x_3} \\ \dfrac{\partial u_2}{\partial x_1} & \dfrac{\partial u_2}{\partial x_2} & \dfrac{\partial u_2}{\partial x_3} \\ \dfrac{\partial u_3}{\partial x_1} & \dfrac{\partial u_3}{\partial x_2} & \dfrac{\partial u_3}{\partial x_3} \end{bmatrix} \tag{4.3.10}$$

The gradient of the displacement vector can be expressed as the sum of symmetric and antisymmetnc or skew-symmetric parts,

$$\text{grad } \mathbf{u} = \begin{bmatrix} \varepsilon_{11} & \varepsilon_{12} & \varepsilon_{13} \\ \varepsilon_{21} & \varepsilon_{22} & \varepsilon_{23} \\ \varepsilon_{31} & \varepsilon_{32} & \varepsilon_{33} \end{bmatrix} + \begin{bmatrix} 0 & \omega_{12} & \omega_{13} \\ \omega_{21} & 0 & \omega_{23} \\ \omega_{31} & \omega_{32} & 0 \end{bmatrix} \tag{4.3.11}$$

where

$$\begin{aligned} \varepsilon_{ij} &= \varepsilon_{ji} = \tfrac{1}{2}(u_{i,j} + u_{j,i}) \\ \omega_{ij} &= -\omega_{ji} = \tfrac{1}{2}(u_{j,i} - u_{i,j}) \end{aligned} \tag{4.3.12}$$

The symmetric part of the deformation gradient is the infinitesimal strain tensor, and the skew-symmetric part is the rigid-body rotation tensor.

The infinitesimal strain tensor is frequently written with respect to rectangular coordinates with axes x, y, z and displacement components u, v, w as

$$\begin{aligned} \varepsilon_{xx} &= \frac{\partial u}{\partial_y} & \varepsilon_{xy} &= \frac{1}{2}\left(\frac{\partial u}{\partial y} + \frac{\partial v}{\partial x}\right) \\ \varepsilon_{yy} &= \frac{\partial v}{\partial y} & \varepsilon_{yz} &= \frac{1}{2}\left(\frac{\partial w}{\partial y} + \frac{\partial v}{\partial z}\right) \\ \varepsilon_{zz} &= \frac{\partial w}{\partial z} & \varepsilon_{yz} &= \frac{1}{2}\left(\frac{\partial w}{\partial y} + \frac{\partial v}{\partial z}\right) \end{aligned} \tag{4.3.13}$$

The other commonly encountered definition for the strain is the engineering strain:

$$[\gamma_{ij}] = \begin{bmatrix} \varepsilon_{11} & \gamma_{12} & \gamma_{13} \\ \gamma_{12} & \varepsilon_{22} & \gamma_{23} \\ \gamma_{13} & \gamma_{23} & \varepsilon_{33} \end{bmatrix}$$

The engineering and tensor normal strain components are equal, $\gamma_{11} = \varepsilon_{11}$, $\gamma_{22} = \varepsilon_{22}$, and $\gamma_{33} = \varepsilon_{33}$; but engineering shear strain components γ_{ij} are twice the corresponding tensor shear strain components, $\gamma_{12} = 2\varepsilon_{12}$, $\gamma_{13} = 2\varepsilon_{13}$, and $\gamma_{23} = 2\varepsilon_{23}$. The engineering shear strain matrix is not a tensor and does not transform as a tensor. However, the engineering shear strain components correspond to the physical change in angle (in radians) of base vectors that are orthogonal. For example, the angle between two lines initially parallel to x_1 and x_2 is $\pi/2 - \gamma_{12}$ after deformation.

Strain is also frequently written in vector form similar to stress, that is,

$$\{\varepsilon\} = \begin{Bmatrix} \varepsilon_{11} \\ \varepsilon_{22} \\ \varepsilon_{33} \\ \varepsilon_{12} \\ \varepsilon_{13} \\ \varepsilon_{23} \end{Bmatrix} = \begin{Bmatrix} \varepsilon_{11} \\ \varepsilon_{22} \\ \varepsilon_{33} \\ \gamma_{12}/2 \\ \gamma_{13}/2 \\ \gamma_{23}/2 \end{Bmatrix} \quad \text{or} \quad \{\varepsilon\} = \begin{Bmatrix} \varepsilon_{11} \\ \varepsilon_{22} \\ \varepsilon_{33} \\ \varepsilon_{23} \\ \varepsilon_{31} \\ \varepsilon_{12} \end{Bmatrix} = \begin{Bmatrix} \varepsilon_{11} \\ \varepsilon_{22} \\ \varepsilon_{33} \\ \gamma_{23}/2 \\ \gamma_{31}/2 \\ \gamma_{12}/2 \end{Bmatrix} \tag{4.3.14}$$

Two engineering shear vectors can be defined like the tensor strain in equation 4.3.14. The strain energy of a linear elastic body is frequently written using the engineering strain vector because it has three fewer terms than the tensor strain formulation, that is,

$$U = \tfrac{1}{2}\int_V \{\sigma\}^T\{\gamma\}\, dV = \tfrac{1}{2}\int_V \sigma_{ij}\varepsilon_{ij}\, dV \tag{4.3.15}$$

where $\{\gamma\}$ is an engineering strain vector with components defined similar to $\{\sigma\}$.

4.3.2 Principal Strains and Invariants of the Strain Tensor

Analogous to the properties for stress, principal planes and principal values also exist for the strain tensor. There are three mutually orthogonal planes that have zero shear strains. The normal strains on these planes are the principal strains, and the directions of the corresponding normal vectors are the principal strain directions or principal strain axes. It can be proven for isotropic materials (polycrystaline metals) that the principal directions of stress are the same as the principal directions of strain.

The characteristic equation for strain is

$$\varepsilon^3 - I_1'\varepsilon^2 + I_2'\varepsilon - I_3' = 0 \tag{4.3.16}$$

and the invariants of the strain tensor are

$$
\begin{aligned}
I_1' &= \varepsilon_{11} + \varepsilon_{22} + \varepsilon_{33} \\
I_2' &= \begin{vmatrix} \varepsilon_{11} & \varepsilon_{12} \\ \varepsilon_{21} & \varepsilon_{22} \end{vmatrix} + \begin{vmatrix} \varepsilon_{22} & \varepsilon_{23} \\ \varepsilon_{32} & \varepsilon_{33} \end{vmatrix} + \begin{vmatrix} \varepsilon_{11} & \varepsilon_{13} \\ \varepsilon_{31} & \varepsilon_{33} \end{vmatrix} \\
I_3' &= \begin{vmatrix} \varepsilon_{11} & \varepsilon_{12} & \varepsilon_{13} \\ \varepsilon_{21} & \varepsilon_{22} & \varepsilon_{23} \\ \varepsilon_{31} & \varepsilon_{32} & \varepsilon_{33} \end{vmatrix}
\end{aligned}
\qquad (4.3.17)
$$

The three roots of the cubic equation are the principal strains ε_1, ε_2, ε_3. The strain invariants I_1', I_2', I_3' may also be written in terms of the principal strains:

$$
\begin{aligned}
I_1' &= \varepsilon_1 + \varepsilon_2 + \varepsilon_3 \\
I_2' &= \varepsilon_2\varepsilon_2 + \varepsilon_1\varepsilon_3 + \varepsilon_2\varepsilon_3 \\
I_3' &= \varepsilon_1\varepsilon_2\varepsilon_3
\end{aligned}
\qquad (4.3.18)
$$

The principal strains and principal strain directions are determined in the same way as the principal stresses and their directions.

4.3.3 Strain Compatibility Conditions

Any three continuously differentiable functions satisfying equation 4.3.4 can be used to define the three displacement components. These functions can then be used in equations 4.3.7, 4.3.8, or 4.3.9 to define the strain field. Therefore, three displacement functions can be used to totally define the six strain components.

The reverse process is not possible. Any six strain functions are not expected to give the total displacement field for two reasons:

1. The material elements in the deformed body are assumed to fit together with no voids, cracks, gaps, or overlap. Thus the strain field must give three single-valued and continuous displacement functions.
2. It is not possible to recover the rigid-body displacements from the strain components. Recall that strain was defined to eliminate the rigid-body motions in equation 4.3.5. The rigid-body displacements must be evaluated separately and added to the displacement components determined from the strain to obtain the total displacements.

These two conditions are described mathematically by defining the total displacement, $\mathbf{u}(x, y, z)$, as the sum of two terms:

$$\mathbf{u}^T(x, y, z) = \int_{P_0}^{P} d\mathbf{u} + \mathbf{u}^{\mathrm{RB}}(x_0, y_0, z_0)$$

The integral represents the displacement of point P relative to point P_0 that results from the deformation or strain field in the body, and $u^{\mathrm{RB}}(x_0, y_0, z_0)$ represents the rigid-body motion (translation and rotation) of point P_0 relative to a fixed orgin. Condition 1 will be satisfied if any path between P_0 and P is used to evaluate the integral, thus the integral must be independent of path and $d\mathbf{u}$ must be an exact differential. This property can be used to establish three independent partial differential equations between the three displacement components. The equations can then be used to establish three restrictions on the strain field. The results are the equations of strain compatibility, or the restrictions on strain components that guarantee a single-valued and continuous displacement field. The six strain compatibility conditions for the infinitesimal strain tensor are

$$\begin{aligned}
&\frac{\partial^2 \varepsilon_x}{\partial y^2} + \frac{\partial^2 \varepsilon_y}{\partial x^2} = 2\frac{\partial^2 \varepsilon_{xy}}{\partial x\, \partial y} \qquad \frac{\partial}{\partial x}\left(-\frac{\partial \varepsilon_{yz}}{\partial x} + \frac{\partial \varepsilon_{xz}}{\partial y} + \frac{\partial \varepsilon_{xy}}{\partial z}\right) = \frac{\partial^2 \varepsilon_x}{\partial y\, \partial z} \\
&\frac{\partial^2 \varepsilon_y}{\partial z^2} + \frac{\partial^2 \varepsilon_z}{\partial y^2} = 2\frac{\partial^2 \varepsilon_{yz}}{\partial y\, \partial z} \qquad \frac{\partial}{\partial y}\left(-\frac{\partial \varepsilon_{xz}}{\partial y} + \frac{\partial \varepsilon_{xy}}{\partial z} + \frac{\partial \varepsilon_{yz}}{\partial x}\right) = \frac{\partial^2 \varepsilon_y}{\partial x\, \partial z} \\
&\frac{\partial^2 \varepsilon_x}{\partial z^2} + \frac{\partial^2 \varepsilon_z}{\partial x^2} = 2\frac{\partial^2 \varepsilon_{xz}}{\partial x\, \partial z} \qquad \frac{\partial}{\partial z}\left(-\frac{\partial \varepsilon_{xy}}{\partial z} + \frac{\partial \varepsilon_{yz}}{\partial x} + \frac{\partial \varepsilon_{xz}}{\partial y}\right) = \frac{\partial^2 \varepsilon_z}{\partial x\, \partial y}
\end{aligned} \tag{4.3.19}$$

There are only three linearly independent compatibility equations; however, the six equations above are generally used because they are simpler than three fourth-order partial differential equations.

For the special case when the strain field is uniform, the relative displacement or rotation between two points can be determined from the properties of the strain tensor. The strain ε_{aa} along any line defined by arbitrary unit vector $\hat{\mathbf{a}}$ is given by

$$\varepsilon_{aa} = \{a\}[\varepsilon]\{a\} = [a_1 \quad a_2 \quad a_3]\begin{bmatrix} \varepsilon_{11} & \varepsilon_{12} & \varepsilon_{13} \\ \varepsilon_{12} & \varepsilon_{22} & \varepsilon_{23} \\ \varepsilon_{13} & \varepsilon_{23} & \varepsilon_{33} \end{bmatrix}\begin{bmatrix} a_1 \\ a_2 \\ a_3 \end{bmatrix} \tag{4.3.20}$$

and the elongation of a line λ units long is $\lambda \varepsilon_{aa}$. The rotation of the unit line defined by $\hat{a}$ in the direction of $\hat{n}$, a unit vector that is normal to $\hat{a}$, is the shear strain ε_{an}, or

$$\varepsilon_{an} = [\hat{n}][\varepsilon][\hat{a}] = [n_1 \quad n_2 \quad n_3]\begin{bmatrix} \varepsilon_{11} & \varepsilon_{12} & \varepsilon_{13} \\ \varepsilon_{12} & \varepsilon_{22} & \varepsilon_{23} \\ \varepsilon_{13} & \varepsilon_{23} & \varepsilon_{33} \end{bmatrix}\begin{bmatrix} a_1 \\ a_2 \\ a_3 \end{bmatrix} \quad \text{rad} \tag{4.3.21}$$

The engineering shear strain $\gamma_{an} = 2\varepsilon_{an}$ is the physical change in angle of the line $\hat{a}$, and the displacement in the direction of $\hat{n}$ of the point λ units down the line is $2\lambda\varepsilon_{ns}$.

Example Problem 4.4: The deformations in a body can be defined as

$$a_1 = x_1 + \alpha x_2^2, \qquad a_2 = x_2, \qquad a_3 = x_3 - \alpha x_2^2$$

where α is a constant that is not necessarily small, and $\mathbf{x}$ and $\mathbf{a}$ are the initial and deformed coordinates, respectively. Determine the Green and Almansi strain tensors. Show that for small displacements defined by $\mathbf{x} \approx \mathbf{a}$ and $\alpha^2 << \alpha$, both tensors reduce to the same small strain tensor.

SOLUTION: Since the Green strain tensor is determined from the displacements as a function of the initial coordinates x, it is necessary to write the displacements as

$$u_1 = a_1 - x_1 = \alpha x_2^2, \qquad u_2 = a_2 - x_2 = 0, \qquad u_3 = a_3 - x_3 = -\alpha x_2^2$$

The displacement gradient can now be determined from the displacements as

$$[\nabla \mathbf{u}(\mathbf{x})] = \begin{bmatrix} \frac{\partial u_1}{\partial x_1} & \frac{\partial u_1}{\partial x_2} & \frac{\partial u_1}{\partial x_3} \\ \frac{\partial u_2}{\partial x_1} & \frac{\partial u_2}{\partial x_2} & \frac{\partial u_2}{\partial x_3} \\ \frac{\partial u_3}{\partial x_1} & \frac{\partial u_3}{\partial x_2} & \frac{\partial u_3}{\partial x_3} \end{bmatrix} = \begin{bmatrix} 0 & 2\alpha x_2 & 0 \\ 0 & 0 & 0 \\ 0 & -2\alpha x_2 & 0 \end{bmatrix}$$

Writing the Green strain in matrix notation $2[E(\mathbf{x})] = [\nabla \mathbf{u}] + [\nabla \mathbf{u}]^T + [\nabla \mathbf{u}]^T[\nabla \mathbf{u}]$ and using the displacement gradient matrix gives:

$$2[E(\mathbf{x})] = \begin{bmatrix} 0 & 2\alpha x_2 & 0 \\ 2\alpha x_2 & 0 & -2\alpha x_2 \\ 0 & -2\alpha x_2 & 0 \end{bmatrix} + \begin{bmatrix} 0 & 0 & 0 \\ 0 & 8\alpha^2 x_2^2 & 0 \\ 0 & 0 & 0 \end{bmatrix}$$

The Almansi strain is determined from the displacements as a function of the coordinates $\mathbf{a}$ in the deformed configuration. Inverting the mapping functions and calculating the displacements gives

$$x_1 = a_1 - \alpha a_2^2, \qquad x_2 = a_2, \qquad x_3 = a_3 + \alpha a_2^2$$

then

$$u_1 = a_1 - x_1 = \alpha a_2^2, \qquad u_2 = a_2 - x_2 = 0, \qquad u_3 = a_3 - x_3 = -\alpha a_2^2$$

The displacement gradient can be determined as a function of the vector **a**,

$$[\nabla \mathbf{u}(\mathbf{a})] = \begin{bmatrix} \dfrac{\partial u_1}{\partial a_1} & \dfrac{\partial u_1}{\partial a_2} & \dfrac{\partial u_1}{\partial a_3} \\ \dfrac{\partial u_2}{\partial a_1} & \dfrac{\partial u_2}{\partial a_2} & \dfrac{\partial u_2}{\partial a_3} \\ \dfrac{\partial u_3}{\partial a_1} & \dfrac{\partial u_3}{\partial a_2} & \dfrac{\partial u_3}{\partial a_3} \end{bmatrix} = \begin{bmatrix} 0 & 2\alpha x_2 & 0 \\ 0 & 0 & 0 \\ 0 & -2\alpha x_2 & 0 \end{bmatrix}$$

The Almansi strain tensor, $2[e(\mathbf{a})] = [\nabla \mathbf{u}] + [\nabla \mathbf{a}]^T - [\nabla \mathbf{u}]^T[\nabla \mathbf{u}]$, is then

$$2[e(\mathbf{a})] = \begin{bmatrix} 0 & 2\alpha a_2 & 0 \\ 2\alpha a_2 & 0 & -2\alpha a_2 \\ 0 & -2\alpha a_2 & 0 \end{bmatrix} - \begin{bmatrix} 0 & 0 & 0 \\ 0 & 8\alpha^2 a_2^2 & 0 \\ 0 & 0 & 0 \end{bmatrix}$$

Upon imposing the conditions that $\mathbf{x} \approx \mathbf{a}$ and $\alpha^2 << \alpha$, both tensors reduce to the same small strain tensor, which is the first matrix in $2[e(a)]$ or $2[E(x)]$.

Example Problem 4.5: The uniform state of strain in a body in the neighborhood of a point, P, is given as

$$[\varepsilon] = \begin{bmatrix} 5 & 3 & 0 \\ 3 & 4 & -1 \\ 0 & -1 & 2 \end{bmatrix} \times 10^{-4}$$

Find the elongation per unit of length in the direction of the line $2e_1 + 2e_2 + e_3$ and the rotation of that line toward or away from a second line, $3e_1 - 4e_3$, emanating from the same point P.

SOLUTION: First note that the elongation per unit length is strain. Begin by writing the two lines as the unit vectors

$$\hat{e}_a = \tfrac{2}{3}e_1 + \tfrac{2}{3}e_2 + \tfrac{1}{3}e_3 \qquad \text{and} \qquad \hat{e}_b = \tfrac{3}{5}e_1 - \tfrac{4}{5}e_3$$

The displacement vector $\mathbf{u}_a$ one unit from point P in the direction $\hat{\mathbf{e}}_a$, as shown in Figure 4.3.2, is determined from

$$\{u_a\} = [\varepsilon]\{\hat{e}_a\} = \frac{1}{3} \times 10^{-4} \begin{bmatrix} 5 & 3 & 0 \\ 3 & 4 & -1 \\ 0 & -1 & 2 \end{bmatrix} \begin{bmatrix} 2 \\ 2 \\ 1 \end{bmatrix} = \frac{1}{3} \times 10^{-4} \begin{bmatrix} 16 \\ 13 \\ 0 \end{bmatrix}$$

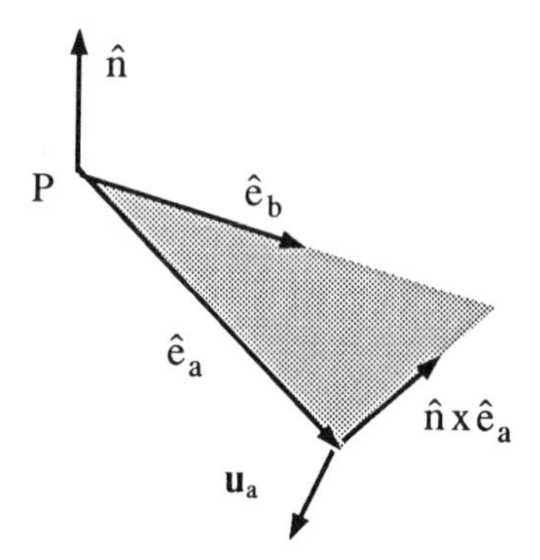

Figure 4.3.2 Definition of the vectors used in Example Problem 4.5.

The projection of $\mathbf{u}_a$ in the direction $\hat{\mathbf{e}}_a$ is the unit of elongation of the line $\hat{\mathbf{e}}_a$ or the normal strain in the $\hat{\mathbf{e}}_a$ direction. This strain is given by

$$\varepsilon_{aa} = u_a \cdot \hat{e}_a = \frac{1}{9} \times 10^{-4}[16 \quad 13 \quad 0]\begin{bmatrix} 2 \\ 2 \\ 1 \end{bmatrix} = 6.44 \times 10^{-4}$$

The rotation α of the line $\hat{\mathbf{e}}_a$ in the direction $\hat{\mathbf{e}}_b$ is the component of $\mathbf{u}_a$ that is normal to $\hat{\mathbf{e}}_a$ and in the plane defined by $\hat{\mathbf{n}} = \hat{\mathbf{e}}_a \times \hat{\mathbf{e}}_b$. Let us start by finding unit vector $\hat{\mathbf{n}}$ normal to the plane of $\hat{\mathbf{e}}_a$ and $\hat{\mathbf{e}}_b$ from

$$\hat{n} = \hat{e}_a \times \hat{e}_b = \begin{vmatrix} \hat{e}_1 & \hat{e}_2 & \hat{e}_3 \\ \frac{2}{3} & \frac{2}{3} & \frac{1}{3} \\ \frac{3}{5} & -\frac{4}{5} & 0 \end{vmatrix} = \frac{1}{15}\,[4\hat{e}_1 \quad 3\hat{e}_2 \quad -14\hat{e}_3]$$

where $\hat{e}_i$ are the base vectors. A unit vector normal to $\hat{\mathbf{e}}_a$ and in the plane of $\hat{\mathbf{e}}_a$ and $\hat{\mathbf{e}}_b$ is given by

$$\hat{n} \times \hat{e}_a = \begin{vmatrix} \hat{e}_1 & \hat{e}_2 & \hat{e}_3 \\ \frac{4}{15} & \frac{3}{15} & -\frac{14}{15} \\ \frac{2}{3} & \frac{2}{3} & \frac{1}{3} \end{vmatrix} = \frac{1}{45}\,[-25\hat{e}_1 \quad -32\hat{e}_2 \quad -2\hat{e}_3]$$

The desired rotation is calculated from

$$a \approx \tan a = (\hat{n} \times \hat{e}_a) \cdot u_a = \frac{10^{-4}}{135}\,[-25 \quad -32 \quad -2]\begin{bmatrix} 16 \\ 13 \\ 0 \end{bmatrix}$$

$$= -6.04 \times 10^{-4} \text{ rad}$$

The minus sign indicates that the rotation of $\hat{\mathbf{e}}_a$ is away from $\hat{\mathbf{e}}_b$, as shown in Figure 4.3.2.

4.4 DEVIATORIC STRESS AND STRAIN

The inelastic response of metals is generally assumed to result primarily from planer slip; thus most constitutive models for metals depend on shear stress and are independent of the mean stress or hydrostatic stress. It has also been observed the inelastic deformation does not produce a significant increase in volume. These two effects are examined in this section, and the deviatoric stress and strain are introduced as a method to implement these observations in constitutive modeling.

4.4.1 Assumptions of Plasticity

In this section the fundamental assumptions of plasticity are examined more closely. The assumption of constant material volume during inelastic deformation, although not exact, is well justified. It was shown in Chapter 1 that inelastic deformation of metals produces defects in the crystal lattice structure. These defects alter the local crystal lattice and produce small voids or vacancies. This leads to slight increases in the material volume which are difficult to observe in mechanical experiments. An example of nearly incompressible behavior is illustrated in Figure 4.4.1. These results show a slight volume increase for several steels loaded past the yield point in tension. Since the plastic volumetric strain is about 100 times smaller than the axial strain, it is negligible compared to the other strain components.

Metals at low temperature exhibit elastic behavior up to a critical stress, after which slip occurs and permanent deformation results. Under a general state of stress the limit of elastic behavior is defined by a yield criterion that is a function of all the stress components. The most commonly used yield criteria for isotropic metals is based on the assumption that the yield stress and subsequent inelastic flow are independent of the hydrostatic or mean stress. This assumption appears justified for many metals at low mean stress, such as a tensile test when the stress vector is $[\sigma_i] = [\sigma \quad 0 \quad 0 \quad 0 \quad 0 \quad 0]$ and the mean stress is $\sigma_m = (\sigma_{11} + \sigma_{22} + \sigma_{33})/3 = \sigma/3$. For example, the fatigue loop in Figure 4.4.2 from a symmetric strain-controlled experiment has nearly symmetric tensile and compressive peak stresses. However, this is not always true. Yielding and plastic flow of metals can depend on mean stress, as illustrated in Figure 4.4.3 for aged and unaged maraging steel. The experiments show that there is a difference in the tensile and compressive response of the maraging steels, and the data also show that when large mean stresses are present, such as in deep-water structures, the effect of mean stress is significant. Thus it appears that the classical assumptions of metals plasticity are not univerally applicable to all materials and all loading conditions.

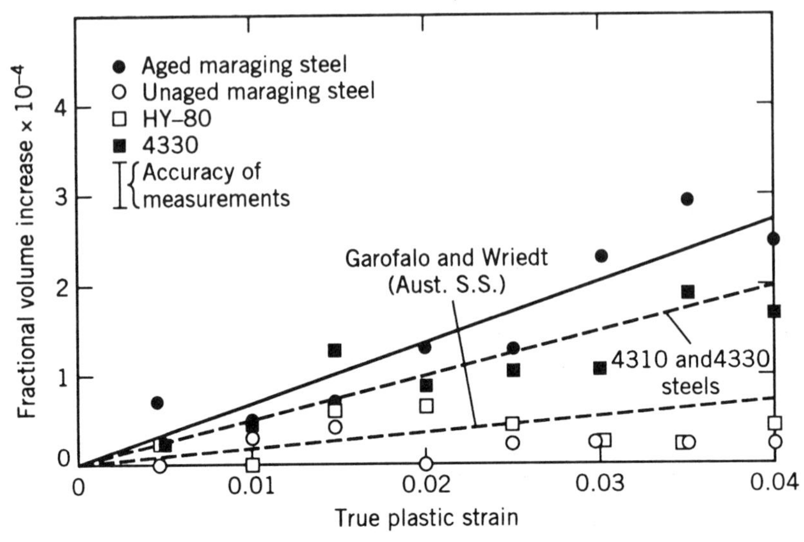

Figure 4.4.1 Plastic volume expansion in four steels as a function of plastic strain. (After Spitzig, Sober, and Richmond, 1976; reproduced with permission of The Minerals, Metals & Materials Society.)

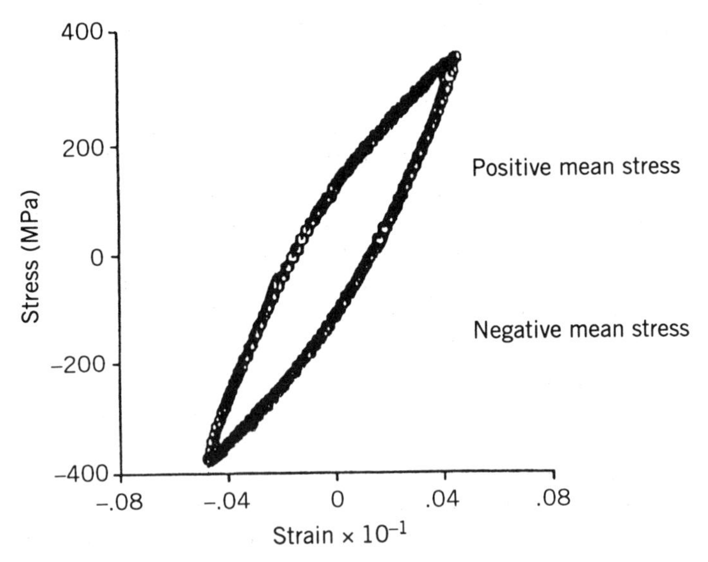

Figure 4.4.2 The cyclic response of a nickel-based alloy is nearly symmetric in tension and compression.

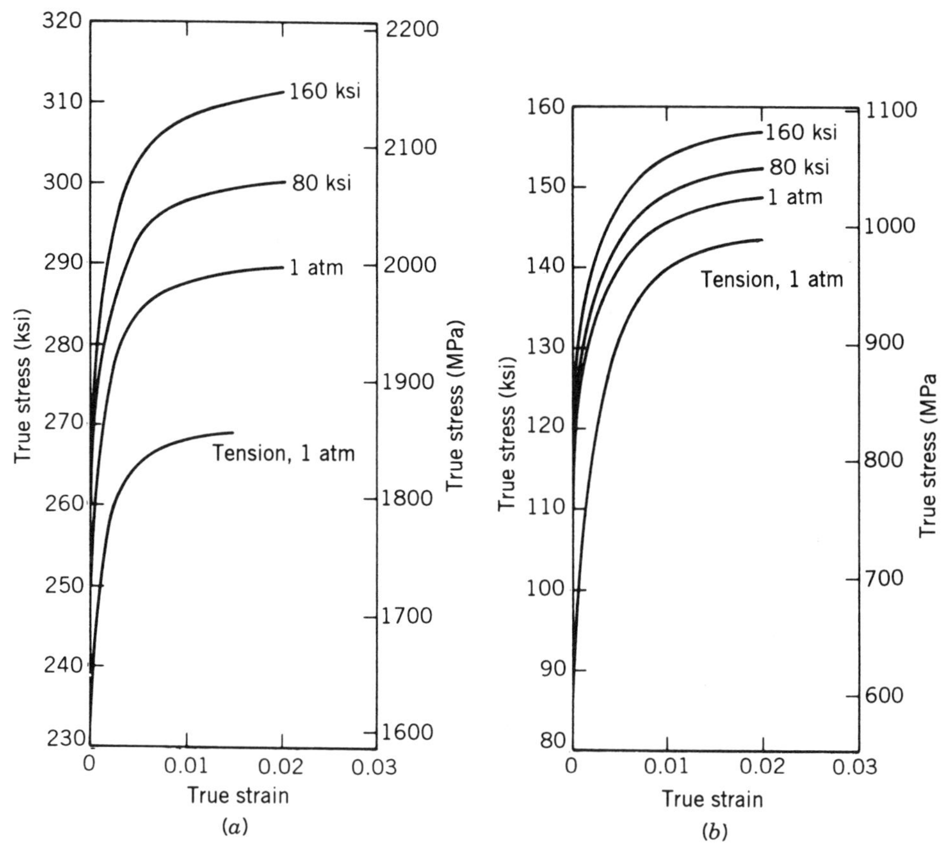

Figure 4.4.3 Effect of hydrostatic pressure on the compressive stress–strain response of aged (*left*) and unaged (*right*) maraging steel. One tensile curve is included for comparison. (From Spitzig, Sober, and Richmond, 1976; reproduced with permission of The Minerals, Metals & Materials Society.)

4.4.2 Deviatoric Stress Tensor

The deviatoric stress is defined to be the difference in the stress and the mean stress. The deviatoric stress tensor s_{ij} is defined by

$$s_{ij} = \sigma_{ij} - \tfrac{1}{3}\sigma_{kk}\delta_{ij} = \sigma_{ij} - \sigma_m \delta_{ij}$$

and

$$\sigma_m = \tfrac{1}{3}(\sigma_{11} + \sigma_{22} + \sigma_{33}) = \tfrac{1}{3}I_1 \tag{4.4.1}$$

where σ_m is the mean or hydrostatic stress. In expanded notation the deviatoric stress tensor is given by

$$S_{ij} = \begin{bmatrix} S_{11} & S_{12} & S_{13} \\ S_{21} & S_{22} & S_{23} \\ S_{31} & S_{32} & S_{33} \end{bmatrix} = \begin{bmatrix} \sigma_{11} - \sigma_m & \sigma_{12} & \sigma_{13} \\ \sigma_{21} & \sigma_{22} - \sigma_m & \sigma_{23} \\ \sigma_{31} & \sigma_{32} & \sigma_{33} - \sigma_m \end{bmatrix} \tag{4.4.2}$$

The directions of the principal axes are not changed by subtracting the hydrostatic stress. The principal deviatoric stresses and the invariants of the deviatoric stress tensor are defined in the same way as the stress tensor. The first invariant of the deviatoric stress is zero by definition; that is, $J_1 = S_{11} + S_{22} + S_{33} = 0$. The second and third invariants are

$$\begin{aligned} J_2 &= \tfrac{1}{2}(S_{11}^2 + S_{22}^2 + S_{33}^2) + \sigma_{12}^2 + \sigma_{13}^2 + \sigma_{23}^2 \\ &= \tfrac{1}{6}[(\sigma_{11} - \sigma_{22})^2 + (\sigma_{22} - \sigma_{33})^2 + (\sigma_{33} - \sigma_{11})^2] + \sigma_{12}^2 + \sigma_{13}^2 + \sigma_{23}^2 \\ J_3 &= \begin{vmatrix} S_{11} & S_{12} & S_{13} \\ S_{21} & S_{22} & S_{23} \\ S_{31} & S_{23} & S_{33} \end{vmatrix} \end{aligned} \tag{4.4.3}$$

The second invariant is widely used in constitutive modeling. The second and third invariants of the deviatoric stress tensor can be written as a function of principal deviatoric stresses S_1, S_2, and S_3 in the form

$$J_2 = \tfrac{1}{2}(S_1^2 + S_2^2 + S_3^2) \qquad \text{and} \qquad J_3 = S_1 S_2 S_3 \tag{4.4.4}$$

and the stress invariants as

$$J_2 = \tfrac{1}{3}(I_1^2 + 3I_2) \qquad \text{and} \qquad J_3 = \tfrac{1}{27}(2I_1^3 + 9I_1 I_2 + 27I_3) \tag{4.4.5}$$

In constitutive modeling the second and third invariants of the deviatoric stress tensor frequently occur in indicial notation as

$$J_2 = \tfrac{1}{2}S_{ij}S_{ij} \qquad \text{and} \qquad J_3 = \tfrac{1}{3}S_{ij}S_{jk}S_{ki} \tag{4.4.6}$$

4.4.3 Deviatoric Strain Tensor

The process of inelastic deformation in metals is usually assumed to be an incompressible process; thus the volumetric inelastic strains are assumed to be negligible. It is convenient to introduce this assumption by defining the deviatoric strain tensor, e_{ij}, as the difference between the strain tensor and the mean strain ε_m,

$$\begin{aligned} e_{ij} &= \varepsilon_{ij} - \tfrac{1}{3}\varepsilon_{kk}\delta_{ij} = \varepsilon_{ij} - \varepsilon_m \delta_{ij} \\ \varepsilon_m &= \tfrac{1}{3}(\varepsilon_{11} + \varepsilon_{22} + \varepsilon_{33}) = \tfrac{1}{3}I_1' \end{aligned} \tag{4.4.7}$$

The quantity $(\varepsilon_{11} + \varepsilon_{22} + \varepsilon_{33})$ is approximately the volumetric strain for small strain. In expanded notation the deviatoric strain is given by

$$e_{ij} = \begin{bmatrix} e_{11} & e_{12} & e_{13} \\ e_{21} & e_{22} & e_{23} \\ e_{31} & e_{32} & e_{33} \end{bmatrix} = \begin{bmatrix} \varepsilon_{11} - \varepsilon_m & \varepsilon_{12} & \varepsilon_{13} \\ \varepsilon_{21} & \varepsilon_{22} - \varepsilon_m & \varepsilon_{23} \\ \varepsilon_{31} & \varepsilon_{32} & \varepsilon_{33} - \varepsilon_m \end{bmatrix} \tag{4.4.8}$$

The directions of the principal deviatoric strain axes are in the same directions as the principal strain axes. The two nonzero invariants of the deviatoric strain tensor are similar to the deviatoric stress tensor, and the invariants of the deviatoric strain tensor can also be written in terms of the principal deviatoric strains e_1, e_2, e_3, as shown for stress in equation 4.4.4. The representations in equations 4.4.5 and 4.4.6 are also valid for the deviatoric strain invariants.

Application of the constant-volume assumption to inelasticity is achieved by writing the total strain tensor as the sum of elastic and inelastic components and representing the inelastic strain by equation 4.4.7, that is,

$$\varepsilon_{ij} = \varepsilon_{ij}^{\mathrm{E}} + \varepsilon_{ij}^{\mathrm{I}} = \varepsilon_{ij}^{\mathrm{E}} + \varepsilon_m^{\mathrm{I}}\delta_{ij} + e_{ij}^{\mathrm{I}}$$

But since the mean strain is negligible, $\varepsilon_m^{\mathrm{I}} \approx 0$, the inelastic strain tensor is equal to the deviatoric inelastic strain tensor, $\varepsilon_{ij}^{\mathrm{I}} = e_{ij}^{\mathrm{I}}$. Therefore, the inelastic strain tensor, $\varepsilon_{ij}^{\mathrm{I}}$, is a deviatoric tensor.

4.5 LINEAR ELASTIC MATERIALS

If a loaded body returns to its initial configuration after unloading, the material in the body is said to behave elastically. Elastic deformation in metals is a result of stretching of bonds between atoms in a lattice structure without any permanent change in the relative positions of the atoms. Even when inelastic strains are present, the elastic stretching of bonds is recoverable on unloading. The elastic strain is not influenced by the inelastic deformation and is a single-valued, linear function of stress alone. The most general linear relationship between the stress and strain tensors is expressed by Hooke's law, that is,

$$\sigma_{ij} = C_{ijkl}\varepsilon_{kl} \tag{4.5.1}$$

where ε_{kl} is the elastic strain tensor and C_{ijkl} is the fourth-order tensor of elastic constants. Since every stress component is a linear function of every strain component, equation 4.5.1 is valid for the most general class of anisotropic elastic materials.

The fourth-order tensor of elastic constants, C_{ijkl}, contains a total of 81 components. Since σ_{ij} and ε_{kl} are symmetric, the number of independent elastic constants can be reduced. In addition, equation 4.5.1 can be derived from the strain energy density (equation 4.3.15) as a potential function of the engineering strain, and since the order of differentiation can be changed, it is possible to prove that C_{ijkl} is symmetric in the first and last pairs of indices. Therefore, it follows that

$$C_{ijkl} = C_{jikl} = C_{ijlk} = C_{klij} \tag{4.5.2}$$

where there are at most 21 independent constants for a general anisotropic elastic solid. The elastic stress–strain relations can be written as a matrix equation if stress and strain are expressed as vectors as defined in equations 4.2.11 and 4.3.14. Thus the following equivalant representation can be used for a general anisotropic elastic solid:

$$\begin{Bmatrix} \sigma_{xx} \\ \sigma_{yy} \\ \sigma_{xx} \\ \tau_{xy} \\ \tau_{xz} \\ \tau_{yz} \end{Bmatrix} = \begin{bmatrix} C_{11} & C_{12} & C_{13} & C_{14} & C_{15} & C_{16} \\ C_{21} & C_{22} & C_{23} & C_{24} & C_{25} & C_{26} \\ C_{31} & C_{32} & C_{33} & C_{34} & C_{35} & C_{36} \\ C_{41} & C_{42} & C_{43} & C_{44} & C_{45} & C_{46} \\ C_{51} & C_{52} & C_{53} & C_{54} & C_{55} & C_{56} \\ C_{61} & C_{62} & C_{63} & C_{64} & C_{65} & C_{66} \end{bmatrix} \begin{Bmatrix} \varepsilon_{xx} \\ \varepsilon_{yy} \\ \varepsilon_{zz} \\ \gamma_{xy} \\ \gamma_{xz} \\ \gamma_{yz} \end{Bmatrix} \tag{4.5.3}$$

where the last symmetry in equation 4.5.2 requires that the elastic stiffness matrix C_{pq} is symmetric with 21 independent components. It is necessary to use engineering strain in equation 4.5.3 to maintain the symmetery of C_{pq}. A material having 21 independent elastic constants is a fully anisotropic material. In this case the base vectors used to establish equation 4.5.3 must be defined and fixed in the material. When the material matrix is fully populated as in equation 4.5.3, there is coupling between the shear and normal terms and within the shear terms. This means that a single shear strain can produce three normal stresses and three shear strains. Most materials have far fewer than 21 elastic constants.

4.5.1 Isotropic Materials

If the elastic response is the same for loading in any direction of the material, the material is isotropic. Metals with a large number of small randomly oriented grains are examples of materials that can be assumed to be isotropic. The isotropic stiffness tensor C_{ijkl} has only two independent scalar constants and can be written

$$C_{ijkl} = \lambda \delta_{ij} \delta_{kl} + \mu(\delta_{ik} \delta_{jl} + \delta_{il} \delta_{jk}) \tag{4.5.4}$$

The elastic parameters λ and μ are the Lamé constants. The parameter μ is frequently referred to as the engineering shear modulus and is designated as G. Substituting equation 4.5.4 into 4.5.1 gives

$$\sigma_{ij} = \lambda \varepsilon_{kk} \delta_{ij} + 2\mu \varepsilon_{ij} \tag{4.5.5}$$

It is common to express the elastic stiffness matrix in equation 4.5.3 using the elastic modulus E and Poisson's ratio ν:

$$[C] = E^* \begin{bmatrix} 1-\nu & \nu & \nu & 0 & 0 & 0 \\ \nu & 1-\nu & \nu & 0 & 0 & 0 \\ \nu & \nu & 1-\nu & 0 & 0 & 0 \\ 0 & 0 & 0 & \frac{1-2\nu}{2} & 0 & 0 \\ 0 & 0 & 0 & 0 & \frac{1-2\nu}{2} & 0 \\ 0 & 0 & 0 & 0 & 0 & \frac{1-2\nu}{2} \end{bmatrix} \tag{4.5.6}$$

where $E^* = E/(1 + \nu)(1 - 2\nu)$.

The bulk modulus, $K = \sigma_{ii}/(3\varepsilon_{kk})$, is used when the elastic constitutive equation is written as a function of the deviatoric and mean stress and strain, that is,

$$S_{ij} = 2\mu\varepsilon_{ij} \quad \text{and} \quad \sigma_m = K\varepsilon_{kk} \tag{4.5.7}$$

The relationships between the elastic constants used for isotropic materials are summarized in Table 4.5.1.

The elastic constants defined in Table 4.5.1 are ordinarily a function of temperature, as shown in Figure 4.5.1. The elastic modulus (Figure 4.5.1*a*) shows a relatively smooth variation with temperature up to 1800°F. These data were used to extrapolate the value of the elastic modulus to higher temperatures, but a subsequent test at 2100°F indicated that the extrapolation was not accurate. The Poisson ratio (Figure 4.5.1*b*) was not determined at 2100°F because it was felt that extrapolation to $\nu = 0.5$ (for an incompressible fluid) at the melting temperature would not present serious errors. In addition to the elastic parameters, the thermal coefficient of expansion is given in Figure 4.5.1*c*. Notice that the value of the coefficient of expansion almost doubles between room temperature and melting for René 80, and the relationship between temperature and the coefficient is nonlinear. This experience suggests that extrapolating a data set can lead to significant errors, and this practice is not recommended.

The elasticity equations can be used to determine the response as a function of temperature if the parameters are written as a function of temperature and

Table 4.5.1 Relationships Between the Five Elastic Constants

	λ, G	K, G	G, ν	E, ν	E, G
λ		$K - \dfrac{2G}{3}$	$\dfrac{2G\nu}{1-2\nu}$	$\dfrac{\nu E}{(1+\nu)(1-2\nu)}$	$\dfrac{G(E-2G)}{3G-E}$
$G = \mu$				$\dfrac{E}{2(1+\nu)}$	
K	$\lambda + \dfrac{2G}{3}$		$\dfrac{2G(1+\nu)}{3(1-2\nu)}$	$\dfrac{E}{3(1-2\nu)}$	$\dfrac{EG}{3(3G-E)}$
E	$\dfrac{G(3\lambda+2G)}{\lambda+G}$	$\dfrac{9KG}{3K+G}$	$2G(1+\nu)$		
ν	$\dfrac{\lambda}{2(\lambda+G)}$	$\dfrac{3K-2G}{GK+2G}$			$\dfrac{E}{2G} - 1$

thermal expansion is included. Letting $\alpha(T)$ represent the thermal expansion from the reference temperature T_0, the thermoelastic response is then

$$\varepsilon_{ij} = \frac{1+\nu(T)}{E(T)}\sigma_{ij} - \frac{\nu(T)}{E(T)}\sigma_{kk}\delta_{ij} + \alpha(\mathrm{T})(T - T_0)\delta_{ij} \tag{4.5.8}$$

The inverted elasticity equations can be written as

$$\sigma_{ij} = \lambda(T)[\varepsilon_{kk} - 3\alpha(T)(T - T_0)]\delta_{ij} + 2G(T)[\varepsilon_{ij} - \alpha(T)(T - T_0)\delta_{ij}] \tag{4.5.9}$$

using the tensor definitions of the elastic material parameters.

4.5.2 Materials with Cubic Symmetry

Single metallic crystals with face-centered cubic or body-centered cubic lattice structure exhibit anisotropic elastic response. Until recently, cubic symmetric materials were mostly of academic interest since no real structures were made from single metallic crystals. Currently, single-crystal nickel-based superalloys are being used for turbine blades in gas and steam turbine engines and other high-temperature structures. The inelastic behavior of these alloys will be discussed in Chapter 8. This section is directed toward establishing a representation for the elastic response with cubic symmetery and developing a method to determine the elastic constants for metals with cubic symmetry.

Materials with cubic symmetry can be shown to have only three independent elastic constants. The three independent elastic constants are desig-

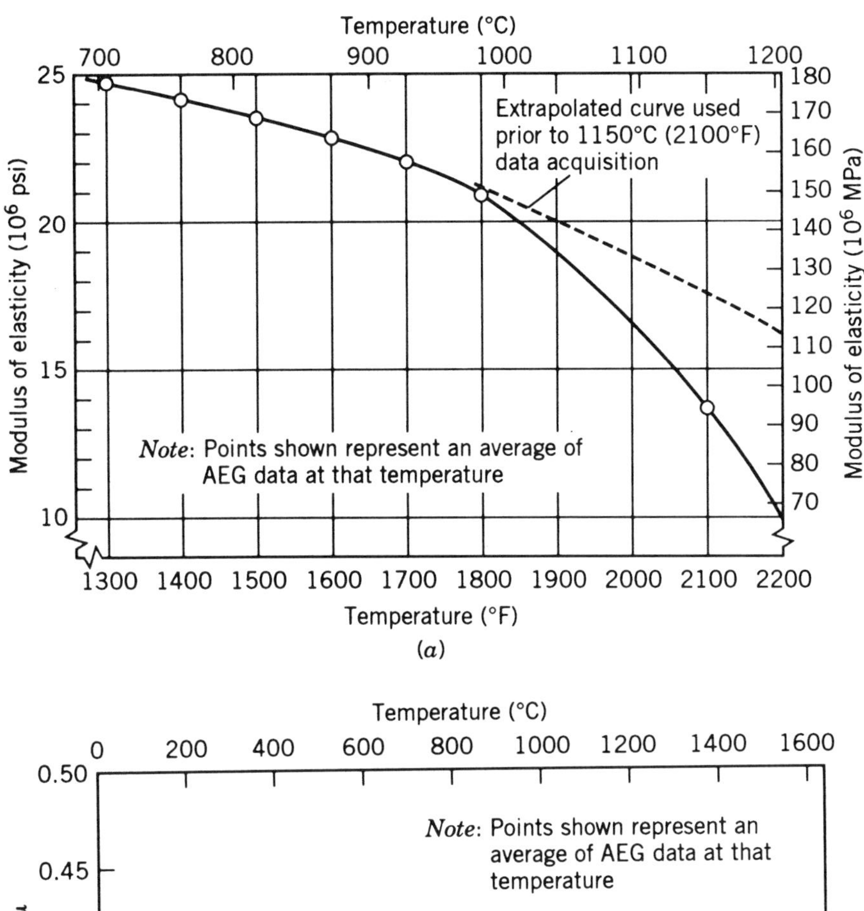

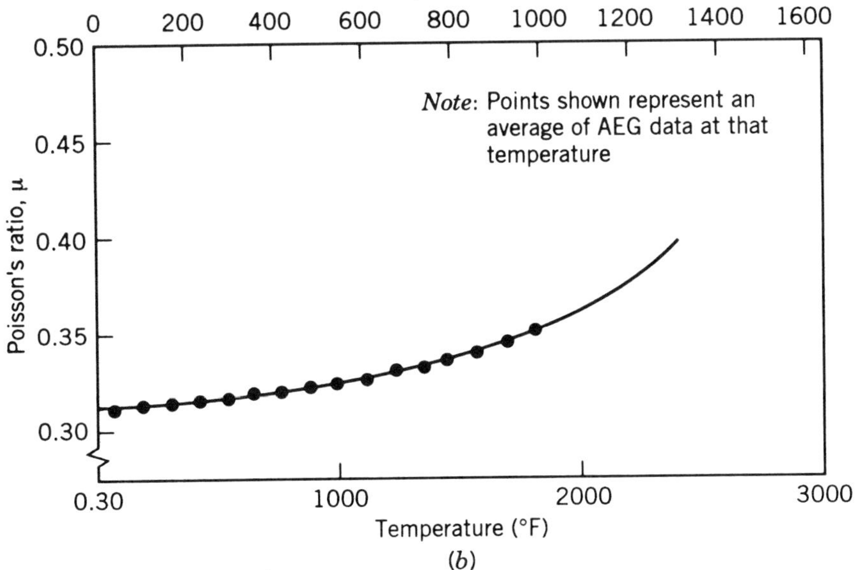

Figure 4.5.1 Variation of the (*a*) elastic modulus, (*b*) Poisson ratio, and (*c*) thermal coefficient of expansion as a function of temperature for René 80. (After McKnight, Laflen, and Spamer, 1982.)

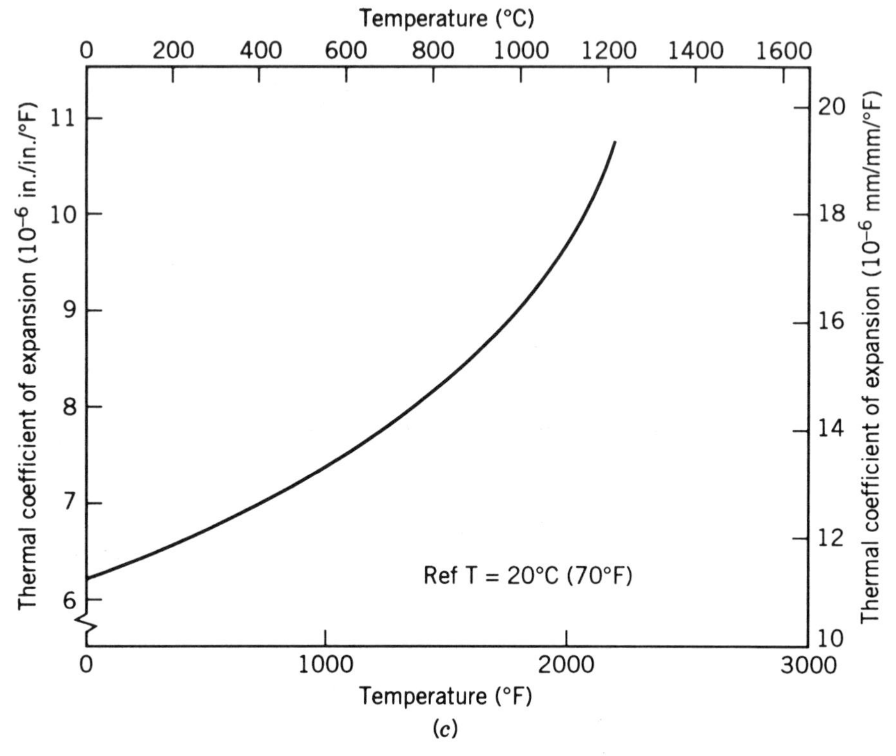

(c)

Figure 4.5.1 (*Continued*)

nated as the elastic modulus, shear modulus, and Poisson ratio. The elastic compliance matrix for principal material directions is

$$\begin{Bmatrix} \varepsilon_{xx} \\ \varepsilon_{yy} \\ \varepsilon_{zz} \\ \gamma_{xy} \\ \gamma_{xz} \\ \gamma_{yz} \end{Bmatrix} = \begin{bmatrix} \dfrac{1}{E} & \dfrac{-\nu}{E} & \dfrac{-\nu}{E} & 0 & 0 & 0 \\ \dfrac{-\nu}{E} & \dfrac{1}{E} & \dfrac{-\nu}{E} & 0 & 0 & 0 \\ \dfrac{-\nu}{E} & \dfrac{-\nu}{E} & \dfrac{1}{E} & 0 & 0 & 0 \\ 0 & 0 & 0 & \dfrac{1}{G} & 0 & 0 \\ 0 & 0 & 0 & 0 & \dfrac{1}{G} & 0 \\ 0 & 0 & 0 & 0 & 0 & \dfrac{1}{G} \end{bmatrix} \begin{Bmatrix} \sigma_{xx} \\ \sigma_{yy} \\ \sigma_{xx} \\ \tau_{xy} \\ \tau_{xz} \\ \tau_{yz} \end{Bmatrix} \tag{4.5.10}$$

This matrix appears identical to the matrix of elastic constants for isotropic materials. However, the elastic constants E, ν, and G are all independent, and the components of the matrix are a fuction of orientation. Equation 4.5.10 is valid only when the loading is in the principal directions of the material or parallel to the edges of the FCC or BCC cubic lattice shown in Figure 1.2.6. Therefore, it is necessary to establish a method to transform equation 4.5.10 to other orientations. The results will be used to determine the elastic constants from three tensile tests on test samples with arbitrary orientation relative to the principal axes of the material, and determine the compliance matrix in other orientations.

Let us establish the elastic stress–strain response of a test specimen with the arbitrary orientation $\mathbf{e}_1''$ relative to the crystal axes as shown in Figure 4.5.2. The vector $\mathbf{e}_1''$ is oriented in space by two sucessive rotations. First, as shown in Figure 4.5.2*b*, is a positive rotation θ about the $\mathbf{e}_3$ axis so that the $\mathbf{e}_i$ basis in the principal direction of the material is transformed into the $\mathbf{e}_i'$ basis. Second is a negative rotation ϕ about $\mathbf{e}_2'$, transforming $\mathbf{e}_1'$ into $\mathbf{e}_1''$ as shown in Figure 4.5.2*c*. The two tranformation matrices are

$$\begin{bmatrix} \mathbf{e}_1' \\ \mathbf{e}_2' \\ \mathbf{e}_3' \end{bmatrix} = \begin{bmatrix} \cos\theta & \sin\theta & 0 \\ -\sin\theta & \cos\theta & 0 \\ 0 & 0 & 1 \end{bmatrix} \begin{bmatrix} \mathbf{e}_1 \\ \mathbf{e}_2 \\ \mathbf{e}_3 \end{bmatrix} \tag{4.5.11a}$$

$$\begin{bmatrix} \mathbf{e}_1'' \\ \mathbf{e}_2'' \\ \mathbf{e}_3'' \end{bmatrix} = \begin{bmatrix} \cos\phi & 0 & \sin\phi \\ 0 & 1 & 0 \\ -\sin\theta & 0 & \cos\phi \end{bmatrix} \begin{bmatrix} \mathbf{e}_1' \\ \mathbf{e}_2' \\ \mathbf{e}_3' \end{bmatrix} \tag{4.5.11b}$$

The constitutive equation 4.5.10 relative to the principal axes of the crystal can be written in matrix notation as

$$\{\gamma\} = [S]\{\sigma\} \tag{4.5.12}$$

where $\{\gamma\}$ is the engineering strain. Define a second coordinate system not in the principal direction of the material such that the tensor stress, $\{\sigma'\}$, and strain, $\{\varepsilon\}$, transform as

$$\{\sigma'\} = [Q]\{\sigma\} \qquad \text{and} \qquad \{\varepsilon'\} = [Q]\{\varepsilon\} \tag{4.5.13}$$

where $[Q]$ is a six-dimensional transformation matrix that will be defined later. However, the engineering and tensor strains are related by

$$\{\varepsilon\} = [R]^{-1}\{\gamma\} \qquad \text{and} \qquad \{\gamma\} = [R]\{\varepsilon\} \tag{4.5.14}$$

where $[R]$ is the Reuter matrix

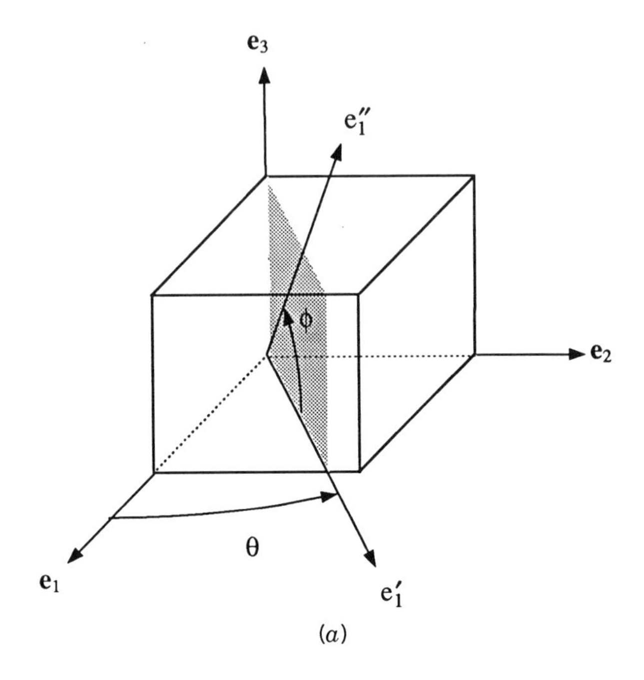

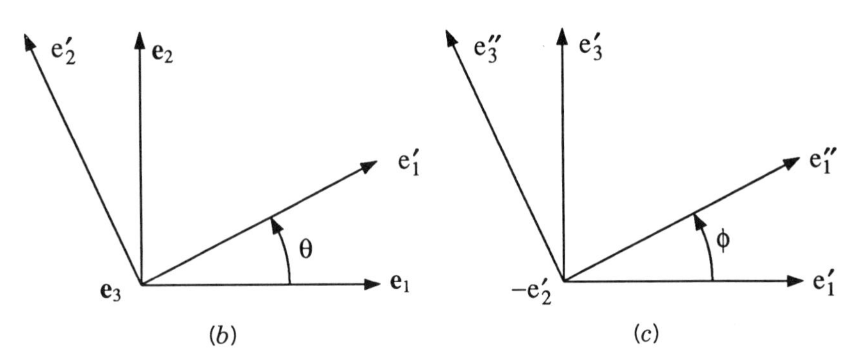

Figure 4.5.2 Definition of two successive rotations defining the three sets of coordinate axes x_i, x_i', and x_i''.

$$[R] = \begin{bmatrix} 1 & 0 & 0 & 0 & 0 & 0 \\ 0 & 1 & 0 & 0 & 0 & 0 \\ 0 & 0 & 1 & 0 & 0 & 0 \\ 0 & 0 & 0 & 2 & 0 & 0 \\ 0 & 0 & 0 & 0 & 2 & 0 \\ 0 & 0 & 0 & 0 & 0 & 2 \end{bmatrix}$$

The engineering strain in the rotated coordinate system, $\{\gamma'\}$, can now be written as

$$\begin{aligned}[\gamma'] &= [R][\varepsilon'] = [R][Q][\varepsilon] = [R][Q][R]^{-1}[\gamma] \\ &= [R][Q][R]^{-1}[S][\sigma] = [R][Q][R]^{-1}[S][Q]^{-1}[\sigma'] \\ &\equiv [S'][\sigma']\end{aligned}$$

where the rotated compliance matrix $[S']$ is given as

$$[S'] = [R][Q][R]^{-1}[S][Q]^{-1} \tag{4.5.15}$$

The transformation matrix $[Q]$ can be established by expanding the transformation rule defined by equation 4.1.15*a* in a six-dimensional vector with the components arranged as shown in equation 4.5.10. This calculation gives

$$\begin{Bmatrix} \sigma'_{11} \\ \sigma'_{22} \\ \sigma'_{33} \\ \sigma'_{12} \\ \sigma'_{13} \\ \sigma'_{23} \end{Bmatrix} = [Q] \times \begin{Bmatrix} \sigma_{11} \\ \sigma_{22} \\ \sigma_{33} \\ \sigma_{12} \\ \sigma_{13} \\ \sigma_{23} \end{Bmatrix}$$

where $[Q]$ is given by

$$[Q] = \begin{bmatrix} a_{11}^2 & a_{12}^2 & a_{13}^2 & 2a_{12}a_{11} & 2a_{13}a_{11} & 2a_{12}a_{13} \\ a_{21}^2 & a_{22}^2 & a_{23}^2 & 2a_{22}a_{21} & 2a_{21}a_{23} & 2a_{22}a_{23} \\ a_{31}^2 & a_{32}^2 & a_{33}^2 & 2a_{32}a_{31} & 2a_{31}a_{33} & 2a_{32}a_{33} \\ a_{11}a_{21} & a_{12}a_{22} & a_{13}a_{23} & a_{12}a_{21}+a_{11}a_{22} & a_{11}a_{23}+a_{13}a_{21} & a_{22}a_{33}+a_{12}a_{23} \\ a_{31}a_{11} & a_{32}a_{12} & a_{33}a_{13} & a_{32}a_{11}+a_{31}a_{12} & a_{31}a_{13}+a_{11}a_{33} & a_{13}a_{22}+a_{33}a_{12} \\ a_{21}a_{31} & a_{22}a_{32} & a_{23}a_{33} & a_{22}a_{31}+a_{21}a_{32} & a_{21}a_{33}+a_{23}a_{31} & a_{22}a_{33}+a_{23}a_{32} \end{bmatrix}$$

Evaluating the transformation matrix for a rotation about the x_3 axis as described in equation 4.5.11*a* gives

$$[Q] = \begin{bmatrix} \mathrm{c}^2\theta & \mathrm{s}^2\theta & 0 & 2\mathrm{c}\theta\,\mathrm{s}\theta & 0 & 0 \\ \mathrm{s}^2\theta & \mathrm{c}^2\theta & 0 & -2\mathrm{c}\theta\,\mathrm{s}\theta & 0 & 0 \\ 0 & 0 & 1 & 0 & 0 & 0 \\ -\mathrm{c}\theta\,\mathrm{s}\theta & \mathrm{c}\theta\,\mathrm{s}\theta & 0 & \mathrm{c}^2\theta - \mathrm{s}^2\theta & 0 & 0 \\ 0 & 0 & 0 & 0 & \mathrm{c}\theta & -\mathrm{s}\theta \\ 0 & 0 & 0 & 0 & \mathrm{s}\theta & \mathrm{c}\theta \end{bmatrix}$$

where c and s represent cosine and sine, respectively. The inverse transformation is obtained by letting $[Q^{-1}] = [Q(-\theta)]$; therefore,

$$[Q^{-1}] = \begin{bmatrix} c^2\theta & s^2\theta & 0 & -2c\theta\, s\theta & 0 & 0 \\ s^2\theta & c^2\theta & 0 & 2c\theta\, s\theta & 0 & 0 \\ 0 & 0 & 1 & 0 & 0 & 0 \\ c\theta\, s\theta & -c\theta\, s\theta & 0 & c^2\theta - s^2\theta & 0 & 0 \\ 0 & 0 & 0 & 0 & c\theta & s\theta \\ 0 & 0 & 0 & 0 & -s\theta & c\theta \end{bmatrix}$$

It can also be shown that $[R][Q][R]^{-1} = [Q]^{-T}$, so the transformed compliance matrix can be expressed as

$$[S'] = [Q]^{-T}[S][Q]^{-1} \tag{4.5.16}$$

The compliance matrix can be determined in the $\mathbf{e}_i'$ basis by substituting $[Q]^{-T}$ and $[Q]^{-1}$ into equation 4.5.16. This calculation produces

$$\begin{Bmatrix} \varepsilon'_{xx} \\ \varepsilon'_{yy} \\ \varepsilon'_{zz} \\ \gamma'_{xy} \\ \gamma'_{xz} \\ \gamma'_{yz} \end{Bmatrix} = \begin{bmatrix} S'_{11} & S'_{12} & \dfrac{-\nu}{E} & S'_{14} & 0 & 0 \\ S'_{12} & S'_{11} & \dfrac{-\nu}{E} & S'_{14} & 0 & 0 \\ \dfrac{-\nu}{E} & \dfrac{-\nu}{E} & \dfrac{1}{E} & 0 & 0 & 0 \\ S'_{14} & S'_{14} & 0 & S'_{44} & 0 & 0 \\ 0 & 0 & 0 & 0 & \dfrac{1}{G} & 0 \\ 0 & 0 & 0 & 0 & 0 & \dfrac{1}{G} \end{bmatrix} \begin{Bmatrix} \sigma'_{xx} \\ \sigma'_{yy} \\ \sigma'_{xx} \\ \tau'_{xy} \\ \tau'_{xz} \\ \tau'_{yz} \end{Bmatrix} \tag{4.5.17}$$

The components in the compliance matrix are

$$S'_{11} = \frac{1}{E'} = \frac{1}{E}(\cos^4\theta + \sin^4\theta) + \left(\frac{1}{G} - \frac{2\nu}{E}\right)\cos^2\theta\sin^2\theta$$

$$S'_{12} = \frac{-\nu'}{E'} = \left(\frac{2}{E} - \frac{1}{G}\right)\cos^2\theta\sin^2\theta - \frac{\nu}{E}(\cos^4\theta + \sin^4\theta)$$

$$S'_{14} = \frac{\eta'}{E'} = \left(\frac{1}{G} - 2\,\frac{1+\nu}{E}\right)\cos\theta\sin\theta(\cos^2\theta - \sin^2\theta)$$

$$S'_{44} = \frac{1}{G'} = 8\left(\frac{1+\nu}{E}\right)\cos^2\theta\sin^2\theta + \frac{1}{G}(\cos^2\theta - \sin^2\theta)^2$$

Equation 4.5.17 shows that when the material is loaded with a tensile stress in the $\mathbf{e}_3' = \mathbf{e}_3$ direction, $\overline{\sigma}' = [0 \quad 0 \quad \sigma_0 \quad 0 \quad 0 \quad 0]^T$, the transverse strains are independent of orientation and the test specimen remains round. If a tensile specimen is loaded in the $\mathbf{e}_1'$ direction, $\overline{\sigma}' = [\sigma_0 \quad 0 \quad 0 \quad 0 \quad 0 \quad 0]^T$, the diametral response does depend upon orientation. Figure 4.5.3 illustrates the change in some of the elastic constants for rotation about the principal axis X_3 for a René N4 single crystal ($E = 19 \times 10^6$ psi, $G = 18 \times 10^6$ psi, $\nu = 0.36$). The elastic modulus, E', and shear modulus, G', vary with orientation and Poisson's ratio, ν', even changes sign. There is coupling between the shear and normal stress and strain components defined by the Lekhnitski coefficients of mutual influence, η' from S_{14}' in equation 4.5.17. In any arbitrary orientation, $\mathbf{e}_j''$, the matrix of elastic constants, can become fully populated and include shear stress–shear strain coupling. The shear stress–shear strain coupling constants are called the Chentsov coefficients. Thus with only three independent elastic constants, the elastic response of a cubic symmetric material is significantly more complex than the elastic response of an isotropic material.

Axial compliance in the $\mathbf{e}_1''$ direction can be determined so that $\varepsilon_{11}' = S_{11}''\sigma_0$ for a tensile stress vector $\overline{\sigma}'' = [\sigma_0 \quad 0 \quad 0 \quad 0 \quad 0 \quad 0]^T$. Following the same

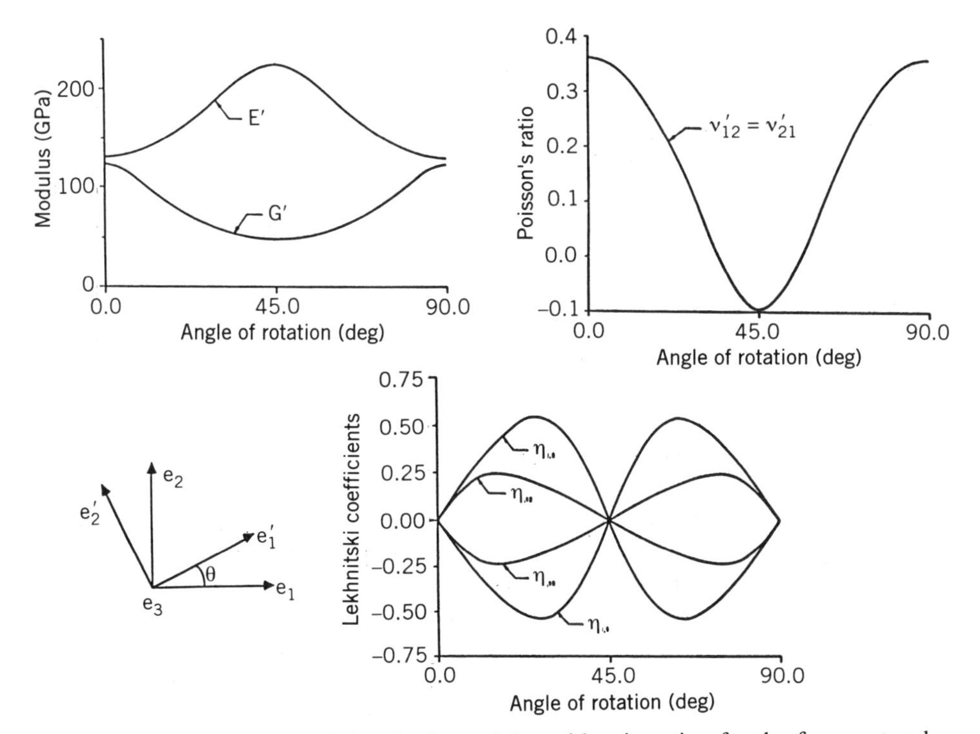

Figure 4.5.3 Variation of the elastic modulus with orientation for the face-centered cubic alloy René N4 at room temperature.

steps for the second rotation, equation 4.5.11b, the compliance of a tensile specimen in the $\mathbf{e}_1''$ direction can be found from

$$\begin{aligned} S''_{11} &= \frac{1}{E''} \\ &= \frac{1}{E}[\cos^4\phi(\cos^4\theta + \sin^4\theta) + \sin^4\phi] + \left(\frac{1}{G} - \frac{2\nu}{E}\right) \\ &\quad \times [\cos^4\phi \cos^2\theta \sin^2\theta + \cos^2\phi \sin^2\phi] \end{aligned} \tag{4.5.18}$$

Since there are only two independent material parameters in S'_{11} and S''_{11}, from equations 4.5.18 and 4.5.17, respectively,

$$\frac{1}{E} \quad \text{and} \quad \left(\frac{1}{G} - \frac{2\nu}{E}\right)$$

tensile compliance measurements from the principal orientation and two other orientations cannot be used to determine the three elastic parameters E, G, and ν. Recalling that a round specimen remains round when loaded in one of the three principal directions, a tensile test in a principal direction can be used to measure E and ν. A second tensile test in an arbitrary onentation then gives the shear modulus G. The accuracy of the parameters depends on the reliability of the orientation measurements.

These results were established for evaluating the elastic constants for materials with cubic symmetry. Transforming the stiffness or compliance matrices in structural calculations is awkward. It may be easier to transform stress or strain from the global coordinates to the principal coordinates of the material, use the constitutive equation in the principal coordinates of the material, and then transform the result back to the global coordinates. Care should be exercised when using both engineering and tensor strain.

4.6 PRINCIPLE OF OBJECTIVITY

The Principle of Objectivity or Material Frame Indifference states that constitutive equations must be invariant under a change of reference. The principle embodies both spatial and temporal reterence changes. Consider, for example, two observers in two different laboratories at two different times who observe the same experiment. It is expected that the results should be identical and not depend on the location and time of the experiment. Thus the observed response must be invariant under a time and spatial change of reference frame.

In the formulation of constitutive equations it is necessary to employ variables that are observer independent or objective. For example, the position and velocity of a particle will be different in different reference frames. These quantities are not objective. However, the distance between two points and the angle between two lines are independent of the rigid-body motion of the observer's frame of reference. These quantities are objective. Newton's laws of motion are valid only in a special frame of reference; thus they are not objective.

Let us establish the rules for objectivity. Define the rectangular reference frame F to be in relative rigid-body motion with respect to frame F'. Since the frames are in rigid motion relative to each other, coordinates x'_i in frame F' can be related to coordinates x_i in frame F by

$$x'_k(t') = Q_{kl}(t)x_l(t) + b_k(t) \qquad \text{and} \qquad t' = t + \lambda \tag{4.6.1}$$

where λ is a constant allowing a different time origin in frame F' and $Q_{kl}(t)$ and $b_k(t)$ are functions of time only. The matrix $Q_{kl}(t)$ is a proper orthogonal transform that satisfies equation 4.1.13, and $b_k(t)$ represents the translation of the orgin of F to F'. Therefore, any a_k vector in frame F will appear as $a'_i = Q_{ik}a_k$ in frame F'. Two motions $x_k\ (X, t)$ and $x'_k\ (X, t')$ are objectively equivalent if

$$x'_k(X, t') = Q_{kl}(t)x_l(X, t) + b_k(t) \qquad \text{and} \qquad t' = t + \lambda \tag{4.6.2}$$

These two motions differ only relative to the reference frame, and time and can be made to coincide with a rigid-body motion and time shift. Thus any vector $\mathbf{a}_k$, or any second-order tensor s_{ij}, is objective if it obeys the appropriate tensor transformation law for all times,

$$\mathbf{a}'_j(X, t') = Q_{jk}(t)a_k(X, t) \qquad \text{and} \qquad s'_{ij}(X, t) = Q_{ik}(t)Q_{jl}(t)s_{kl}(X, t) \tag{4.6.3}$$

All vectors and tensors that are independent of time are objective. Time-dependent quantities are not always objective. Consider, for example, the velocity vector. Differentiating equation 4.6.2 gives

$$\mathbf{v}'_k = \frac{dx'_k}{dt} = Q_{kl}\frac{dx_l}{dt} + \frac{dQ_{kl}}{dt}x_l + \frac{db_k}{dt} = Q_{kl}v_l + \frac{dQ_{kl}}{dt}x_l + \frac{db_k}{dt} \tag{4.6.4}$$

which is not of the form of equation 4.6.3. The extra term arises since velocity would a different in a moving reference frame. Similarly, acceleration is not objective.

Malvern (1969) and Eringen (1967) have evaluated several vectors and tensors for objectivity. The spatial part of the principle of objectivity is not important when small rotations are involved. Stress rate, strain rate, and spin (vorticity) are not objective tensors for arbitrary motions, so constitutive equations for finite deformations that involve these variables do not in general satisfy objectivity. Special forms for these tensors have been developed that do satisfy the principle of objectivity, and details are available in most continuum mechanics books.

The time translation criteria can be stated more explicitly by considering two stress histories $\sigma_{ij}(\tau)$ defined on the time interval $[0, t]$ and $\sigma'_{ij}(\tau)$ defined on the interval $[\lambda, \lambda + t]$. The strain histories corresponding to $\sigma_{ij}(\tau)$ and $\sigma'_{ij}(\tau)$ are $\varepsilon_{ij}(\tau)$ and $\varepsilon'_{ij}(\tau)$, respectively. If the two stress histories are identical over their time domains, $\sigma_{ij}(\tau + \lambda) = \sigma'_{ij}(\tau)$ for all τ on $[0, t]$, then objectivity requires that $\varepsilon_{ij}(\tau + \lambda) = \varepsilon'_{ij}(\tau)$ for all τ on $[0, t]$. All constitutive equations are expected to satisfy this requirement. It will be shown in Section 5.10 that not all models satisfy this test, and the properties of time as a constitutive variable will be examined further.

4.7 SUMMARY

The deviatoric stress and strain tensors are key variables that are used in the formulation of constitutive equations. It is generally assumed that plastic flow or slip is independent of the mean stress; thus the deviatoric stress is used rather than the total stress. This assumption is usually valid except for extreme pressure loads, such as in deep-water structures. Since deformation occurs primarily by slip, permanent changes in material volume due to the development of point defects and vacancies are small and can be neglected. Thus deviatoric inelastic (plastic or creep) strain is used rather than total inelastic strain; however, elastic volume strains are not usually neglected. The second invariant of the deviatoric stress tensor generally arises in the formulation of three-dimensional constitutive equations for metal plasticity; thus many of the representations for the invariant J_2 in equation 4.4.3 will be used.

The constitutive equations of linear elasticity are part of every inelasticity model and the elastic strain is added to the inelastic strain to obtain the total strain. The linear strain can be used for values up to about three times the yield strain, and nonlinear strain measures are important above 5 or 10% strain. The elastic parameters depend on temperature but do not change due to inelastic deformation. There are two independent elastic parameters for isotropic (polycrystalline) metals, but single crystals with cubic symmetry have three independent parameters. These parameters are usually the elastic modulus, shear modulus, and Poisson ratio in the principal directions of the material (parallel to the edges of the cubic lattice). The stress and strain tensors must be transformed to the principal directions of the material to use these three parameters, or the constitutive matrix must be rotated to the frame of the applied loading.

All constitutive equations must satisfy the principal of objectivity. The spatial part of the principal of objectivity is satisfied when rigid-body rotations are small and can be neglected. Thus constitutive equations that are formulated for small strain and the linear strain measure satisfy the spatial part of the principal of objectivity. However, it will be shown in Section 5.10 that not all constitutive equations satisfy the temporal or time part of the principal of objectivity. Additional information on tensors, stress and strain, and the field equations of mechanics is available from the references.

REFERENCES

Boresi, A. P., and P. P. Lynn (1974). *Elasticity in Engineering Mechanics,* Prentice Hall, Englewood Cliffs, NJ.

Borisenko, A. I., and I. E. Tarapov (1979). *Vector and Tensor Analysis with Applications,* Dover Publications, New York.

Dieter, G. E. (1986). *Mechanical Metallurgy,* 3rd ed., McGraw-Hill, New York.

Eringen, A. C. (1967). *Mechanics of Continua,* Wiley, New York.

Fung, Y. C. (1977). *A First Course in Continuum Mechanics,* 2nd ed., Prentice Hall, Englewood Cliffs, NJ.

Liu, K. C., and W. L. Greenstreet (1976). Experimental studies to examine the behavior of metal alloys used in nuclear structures in *Constitutive Equations in Viscoplasticity,* ASME Applied Mechanics Division Publication 20, ASME, New York, p. 43.

Malvern, L. E. (1969). *Introduction to the Mechanics of a Continuous Medium,* Prentice Hall, Englewood Cliffs, NJ.

McKnight, R. W., J. H. Laflen, and G. T. Spamer (1982). Turbine blade tip durability analysis, *NASA CR 165268.*

Selby, S. M. (1972). *CRC Standard Mathematical Tables,* 20th ed., The Chemical Ruber Company, Cleveland.

Sokolnikoff; I. F. (1956). *Mathematical Theory of Elasticity,* 2nd ed., McGraw-Hill, New York.

Spitzig, W. A., R. J. Sober, and O. Richmond (1976). The effect of hydrostatic pressure on the deformation behavior of maraging HY-80 steels and its implications on plasticity theory, *Metallurgical Transactions,* Vol. 7A, pp. 1703–1710.

Synge, J. L. (1978), *Tensor Calculus,* Dover Publications, New York.

Truesdell, C., and R. A. Toupin (1960). The classical field theories, *Encyclopedia of Physics,* W. Flugge, ed., Vol. 3.1, Springer-Verlag, Berlin.

Yang, S. (1984). Elastic constants of a monocrystalline nickel-base superalloy, *Metallurgical Transactions,* Vol. 16A, No. 4, p. 661.

PROBLEMS

4.1 Show that the Kronecker delta defined by equation 4.1.4 is an isotropic tensor. That is, prove that the components are the same for an arbitrary change of basis. Also show that $\delta_{ij}\delta_{kl}$ is an isotropic tensor.

4.2 Write equations 4.2.12, 4.3.7, and 4.4.1 as matrix equations in a three-dimensional space. Write equations 4.2.15 and 4.3.19 in indical notation.

4.3 Compute the invariants, the principal stresses, and the associated principal axes for the following states of stress. Compute the transformation matrices and verify the results by rotating the stress tensor from the global coordinate axes into the principal directions.

$$\textbf{(a)}\ \sigma_{ij} = \begin{bmatrix} -3 & -6 & 0 \\ -6 & 6 & 0 \\ 0 & 0 & 2 \end{bmatrix} \qquad \textbf{(b)}\ \sigma_{ij} = \begin{bmatrix} 7 & 2 & 0 \\ 2 & 6 & -2 \\ 0 & -2 & 5 \end{bmatrix}$$

$$\textbf{(c)}\ \sigma_{ij} = \begin{bmatrix} -10 & 2 & 3 \\ 2 & -5 & 4 \\ 3 & 4 & 10 \end{bmatrix}$$

4.4 For the stress tensors given in Problem 4.3, compute the deviatoric stress tensor, invariants of the deviatoric stress tensor, principal deviatoric stresses, and principal deviatoric directions for each.

4.5 An octahedral plane is defined to be a plane whose normal is at equal angles from each of the principal axes. There are eight such planes with normals, as shown in Figure 1.2.8.

$$\mathbf{n} = (n_1, n_2, n_3) = \frac{1}{\sqrt{3}}(\pm 1, \pm 1, \pm 1)$$

(a) Compute the normal stress and shear stress on an octahedral plane in terms of the principal stresses.

(b) Compute the octahedral normal stress and octahedral shear stress for the stress tensors given in Problem 4.3.

4.6 Verify the development of the Almansi and Green strain tensors, equations 4.3.7 and 4.3.8, respectively. Write the results in matrix notation.

4.7 For small displacement theory, determine if the following displacement components (u, v, w) relative to Cartesian coordinate axes (x, y, z) satisfy the compatibility conditions.

(a) $u = 2yx, \quad v = 4xz, \quad w = xy$

(b) $u = 3x^2 + 4xy + 5, \quad v = 4y^2 + 5x + 6z, \quad w = 6z^3$

4.8 Compute the inverse of the matrix of elastic constants for an isotropic material given by equation 4.5.6.

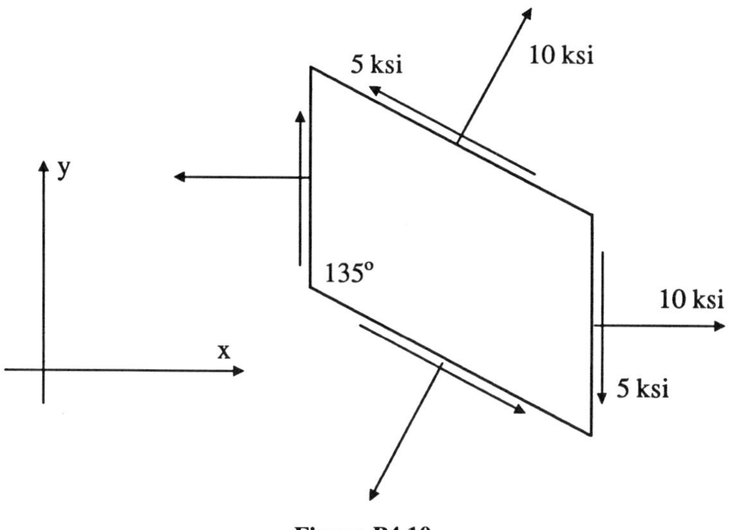

Figure P4.10

4.9 Given the elastic strain components

$$\varepsilon_{11} = 36 \times 10^{-6} \qquad \varepsilon_{22} = 40 \times 10^{-6} \qquad \varepsilon_{33} = 25 \times 10^{-6}$$

$$\varepsilon_{12} = 12 \times 10^{-6} \qquad \varepsilon_{23} = 0 \qquad \varepsilon_{31} = 30 \times 10^{-6}$$

Find the corresponding stresses for $\lambda = 17.3 \times 10^6$ psi and $\mu = 11.5 \times 10^6$ psi.

4.10 A thin plate is in a state of stress as shown in Figure P4.10. Find the stress tensor in the x–y coordinate system.

4.11 The strength-of-materials solution for the stress in the wedge-shaped plate shown in Figure P4.11 is $\sigma_{11} = \sigma_0(a/x)$ with $\sigma_{22} = \sigma_{33} = \sigma_{12} = \sigma_{23} = \sigma_{31} = 0$. Does this solution satisfy the equations of elasticity?

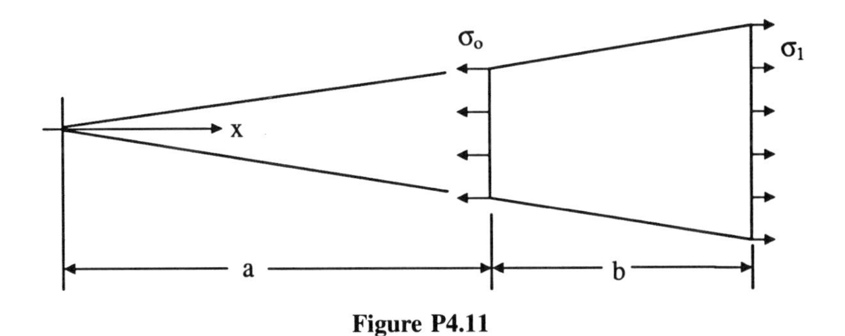

Figure P4.11

4.12 Show that

$$T = \begin{bmatrix} 5 & -10 & 8 \\ -10 & 2 & 2 \\ 8 & 2 & 11 \end{bmatrix} \quad \text{and} \quad T' = \begin{bmatrix} -9 & 0 & 0 \\ 0 & 9 & 0 \\ 0 & 0 & 18 \end{bmatrix}$$

are the same tensor. Find the matrix that can be used to transform T into T'.

4.13 Let S_{ij} be symmetric and T_{ij} be skew symmetric. Expand and reduce $S_{ij}T_{ij}$ to the lowest number of terms.

4.14 The stress at point P is given by

$$\sigma_{11} = 5 \qquad \sigma_{22} = 0 \qquad \sigma_{33} = 0$$

$$\sigma_{12} = a \qquad \sigma_{23} = b \qquad \sigma_{31} = -a$$

where a and b are unknown. At point P it is also known that the second principal stress $\sigma_2 = 2$ ($\sigma_1 > \sigma_2 > \sigma_3$) and that the absolute maximum shear stress is 5.5. Determine σ_1 and σ_3.

4.15 A 2-in. steel cube ($E = 30 \times 10^6$, $\nu = 0.3$) is subjected to the uniform state of plane stress $\sigma_{11} = 6 \times 10^3$ psi, $\sigma_{12} = 2 \times 10^3$ psi, $\sigma_{22} = -3 \times 10^3$ psi. Find the change in length of the face diagonal AB shown in Figure P4.15.

4.16 Find the proper orthogonal transform that carries the **x** coordinates into the **x**$'$ coordinates. The **x**$'$ coordinates are obtained by a sequence of two rotations: first is a 30° right hand rotation about the $-x_3$ axis followed by a 45° right hand rotation about the rotated x_1 axis.

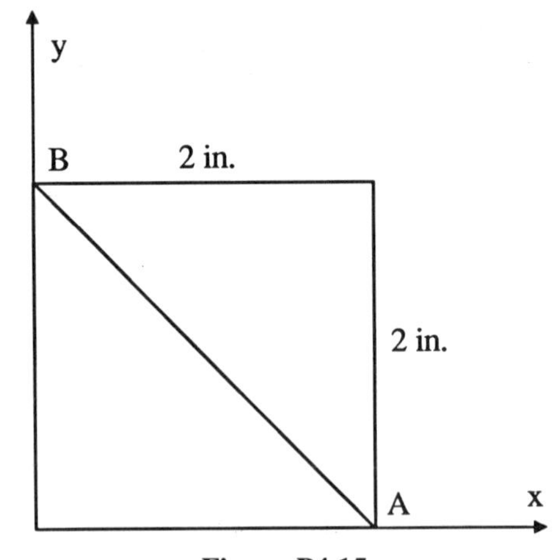

Figure P4.15

4.17 Does the following stress distribution satisfy equilibrium in the absence of body forces?

$$\sigma_{11} = 3x^2 + 4xy - 8y^2 \qquad \sigma_{22} = 2x^2 + xy + 3y^2$$

$$\sigma_{12} = -\tfrac{1}{2}x^2 - 6xy - 2y^2 \qquad \sigma_{13} = \sigma_{23} = \sigma_{33} = 0$$

4.18 The most general form for a constitutive equation for nonlinear isotropic solids can be written as $\sigma_{ij} = f_0(I)\delta_{ij} + f_1(I)\varepsilon_{ij} + f_2(I)\varepsilon_{ik}\varepsilon_{kj}$, where f_0, f_1, and f_2 are functions of the strain invariants, where I is symbolic for the three invariants of strain I_1, I_2, I_3.

(a) This representation was derived from a power series of strain. Show that $\varepsilon_{ik}\varepsilon_{kl}\varepsilon_{li}$ can be written as $\lambda\varepsilon_{ik}\varepsilon_{kj}$, where λ is a function of the first invariant. (*Hint:* Expand $\varepsilon_{ik}\varepsilon_{kl}\varepsilon_{li}$ in the principal coordinate system of strain.

(b) Show that principal directions of stress and strain coincide.

(c) What is a general representation for a second-order solid? Reduce this result for second-order incompressible materials such as rubber.

(d) Show that this result reduces to the linear elastic equation (Hooke's law).

4.19 Small rigid-body rotations can be described by the proper orthogonal transform $\mathbf{Q} = \mathbf{I} + \alpha\mathbf{Q}'$, where α approaches zero as the angle of rotation approaches zero. Show that for two small successive rotations Q_1 and Q_2, the final position is independent of the order in which the rotations are performed.

4.20 The rate of deformation and spin (vorticity) tensors are obtained by expanding the velocity gradient into symmetric and skew-symmetric tensors, similar to equation 4.3.11. Show that the rate of deformation tensor is objective but that the spin tensor is not. How does this result relate to the strain rate and rate of rotation tensors obtained by differentiating equation 4.3.12 with respect to time?

4.21 Show that the second invariant of the deviatoric stress tensor can be written in any of the following forms:

$$J_2 = \tfrac{1}{3}(I_1^2 - I_2) = \tfrac{1}{2}S_{ij}S_{ij}$$

$$= \tfrac{1}{2}[(\sigma_1 - \sigma_m)^2 + (\sigma_2 - \sigma_m)^2 + (\sigma_3 - \sigma_m)^2]$$

$$= \tfrac{1}{6}[(\sigma_{11} - \sigma_{22})^2 + (\sigma_{22} - \sigma_{33})^2 + (\sigma_{33} - \sigma_{11})^2$$

$$+ 6(\sigma_{12}^2 + \sigma_{23}^2 + \sigma_{31}^2)]$$

4.22 In face-centered cubic crystal structures the octahedral plane, as defined in Problem 4.5, is a primary slip plane. Show that the maximum shear stress on the octahedral plane, σ_{oct} the octahedral shear stress, is related to the second invariant of the deviatoric stress tensor by

$$J_2 = \tfrac{3}{2}\sigma_{oct}^2$$

4.23 Equation 4.1.11 gives the matrix equivalent for several tensor equations assuming that vectors are represented by row matrices. Write the equivalent matrix expressions if vectors are represented as column matrices.

CHAPTER 5

Yield Surface Plasticity and Classical Creep Modeling

The traditional approach for predicting the inelastic response of metals to loading utilizes yield surface plasticity at low temperatures and classical creep methods at high temperatures. This approach is based on the assumption that low- and high-temperature inelastic strains are independent and additive. It is also assumed that the plastic strain history does not affect the creep response and the creep strain history does not affect the plastic response. This assumption has not been proven experimentally and in recent years much effort has been devoted to developing "creep–fatigue" (plasticity) interaction rules. Unfortunately, these interaction rules have not been particularly successful and the state-variable approach, presented in Chapters 6 to 8, has emerged as a more successful method for modeling high-temperature time-dependent inelastic deformation. However, yield surface plasticity and classical creep models are valuable for modeling certain classes of materials and loading conditions. Furthermore, these models require relatively few input data and are widely available in numerous general-purpose commercial finite-element codes. In this chapter we present the elements of yield surface plasticity and classical creep modeling relevant to metals. No attempt is made to model creep–fatigue interaction with these approaches.

Plasticity in metals occurs at low temperatures when stress is above the critical resolved shear stress. Slip is the dominant deformation mechanism, but twinning may also be present. Creep in metals occurs at high temperatures and stress is below the critical resolved shear stress. Deformation occurs by dislocation glide, dislocation creep, and bulk and grain boundary diffusion.

A. YIELD SURFACE PLASTICITY

Yield surface plasticity models were developed for metals at relatively low temperatures where the relationship between stress and plastic strain does not depend on the strain rate, and the onset of plastic (inelastic) deformation occurs at a well-defined stress level or yield point. Yield surface plasticity models consist of a criterion for yielding, a loading criterion, a plastic flow equation, and a hardening rule. The yield criterion defines the limit of elastic behavior throughout the loading history for a general state of stress. The loading criterion determines if a stress or strain increment will produce elastic or inelastic straining when the current stress state is at the limit of elastic behavior. The flow equation relates the plastic strain increment tensor to the stress state and loading increment. The hardening rule is used to predict changes in the yield criterion and flow equation as a result of inelastic straining.

5.1 CRITERIA FOR INITIAL YIELD

The initial yield of metals in uniaxial loading is usually defined as the stress for which some prescribed value of permanent plastic strain is produced (typically, 0.2%). The yield stress in tension or compression is therefore readily available from a simple uniaxial test. The yield criterion or yield function is used to determine the elastic limit for a general three-dimensional state of stress. The initial uniaxial yield stress is used to calibrate the yield function for a particular material and temperature.

The initial yield function or yield surface for a general state of stress, σ_{ij}, may be written as

$$f(\sigma_{ij}) = k \qquad \text{or} \qquad F(\sigma_{ij}, k) = 0 \tag{5.1.1}$$

The yield function is frequently referred to as a yield surface since equation 5.1.1 defines a surface in a six-dimensional stress space. The value of the parameter k in equation 5.1.1 is a simple function of the uniaxial yield stress and may depend on temperature. The subsequent yield surface (after initial yield), discussed in Section 5.3, may also depend on variables that are a function of the plastic deformation history.

The yield or onset of plastic straining is typically assumed to occur if a calculated "trial" value of the yield function is greater than a prescribed value, $f(\sigma_{ij}) > k$. The trial value of the yield function is computed from a "trial stress" based on the assumption of elastic straining. However, it is not permissible to have stress states outside the yield surface. Plastic straining and changes in the yield surface will occur such that the stress remains on the current yield surface. Therefore, defining the conditions for plastic flow for a loading history can be more involved than using a simple inequality. Models for the evolution of the yield surface are presented in Section 5.3 and the loading criteria that produce plastic flow are discussed in Section 5.4.

For polycrystalline metals it is reasonable to assume that the initial yield behavior of the material is isotropic. In this case the yield function may be written as a function of the principal stresses σ_1, σ_2, σ_3 or the principal stress invariants I_1, I_2, I_3.

$$f_1(\sigma_1, \sigma_2, \sigma_3) = f_2(I_1, I_2, I_3) = k \tag{5.1.2}$$

For metals the assumption is also usually made that yielding is independent of hydrostatic pressure or mean stress, at least up to moderate pressures. These assumptions were discussed in Chapter 4. Thus the yield function is not a function of the first invariant of the stress tensor and may be written in the form

$$f_3(J_2, J_3) = k \tag{5.1.3}$$

Equation 5.1.3 represents the general form of a yield function that satisfies the assumptions of isotropy and plastic incompressibility.

A graphical representation for the yield function was developed by Westergaard (1920) by representing a general three-dimensional state of stress as a stress vector

$$\boldsymbol{\sigma} = [\sigma_1, \sigma_2, \sigma_3] \tag{5.1.4}$$

in the space of the principal stresses as shown in Figure 5.1.1. The state of stress, $\boldsymbol{\sigma}$, can be resolved into hydrostatic, $\boldsymbol{\sigma}_h = [\sigma_m, \sigma_m, \sigma_m]$, and deviatoric components, $\mathbf{S} = \boldsymbol{\sigma} - \boldsymbol{\sigma}_h$, where the mean stress is defined as $\sigma_m = (\sigma_1 + \sigma_2 + \sigma_3)/3$. Combining these results, the deviatoric stress is

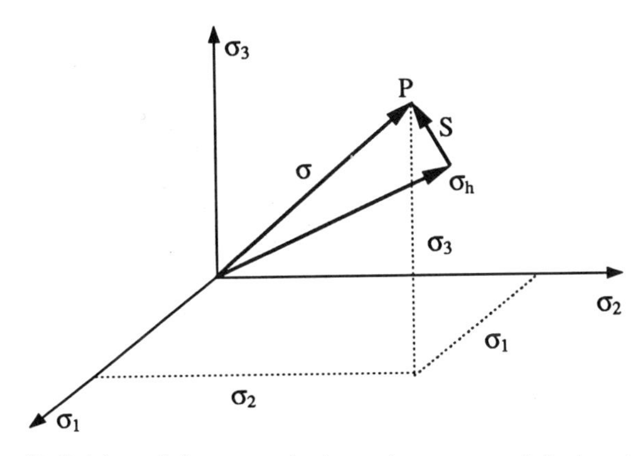

Figure 5.1.1 Definition of the stress, hydrostatic stress, and deviatoric stress vectors in the Haigh–Westergard principal stress space.

$$\mathbf{S} = \boldsymbol{\sigma} - \boldsymbol{\sigma}_h = [(\sigma_1 - \sigma_m),(\sigma_2 - \sigma_m),(\sigma_3 - \sigma_m)] \tag{5.1.5}$$

The magnitude squared of the deviatoric stress vector is written as

$$\begin{aligned}\mathbf{S} \cdot \mathbf{S} = S^2 &= (\sigma_1 - \sigma_m)^2 + (\sigma_2 - \sigma_m)^2 + (\sigma_3 - \sigma_m)^2 \\ &= \tfrac{1}{3}[(\sigma_1 - \sigma_2)^2 + (\sigma_2 - \sigma_3)^2 + (\sigma_3 - \sigma_1)^2] = 2J_2\end{aligned} \tag{5.1.6}$$

The magnitude of the deviatoric stress is related to the second invariant of the deviatoric stress tensor, $S = \sqrt{2J_2}$, and is proportional to the octahedral shear stress, the shear stress on the octahedral plane (Problem 4.22). The deviatoric stress vector is also normal to the hydrostatic stress line since

$$\begin{aligned}\mathbf{S} \cdot \boldsymbol{\sigma}_h &= (\sigma_1 - \sigma_m)\sigma_m + (\sigma_2 - \sigma_m)\sigma_m + (\sigma_3 - \sigma_m)\sigma_m \\ &= (\sigma_1 + \sigma_2 + \sigma_3 - 3\sigma_m)\sigma_m = 0\end{aligned} \tag{5.1.7}$$

Assume that the stress, $\boldsymbol{\sigma}$, defined in equation 5.1.4 is a three-dimensional state of stress at point P in Figure 5.1.1 that just produces yield. Next define a second stress, σ', at point P' as

$$\boldsymbol{\sigma}' = \boldsymbol{\sigma} + a\boldsymbol{\sigma}_h \tag{5.1.8}$$

that is equal to the stress $\boldsymbol{\sigma}$ but with the addition of an arbitrary hydrostatic stress $a\boldsymbol{\sigma}_h$. Since it is assumed that yield does not depend on the hydrostatic stress, the second stress $\boldsymbol{\sigma}'$ will also just produce yield. Therefore, the two vectors $\boldsymbol{\sigma}$ and $\boldsymbol{\sigma}'$ define a line that is parallel to the hydrostatic line. This

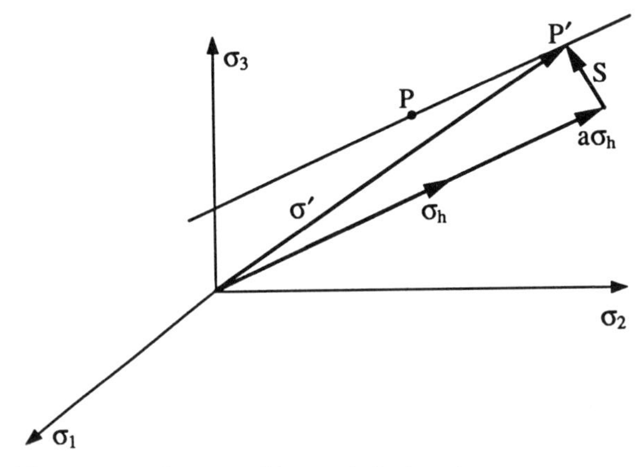

Figure 5.1.2 Two states of stress with equal deviatoric stress components but different hydrostatic stress components.

line contains a set of stress values that just produce yield, as shown in Figure 5.1.2. The deviatoric stress vector, **S**, is identical for all the values of stress on this line that just produce yield. A second deviatoric stress vector, **S**′, that just produces yield will define a second line parallel to the hydrostatic line. Therfore, a set of stress vectors, $\mathbf{S}_i$, that just produce yield lie on a right cylindrical surface (not necessarily circular) that is the yield surface. Points inside the surface correspond to elastic deformation, whereas points on the surface are at the limit of elastic behavior. The hydrostatic line is the axis of the yield surface.

The shape of the cross section of the cylinder can be established by examining the stresses in the plane normal to the hydrostatic line. This plane is usually called the pi plane. The projection of the three stress axes on the pi plane show that stresses plotted on the plane are $\sqrt{\frac{2}{3}}$ of their nominal value. Define x_1 and x_2 as the pair of orthogonal axes in the pi plane oriented as shown in Figure 5.1.3. The coordinates of the projection of an arbitrary state of stress in the pi plane are given by

$$x_1 = \sqrt{\tfrac{2}{3}}\,\sigma_2 \cos 30 - \sqrt{\tfrac{2}{3}}\,\sigma_1 \cos 30 = \sqrt{\tfrac{1}{2}}\,(\sigma_2 - \sigma_1)$$

$$x_2 = \sqrt{\tfrac{2}{3}}\,\sigma_3 - \sqrt{\tfrac{2}{3}}\,\sigma_1 \sin 30 - \sqrt{\tfrac{2}{3}}\,\sigma_2 \sin 30 = \sqrt{\tfrac{1}{6}}\,(2\sigma_3 - \sigma_1 - \sigma_2) \tag{5.1.9}$$

It is also possible to show that $x_1^2 + x_2^2 = S^2$.

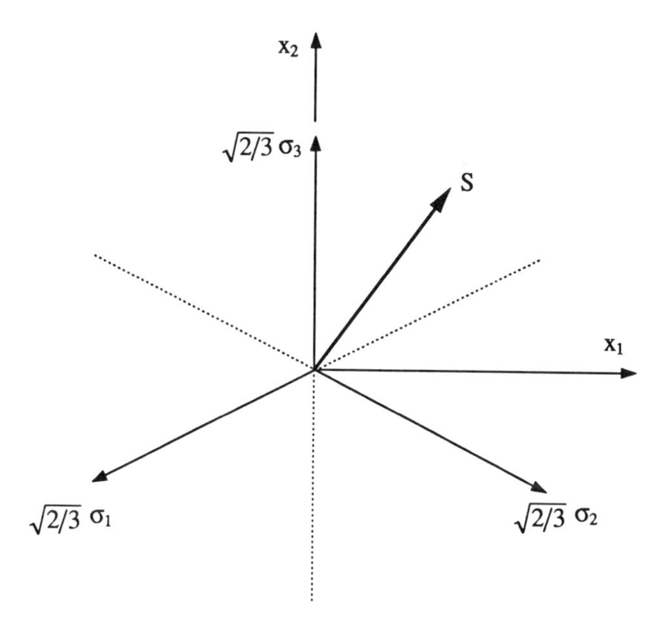

Figure 5.1.3 Projection of the three-dimensional stress vector onto two orthogonal axes in the pi plane.

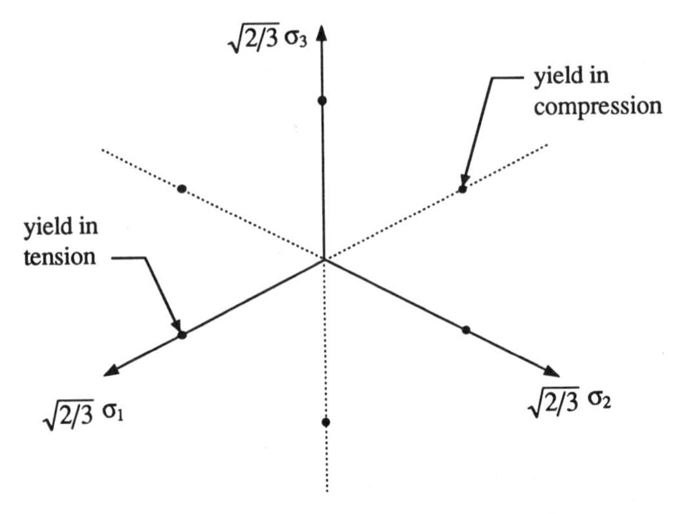

Figure 5.1.4 Six points on the yield surface perimeter for a polycrystalline material that is isotropic and has the same value for yield in tension and compression.

Since it is assumed that metals that deform by slip have the same yield stress in tension and compression and that polycrystalline metals are isotropic, the right cylinder must pass through six points with the value $\pm\sqrt{\frac{2}{3}}\,\sigma_y$. These points are shown on the pi plane in Figure 5.1.4 on the three principal stress axes. The von Mises (1913) yield criterion is a right circular cylinder that passes through these six points and the Tresca or maximum shear stress criterion is defined by the hexagon inside the circular cylinder, as shown in Figure 5.1.5. It will be shown in Section 5.4.3 that the yield surface for

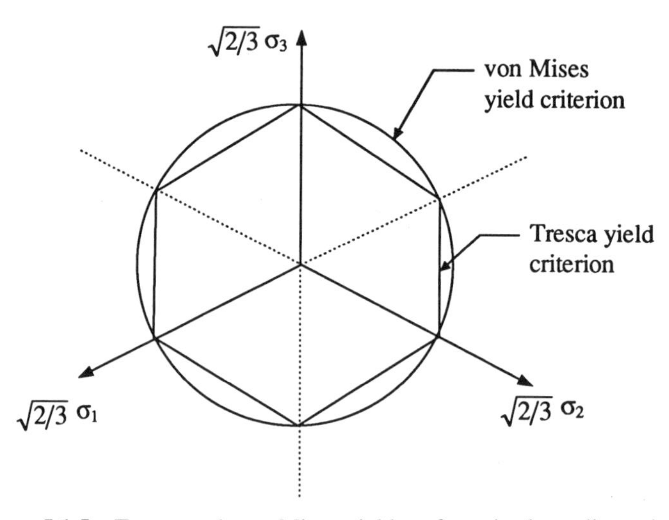

Figure 5.1.5 Tresca and von Mises yield surfaces in three dimensions.

hardening materials must be a right cylinder with a cross section no larger than the hexagon outside the von Mises circular cylinder. Thus, for hardening materials, the initial yield is defined by a deviatoric stress vector that is between the inner and outer hexagons shown in Figure 5.1.6. The von Mises criterion is near the average of the acceptable values for the initial yield stress. The mathematical representation of the von Mises yield criterion can be established by requiring the radius of the circle to pass through the six yield points in the pi plane, that is,

$$|\mathbf{S}| = \sqrt{\tfrac{2}{3}}\, \sigma_y \tag{5.1.10}$$

Using equation 5.1.6 with equation 5.1.10 gives

$$\sqrt{\tfrac{1}{2}[(\sigma_1 - \sigma_2)^2 + (\sigma_2 - \sigma_3)^2 + (\sigma_3 - \sigma_1)^2]} = \sigma_y \tag{5.1.11}$$

The von Mises criterion for initial yield for a general state of stress can be written as

$$(\sigma_x - \sigma_y)^2 + (\sigma_y - \sigma_z)^2 + (\sigma_z - \sigma_x)^2 + 6\tau_{xy}^2 + 6\tau_{yz}^2 + 6\tau_{zx}^2 = 2\sigma_y^2 \tag{5.1.12}$$

or

$$J_2 = \frac{\tau_y^2}{3}$$

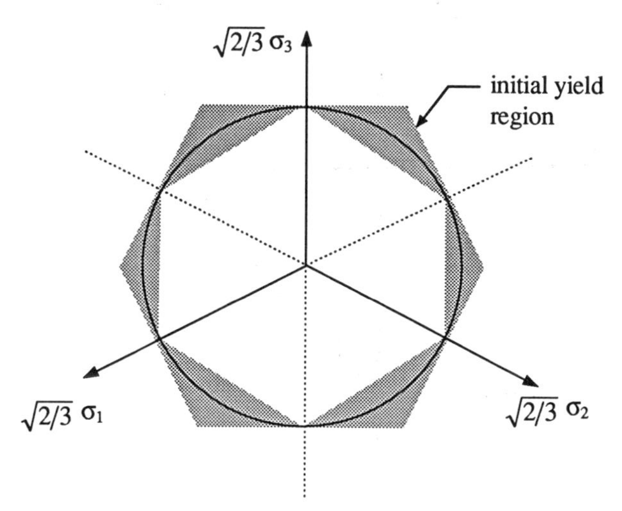

Figure 5.1.6 Von Mises and Tresca yield criteria in the pi plane. The outer hexagon is the maximum value of the yield stress for strain-hardening material according to Drucker's postulate discussed in Section 5.2.3.

Yielding occurs for the Tresca yield criterion, which is the same as the inner hexagon, when the maximum shear stress reaches a critical value

$$\max(|\sigma_1 - \sigma_3|, |\sigma_2 - \sigma_3|, |\sigma_1 - \sigma_2|) = \sigma_y \tag{5.1.13}$$

The Tresca criterion for metals in a multiaxial state of stress was proposed in 1864. It is identical to the von Mises criterion for uniaxial loading but is more conservative for any other stress state.

Many problems approximately satisfy the condition of plane stress where one principal stress is zero. For plane stress problems it is convenient to define yield in a plane with two orthogonal principal stress axes. The Tresca criterion in this space is a hexagon, and the von Mises criterion is an ellipse that is rotated 45°, as shown in Figure 5.1.7. In this stress subspace with $\sigma_3 = 0$, the Tresca yield criterion is a hexagon defined by

$$\max(|\sigma_1|, |\sigma_2|, |\sigma_1 - \sigma_2|) = \sigma_y \tag{5.1.14}$$

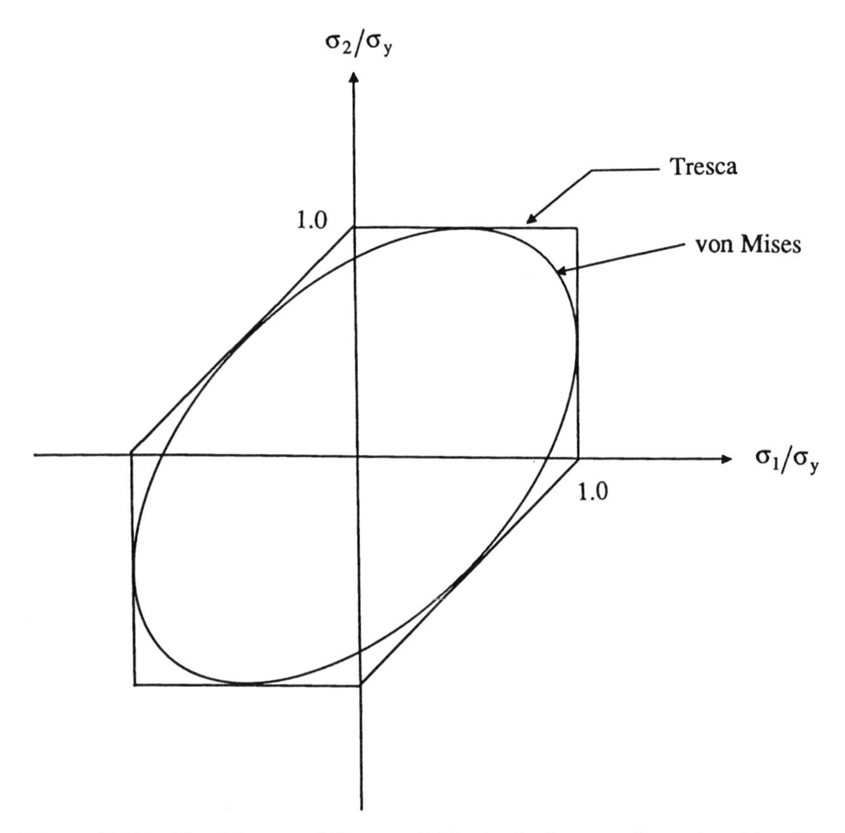

Figure 5.1.7 Von Mises and Tresca yield criteria for two-dimensional loading.

The von Mises yield criterion is an ellipse in the σ_1, σ_2 plane and is given by the equation

$$\sigma_1^2 + \sigma_1\sigma_2 + \sigma_2^2 = \sigma_y^2 \tag{5.1.15}$$

Combined tension torsion stress states may conveniently be represented on a plot with normal stress, σ, and shear stress, τ, axes as shown in Figure 5.1.8. This particular plot is useful for comparing tension torsion test data with theoretical predictions. The Tresca yield criterion in the tension torsion subspace is an ellipse given by

$$\sigma^2 + 4\tau^2 = \sigma_y^2 \tag{5.1.16}$$

and the von Mises criterion is an ellipse given by

$$\sigma^2 + 3\tau^2 = \sigma_y^2 \tag{5.1.17}$$

Equations 5.1.16 and 5.1.17 show that the Tresca and von Mises criteria do not agree in pure torsion (shear).

There have been numerous experimental programs to verify the three-dimensional yield criteria developed by von Mises and Tresca. The von Mises criterion has been shown to be more accurate for a large number of polycrystalline metals. An example is shown in Figure 5.1.9 for annealed 304 stainless steel. The Tresca criterion is more conservative and tends to represent a lower bound for yield. It is also very simple to apply.

Experiments to verify the yield surface for combined loading are not easy to conduct. On loading along some stress path experimentally it is necessary

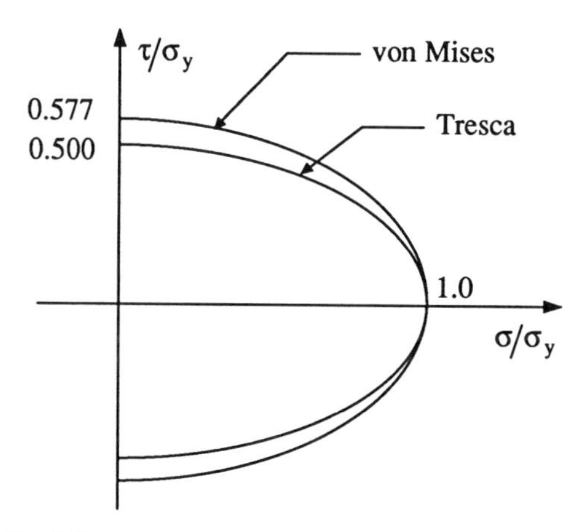

Figure 5.1.8 Von Mises and Tresca yield criteria for tension–torsion loading.

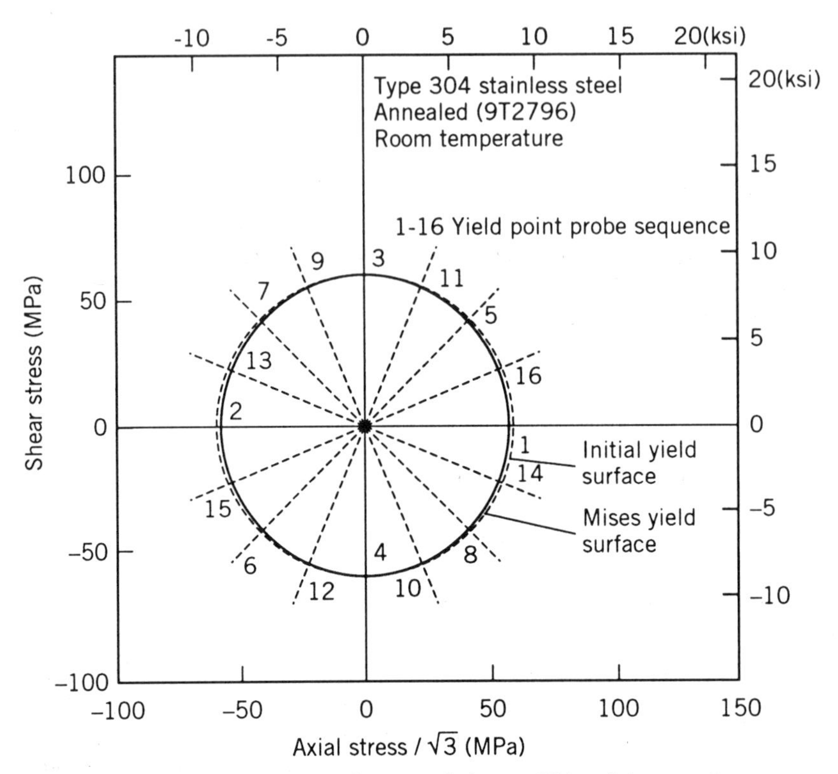

Figure 5.1.9 Initial yield surface for annealed type 304 stainless steel at room temperature compared to the von Mises yield criterion. The data are plotted against $\sigma/\sqrt{3}$ to map the von Mises yield surface into a circle. (After Liu and Greenstreet, Constitutive Equations in Viscoplasticity, pg. 43, 1976. Copyright © The American Society of Mechanical Engineers, 345 East 47th St., New York, NY 10017. Used with permission.)

to load the material beyond yield in order to detect the onset of plastic flow. The plastic deformation during loading to find the yield stress also hardens the material. After unloading and reloading several times to find multiple points on the yield surface, the material can change significantly. Thus the material at the end of the test may not be the same as the material at the beginning of the test. The effect of testing on the definition of the yield surface was demonstrated by Williams and Svensson (1970). Figure 5.1.10 shows how the proof strain (amount of strain used to detect yield) affects the yield stress in a tension torsion test.

As temperature increases, the uniaxial yield point for metals decreases. This is not surprising when consideration is given to the role of thermally activated deformation mechanisms discussed in Part One. For multiaxial states of stress this may be interpreted as a decrease in the size of the yield surface with temperature. A typical example of the temperature dependence of the initial yield surface is shown in Figure 5.1.11 for annealed 304 stainless steel.

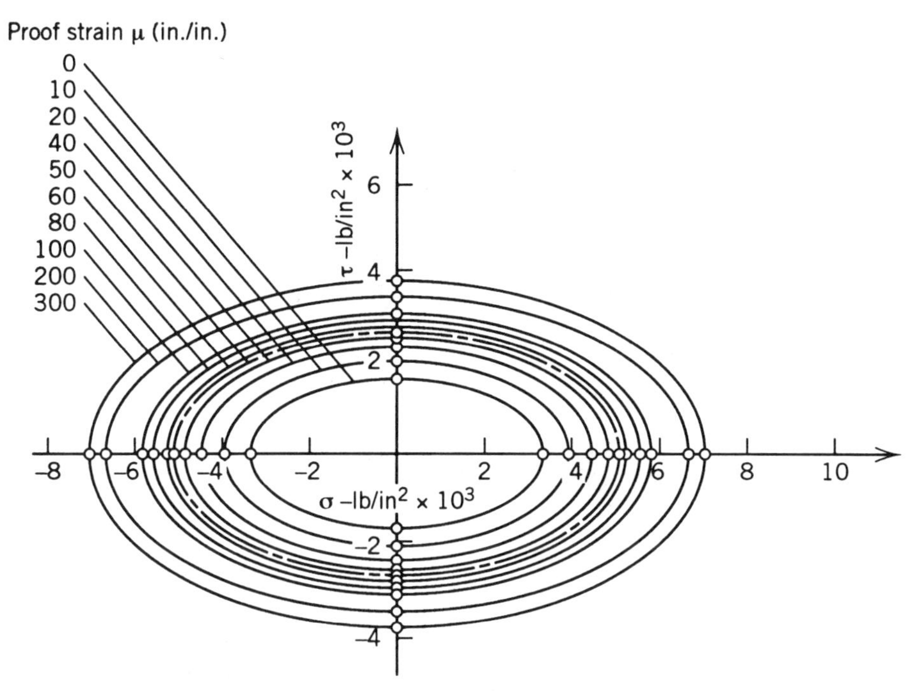

Figure 5.1.10 Effect of the yield stress definition on the initial yield surface of 1100-F aluminum alloy at room temperature. The effective proof strain used in this study is defined by $\hat{\varepsilon}^P = \sqrt{(\varepsilon^P)^2 + (\gamma^P)^2/3}$, where ε^P and γ^P are the tensile and torsional plastic strains, respectively. (From Williams and Svensson, 1970; reprinted by permission of the Council of the Institution of Mechanical Engineers from the *Journal of Strain Analysis.*)

5.2 ELASTIC-PLASTIC FLOW EQUATION

If a hardening material is loaded to a state of stress on the yield surface, and if an additional loading increment produces a trial stress outside the starting yield surface, plastic straining results and the yield surface changes. In general, it is impossible to have a stress state that is outside the yield surface. During application of the load increment the yield surface moves or changes shape such that the current stress point always lies on the surface. Thus the development of the theory requires three more elements: a rule to determine the plastic strain increment, a rule to determine how the yield surface changes and a loading criterion to decide if plastic strain occurs.

The total strain response is written as the sum of the elastic strain, which is given in Section 4.5, and the plastic strain in the form

$$\varepsilon_{ij} = \varepsilon_{ij}^e + \varepsilon_{ij}^p = \varepsilon_{ij}^e + \int d\varepsilon_{ij}^p \tag{5.2.1}$$

The plastic strain is the integral of the plastic strain increments that are de-

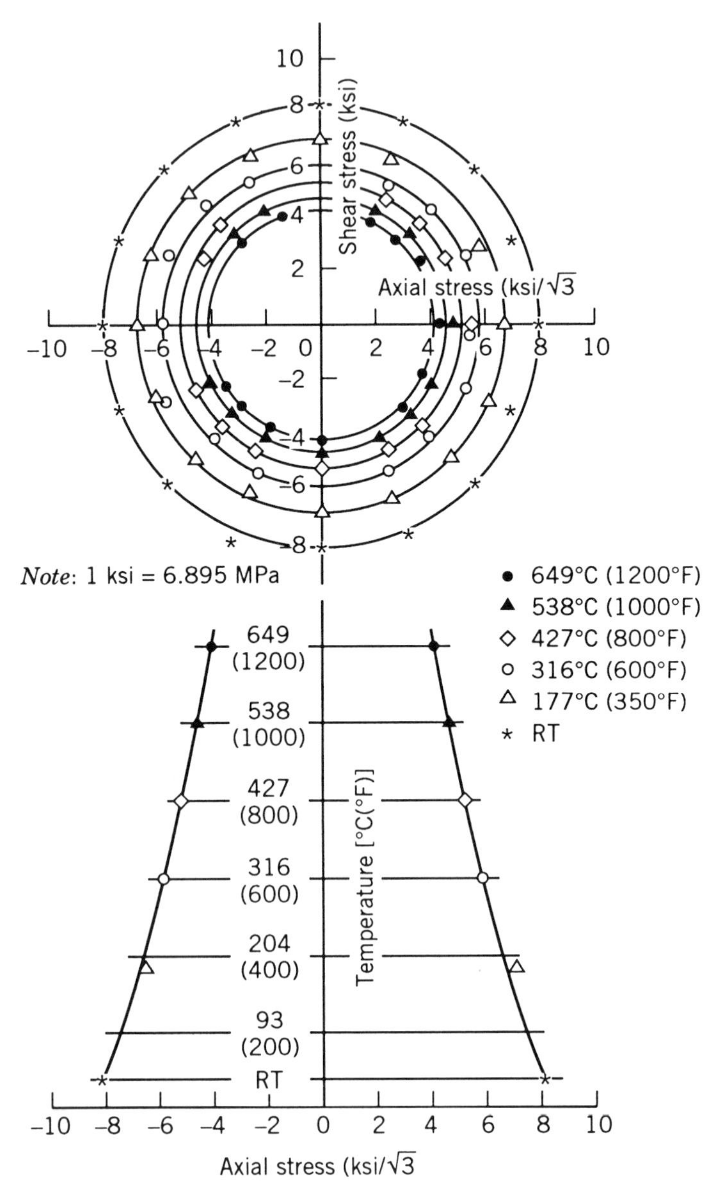

Figure 5.1.11 Effect of temperature on the initial yield stress of annealed type 304 stainless steel. (After Liu and Greenstreet, Constitutive Equations in Viscoplasticity, p. 44, 1976. Copyright © The American Society of Mechanical Engineers, 345 East 47th St., New York, NY 10017. Used with permission.)

veloped during the loading history. Therefore, it is necessary to determine the plastic strain increments that result from the history of the stress or stress increments. The flow law was put on a firm theoretical foundation by Drucker (1950, 1951) by establishing a relationship between the yield surface and plastic strain increments for hardening materials. These results can be used to derive the Prandtl (1924)–Reuss (1930) plastic flow equation.

5.2.1 Drucker's Postulate

Suppose a body is loaded such that the stress and strain, σ_{ij}^* and ε_{ij}^*, satisfy equilibrium and compatibility throughout the loading history. Then a load increment is applied that produces a further stress increment, elastic strain increment, and plastic strain increment. Drucker (1950) assumed for hardening materials that the body will remain in equilibrium, and that

1. Positive work is done during the application of the load increment.
2. Work done during application and removal of the load increment is positive or zero.

The first statement can be expressed mathematically as

$$d\sigma_{ij}\, d\varepsilon_{ij} > 0 \tag{5.2.2}$$

The second statement requires that nonnegative plastic work be done during a loading and unloading cycle, that is,

$$d\sigma_{ij}(d\varepsilon_{ij} - d\varepsilon_{ij}^e) = d\sigma_{ij}\, d\varepsilon_{ij}^p \geq 0 \tag{5.2.3}$$

The quantity $d\sigma_{ij}\, d\varepsilon_{ij}^p$ is the scalar product between the stress increment, $d\sigma_{ij}$, and plastic strain increment, $d\varepsilon_{ij}^p$. Equation 5.2.3 implies that the angle between the stress increment and plastic strain increment vectors must be acute. Drucker defined a material that satisfies equation 5.2.2 as a plastically stable material. If there is no elastic straining during the loading cycle, the material is perfectly plastic or neutrally stable and equation 5.2.3 will be an equality. A plastically stable material implies that the slopes of the stress–strain and stress–plastic strain tensile curves are positive, as shown in Figure 5.2.1.

5.2.2 Yield Surface as a Plastic Potential Function

Before establishing a general structure for the plastic flow equation it is helpful to extend a few ideas from vector algebra to a nine-dimensional vector space. Consider any two second-order tensors a_{ij} and b_{ij}. It is possible to define nine-dimensional vectors from components of the tensors as

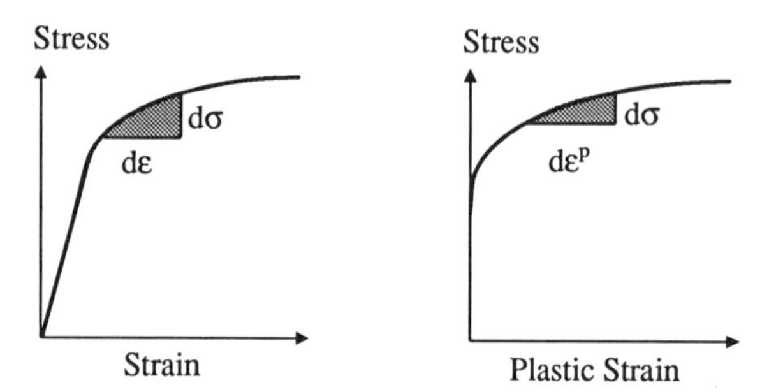

Figure 5.2.1 Stress–strain and stress–plastic strain showing Drucker's postulate for a tensile test.

$$a = [a_{11}, a_{12}, a_{13}, a_{21}, \ldots, a_{33}]$$
$$b = [b_{11}, b_{12}, b_{13}, b_{21}, \ldots, b_{33}] \tag{5.2.4}$$

The scalar product of the two vectors is defined by

$$\mathbf{a} \bullet \mathbf{b} = a_{ij} b_{ij}$$

If $\mathbf{a}$ is parallel to $\mathbf{b}$, then $\mathbf{a} = \lambda \mathbf{b}$, where λ is a scalar. If $\mathbf{a} \bullet \mathbf{b} = 0$, then (1) $\mathbf{a} = 0$ or $\mathbf{b} = 0$ or (2) $\mathbf{a} \neq 0$, $\mathbf{b} \neq 0$ and $\mathbf{a}$ is normal to $\mathbf{b}$.

The differential of a function $f(\mathbf{a})$ can also be written in the nine-dimensional vector space as

$$df = \frac{\partial f}{\partial a_{ij}} da_{ij} = \left[\frac{\partial f}{\partial a_{11}}, \frac{\partial f}{\partial a_{12}}, \frac{\partial f}{\partial a_{13}}, \ldots, \frac{\partial f}{\partial a_{33}} \right] \cdot [da_{11}, da_{12}, da_{13}, \ldots, da_{33}] = \text{grad } f \cdot d\mathbf{a} \tag{5.2.5}$$

On any surface defined by f = constant, grad f is a vector normal to the surface and $d\mathbf{a}$ is a vector tangent to the surface.

Drucker established a relationship between the plastic strain increments and the yield surface by assuming:

1. The material is a hardening material that satisfies the conditions expressed in equations 5.2.2 and 5.2.3.
2. A yield surface exists for the material in the form of equation 5.1.1*a* where f and k both depend on the deformation history.

Assume that a differential stress increment from a point on the yield surface can be resolved into components that are normal, $d\sigma_{ij}^N$, and tangent, $d\sigma_{ij}^T$, to the yield surface as shown in Figure 5.2.2. The total stress increment can be then written as

$$d\sigma_{ij} = d\sigma_{ij}^T + d\sigma_{ij}^N \tag{5.2.6}$$

For hardening materials Drucker further assumed that the stress increment $d\sigma_{ij}^T$ does not produce any plastic strain since it is tangent to the yield surface. The incremental change in the yield surface df is given from equation 5.1.1*a* by

$$df = \frac{\partial f}{\partial \sigma_{ij}} d\sigma_{ij} = \frac{\partial f}{\partial \sigma_{ij}} d\sigma_{ij}^T + \frac{\partial f}{\partial \sigma_{ij}} d\sigma_{ij}^N \tag{5.2.7}$$

Since $\partial f / \partial \sigma_{ij}$ is normal to $d\sigma_{ij}^T$, the scalar product $(\partial f / \partial \sigma_{ij}) d\sigma_{ij}^T = 0$. The vector $\partial f / \partial \sigma_{ij}$ is parallel to $d\sigma_{ij}^N$, so $d\sigma_{ij}^N = (\partial f / \partial \sigma_{ij}) da$, where da is a positive differential since the normal vector must be in the outward direction. The differential change of the yield surface becomes

$$df = \frac{\partial f}{\partial \sigma_{ij}} d\sigma_{ij} = 0 + \frac{\partial f}{\partial \sigma_{ij}} d\sigma_{ij}^N = \frac{\partial f}{\partial \sigma_{ij}} \frac{\partial f}{\partial \sigma_{ij}} da \equiv g(\sigma)\, da > 0 \tag{5.2.8}$$

since $(\partial f / \partial \sigma_{ij})(\partial f / \partial \sigma_{ij}) = g(\sigma)$ is a positive definite quantity. Thus $f + df$ satisfies the initial assumptions that plastic flow exists and a stress increment $d\sigma_{ij}^T$ alone will not produce plastic flow since df would be zero.

For a hardening material, equation 5.2.3 becomes

$$d\sigma_{ij}\, d\varepsilon_{ij}^p = (d\sigma_{ij}^N + d\sigma_{ij}^T)\, d\varepsilon_{ij}^p > 0 \tag{5.2.9}$$

Since $d\sigma_{ij}^T$ is tangent to the yield surface and does not produce plastic straining, it can be replaced by $Cd\sigma_{ij}^T$, so that

$$(d\sigma_{ij}^N + Cd\sigma_{ij}^T)\, d\varepsilon_{ij}^p > 0 \tag{5.2.10}$$

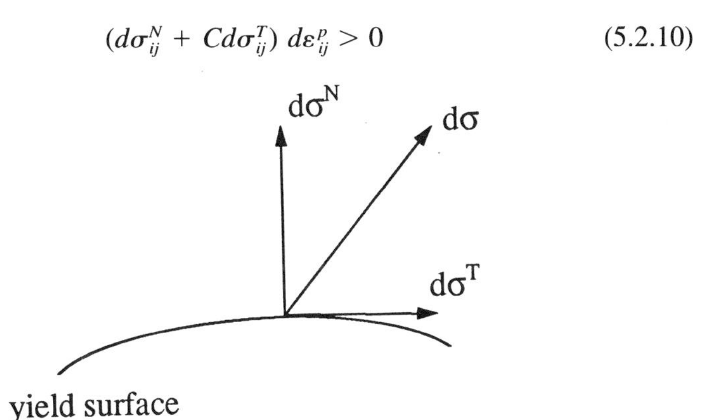

Figure 5.2.2 Definition of normal and tangential stress increment components.

where C is an arbitrary constant. This inequality can be satisfied for arbitrary values of C only if $d\sigma_{ij}^T \, d\varepsilon_{ij}^p = 0$. Therefore, $d\varepsilon_{ij}^p$ and $d\sigma_{ij}^T$ must be normal and $d\varepsilon_{ij}^p$ can be written as

$$d\varepsilon_{ij}^p = d\lambda \frac{\partial f}{\partial \sigma_{ij}} \tag{5.2.11}$$

where $d\lambda$ is a scalar. Thus it follows that:

1. The plastic strain vector is normal to the yield surface for strain-hardening materials.
2. The yield surface is a potential function that can be used to derive the plastic strain increments.

Von Mises (1928) postulated that the plastic strain increments are derivable from a plastic potential function, $g(\sigma_{ij})$, which is a scalar function of the components of the stress tensor:

$$d\varepsilon_{ij}^p = d\lambda \frac{\partial g}{\partial \sigma_{ij}} \tag{5.2.12}$$

The proportionality constant, $d\lambda$, is a nonnegative scalar quantity that is nonzero only when the current state of stress is on the yield surface and the conditions for loading are satisfied. The proportionality constant depends on the slope of the stress–strain curve and the current stress or strain increment. If the plastic potential function is the same as the yield function, $g = f$, the plastic strain increments are associated with the yield surface. Then the flow rule, such as Drucker's formulation, is called an associated flow rule.

5.2.3 Yield Surface Convexity

Drucker's concept of hardening is broader than that stated in equation 5.2.3. The definition can also be applied to finite stress steps. Consider, for example, a loading and unloading sequence that is made up of a finite stress step and a differential stress step. The initial stress state σ_{ij}^* may be inside or on the yield surface. The material is then loaded to a stress state, σ_{ij}, which is on the yield surface as shown in Figure 5.2.3. It is then loaded into the plastic region with an additional stress increment, $d\sigma_{ij}$, which produces a plastic strain increment, $d\varepsilon_{ij}^p$. The stress is then returned to the initial state σ_{ij}^*. The work done during the loading and unloading cycle can be approximated by

$$dW \simeq (\sigma_{ij} - \sigma_{ij}^*) \, d\varepsilon_{ij}^p + d\sigma_{ij} \, d\varepsilon_{ij}^p \geq 0 \tag{5.2.13}$$

The quantity $(\sigma_{ij} - \sigma_{ij}^*)$ can always be made larger than $d\sigma_{ij}$; therefore,

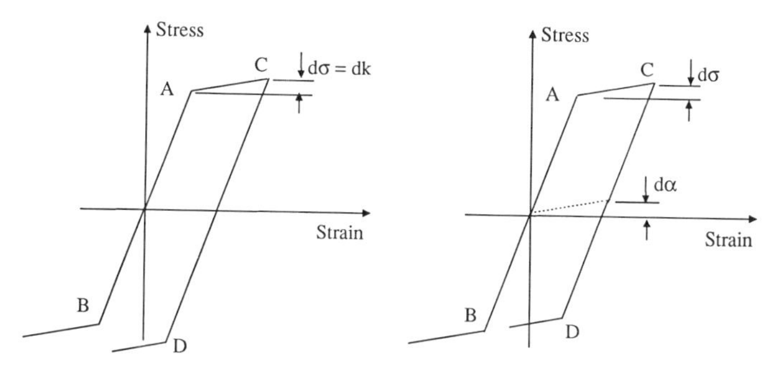

Figure 5.2.3 Yield surfaces that do (*left*) and do not (*right*) satisfy Drucker's assumption for hardening.

$$(\sigma_{ij} - \sigma_{ij}^*)\, d\varepsilon_{ij}^p \geq 0 \tag{5.2.14}$$

Superimposing the plastic strain and stress axes, the vectors $(\sigma_{ij} - \sigma_{ij}^*)$ and $d\varepsilon_{ij}^p$ must form an acute angle (Figure 5.2.3). Since the choice of the stress, σ_{ij}^*, inside the yield surface is arbitrary, equation 5.2.14 must hold for all σ_{ij}^* and $d\varepsilon_{ij}^p$. For the inequality of equation 5.2.14 to be satisfied, the plastic strain increment must be normal to the yield surface and the yield surface must be convex (i.e., the surface must always be inside a plane tangent to the surface). This conclusion is illustrated in Figure 5.1.6 for the initial yield surface. Since all surfaces must pass through the six uniaxial yield points, the outer hexagon is the limiting surface that satisfies equation 5.2.14.

5.2.4 Prandtl–Reuss Flow Equation

The flow rule associated with the von Mises yield function is the Prandtl–Reuss equation. Substituting $f = J_2$ into equation 5.2.11 gives

$$d\varepsilon_{ij}^p = d\lambda \frac{\partial J_2}{\partial \sigma_{ij}} \tag{5.2.15}$$

Note that for the von Mises criterion, (which is independent of mean stress) the derivatives of the yield function with respect to the total stress components are the same as the derivatives of the yield function with respect to the deviatoric stress components:

$$d\varepsilon_{ij}^p = d\lambda \frac{\partial J_2}{\partial \sigma_{ij}} = d\lambda \frac{\partial J_2}{\partial S_{ij}}$$

Thus $d\varepsilon_{11}^p$, for example, can be evaluated as

$$d\varepsilon_{11}^{p} = d\lambda \frac{\partial}{\partial \sigma_{11}} \{\tfrac{1}{6}[(\sigma_{11} - \sigma_{22})^2 + (\sigma_{22} - \sigma_{33})^2 + (\sigma_{33} - \sigma_{11})^2] + \sigma_{12}^2 + \sigma_{13}^2 + \sigma_{23}^2\} = d\lambda\, S_{11} \quad (5.2.16)$$

The Prandtl–Reuss equation can be written as

$$d\varepsilon_{ij}^{p} = d\lambda\, S_{ij} \quad (5.2.17)$$

or

$$d\lambda = \frac{d\varepsilon_{11}^{p}}{S_{11}} = \frac{d\varepsilon_{22}^{p}}{S_{22}} = \frac{d\varepsilon_{33}^{p}}{S_{33}} = \frac{d\varepsilon_{12}^{p}}{S_{12}} = \frac{d\varepsilon_{13}^{p}}{S_{13}} = \frac{d\varepsilon_{23}^{p}}{S_{23}} \quad (5.2.18)$$

Equations 5.2.17 and 5.2.18 show that the plastic strain increments depend on the deviatoric stress, not the total stress or the stress increments. Further, the plastic strain increments and deviatoric stress are parallel, and the principal axes of stress and plastic strain increments coincide. The magnitude of the plastic strain increments depends on the parameter $d\lambda$.

The parameter $d\lambda$ can be found by multiplying equation 5.2.17 by the plastic strain increment to get

$$\tfrac{1}{2} d\varepsilon_{ij}^{P} d\varepsilon_{ij}^{P} = \tfrac{1}{2} d\varepsilon_{ij}^{P} S_{ij}\, d\lambda = \tfrac{1}{2} S_{ij} S_{ij}\, d\lambda^2 \quad (5.2.19)$$

The von Mises effective stress is defined as

$$\sigma_e = \sqrt{3J_2} \quad (5.2.20)$$

and the effective plastic strain increment is defined as

$$d\varepsilon_e^{p} = \sqrt{\tfrac{2}{3}\, d\varepsilon_{ij}^{p}\, d\varepsilon_{ij}^{p}} \quad (5.2.21)$$

Solving equation 5.2.19 for $d\lambda$ and using the definitions of the von Mises effective stress and the effective plastic strain increment gives

$$d\varepsilon_{ij}^{p} = \frac{3 d\varepsilon_e^{p}}{2\sigma_e} S_{ij} = \frac{\sqrt{3}}{2}\, d\varepsilon_{ij}^{p} \frac{S_{ij}}{\sqrt{J_2}} \quad (5.2.22)$$

which are two common expressions for the Prandtl–Reuss flow equation. The factors $\sqrt{3}$ and $\sqrt{\tfrac{2}{3}}$ in the effective stress and effective plastic strain increment equations are chosen so that the effective stress and effective plastic strain increments are equal to the uniaxial stress and uniaxial plastic strain increment in a tensile test.

5.3 HARDENING RULES

For hardening materials, the Tresca and von Mises yield criteria presented in Section 5.1 are valid only for initial yield. The yield surface changes as a result of the hardening that develops during the history of plastic deformation. There are two basic models for hardening. Isotropic hardening is based on the assumption that the center of the yield surface remains fixed and the surface expands without changing shape. Kinematic hardening is based on the assumption that the yield surface translates in the six dimensional stress space but does not change size or shape. Mixed or combined hardening models have also been developed which include both a yield surface expansion and translation. Softening behavior is not included in Drucker's postulate and is generally excluded from the theory of yield surface plasticity. Experimental observations show that the shape, size, and orientation of the yield surface can be very complex during a multiaxial loading history. The discussion of hardening begins with the definitions of the plastic and tangent moduli for uniaxial data.

5.3.1 Definition of the Plastic and Tangent Moduli

Experimentally, the tensile or cyclic stress strain response is the basic information used in most multiaxial hardening and plastic flow equations. There are only two independent material constants for isotropic elasticity. Similarly, for isotropic plasticity of polycrystalline metals there are two scalar material functions. Since there is almost no plastic volume change, only one independent scalar material function is required to model three-dimensional plastic deformations. Thus the challenge of three-dimensional modeling is to use the information from the uniaxial monotonic or cyclic tests to predict the three-dimensional plastic response of metals.

The slope of the stress–total strain curve at any point is the tangent modulus, E_t, as shown in Figure 5.3.1. The elastic modulus, E, is defined as the slope of the stress–elastic strain curve and the slope of the stress–plastic strain curve is the plastic modulus, E_p. The stress increment, $d\sigma$, is related to the total strain increment, $d\varepsilon$, the elastic strain increment, $d\varepsilon^e$, and the plastic strain increment, $d\varepsilon^p$, by

$$d\sigma = E_t\, d\varepsilon = E\, d\varepsilon^e = E_p\, d\varepsilon^p \tag{5.3.1}$$

Since $d\varepsilon = d\varepsilon^e + d\varepsilon^p$, equation 5.3.1 can be rearranged to give the relationship between the elastic, tangent, and plastic moduli as

$$\frac{1}{E_t} = \frac{1}{E} + \frac{1}{E_p} \tag{5.3.2}$$

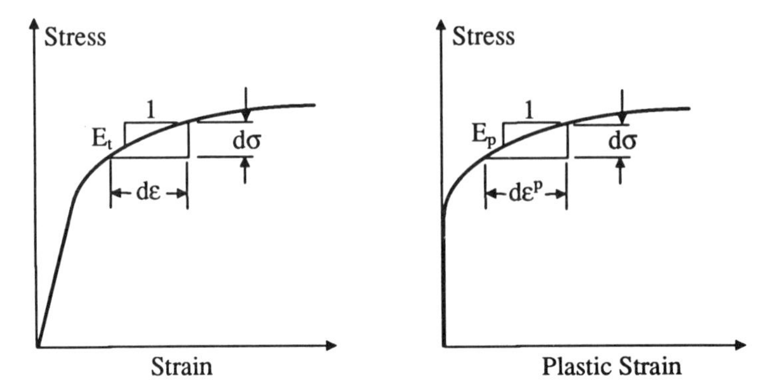

Figure 5.3.1 Schematic diagram showing the definitions of the tangent and plastic moduli.

In uniaxial response, the material is a hardening material when the tangent modulus, E_t, is greater than zero, it is elastic–perfectly plastic if the tangent modulus is zero, and it is a softening material if the tangent modulus is less than zero.

The yield stress and plastic modulus for isotropic hardening materials is generally assumed to be a function of the accumulated plastic work or the integrated plastic strain. Thus two definitions of a hardening parameter, κ, are introduced. The hardening parameter for a plastic work-hardening material is defined as

$$\kappa = W_p = \int \sigma_{ij}\, d\varepsilon_{ij}^p \tag{5.3.3}$$

The integral is evaluated along the stress–strain path for the loading history of the material. For a strain-hardening material the hardening parameter may be expressed as the integral of the effective plastic strain increments, $d\varepsilon_e^p$, or in terms of the components of the plastic strain increment tensor, $d\varepsilon_{ij}^p$:

$$\kappa = \varepsilon_e^p = \int d\varepsilon_e^p = \int \sqrt{\tfrac{2}{3}\, d\varepsilon_{ij}^p\, d\varepsilon_{ij}^p} \tag{5.3.4}$$

The strain-hardening model is used most frequently for metals. The axial plastic strain increment in a tensile test can easily be interpreted as an effective plastic strain increment.

5.3.2 Isotropic Hardening

The isotropic hardening model is based on the assumption that the yield surface expands uniformly in stress space as yielding occurs. The surface does not translate or change shape. The isotropic hardening assumption for an elastic-linear hardening material is illustrated in Figure 5.3.2*a*. Initially, the tensile and compressive yield stresses at points *A* and *B* are equal in magnitude. Once the stress has exceeded yield in tension, the yield stress increases uniformly in both tension and compression. This is illustrated by the loading path from *A* to *C*, followed by an unload to zero stress and a compressive loading to the new compressive yield point *D*. The material remains isotropic after yielding and the new tensile and compressive yield stresses are equal in magnitude throughout the deformation history.

The three-dimensional yield surface size is characterized by the yield stress, which is a function of the scalar isotropic hardening parameter κ. Either the plastic work for work-hardening materials, (equation 5.3.3) or the effective plastic strain increment for strain-hardening materials (equation 5.3.4) can be used. The general form of the isotropic hardening rule can be written as

$$f(\sigma_{ij}) = k(\kappa) \tag{5.3.5}$$

where $f(\sigma_{ij})$ is the von Mises, Tresca, or some other function that defines the shape and orientation of the yield surface. The parameter $k(\kappa)$ defines the size of the yield surface as a function of the deformation history. For example, if the uniaxial tensile response of a material is characterized by the Ramberg–Osgood model in engineering variables (equation 2.3.7), the uniaxial plastic strain is replaced by the effective accumulated plastic strain, κ, so that

$$k(\kappa) = K(\kappa)^n \equiv \sigma_{y\ \text{current}} \tag{5.3.6}$$

which is the current value of the yield stress in isotropic hardening. The Ramberg–Osgood model can be replaced by any other model or the tensile

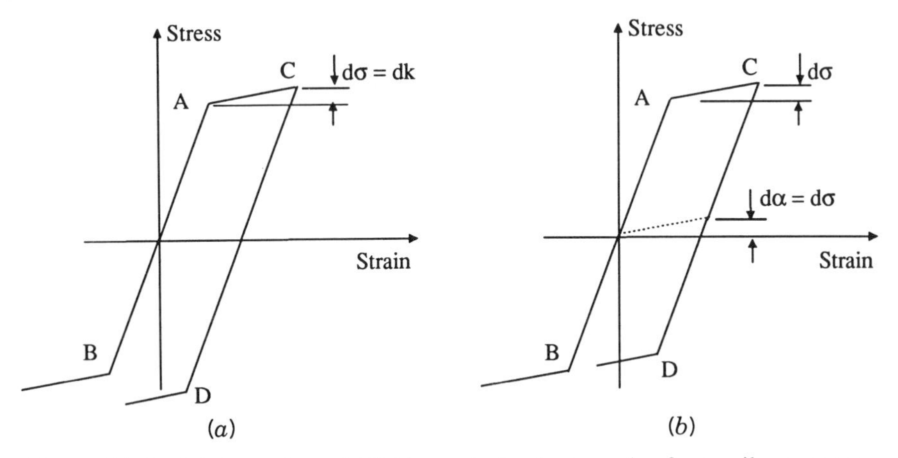

Figure 5.3.2 (*a*) Isotropic and (*b*) kinematic hardening rules for tensile response.

curve to determine the current yield stress. Figure 5.3.3 shows the expansion of the yield surface in a multiaxial stress space for the von Mises and Tresca yield surfaces.

When performing numerical calculations with the isotropic hardening model the parameter κ is used as a state variable (a variable that defines the current amount of hardening due to the loading history). It is an effective scalar measure of the dislocation network and net hardening that has been

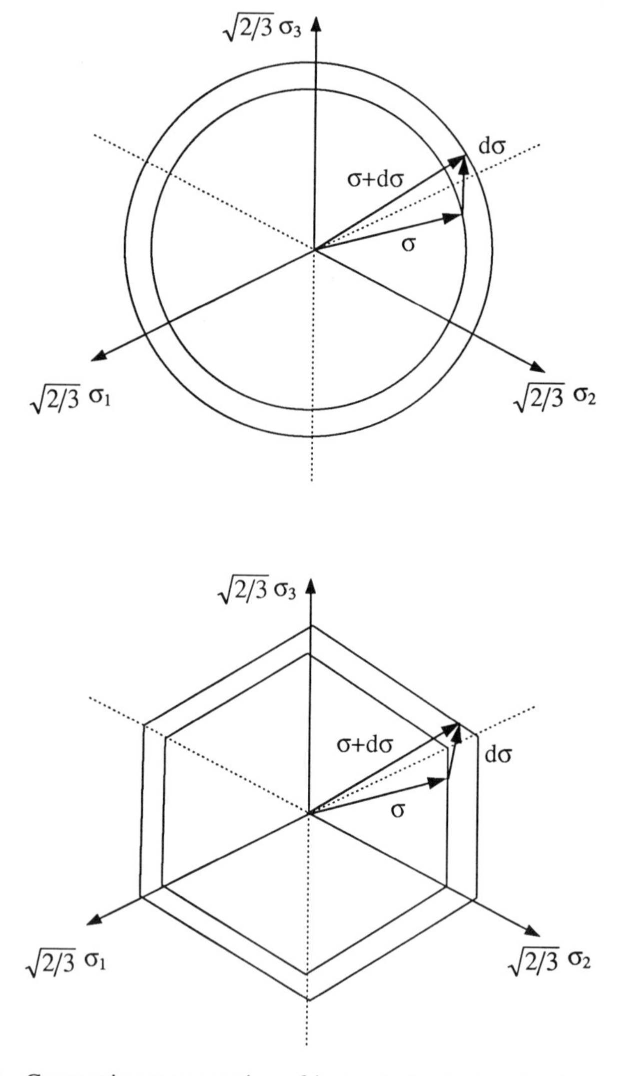

Figure 5.3.3 Geometric representation of isotropic hardening for the von Mises and Tresca yield surfaces in the pi plane.

developed. This type of interpretation is extremely important for numerical simulations that involve changes in temperature or strain rate. For example, Figure 5.3.4 shows the tensile response of a metal at two different temperatures. Consider a material that is loaded in tension at temperature T_1, then the temperature is increased instantaneously a small amount to T_2 for the subsequent loading. The tensile response immediately after the temperature change to T_2 is determined by assuming that the strain-hardened state and associated dislocation structure is not changed by the instantaneous temperature change. Thus the material states at T_1 and T_2 are assumed to be identical. This is correct if the deformation mechanisms at temperatures T_1 and T_2 are the same and if recovery (or annealing) does not occur during the temperature change.

The isotropic hardening model is relatively simple, but unfortunately not very realistic for cyclic loading of metals. The isotropic hardening model may be quite acceptable for the special case of no cyclic loading.

5.3.3 Kinematic Hardening

Kinematic hardening rules were developed to model the Bauschinger effect described in Section 2.5. The model is based on the assumption that the yield surface translates in stress space but does not change size, shape, or orientation as yielding occurs. The kinematic hardening assumption for a uniaxial loading history is illustrated in Figure 5.3.2*b*. Since the material is initially isotropic, the initial tensile and compressive yield stresses are equal in magnitude. The initial yield in tension is at point *A* and the initial yield in compression is at point *B*. On loading past the initial yield point in tension the magnitude of the yield stress in compression is assumed to decrease so that the elastic stress range from the tensile to the compressive yield remains unchanged. This is illustrated by the loading path from point *A* to *C* followed

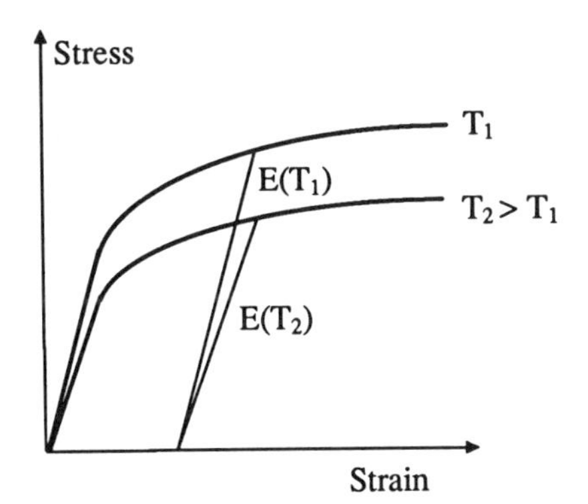

Figure 5.3.4 Use of plastic strain as a state variable.

by an unload to zero stress and a compressive loading to the new compressive yield point *D*. The material is no longer isotropic after yielding since the tensile and compressive yield stresses are different.

The motion of the yield surface in a multiaxial stress space is shown in Figure 5.3.5. The position of the yield surface center in stress space is given by the back stress tensor, α_{ij}. The back stress is similar mathematically to the back stress discussed in Section 1.3.1 resulting from dislocation pileups. Initially zero, the back stress is a function of the deformation history. The difference $\sigma_{ij} - \alpha_{ij}$ is sometimes called the overstress, reduced stress, or shift stress. The general expression for a yield function with kinematic hardening may be written as

$$f(\sigma_{ij} - \alpha_{ij}) = k \tag{5.3.7}$$

The parameter k is a simple function of the initial yield stress. It may be a function of temperature but is unaffected by the kinematic hardening. As an example, the von Mises yield function with kinematic hardening after yielding will be

$$f = \tfrac{1}{2}(S_{ij} - \alpha_{ij})(S_{ij} - \alpha_{ij}) = k^2 \tag{5.3.8}$$

The associated flow equation for a kinematic hardening von Mises material is obtained by substituting equation 5.3.8 for the yield function into equation 5.2.11, that is,

$$d\varepsilon_{ij}^p = d\lambda(S_{ij} - \alpha_{ij}) \tag{5.3.9}$$

The evolution of the back stress, α_{ij}, depends on the plastic strain or plastic work history. The earliest model for the evolution of the back stress is due to Prager (1956), and was subsequently modified by Ziegler (1959). The Pra-

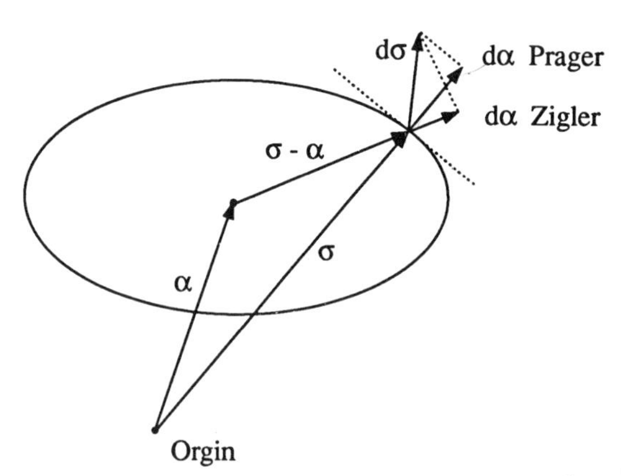

Figure 5.3.5 Definition of Prager and Zigler kinematic hardening rules.

ger hardening assumption is that during yielding the back stress increment $d\boldsymbol{\alpha}$ is equal to the component of $d\boldsymbol{\sigma}$ in the direction normal to the yield surface (Figure 5.3.5). Since the plastic strain increment is also normal to the yield surface, $\mathbf{n}$, the increment $d\boldsymbol{\alpha}$ can be written as

$$d\boldsymbol{\alpha} = [\mathbf{n} \cdot d\boldsymbol{\sigma}]\mathbf{n} = \left[d\sigma \cdot \frac{d\boldsymbol{\varepsilon}^p}{|d\boldsymbol{\varepsilon}^p|} \right] \frac{d\boldsymbol{\varepsilon}^p}{|d\boldsymbol{\varepsilon}^p|} \tag{5.3.10}$$

The Prager definition of $d\alpha$ can also be expressed as a function of the plastic modulus. If $\boldsymbol{\sigma}$, $\boldsymbol{\alpha}$, and $\boldsymbol{\varepsilon}^p$ are defined as the uniaxial stress, back stress, and plastic strain respectively, then

$$\tfrac{3}{2}\, d\boldsymbol{\alpha} = d\boldsymbol{\sigma} = E_p d\boldsymbol{\varepsilon}^p \tag{5.3.11}$$

The $\frac{3}{2}$ factor arises because α_{ij} must be a deviatoric tensor, so that $d\varepsilon_{ii}^p = 0$ in equation 5.3.9. Replacing the uniaxial plastic strain increment by the plastic strain increment vector, it follows that the Prager hardening rule is

$$d\boldsymbol{\alpha} = \tfrac{2}{3}\, E_p\, d\boldsymbol{\varepsilon}^p \tag{5.3.12}$$

The plastic strain, deviatoric stress, and back-stress tensors in uniaxial loading in equation 5.3.9 are defined as

$$[d\varepsilon_{ij}^p] = \begin{bmatrix} d\varepsilon^p & 0 & 0 \\ 0 & -\frac{1}{2}\, d\varepsilon^p & 0 \\ 0 & 0 & -\frac{1}{2}\, d\varepsilon^p \end{bmatrix} \quad \text{and}$$

$$[S_{ij} - \alpha_{ij}] = \begin{bmatrix} \frac{2}{3}\,\sigma - \alpha & 0 & 0 \\ 0 & -\frac{1}{3}\,\sigma + \frac{1}{2}\,\alpha & 0 \\ 0 & 0 & -\frac{1}{3}\,\sigma + \frac{1}{2}\,\alpha \end{bmatrix} \tag{5.3.13}$$

to satisfy that requirement of no plastic volume change.

The Prager hardening model is not always defined. Yield surfaces such as the Tresca hexagon do not have defined normal vectors at the corners, as shown in Figure 5.3.6. Furthermore, bodies that experience yielding in a stress subspace using the Prager model may produce a translation of the yield surface that is not in the stress subspace. For example, consider a body loaded in plane stress. The out-of-plane normal stress is zero but the out-of-plane normal plastic strain increment is not zero. The out-of-plane plastic strain increment produces an out-of-plane increment $d\boldsymbol{\alpha}$ that is out of the plane stress subspace. These objections are avoided by Ziegler's modification to the Prager hardening model.

The Ziegler hardening model is based on the assumption that the back-stress increment $d\boldsymbol{\alpha}$ is in the direction of the shift stress, $\sigma_{ij} - \alpha_{ij}$. Geometrically, this is in the direction from the current yield surface center to the

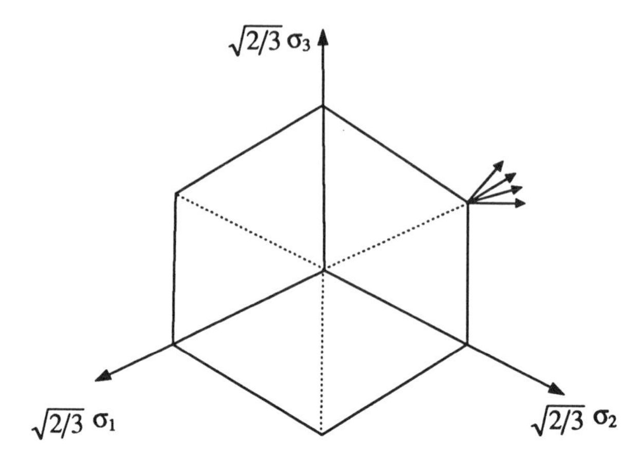

Figure 5.3.6 Normal vectors associated with the Tresca yield criterion are not defined everywhere.

current stress state (Figure 5.3.5). The expression for the back-stress increment with the Ziegler model is given by

$$d\boldsymbol{\alpha} = \mu(\boldsymbol{\sigma} - \boldsymbol{\alpha}) \tag{5.3.14}$$

where μ is a material parameter that must be determined from data.

5.3.4 Mixed Hardening

A hardening model that includes both translation and expansion of the yield surface was proposed by Hodge (1957). This is derived by combining the isotropic yield function (equation 5.3.5) and the kinematic yield function (equation, 5.3.7):

$$f(\sigma_{ij} - \alpha_{ij}) = k(\kappa) \tag{5.3.15}$$

Hodge's mixed hardening model is based on the assumption that the plastic strain increment may be linearly decomposed into components that produce kinematic hardening, $d\varepsilon_{ij}^{k}$, and isotropic hardening, $d\varepsilon_{ij}^{i}$.

$$d\varepsilon_{ij}^{p} = d\varepsilon_{ij}^{k} + d\varepsilon_{ij}^{i} \tag{5.3.16}$$

The ratio of isotropic to kinematic hardening is defined by a mixed hardening parameter, M, which must be obtained from experimental observations. The mixed hardening parameter may have values from 0 to 1. For pure kinematic hardening, $M = 0$, and for pure isotropic hardening $M = 1$. The part of the plastic strain increment that produces isotropic hardening is given by

$$d\varepsilon_{ij}^{i} = M\, d\varepsilon_{ij}^{p} \tag{5.3.17}$$

The part of the plastic strain increment that produces kinematic hardening is

$$d\varepsilon_{ij}^{k} = (1 - M)\, d\varepsilon_{ij}^{p} \tag{5.3.18}$$

Either a Prager or a Ziegler model may be used for kinematic hardening. The isotropic hardening parameter for strain hardening with the mixed hardening model is

$$\kappa = \int \sqrt{\tfrac{2}{3}\, d\varepsilon_{ij}^{i}\, d\varepsilon_{ij}^{i}} = \int \sqrt{\tfrac{2}{3} M^2\, d\varepsilon_{ij}^{p}\, d\varepsilon_{ij}^{p}} = M \int d\varepsilon_{e}^{p} = M\varepsilon_{e}^{p} \tag{5.3.19}$$

and for work-hardening material the isotropic hardening parameter is

$$\kappa = \int \sigma_{ij}\, d\varepsilon_{ij}^{i} = M \int \sigma_{ij}\, d\varepsilon_{ij}^{p} = MW_p \tag{5.3.20}$$

Using the Prager hardening model, the back-stress increment is determined from equation 5.3.12 as

$$d\alpha_{ij} = \tfrac{2}{3} E_p\, d\varepsilon_{ij}^{k} = \tfrac{2}{3} E_p (1 - M)\, d\varepsilon_{ij}^{p} \tag{5.3.21}$$

A similar representation can be developed for the Ziegler model.

5.3.5 Experimental Observations of Subsequent Yield Surfaces

As seen in Figure 5.1.10, the yield stress for a material is defined as the stress level at which a prescribed (but somewhat arbitrary) amount of plastic strain is produced. The plastic strain used to define the onset of yield is called the proof strain. As a result, the yield surface size, shape, and orientation depends on the proof strain used to define the yield point. The yield surface with the smallest proof strain is closest to the true yield surface that defines the stress required to initiate significant slip. Many early experiments, Lode (1926) and Taylor and Quinney (1931), in yield surface plasticity did not recognize the sensitivity of the yield surface to proof strain. However, more recent results show some different trends.

Figure 5.3.7 shows the effect of the proof strain at yield on the size and shape of the yield surface. The tension torsion yield surfaces were obtained after a 1% previous strain in tension. Figure 5.3.8 shows the initial yield surface and the yield surface after two different prestress loading conditions at four temperatures. The preloads were applied at room temperature. Phillips and Tang (1972) also reported a number of yield surfaces showing results similar to Figure 5.3.7. These results tend to indicate that the size, shape, and orientation of the yield surface can change dramatically during a load history, and the simple isotropic, kinematic, and mixed hardening rules presented in

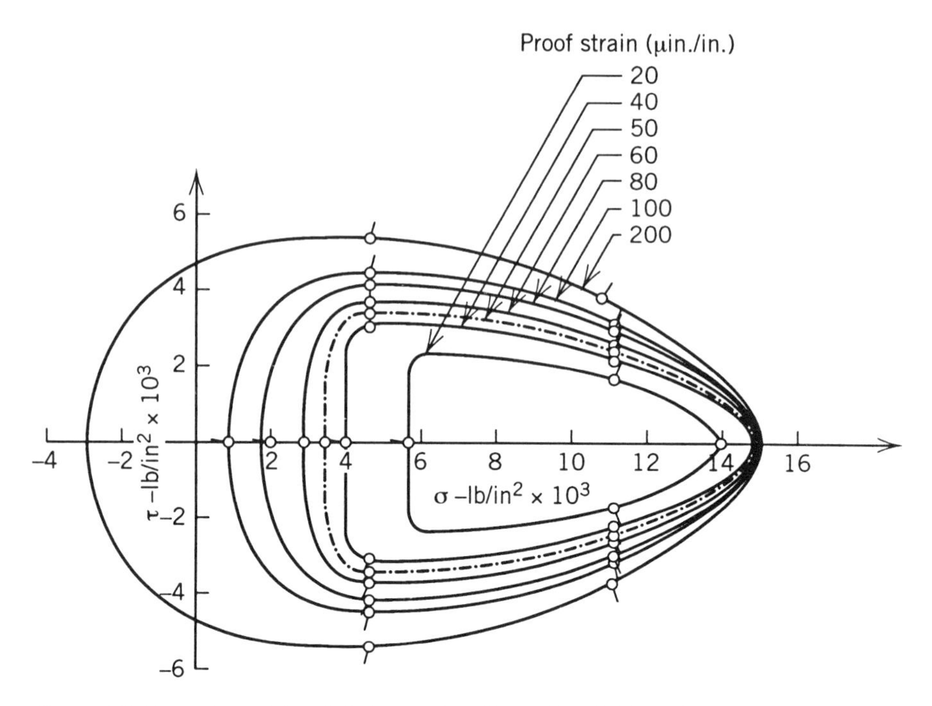

Figure 5.3.7 Effect of the definition of yield on the yield surface for an aluminum alloy following 1% tensile prestrain. (From Williams and Svensson, 1970; reprinted by permission of the Council of the Institution of Mechanical Engineers from the *Journal of Strain Analysis*.)

Sections 5.3.2 to 5.3.4 are to be viewed as rough approximations. However, the von Mises model for initial yield is generally an excellent representation of the actual response.

Next, it is useful to consider experimental evidence concerning the directions of the deviatoric stress and plastic strain increment vectors. However, to appreciate the data fully, it is helpful to explain the difference between proportional and non proportional multiaxial loading. Under strain control testing, proportional loading is defined by

$$\varepsilon_{ij} = g(t)E_{ij} \tag{5.3.22}$$

where $g(t)$ is a scalar function of time and E_{ij} is a constant tensor that is independent of time. The ratios of the strain tensor components are constant throughout the loading history. In structures with nonproportional loading the components of the strain tensor vary independently with time. An example of strain control that can produce proportional or nonproportional tension torsion loading is given by

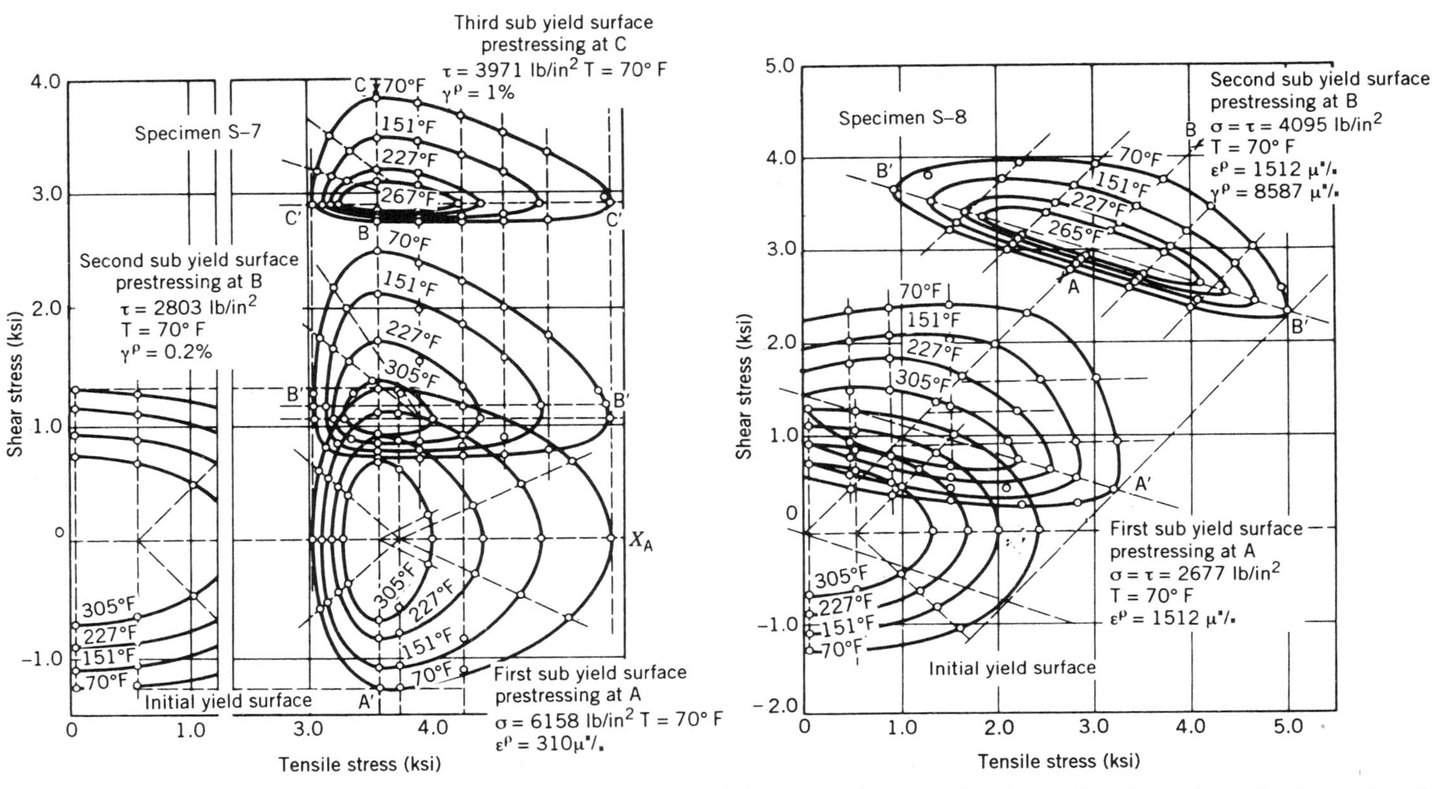

Figure 5.3.8 Effect of prestrain and temperature on two sequential yield surfaces of aluminum. (Reprinted from the International Journal of Solids and Structures, Phillips and Tang, "The Effect of Loading Path on the Yield Surface at High Temperature," pg. 463, 1972, with the kind permission from Elsevier Science Ltd.)

$$\varepsilon_{11} = \varepsilon_a \sin \tilde{\omega} \qquad \text{and} \qquad \varepsilon_{12} = \gamma_a \sin(\tilde{\omega} t + \beta) \tag{5.3.23}$$

The parameter $\tilde{\omega}$ is the cyclic frequency, β is a phase shift, and ε_a and γ_a are the normal and shear strain amplitudes, respectively. If $\beta = 0$, the test is proportional, and if $\beta = \pi/2$, the test is a 90° out-of-phase nonproportional test. The amount of nonproportionality can be adjusted by varying the phase-shift angle. Proportional and nonproportional tests correspond to much different physical conditions. In proportional loading the axes of the maximum shear stress are constant throughout the load history and slip occurs on a fixed set of slip planes. A tensile test is a test with a proportional load history. In nonproportional loading the active slip planes change during the loading history. Nonproportional deformation is much more complicated than proportional deformation because there are more dislocation intersections. The load paths that produced the yield surface results in Figures 5.3.7 and 5.3.8 are nonproportional load paths.

Tension torsion tests were conducted by Liu and Greenstreet (1976) on type 304 stainless steel on five different specimens. Each specimen was loaded in a tension torsion stress space along five equally spaced radial loading paths in the first and third quadrants, as shown in Figure 5.3.9. The initial yield stresses were found to agree with the von Mises yield criterion. The plastic strain trajectories were determined from the experimental data and found to be parallel to the stress paths and normal to the initial yield surface. Each test in this program was a proportional test.

Examples of the response of type 304 stainless steel to nonproportional load histories were given by McDowell (1983). The strain control tension torsion histories for two tests with experimentally determined yield stress results are shown in Figure 5.3.10. The short vectors are in the direction of the deviatoric stress vectors. The longer vectors are parallel to the inelastic strain rate vectors at the same point. It is clear from these experiments that the angle between the deviatoric stress and inelastic strain rate vectors change continuously throughout the cycle. Thus it appears that the plastic strain increment vector is not normal to the yield surface and is not parallel to the deviatoric stress vector in nonproportional loading. It does appear that these conditions are satisfied in proportional loading.

5.4 LOADING CRITERIA

For a hardening material loading occurs when the current stress state is on the yield surface and an additional stress increment, $d\sigma_{ij}$, produces plastic straining. During loading the yield surface will change such that the stress state will remain on the yield surface. Unloading occurs when the current stress state is on the yield surface and a stress increment moves the stress inside the yield surface and produces no plastic straining. Neutral loading occurs when the stress increment is tangent to the yield surface.

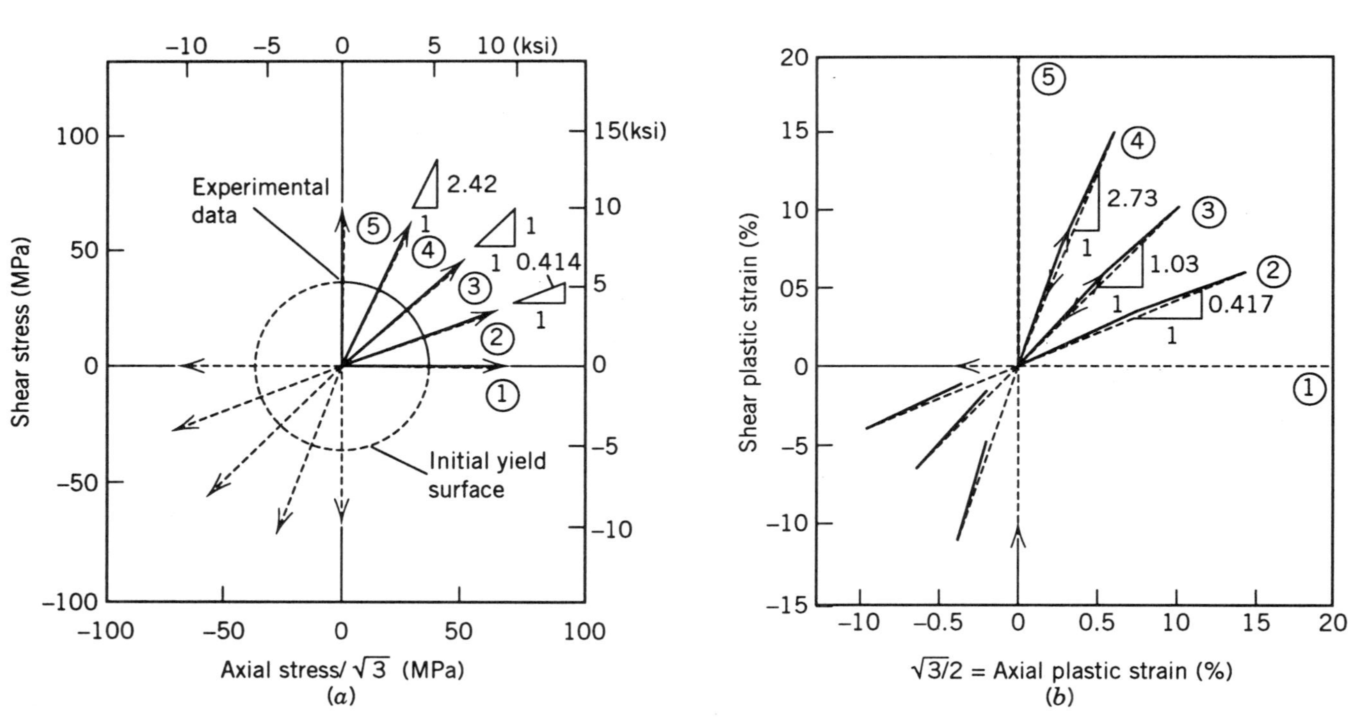

Figure 5.3.9 Radial loading paths and resulting plastic strain directions for the initial yield of type 304 stainless steel at room temperature. (After Liu and Greenstreet, Constitutive Equations in Viscoplasticity, pg. 46 and 47, 1976. Copyright © The American Society of Mechanical Engineers, 345 East 47th St., New York, NY 10017. Used with permission.)

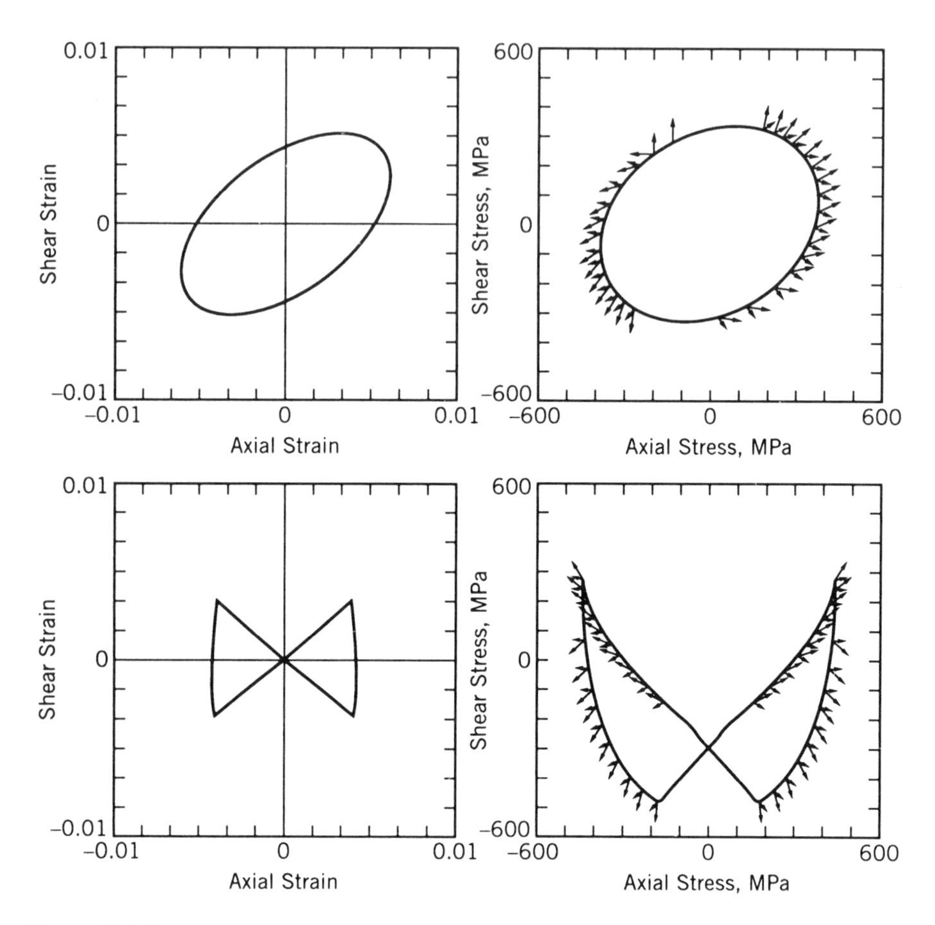

Figure 5.3.10 Variation in the direction of the inelastic strain rate (long vectors) and the deviatoric stress (short vectors) in a nonproportional test on type 304 stainless steel at room temperature. The input strain histories (*a*) and (*b*) correspond to (*c*) and (*d*), respectively. (After McDowell, Journal of Engineering Materials and Technology, V107, pg. 309, 1985. Copyright © The American Society of Mechanical Engineers, 345 East 47th St., New York, NY 10017. Used with permission.)

The yield surface can be used to determine if plastic flow will result from the application of an increment of stress, $d\sigma_{ij}$. A trail value of the yield function can be computed by applying an increment in stress, $d\sigma_{ij}$, while holding the back stress, α_{ij}, and yield surface size, k, constant. Plastic flow will result from the stress increment if

$$f((\sigma_{ij} + d\sigma_{ij}) - \alpha_{ij}) > k \tag{5.4.1}$$

However, it is not possible to have stress states outside the yield surface. Plastic straining will result and the yield surface will change due to hardening as the material is deformed. Equation 5.4.1 is often useful, but a more complete analysis of the loading criteria is required to produce a consistency condition between the stress and strain increments during loading.

Loading occurs and plastic strain will result when the stress is at a point on the yield surface and an additional stress increment is directed outward from the yield surface. Since the normal to the yield surface in equation 5.1.1*b* may be written mathematically as $\partial F/\partial \sigma_{ij}$, the loading criterion is given by

$$F(\sigma_{ij}, k) = 0 \qquad \text{and} \qquad \frac{\partial F}{\partial \sigma_{ij}}\, d\sigma_{ij} > 0 \tag{5.4.2}$$

Plastic straining will occur when equation 5.4.2 is satisfied.

Unloading occurs when the stress state is on the yield surface and the stress increment is directed inward from the yield surface,

$$F(\sigma_{ij}, k) = 0 \qquad \text{and} \qquad \frac{\partial F}{\partial \sigma_{ij}}\, d\sigma_{ij} < 0 \tag{5.4.3}$$

Elastic straining will occur when equation 5.4.3 is satisfied.

Neutral loading occurs when the stress state is on the yield surface and the stress increment is tangent to the yield surface,

$$F(\sigma_{ij}, k) = 0 \qquad \text{and} \qquad \frac{\partial F}{\partial \sigma_{ij}}\, d\sigma_{ij} = 0 \tag{5.4.4}$$

No plastic straining occurs during neutral loading conditions.

For an elastic–perfectly plastic material the yield surface does not change and it is impossible to increase the stress beyond the initial yield point. In this case, for loading, any stress increment must be tangent to the yield surface, so the loading criterion is defined by equation 5.4.4. The unloading criterion is still given by equation 5.4.3 and neutral loading is not defined.

During loading the stress state must remain on the yield surface. This condition is a consistency condition that must be satisfied. It can be stated mathematically by using the yield surface as given by the second representation in equation 5.1.1 and assuming that $k(\kappa)$ is a function of the plastic strain. Since the stress remains on the yield surface, $dF = 0$, or

$$dF = \frac{\partial F}{\partial \sigma_{ij}}\, d\sigma_{ij} + \frac{\partial F}{\partial \varepsilon_{ij}^{p}}\, d\varepsilon_{ij}^{p} = 0 \tag{5.4.5}$$

Note that the consistency condition for the case of combined isotropic-kinematic hardening may be expanded to

$$dF = \frac{\partial F}{\partial \sigma_{ij}}\, d\sigma_{ij} + \frac{\partial F}{\partial \alpha_{ij}}\, d\alpha_{ij} + \frac{\partial F}{\partial \kappa}\, d\kappa = 0 \tag{5.4.6}$$

5.5 SPECIFIC REPRESENTATIONS FOR THE FLOW LAW

There are many specific forms for the plastic flow law in the literature. This section contains a description of three basic methods to develop a specific plastic flow rule. In all cases the tensile response data or a model for the tensile response is used. The cyclic stress–strain curve can be substituted for the tensile response for cyclic simulations. In most cases the actual simulations for real structural applications are numerical.

One method to establish a representation for equation 5.2.22 is to replace the effective strain increment by representative stress and plastic strain increments from a tensile curve. That is, if the uniaxial stress plastic strain is defined by

$$\varepsilon^p = g(\sigma) \tag{5.5.1}$$

where $g(\sigma)$ is a material function, the effective stress–plastic strain representation and plastic strain increments are given by

$$\varepsilon_e^p = g(\sigma_e) \qquad \text{and} \qquad d\varepsilon_e^p = \frac{dg}{d\sigma_e}\, d\sigma_e \tag{5.5.2}$$

Recall that the effective stress (equation 5.2.20) and effective plastic strain increment (equation 5.2.21) were defined to reduce to the uniaxial stress and plastic strain increment values. Applying this result to the Ramberg–Osgood equation gives

$$d\varepsilon_e^p = \frac{n}{K}\left(\frac{\sigma_e}{K}\right)^{(1-n)/n} d\sigma_e \tag{5.5.3}$$

Substitution of equation 5.5.3 into equation 5.2.22 gives a specific representation for $d\varepsilon_{ij}^p$. The values of the strength coefficient and strain-hardening exponent are available for many materials in the literature. This approach can be used with any of the tensile response models.

Another approach to determine $d\varepsilon_e^p$ is to introduce the plastic modulus into the representation. This is accomplished by determining a representation for $d\lambda$ in equation 5.2.17. Assume that $d\lambda$ is a function of the second invariant of the deviatoric stress, that is,

$$d\lambda = g(J_2)\, dJ_2 \tag{5.5.4}$$

where $g(J_2)$ is another material function. Prandtl–Reuss equation (5.2.17) can be written for a tensile test as

$$d\varepsilon^p = \frac{2}{3}\sigma\, d\lambda = \frac{1}{E_p}\, d\sigma \quad \text{where} \quad dJ_2 = \frac{2}{3}\sigma\, d\sigma \tag{5.5.5}$$

Combining equations 5.5.4 and 5.5.5 gives

$$g(J_2) = \frac{3}{4J_2E_p} \tag{5.5.6}$$

A final representation can be obtained by substituting the result for $g(J_2)$ into the Prandtl–Reuss equation:

$$d\varepsilon_{ij}^p = \frac{3S_{ij}}{4J_2E_p}\, dJ_2 \tag{5.5.7}$$

where the plastic tangent modulus is a function of the strain-hardening or work-hardening parameter.

Let us next develop a representation for $d\lambda$ in the Prandtl–Reuss equation with kinematic hardening (equation 5.3.9). The first step is to note that the differential of the associated yield surface (equation 5.3.8) becomes

$$(S_{ij} - \alpha_{ij})(dS_{ij} - d\alpha_{ij}) = 0 \tag{5.5.8}$$

since the parameter k (the initial yield stress) is a constant. Using the Prager hardening rule, equation 5.3.11, the equation above becomes

$$(S_{ij} - \alpha_{ij})\, dS_{ij} = (S_{ij} - \alpha_{ij})\, d\alpha_{ij} = \tfrac{2}{3}E_p(S_{ij} - \alpha_{ij})\, d\varepsilon_{ij}^p \tag{5.5.9}$$

Using the flow rule (equation 5.3.9) and the definition of the yield surface (equation 5.3.8) gives

$$(S_{ij} - \alpha_{ij})\, dS_{ij} = \tfrac{2}{3}E_p(S_{ij} - \alpha_{ij})(S_{ij} - \alpha_{ij})\, d\lambda = \tfrac{4}{3}E_pk^2\, d\lambda \tag{5.5.10}$$

which can easily be solved for $d\lambda$. Substituting this value into equation 5.3.9 gives

$$d\varepsilon_{ij}^p = \frac{(S_{kl} - \alpha_{kl})\, dS_{kl}}{\tfrac{4}{3}E_pk^2}(S_{ij} - \alpha_{ij}) \tag{5.5.11}$$

Equation 5.5.11 with the material parameters in Appendix 5.1 was used to calculate the fatigue response of IN 100 at 1300°F as shown in Figure 5.5.1. The response at 1300°F is strain-rate sensitive and the material parameters were determined for tensile tests at three strain rates. The fatigue response, which is at a different strain rate, was determined by interpolation of the tensile parameters with respect to strain rate. The fatigue test was run under

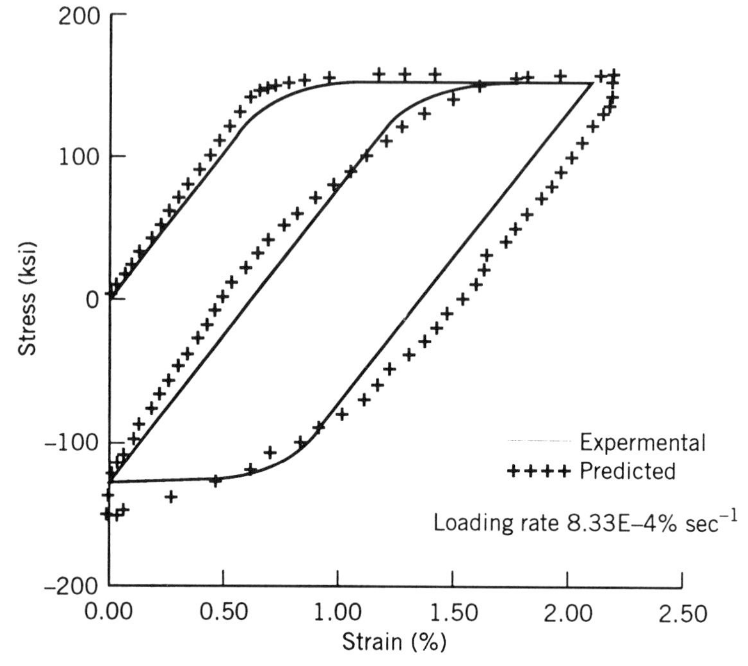

Figure 5.5.1 Comparison of experimental response and equation 5.5.11 using constants determined from tensile test on IN100 at 1350°F at three different strain rates. The IN100 data are given in Appendix 2.3. (Courtesy of Binyu Tian, 1992.)

strain control with a 25-sec hold in compression, during which there was stress relaxation that was not captured by the model. This was expected since the model does not incorporate time or rate effects. This exercise gives some idea of the accuracy of yield surface models for a strain-rate-sensitive uniaxial prediction.

Example Problem 5.1: An elastic–plastic material is assumed to have the tensile response curve shown in Figure 5.5.2, which can be modeled by

$$\varepsilon = \frac{\sigma}{10^7} + \frac{\sigma^3}{10^{15}}$$

The first term characterizes the elastic strain and the second, the plastic strain. Units of stress are psi. Assume that the material obeys the von Mises yield criterion and can be modeled with isotropic hardening. Determine the plastic strain response to the tension torsion cycle shown in Figure 5.5.2 using equation 5.5.7. Sketch the yield surface at the end of each load path.

SOLUTION: Let us begin by introducing a simplified notation

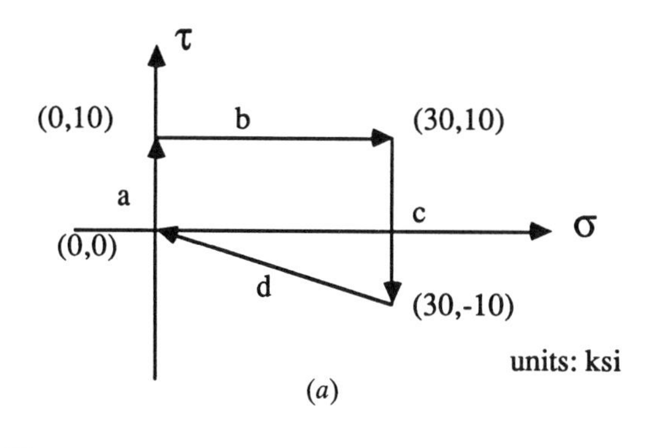

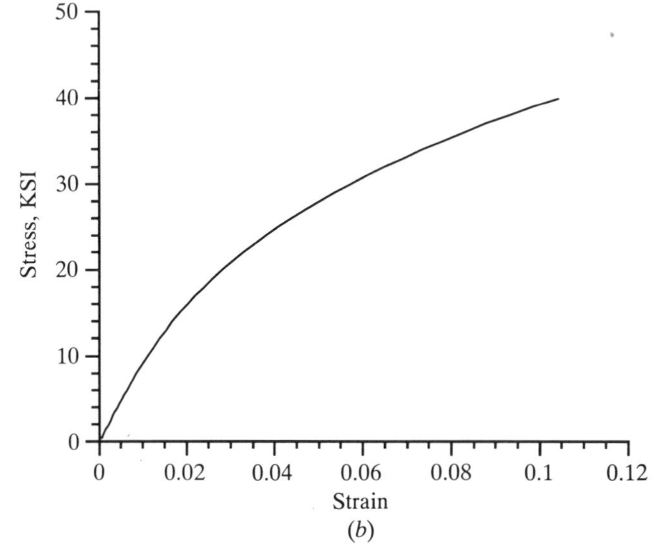

Figure 5.5.2 Definition of the load cycle and tensile model used in Example Problem 5.1.

$$\sigma_{ij} = 0 \quad \text{except} \quad \sigma_{11} = \sigma \quad \text{and} \quad \sigma_{12} = \tau$$

The deviatoric stress and its differential are

$$\begin{aligned} J_2 &= \tfrac{1}{6}[(\sigma_1 - \sigma_2)^2 + (\sigma_2 - \sigma_3)^2 + (\sigma_3 - \sigma_1)^2] + \sigma_{12}^2 + \sigma_{23}^2 + \sigma_{31}^2 \\ &= \tfrac{1}{3}\sigma^2 + \tau^2 \\ dJ_2 &= \tfrac{2}{3}\sigma\, d\sigma + 2\tau\, d\tau \end{aligned}$$

To determine the plastic modulus, differentiate the plastic strain term of the tensile curve and note the relationship between tensile stress and the second invariant of the deviatoric stress in equation 5.2.20:

$$\frac{d\varepsilon^p}{d\sigma} = \frac{3\sigma^2}{10^{15}} = \frac{9J_2}{10^{15}} = \frac{1}{E_p}$$

Substituting the plastic modulus and dJ_2 into equation 5.5.7 gives

$$d\varepsilon_{ij}^p = \frac{27}{4 \times 10^{15}} S_{ij} \left(\frac{2}{3} \sigma \, d\sigma + 2\tau \, d\tau \right)$$

The deviatoric stresses for this problem are

$$S_{11} = \tfrac{2}{3}\sigma, \qquad S_{22} = S_{33} = -\tfrac{1}{3}\sigma, \qquad S_{12} = \tau, \qquad S_{23} = S_{31} = 0$$

The yield surface has to be updated at the end of each load step. Since the stresses are given as part of the problem it is possible to write the loading criteria (equation 5.4.2) as

$$J_2 = \tfrac{1}{3}\sigma^2 + \tau^2 = k \qquad \text{and} \qquad dJ_2 > 0$$

and update k at the end of each load step using the stresses. If the stress were not known, the plastic strain increments would be calculated from equation 5.3.4 and used to calculate the current yield stress.

The equations are now specialized for this particular problem. On path a all the plastic strains are zero except ε_{12}^p. Since the initial yield stress is defined as zero (a special case for this particular model) and $\tau \, d\tau > 0$ on path a, plastic straining will occur. Thus the shear strain at the end of path a is

$$\varepsilon_{12}^p = \frac{27}{4 \times 10^{15}} \int_0^{10^4} \tau(2\tau \, d\tau) = 4.5 \times 10^{-3}$$

The yield stress at the end of path a is

$$k_A = \tfrac{1}{3}\sigma^2 + \tau^2 = (10^4)^2 = 10^8$$

and the intercept of the yield surface with the stress axis, $\tau = 0$, is at 17.3 ksi (Figure 5.5.3*a*). There will be additional plastic strain at the end of path B since $\sigma \, d\sigma > 0$ and the stress at the end of path a is on the yield surface. At the end of path b

$$J_2 = \tfrac{1}{3}\sigma^2 + \tau^2 = \tfrac{1}{3}(3 \times 10^4)^2 + 10^8 \equiv k_B > k_A$$

The normal stress intercept of the yield surface is at 34.6 ksi and the shear stress intercept is at 20 ksi. The axial strain on loading path b is given by

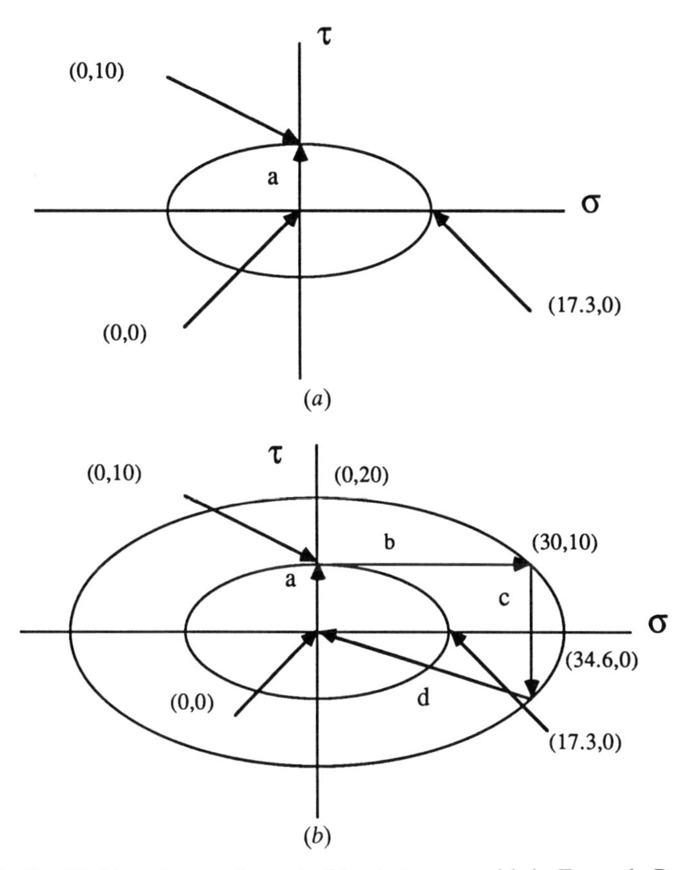

Figure 5.5.3 Yield surface at the end of load steps a and b in Example Problem 5.1.

$$\varepsilon_{11}^{p} = \frac{27}{4 \times 10^{15}} \int_{0}^{3\times 10^{4}} \frac{2}{3}\,\sigma \left(\frac{2}{3}\,\sigma\, d\sigma + 0\right) = 27.0 \times 10^{-3}$$

The transverse plastic strains could be determined from a similar analysis, but since plastic volume must be conserved, it follows that

$$\varepsilon_{22}^{p} = \varepsilon_{33}^{p} = -\tfrac{1}{2}\varepsilon_{11}^{p} = -13.5 \times 10^{-3}$$

There is also a nonzero shear strain that is developed by coupling with the axial strain. This is calculated from

$$\varepsilon_{12}^{p} = 4.5 \times 10^{3} + \frac{27}{4 \times 10^{15}} \int_{0}^{3\times 10^{4}} 10^{4} \left(\frac{2}{3}\,\sigma\, d\sigma\right) = 24.75 \times 10^{-3}$$

From the diagram of the yield surface after loading on path b it can be seen that no plastic strains are developed on paths c and d. The loading criterion is never satisfied because $J_2 < k_B$ except at the end of path c, when $J_2 = k_B$, but $dJ_2 < 0$. The plastic strains could be added to the elastic strains to find the total strain at the end of each load path.

This problem shows the steps required for a plasticity calculation and the coupling that exists in the equations, but it is not a totally typical application. The constitutive equation was picked so that the integrals would be easy to evaluate; consequently, the initial yield stress was zero. In a typical numerical simulation an isotropic hardening parameter (such as plastic work or effective plastic strain) would be computed and used to find the yield stress.

5.6 NUMERICAL TECHNIQUES FOR PLASTICITY

The calculation of a stress or strain history is usually both an incremental and an iterative computational procedure. Nonlinear finite element codes are typically used for the solution of structural problems involving the plasticity of metals. For studying the characteristics of nonlinear material models, stand-alone constitutive codes are often used. These codes are used to solve for the stress or strain history under uniform conditions typically found in test specimens. These simple codes, which are representative of the material subroutines in a finite-element code, are relatively simple to write, economical to run, and useful for the development of more complex finite-element procedures. It is the goal of the current section to present the numerical procedures for stand-alone constitutive codes and provide specifics relevant to plasticity theory (i.e., predicting incremental plastic strains from the plasticity equations). The results of this section are applied to state-variable methods in Section 6.7 and used in Chapter 9 with finite-element methods.

5.6.1 Integration of the Incremental Plasticity Equations

Integration of the incremental equations of plasticity requires the use of finite-length loading steps. For finite-element procedures the strains are computed directly from displacements, and during the load step the strain path is typically assumed to be linear. For a stand-alone constitutive code the stress or strain increments may be input quantities. The most commonly used techniques for integrating the plastic strain increments are the generalized trapezoidal rule, the generalized midpoint rule, or a special case of one of these. The generalized trapezoidal rule for a loading step is

$$\{\Delta\sigma\} = [C](\{\Delta\varepsilon\} - \{\Delta\varepsilon^p\}) \tag{5.6.1}$$

where $[C]$ is the matrix of elastic constants the plastic strain increment going from load point n to $n + 1$ is given by

$$\{\Delta\varepsilon^p\} = \Delta\lambda[(1 - \beta)\{b\}_n + \beta\{b\}_{n+1}] \tag{5.6.2}$$

The plastic multiplier, $\Delta\lambda$, is related to the magnitude of the plastic strain increment for the load step, the flow vector, $\{b\}_n$, gives the direction of plastic flow at the beginning of the increment, and $\{b\}_{n+1}$ gives the direction of the plastic flow at the end of the increment. The flow vector is normal to the plastic potential function and is defined by $\{b\} = \partial g/\partial\{\sigma\}$. For the associated flow rules typically used for metals, the plastic potential function, g, is equal to the yield function, f, as discussed in Section 5.2.2. The parameter β may be given values from 0 to 1. For $\beta = 0$ the generalized trapezoidal rule becomes the forward Euler algorithm and for $\beta = 1$ it reduces to the backward Euler algorithm. Note that if the loading increment is infinitesimal, equation 5.6.2 reduces to equation 5.2.12, and for an associated flow rule, $g = f$, it reduces to equation 5.2.11.

In the beginning of the step the strain increment, $\{\Delta\varepsilon\}$, or stress increment, $\{\Delta\sigma\}$, is generally known or estimated. The plastic multiplier, $\Delta\lambda$, the plastic strain increment, $\{\Delta\varepsilon^p\}$, the yield surface at the end of the increment, f_{n+1}, and the plastic flow vector at the end of the increment, $\{b\}_{n+1}$, are unknown.

For the generalized midpoint rule the plastic strain increment is given by

$$\{\Delta\varepsilon^p\} = \Delta\lambda\{b\}_{n+\beta} \tag{5.6.3}$$

The parameter β in this expression may be given values from 0 to 1. For $\beta = 0$ the flow vector at the beginning of the increment is used and the generalized midpoint rule reduces to the forward Euler algorithm. For $\beta = 1$ the flow vector at the end of increment is used and the generalized midpoint rule reduces to the backward Euler algorithm.

When the forward Euler algorithm is used the solution will "drift" such that the stress state no longer lies on the yield surface. Subincrementing is often used along with a method for correcting the computed stress so that it remains on the yield surface. When loading from a point inside the yield surface using the forward Euler algorithm it is also necessary to divide the stress increment into an elastic part and an elastic–plastic part. Thus it becomes necessary to compute the point at which the stress increment intersects the yield surface.

5.6.2 Backward Euler Algorithm for von Mises Yield Surface and Mixed Hardening

As an example we develop the equations for the backward Euler algorithm with a von Mises yield function, associated flow rule, and a Ziegler–Prager mixed strain-hardening model for a given total strain increment. The von Mises yield function for a kinematic or mixed hardening model is given by equation 5.3.15. The yield stress, σ_y, is a function of the effective plastic strain, ε_e^p, as defined in equation 5.3.19 and a bilinear stress strain curve is used with a plastic modulus, E_p. At the end of the plastic increment the yield function may be written as

$$F_{n+1} = F(\{\sigma\}_{n+1}, \{\alpha\}_{n+1}, (\varepsilon_e^p)_{n+1}) = 0 \tag{5.6.4}$$

The trial state, $F_{(n+1)t}$, at the end of the increment is computed by assuming that the entire strain increment is elastic:

$$F_{(n+1)t} = F(\{\sigma\}_{(n+1)}, \{\alpha\}_n, (\varepsilon_e^p)_n) = 0 \tag{5.6.5}$$

Using the first term of a Taylor series expansion about the trial state produces

$$F_{n+1} = F_{(n+1)t} + \left(\frac{\partial F}{\partial\{\sigma\}}\right)^T_{(n+1)t} \{d\sigma\} + \left(\frac{\partial F}{\partial\{\alpha\}}\right)^T_n \{d\alpha\} + \left(\frac{\partial F}{\partial \varepsilon_e^p}\right)_n d\varepsilon_e^p \tag{5.6.6}$$

Recalling equation 5.1.12 and 5.3.19, equation 5.6.6 may be written

$$f + \{a\}^T\{d\sigma\} - \{a\}^T\{d\alpha\} - \tfrac{4}{9}\sigma_y\overline{\sigma}_e M E_p \, d\lambda = 0 \tag{5.6.7}$$

The subscripts are dropped for simplicity, with the understanding that the derivatives are with respect to successive trial states. The vector $\{\mathbf{a}\}$ is defined by

$$\{\mathbf{a}\} = \frac{\partial F}{\partial\{\sigma\}} = -\frac{\partial F}{\partial\{\alpha\}} \tag{5.6.8}$$

and the last term of equation 5.6.7 is determined by noting that

$$F = J_2 - \frac{\sigma_y^2}{3} = 0 \qquad \text{where} \qquad \sigma_y = \sigma_{yo} + M E_p \varepsilon_e^p$$

and recalling that $d\lambda$ is the magnitude of the plastic strain increment. The effective stress $\overline{\sigma}_e$ in equation 5.6.7 is defined as

$$\overline{\sigma}_e = [\tfrac{3}{2}(S_{ij} - \alpha_{ij})(S_{ij} - \alpha_{ij})]^{1/2} \tag{5.6.9}$$

During the iteration process the total strain increment is held fixed and the stress increment is given by

$$\{d\sigma\} = [C]\{d\varepsilon^E\} = -[C]\{d\varepsilon^p\} = -d\lambda[C]\{a\} \tag{5.6.10}$$

where $[C]$ is the matrix of elastic constants and

$$\{d\alpha\} = d\lambda\{A\} \tag{5.6.11}$$

For Ziegler–Prager mixed hardening the vector {A} is given by

$$\{A\} = E_p(1 - M)(\{\sigma - \alpha\}) \tag{5.6.12}$$

Substituting equations 5.6.10 and 5.6.11 into equation 5.6.7 and solving for the plastic multiplier, $d\lambda$, produces

$$d\lambda = \frac{f}{h} \tag{5.6.13}$$

where h is given by

$$h = \{a\}^T[C]\{a\} + \{a\}^T\{A\} + \tfrac{4}{9}\sigma_y\overline{\sigma}_e ME_p \tag{5.6.14}$$

Once the plastic multiplier, $d\lambda$, is computed, the additional plastic strain increments, back-stress increments, and stress increments may be computed:

$$\{d\varepsilon^p\}_{n+1} = d\lambda\{a\} + \{d\varepsilon^p\}_n \tag{5.6.15}$$

$$\{d\alpha\}_{n+1} = d\lambda\{A\} + \{d\alpha\}_n \tag{5.6.16}$$

$$\{d\sigma\}_{n+1} = -d\lambda[C]\{a\} + \{d\sigma\}_n \tag{5.6.17}$$

After updating the plastic strain increment, back stress increment, and stress increment, a new value for the trial yield function, F, is computed. If the trial yield function is sufficiently close to zero, the iteration is terminated. If convergence is not achieved, return to equation 5.6.6 and proceed with an additional iteration.

The backward Euler algorithm is popular for its effectiveness and simplicity. Details for other algorithms may be found in Owen and Hinton (1980), Crisfield (1991), and Hinton (1992).

B. CLASSICAL CREEP MODELING

Classical creep models have been developed for metals at high temperatures, above about $0.5T_h$, where diffusion and climb are important. The response to loading is rate dependent and the onset of inelastic (creep) deformation occurs at stress levels below the yield stress. Even though there is no yield surface, the theory borrows heavily from the ideas embedded in yield surface plasticity. The creep models presented in Chapter 3 are for uniaxial loading and restricted to constant stress and temperature. Time and strain hardening rules are introduced in this chapter to allow for time-varying stress and temperature histories. The uniaxial creep or creep rate equations are extended to three dimensions by the use of an equivalent creep strain rate and effective stress similar to plasticity. Numerical techniques for integrating the creep rate equations are presented. Finally, the properties of time are discussed for use in constitutive modeling, and it is shown that time is not always an appropriate variable.

5.7 CREEP HARDENING RULES

For practical applications creep does not usually occur at a fixed temperature and stress over the entire life of the part. This being the case, it is not possible to compute the creep strain directly from a uniaxial creep equation like those in Chapter 3. The rules of time hardening and strain hardening were developed to extend the constant stress and temperature creep equations to time-varying stress and temperature histories. The creep strain is determined by integrating the creep strain rate equations for changing stress and temperature conditions. To compute the creep strain rate at a particular instant, the mechanical and thermal loads as well as the material history must be known. Hardening rules specify the state of the material as a function of its loading history.

Computationally, the creep hardening rules serve the same function as the hardening rules in yield surface plasticity theory (i.e., to include loading history effects in the model). In its simplest form a creep hardening rule may be thought of as a method of moving between constant stress and temperature creep curves as the stress and temperature change. Two different hardening rules have been used with classical creep models: time hardening and strain hardening. When the creep rate equations are integrated for conditions of constant stress and temperature, both hardening rules will produce identical results: namely, the uniaxial creep curve for that stress and temperature. Creep equations that model only secondary creep (where the rate is only a function of stress and temperature) also produce results that are the same for both hardening models. For most creep models the uniaxial creep strain, ε^C, is written as a function of the constant stress, σ, constant temperature, T, and time, t, that is

$$\varepsilon^C = \mathfrak{C}(\sigma, T, t) \tag{3.5.1}$$

The creep rate can be determined from equation 5.7.1 as

$$\dot{\varepsilon}^C = \frac{\partial \mathfrak{C}}{\partial \sigma}\dot{\sigma} + \frac{\partial \mathfrak{C}}{\partial T}\dot{T} + \frac{\partial \mathfrak{C}}{\partial t} \tag{5.7.1}$$

The creep rate in the time- and strain-hardening formulations is defined as

$$\dot{\varepsilon}^C = \frac{\partial \mathfrak{C}}{\partial t} \qquad (5.7.2)$$

Equation 5.7.2 neglects the effect of the stress rate and temperature rate on the creep rate. This is one of the shortcomings of classical hardening rules. Theoretically, these rules should be used only for slowly changing stress and temperature histories. The accumulated creep strain is then determined by integrating the creep rate for time-varying stress and temperature histories; that is,

$$\varepsilon^C = \int \dot{\varepsilon}^C \, dt \qquad (5.7.3)$$

For the time-hardening rule the accumulated creep time is used as a material state variable. Creep strain rate is considered to be a function of the accumulated creep time and the stress and temperature. When stress or temperature change instantaneously, the material state (accumulated creep time) does not and the creep strain rate may be computed for the new conditions. This is illustrated graphically in Figure 5.7.1. A material is subjected to stress σ_1 until time t_1 at point A. The stress is then increased to σ_2. The creep rate is then computed from the creep rate equation using stress σ_2 and time t_1. This is equivalent to translating the curve for stress σ_2 at point B to point A as shown in the figure.

The time-hardening rule is very simple to apply for hand calculations since the creep rate equations may be used directly. However, this assumption is particularly poor for situations in which there are large changes of stress or temperature. The use of time as a state variable is also in contradiction of the principle of objectivity discussed in Sections 4.6 and 5.10.

The strain-hardening rule is based on the assumption that the material state is characterized by the accumulated creep strain. For most creep models, the

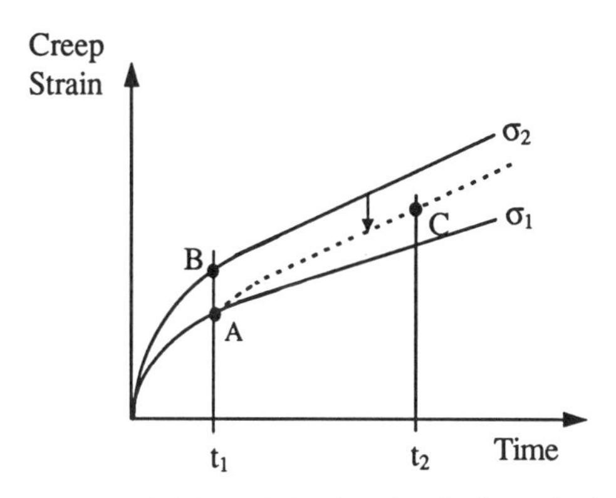

Figure 5.7.1 Definition of the time-hardening rule for creep.

strain-hardening rule requires the calculation of an equivalent time to be used in the creep-rate equation. The equivalent time is computed by solving the uniaxial creep equation for time using the current values of temperature, stress, and accumulated creep strain. Some creep models may be solved analytically for time, while others require an iterative numerical procedure. A few creep models are implicitly strain hardening and the creep rate is expressed as a function of stress, temperature, and the accumulated creep strain

$$\dot{\varepsilon}^C = \dot{\mathfrak{C}}(\sigma, T, \varepsilon^C) \tag{5.7.4}$$

Strain hardening can also be described graphically as shown in Figure 5.7.2. When stress or temperature changes instantaneously, the material state (accumulated creep strain) does not and the creep strain rate may be computed for the new conditions. This is illustrated graphically in Figure 5.7.2. Assume that a material is subjected to stress σ_1 until time t_1 at point A. The stress is then increased to σ_2. An equivalent time is found for the new stress level by moving horizontally along a line of constant creep strain to the creep curve at stress σ_2. The creep rate is then computed from the creep rate equation using stress σ_2 and equivalent time t^E. This is equivalent to translating the curve for stress σ_2 at point B to point A, as shown in Figure 5.7.2.

The strain-hardening rule has generally been accepted as being more accurate than time hardening. However, the calculation of an equivalent time for complicated creep equations may require an iterative numerical solution. The strain-hardening rule is therefore more complex to use in hand calculations but does not present any real difficulty when used in a computer code. Strain hardening is the default model in many computer codes developed for creep analysis. Although less objectionable and somewhat more realistic than time hardening, strain hardening is still very simplistic and does not take into account a recovery mechanism or the effect of stress and temperature rates.

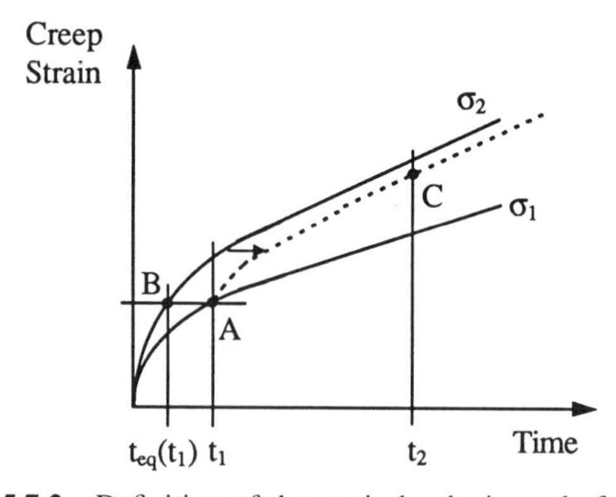

Figure 5.7.2 Definition of the strain-hardening rule for creep.

Time- and strain-hardening rules are both similar to the isotropic hardening assumption in the yield surface plasticity theory and produce extremely poor results for cyclic loading conditions, Techniques have been developed to account for stress reversals or cyclic creep (see, e.g., Kraus, 1980). These methods are somewhat contrived and are not discussed here.

Example Problem 5.2: A material is assumed to obey the Bailey–Norton creep law (equation 3.5.3) with the following constants (for time in hours and stress in psi): $A = 1 \times 10^{-5}$, $\sigma_0 = 10^4$, $m = 2$ to $= 1$, and $n = 0.5$. A bar of the material is subjected to a uniform stress of 2×10^4 psi for 100 h. The stress is then lowered to 1×10^4 psi for an additional 300 h. What is the predicted creep strain in the bar for the time and strain hardening rules?

SOLUTION: For time hardening, the creep strain from equation 3.5.3 at the end of the first 100 h is

$$\varepsilon^C = A\left(\frac{\sigma}{\sigma_0}\right)^m \left(\frac{t}{t_0}\right)^n (1 \times 10^{-5})\left(\frac{2 \times 10^4}{1 \times 10^4}\right)^2 (100)^{0.5} = 4 \times 10^{-4}$$

Rather than using equation 5.7.3, the accumulated creep strain at the end of 400 h is determined by adding the additional strain increment from the second load, 1×10^4 psi, to the strain above, that is,

$$\begin{aligned}\varepsilon^C &= 4 \times 10^{-4} + [\varepsilon^C(\sigma_2, t_2) - \varepsilon^C(\sigma_2, t_1)] \\ &= 4 \times 10^{-4} + (1 \times 10^{-5})\left(\frac{1 \times 10^4}{1 \times 10^4}\right)^2 [(400)^{0.5} - (100)^{0.5}] = 5 \times 10^{-4}\end{aligned}$$

The creep strain at the end of the first 100 h is the same for time hardening and strain hardening (i.e., $\varepsilon^C = 4 \times 10^{-4}$). The equivalent time at the second stress level is determined from the inverse of the creep law:

$$\begin{aligned}t_2^E = \left[A\left(\frac{\sigma}{\sigma_0}\right)^m\right]^{-1/n} (\varepsilon^C)^{1/n} = \left[1 \times 10^{-5}\left(\frac{1 \times 10^4}{1 \times 10^4}\right)^2\right]^{-2} \\ \times (4 \times 10^{-4})^2 = 1600\end{aligned}$$

At the end of the second load increment, 400 h, the creep strain is

$$\varepsilon^C = 4 \times 10^{-4} + [\varepsilon^C(\sigma_2, t_2^E) - \varepsilon^C(\sigma_2, t_1^E)]$$

where the second equivalent time is defined as $t_2^E = t_1^E + 300 = 1900$ h. The accumulated creep strain according to the strain-hardening rule is then

$$\varepsilon^C = 4 \times 10^{-4} + (1 \times 10^{-5})\left(\frac{1 \times 10^4}{1 \times 10^4}\right)^2 [(1900)^{0.5} - (1600)^{0.5}] = 4.358 \times 10^{-4}$$

Keep in mind that for computer-based numerical procedures the strain levels are computed by numerically integrating the creep-rate equations. However, the time and equivalent time are determined in the manner illustrated in the example problem or by an iterative numerical procedure if the creep equation cannot be inverted.

5.8 CREEP FLOW EQUATION

The classical creep approach is based on decomposition of the total strain tensor into elastic, creep, and thermal components:

$$\varepsilon_{ij} = \varepsilon_{ij}^E + \varepsilon_{ij}^C + \varepsilon_{ij}^T \tag{5.8.1}$$

For combined plasticity and creep, or rate-independent and rate-dependent response, the classical approach is to add the plastic strain to equation 5.8.1, to obtain

$$\varepsilon_{ij} = \varepsilon_{ij}^E + \varepsilon_{ij}^P + \varepsilon_{ij}^C + \varepsilon_{ij}^T \tag{5.8.2}$$

This approach clearly neglects the interaction between plasticity and creep. However, experimental evidence (Section 3.5) has shown that plastic and creep strain are not independent. After several years of attempting to model creep–fatigue interaction, most people now use a single inelastic strain, as defined in Chapter 6.

The classical approach to creep modeling of polycrystalline metals is based on developing a general method to extend the uniaxial model to three dimensions. This is achieved by making the following assumptions, which are also typical for yield surface plasticity:

1. The material is isotropic and homogeneous.
2. The response does not depend on mean or hydrostatic stress.
3. Volumetric strain during creep can be neglected.
4. The response is the same in tension and compression.
5. The three-dimensional model must reduce to the correct one-dimensional model.

These assumptions are not totally valid for creep. Even though the macroscopic volume changes are small, recall that Nabarro–Herring creep and

Coble creep are diffusion-driven processes that depend on local changes in volume (see Sections 3.2.2 and 3.2.3). The mean stress can expand or contract the crystal lattice and increase or decrease the rate of diffusion, respectively. Figure 3.3.7 illustrates the degree of tension compression anisotropy that occurs in creep of a typical material. This result clearly shows that the response is not independent of mean stress, nor is creep symmetric in tension and compression.

For modeling purposes these assumptions are satisfied by writing the effective creep rate as a function of the von Mises or effective stress. Similar to the effective plastic strain increment defined in equation 5.2.21 for yield surface plasticity, an effective creep strain rate is defined by

$$\dot{\varepsilon}_e^C = \sqrt{\tfrac{2}{3}\dot{\varepsilon}_{ij}^C\,\dot{\varepsilon}_{ij}^C} \tag{5.8.3}$$

The constant $\sqrt{\frac{2}{3}}$ is chosen so that the effective creep strain rate or creep strain increment is equal to the uniaxial creep strain rate or increment during a uniaxial test. Also recall that the von Mises effective stress is defined such that it is equal to the uniaxial stress in a uniaxial test. Therefore, the uniaxial creep equations discussed in Section 3.5 can also be considered to be a function of the effective stress and strain,

$$\varepsilon_e^C = \mathfrak{C}_e(\sigma_e, T, t) \tag{5.8.4}$$

The creep-rate equation, defined by Equation 5.7.2 or derived from partial differentiation, can also be written as a function of the effective variables in the form

$$\dot{\varepsilon}_e^C = \dot{\mathfrak{C}}_e(\sigma_e, T, t) \qquad \text{or} \qquad \dot{\varepsilon}_e^C = \dot{\mathfrak{C}}_e(\sigma_e, T, \varepsilon_e^C) \tag{5.8.5}$$

for time or strain hardening.

A three-dimensional creep equation can be formulated to satisfy the first four assumptions by using the mathematical structure of Prandtl–Reuss equations of yield surface plasticity. The creep strain rate components are written as a function of the deviatoric stress tensor, S_{ij}, and an effective scalar creep rate, $\dot{\lambda}$, in the form

$$\dot{\varepsilon}_{ij}^C = \dot{\lambda} S_{ij} \tag{5.8.6}$$

The quantity, $\dot{\lambda}$, is similar to the plastic multiplier in yield surface plasticity and may be obtained directly from the effective creep strain rate similar to equation 5.2.22:

$$\dot{\lambda} = \frac{3\dot{\varepsilon}_e^C}{2\sigma_e} \tag{5.8.7}$$

Combining equations 5.8.6 and 5.8.7 produces the creep flow equation as a function of the effective creep strain rate, the current effective stress, and the deviatoric stress tensor:

$$\dot{\varepsilon}_{ij}^{C} = \frac{3\dot{\varepsilon}_{e}^{C}}{2\sigma_{e}} S_{ij} \tag{5.8.8}$$

As an example consider the Bailey–Norton creep law (equation 3.5.3). Taking the partial derivative with respect to time and replacing the stress by the von Mises effective stress, the effective creep strain rate is given by

$$\dot{\varepsilon}_{e}^{C} = nA\left(\frac{\sigma_{e}}{\sigma_{0}}\right)^{m} \frac{t^{n-1}}{t_{0}^{n}} \tag{5.8.9}$$

Once the effective creep strain rate has been established from the uniaxial equation as a function of the von Mises effective stress, the flow rule is used to compute the components of the creep strain rate tensor. Using equation 5.8.8, the components of the creep strain rate, written in vector form, are

$$\begin{Bmatrix} \dot{\varepsilon}_{11}^{C} \\ \dot{\varepsilon}_{22}^{C} \\ \dot{\varepsilon}_{33}^{C} \\ \dot{\varepsilon}_{12}^{C} \\ \dot{\varepsilon}_{13}^{C} \\ \dot{\varepsilon}_{23}^{C} \end{Bmatrix} = \frac{3nA(\sigma_{e}/\sigma_{0})^{m}\, t^{n-1}}{2\sigma_{e} t_{0}^{n}} \begin{Bmatrix} S_{11} \\ S_{22} \\ S_{33} \\ S_{12} \\ S_{13} \\ S_{23} \end{Bmatrix} \tag{5.8.10}$$

5.9 NUMERICAL TECHNIQUES FOR CREEP

The integration of the creep strain rate equations involves special techniques. For many models the creep strain rate can vary dramatically from the high initial rates in primary creep to the much slower rates in secondary creep. In fact, some creep-rate equations give an infinite strain rate at zero time (the Bailey–Norton law, equation 5.8.9, for typical values of $n < 1$). The creep rate is also usually very sensitive to stress and temperature. Therefore, selection of an appropriate time increment to achieve both an economical and accurate solution is crucial. Several effective, yet simple time stepping strategies have been developed. Most of the techniques for integrating the creep rate equations may be applied directly to the state variable equations presented in the remaining chapters.

5.9.1 Integration of the Creep Rate Equations

The most common techniques for integrating the creep rate equations are based on the generalized trapezoidal rule and the generalized midpoint rule.

The generalized trapezoidal rule (equation 5.6.2) for integrating the creep-rate equation over a time step Δt is

$$\{\Delta\varepsilon^c\} = \Delta t[(1 - \beta)\{\dot{\varepsilon}^C\}_t + \beta\{\dot{\varepsilon}^C\}_{t+\Delta t}] \tag{5.9.1}$$

The parameter β may be given values from 0 to 1. Both the forward Euler, $\beta = 0$, and backward Euler, $\beta = 1$, algorithms are first-order accurate (i.e., integration error is proportional to Δt^2). For $\beta = \frac{1}{2}$ (the trapezoidal rule) the integration error is second-order accurate or proportional to Δt^3. The forward Euler method is explicit since all of the information required to integrate the rate equations is known at the beginning of the time step. For $\beta > 0$ the method is implicit and the stress, creep strain rate, and other variables at the end of the step are initially unknown and must be determined.

The generalized midpoint rule is given by

$$\{\Delta\varepsilon^C\} = \Delta t\{\dot{\varepsilon}^C\}_{t+\beta\Delta t} \tag{5.9.2}$$

where the parameter β may take on values from 0 to 1. Like the generalized trapezoidal rule, for $\beta = 0$ the generalized midpoint rule becomes the forward Euler algorithm, and for $\beta = 1$ it becomes the backward Euler algorithm. For $\beta = \frac{1}{2}$, equation 5.9.2 reduces to the second-order accurate midpoint rule, which is generally preferred over the first-order accurate forward and backward Euler algorithms. Once again, except for the special case of $\beta = 0$, the stress, creep strain rate, and so on, are initially unknown at time $t + \beta\,\Delta t$.

Like the generalized trapezoidal and generalized midpoint rules, implicit integration rules are frequently implemented using predictor–corrector methods. The creep strain rate is initially computed with a forward Euler method, Once the creep strain increment has been estimated, the creep rate at some other time in the increment can be computed. Iteration on the creep strain rate at the appropriate point in the time step and the creep strain increment continues until a preselected convergence criterion is satisfied.

5.9.2 Difficulties Encountered Integrating the Creep Strain Rate

Most creep equations are simply curve fits of creep data from tests at several different constant stress and temperature conditions (Chapter 3). The equations usually express the accumulated creep strain as a function of temperature, time, and stress (equation 5.8.4), or less frequently, the creep strain rate as a function of temperature, time, and stress (equation 5.8.5a). To integrate the creep strain rate equations for conditions of varying stress and temperature, both the rate equation and integrated equation are usually necessary. The rate equation is used directly to compute the rates that are numerically integrated. The integrated form of the equation is needed to compute an "equivalent time" for strain hardening and sometimes for estimating the initial creep strain increment if the initial creep rate is not defined.

One of the difficulties encountered in integrating many creep equations with the generalized trapezoidal rule is evaluating the creep strain rate at time zero when the rate is infinite for some models. This represents a special case that must be analyzed separately. One simple way to avoid the problem is to assume an initial nonzero time (for time hardening) or nonzero accumulated creep strain (for strain hardening). Another approach is to assume that the stress and temperature are constant during the first time increment. The initial creep increment may then be computed directly from the integrated form of the creep equation. For some adaptive time-stepping procedures it may be necessary to obtain a numerical value of the initial creep rate. This value can be back-calculated from equation 5.9.1 after the creep increment is computed. This is an approximate procedure, but it will lead to exact results for the special case of constant stress and temperature over the time interval.

For the generalized midpoint rule, the creep strain rate for the first time step in not infinite, but numerical integration during the first time increment may be very inaccurate. An approach similar to that used for the generalized trapezoidal rule is suggested. The creep increment can be computed from the integrated form of the creep equation assuming constant stress and temperature during the increment. If a numerical value for the creep strain rate is desired, it can then be back calculated from equation 5.9.2.

Regardless of the integration procedure it is more accurate and efficient to use smaller time steps when the creep rate is changing rapidly and larger steps when the creep rate is relatively constant.

Implementation of the strain-hardening rule becomes more complicated when the integrated form of the creep equation cannot be solved explicitly for equivalent time. In this case it is necessary to solve numerically for equivalent time. Several methods that have been used effectively for this purpose are Newton's method, the secant method, the bisection method, and various combinations. The basic iterative procedure using these methods for computing the equivalent time t_{eq}, given the accumulated equivalent creep strain ε^C, is as follows:

1. Estimate a lower bound t_l and an upper bound t_u for the equivalent time t^E.
2. Using the integrated form of the creep equation, compute creep strains ε_l^C and ε_u^C associated with the lower and upper time bounds, respectively; that is,

$$\varepsilon_l^C = \mathfrak{C}(\sigma, T, t_l), \qquad \varepsilon_u^C = \mathfrak{C}(\sigma, T, t_u)$$

3. Estimate a new equivalent time t^E from the lower and upper bounds for the equivalent time, t_l and t_u, the associated creep strains, ε_l^C and ε_u^C, and the known value for the accumulated creep strain, ε^C.
4. Using the integrated form of the creep equation, compute the creep strain ε_{eq}^C associated with the estimated equivalent time t^E:

$$\varepsilon_{\text{eq}}^{C} = \mathfrak{C}(\sigma, T, t^{E})$$

5. If the accumulated creep strain ε^C is greater than the creep strain associated with the estimated equivalent time, reset the lower-bound time estimate t_1 equal to the estimated equivalent time t^E:

$$\text{if} \quad \varepsilon^{C} > \varepsilon_{\text{eq}}^{C}, \quad \text{then} \quad t_1 = t^{E}$$

6. If the accumulated creep strain ε^C is less than the creep strain associated with the estimated equivalent time, reset the upper-bound time estimate t_u equal to the estimated equivalent time t^E:

$$\text{if} \quad \varepsilon^{C} < \varepsilon_{\text{eq}}^{C}, \quad \text{then} \quad t_u = t^{E}$$

7. Check for convergence, and proceed with step 2 if not converged.

5.9.3 Adaptive Time-Stepping Procedures

In stand-alone codes used to validate constitutive models, the time step size and numerical efficiency are often of little concern. However, for finite-element simulations where a considerable amount of computation is associated with even a single solution step, it is important to adjust the time step to achieve an economical solution. A number of adaptive strategies have been developed. Among the criteria for adjusting the time step are:

1. The ratio of the creep strain increment to the elastic strain
2. The ratio of the creep strain increment to the total strain
3. The size of the effective creep strain increment
4. The rate of change of the creep strain rate

Sometimes multiple criteria are applied simultaneously. Limits are usually set as to the minimum and maximum allowable time steps and the minimum and maximum ratio of the current time increment to the previous time increment. Within a finite element code the global time step is usually set by the minimum time increment computed at any point in the model.

As an example, consider an adaptive time-stepping criterion based on the ratio of the creep strain increment to the elastic strain, $\Delta\varepsilon^C/\varepsilon^E$. The next time step is calculated from the current time step using

$$\Delta t_{\text{next}} = r \times \frac{\varepsilon_{\text{current}}^{E}}{\Delta\varepsilon_{\text{current}}^{C}} \Delta t_{\text{current}} \tag{5.9.3}$$

The desired ratio of creep strain increment to elastic strain, r, would typically have values in the range 0.1 to 0.5. Equations for other criteria can be developed similarly, as shown in Section 6.1.

Limits setting the maximum allowable change from one time step to the next are also usually part of an adaptive time-stepping procedure. This is required especially when the creep strain rate is changing significantly from one step to the next, such as in the primary creep regime or for rapidly changing stress or temperature conditions. The adaptive time-stepping equations are usually linear functions of key variables. The order of magnitude for typical limiting ratios for minimum and maximum are typically 0.2 to 5. These limits prevent excessive extrapolation of current conditions.

Selecting the initial time step can also be automated based on the initial creep strain rate or the desired initial creep strain increment. Another strategy is to select arbitrarily a very small initial time step and allow the adaptive time-stepping algorithm to adjust the step to an appropriate size.

5.9.4 Predictor–Corrector Procedure for Integrating Creep-Rate Equations

Although there are many possibilities for integrating the creep-rate equations, a specific example may be helpful. Library subroutines can also be used to integrate the creep rate and state variable strain rate equations. For example, ODEPACK (a systematized collection of six ordinary differential equation solvers by Alan C. Hindmarsh, Lawrence Livermore National Laboratory, Livermore, CA 94550, 1987), has been used to integrate stiff rate equations in constitutive models successfully with backward differentiation methods.

The principal steps at a typical time point in the calculation procedure using a predictor–corrector procedure and generalized trapezoidal rule are given below. It is assumed that the values of all variables are known at time t and that the time step Δt has been determined. The total strain increment is known and the creep strain increment and stress at time $t + \Delta t$ are to be determined.

1. Increment the total strains:

$$\{\varepsilon\}_{t+\Delta t} = \{\varepsilon\}_t + \{\Delta\varepsilon\}$$

2. Compute the stress increment and creep strain increment for the first iteration using a forward Euler predictor.

Creep strain increment:	$\{\Delta\varepsilon^C\} = \Delta t\{\dot{\varepsilon}^C\}_t$
Creep strain at $t + \Delta t$:	$\{\varepsilon^c\}_{t+\Delta t} = \{\varepsilon^C\}_t + \{\Delta\varepsilon^C\}_t$
Elastic strain at $t + \Delta t$:	$\{\varepsilon^E\}_{t+\Delta t} = \{\varepsilon\}_{t+\Delta t} - \{\varepsilon^C\}_{t+\Delta t}$
Stress at $t + \Delta t$:	$\{\sigma\}_{t+\Delta t} = [C]\{\varepsilon^E\}_{t+\Delta t}$

For subsequent iterations a generalized trapezoidal rule (equation 5.9.1) can be used to compute the creep strain increment from the creep strain rate at times t and $t + \Delta t$.

3. Compute a new estimate for the creep strain rate at the end of the time increment:

$$\{\dot{\varepsilon}^C\}_{t+\Delta t} = \dot{\mathfrak{C}}(\{\sigma\}_{t+\Delta t}, \{\varepsilon^C\}_{t+\Delta t}, t + \Delta t, T_{t+\Delta t})$$

4. Check for convergence. If not converged, repeat step 2 using the updated value for the creep rate at time $t + \Delta t$. If convergence is satisfied, proceed to step 5.
5. Update the creep strain and initialize creep strain rate for the beginning of the next time step.
6. If the solution is complete exit. If the solution is not complete compute the next time increment and go to step 1.

This procedure would require special processing on the first iteration of the first step to compute the initial creep strain rate.

5.10 TIME AS A VARIABLE

In this section it will be shown that time is not always a good variable for use in constitutive modeling. In particular, it will be shown that when time is present as a variable in a rate equation, such as equation 5.8.8 for the effective creep strain rate, it does not satisfy the time-translation portion of the principle of objectivity (Section 4.6). However, if an elapsed time is used, such as in viscoelasticity, the principle of objectivity is satisfied. However, the resulting representation is a hereditary integral that is not desirable for numerical simulations or evaluation of the material parameters.

Let us begin with a brief review of the viscoelasticity approach to constitutive modeling. Viscoelasticity is a phenomenological approach that was originally developed for inelastic response of polymers. It is widely used in biomechanics to model soft tissue and is occasionally used for metals. There is a large body of literature in the development of viscoelastic models based on mechanical analogies of combinations of linear springs ($\sigma = E\varepsilon$) and dashpots ($\sigma = \eta\dot{\varepsilon}$). There are also a number of representations with nonlinear material functions as integrands or integral series representations. One major difference in viscoelasticity is that the stress is usually related to the total strain. The elastic modulus is introduced by a linear spring in series with other spring and dashpot elements. The reason for introducing viscoelasticity it to show how time enters the constitutive equation. The spring and dashpot analogies will not be used since the resulting models are usually generalized by introducing the creep compliance and stress relaxation functions. Instead, the creep compliance function will be introduced directly.

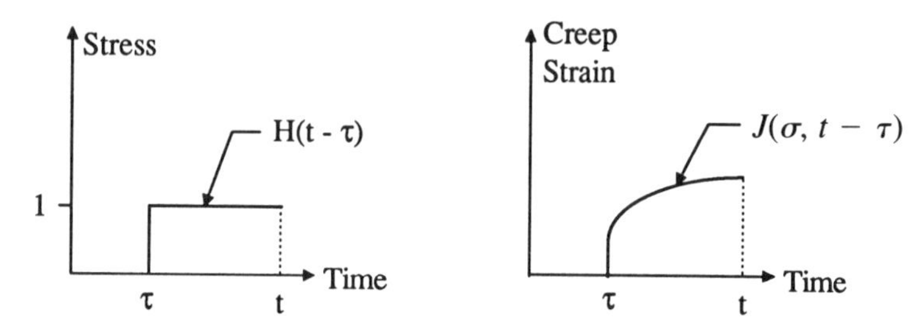

Figure 5.10.1 A step stress history and associated creep strain.

The creep compliance function, $J(\sigma, t - \tau)$, is defined as the total uniaxial strain at time t due to a constant uniaxial stress σ applied at time τ as shown in Figure 5.10.1. The measure $t - \tau$ is the elapsed time during which the stress is acting on the material. The material is a linear viscoelastic material if

$$\frac{J(\sigma, t - \tau)}{\sigma} = J(t - \tau) \tag{5.10.1}$$

where $J(t - \tau)$ is a function of time only and is defined as the strain response to a unit step stress. The desired formulation with elapsed time can be obtained with the linear theory; however, in practice there are very few linear viscoelastic materials.

Viscoelasticity deviates from time and strain hardening in the extension to time-varying stress histories. Let us assume that a stress history, $\sigma(\tau)$, on an interval $[0, t]$, can be partitioned into several steps, as shown in Figure 5.10.2. The stress history is approximated mathematically as

$$\sigma(t) \approx \sigma(0)H(t - 0) + \Delta\sigma_1 H(t - \tau_1) + \Delta\sigma_2 H(t - \tau_2) + \cdots \tag{5.10.2}$$

where $H(t - a)$ is the Heaviside unit step function, defined as

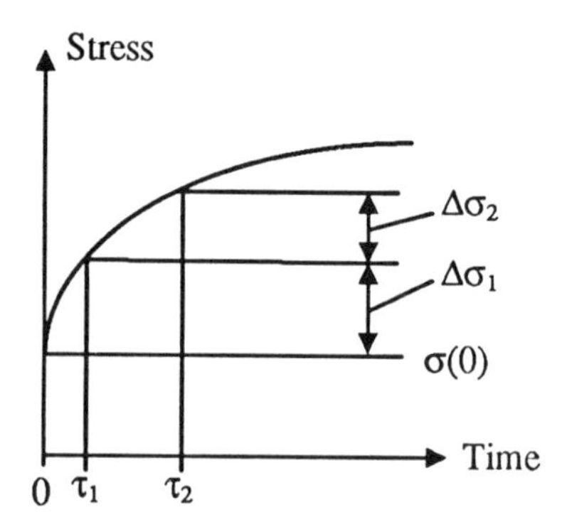

Figure 5.10.2 Decomposition of a stress history into a number of stress increments.

$$H(t - a) = \begin{cases} 0 & \text{for } t < a \\ 1 & \text{for } t > a \end{cases} \tag{5.10.3}$$

Using the definition of the creep compliance function for a linear representation, equation 5.10.1, the total strain at time t is given by

$$\varepsilon(t) \approx \sigma(0)J(t - 0) + \Delta\,\sigma_1\, J(t - \tau_1) + \Delta\sigma_2\, J(t - \tau_2) + \cdots \tag{5.10.4}$$

as shown graphically in Figure 5.10.3. Approximating the stress increment at each step by the stress rate and the time step, $\Delta\sigma_i = \dot{\sigma}_i\, \Delta\tau_i$, Equation 5.10.4 can be rewritten as

$$\varepsilon(t) \approx \sigma(0)J(t) + \dot{\sigma}_1\, \Delta\tau_1\, J(t - \tau_1) + \dot{\sigma}_2\, \Delta\tau_2\, J(t - \tau_2) + \cdots$$

or

$$\varepsilon(t) \approx \sigma(0)J(t) + \sum_1^n J(t - \tau_i)\dot{\sigma}_i\, \Delta\tau_i$$

Replacing the summation by an integral for an infinite number of infinitesimal time steps gives

$$\varepsilon(t) = \sigma(0)J(t) + \int_0^t J(t - \tau)\dot{\sigma}(\tau)\, d\tau \tag{5.10.5}$$

where the first term is the initial response and integral gives the additional

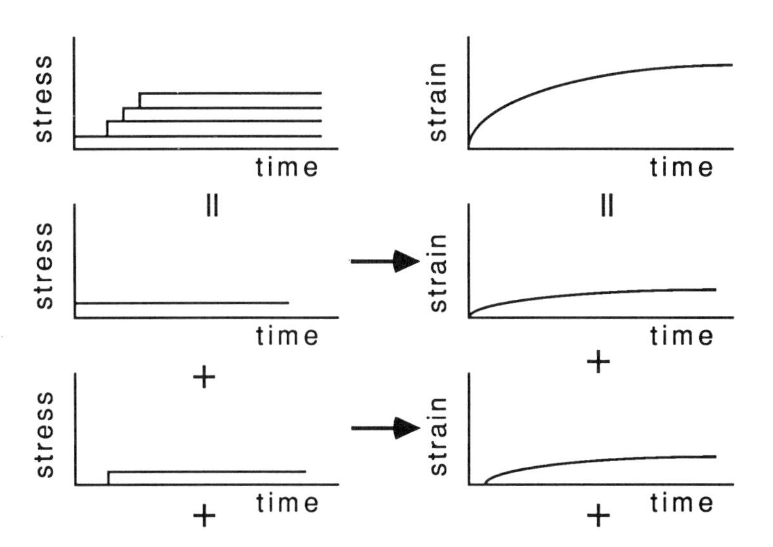

Figure 5.10.3 Linear superposition integral for creep.

strain due to creep strain. The linear viscoelasticity integral is frequently written as

$$\varepsilon(t) = \int_{-\infty}^{t} J(t - \tau)\dot{\sigma}(\tau)\, d\tau \tag{5.10.6}$$

for convenience. The integral is evaluated by separating the initial elastic jump as shown in equation 5.10.5. Nonlinear representations for equation 5.10.6 can be obtained by replacing the kernel function by a general function of stress or viewing integral as the first term of a multiple integral series.

The temporal part of the principle of objectivity presented in Section 4.6 can be stated more explicitly by considering two stress histories: $\sigma(\tau)$ defined on the time interval $[0, t]$, and $\sigma'(\tau)$ defined on the interval $[\lambda, \lambda + t']$. The strain histories corresponding to $\sigma(\tau)$ and $\sigma'(\tau)$ are $\varepsilon(\tau)$ and $\varepsilon'(\tau)$, respectively. If the two stress histories are identical over their respective time domains, the principle of objectivity requires that the corresponding strain histories be identical. That is:

$$\begin{aligned} &\text{if} \quad \sigma(\tau) = \sigma'(\tau + \lambda) \qquad \text{for all } \tau \text{ on } [0, t] \\ &\text{then} \quad \varepsilon(\tau) = \varepsilon'(\tau + \lambda) \qquad \text{for all } \tau \text{ on } [0, t] \end{aligned} \tag{5.10.7}$$

All constitutive equations are expected to satisfy this requirement. Practically, this implies that a specimen can be tested on any day of the week. The relevant time is the elapsed time between the beginning and end of the test, not the time on the clock. The principle of objectivity can be applied to equation 5.10.6 by letting $\sigma(\tau) = \sigma'(\tau + \lambda)$ and finding the resulting strain:

$$\varepsilon(t) = \int_{-\infty}^{t} J(t - \tau)\dot{\sigma}'(\tau + \lambda)\, d\tau \tag{5.10.8}$$

Using the change of variable $\theta = \tau + \lambda$ allows equation 5. 10.8 to be written as

$$\varepsilon(t + \lambda) = \int_{-\infty}^{t+\lambda} J[(t + \lambda) - \theta]\dot{\sigma}'(\theta)\, d\theta \tag{5.10.9}$$

Let us next determine $\varepsilon'(t + \lambda)$ from $\sigma'(\tau)$ by using equation 5.10.6 directly; that is, substitute $\sigma'(\tau)$ for $\sigma(\tau)$ and integrate on $(-\infty, t + \lambda)$ to obtain

$$\varepsilon'(t + \lambda) = \int_{-\infty}^{t+\lambda} J[(t + \lambda) - \theta]\dot{\sigma}'(\theta)\, d\theta \tag{5.10.10}$$

Comparing equations 5.10.9 and 5.10.10 shows that $\varepsilon(t + \lambda) = \varepsilon'(t + \lambda)$ for any value of t; therefore, the principle of objectivity is satisfied.

Let us apply the time-translation criterion to the time-hardening rule of classical creep. Let the uniaxial creep rate be given by an equation of the form

$$\dot{\varepsilon} = f(\sigma)g(t) \tag{5.10.11}$$

where f and g are functions only of stress and time, respectively. This representation of the creep rate is typical of the equations given earlier. The accumulated strain by the time-hardening rule for the two stress histories is then given by

$$\varepsilon(t) = \int_{-\infty}^{t} f[\sigma(\tau)]g(\tau)\, d\tau \tag{5.10.12}$$

and

$$\varepsilon'(t + \lambda) = \int_{-\infty}^{t+\lambda} f[\sigma'(\tau)]g(\tau)\, d\tau \tag{5.10.13}$$

where the lower limit is used for convenience with the assumption that $f[\sigma(\tau)] = 0$ for $\tau < 0$. Introducing $\sigma(\tau + \lambda) = \sigma'(\tau)$ in Equation 5.10.12 gives

$$\varepsilon(t) = \int_{-\infty}^{t} f[\sigma'(\tau + \lambda)]g(\tau)\, d\tau$$

and using the same change of variable, $\theta = \tau + \lambda$, allows the integral to be rewritten as

$$\varepsilon(t + \lambda) = \int_{-\infty}^{t+\lambda} f[\sigma'(\theta)]g(\theta - \lambda)\, d\theta \tag{5.10.14}$$

But $\varepsilon(t + \lambda)$ from equation 5.10.14 does not equal $\varepsilon'(t + \lambda)$ from equation 5.10.13; thus the principle of objectivity is not satisfied for the simple rule of time hardening.

These results suggest that the viscoelasticity formulation should be used rather than time or strain hardening. However, the viscoelasticity integral (equation 5.10.6) has some undesirable properties. First, $\dot{\sigma}(\tau)$ is integrated from 0 to t while $J(t - \tau)$ is integrated from t to 0, This means that if the strain is required at 500 points, for example, the entire integral must be evaluated 500 times. This is not nearly as efficient as adding 500 strain increments for a rule such as time hardening or strain hardening. Next consider evaluation of the material function $J(t)$. If the stress and strain histories are observed in an experiment, equation 5.10.6 is an integral equation that must be solved for the creep compliance function. This can easily be accomplished for linear materials by using the Laplace transformation, but it is not so simple for nonlinear kernel functions or multiple integral series. As a result, the viscoelasticity formulation is not often used for structural simulations because it is hard to find the material parameters and the formulation is not very efficient numerically.

To summarize, elapsed time should be used rather than time in rate equations to satisfy the principle of objectivity. But elapsed time leads to a mathematical structure with some undesirable properties. One way to avoid this dilemma is to avoid time as a variable in rate equations. This does not mean that elapsed time should not appear in constitutive models; it only means that it should not be used with variables that will be integrated if the foregoing difficulties are to be avoided.

5.11 SUMMARY

A useful phase of the study of constitutive models is to review critically the material that was recently presented. Since our primary objective is the development of constitutive models for design applications, it is useful to establish evaluation criteria to identify the attributes and limitations of the models. In this study each model will be tested for computational efficiency, ease in determining material parameters, and accuracy. These criteria have evolved after several years of work with a variety of constitutive models.

It is generally believed by the community that a model must be reasonably efficient for numerical simulations if it is to be useful in engineering design. It is not acceptable to develop a model that cannot be used in a finite-element code, where it is expected to be called thousands of times. The superposition integral associated with the viscoelasticity theory is an example of an unacceptable approach because of its numerical inefficiency. This point was learned the hard way. Numerical efficiency is absolutely essential for modern engineering applications.

Another critical factor controlling the usefulness of a model is the ease in finding the material parameters. It is necessary to be able to find material constants from relatively standard sets of experimental data. The need to run a number of special tests could severely limit the use of a model. The need to solve a set of difficult differential or integral equations to find the material parameters could also make a model useless. It is critical to be able to find the material parameters: the easier the better. It is impossible to carry out engineering design calculations without the material parameters.

Accuracy is desirable but not always necessary. Sometimes it is only necessary to get an estimate of the response and know the sources of possible error. At other times it may be necessary to know material response accurately. For example, a life prediction analysis could be critical for public safety. However, in many life models a 10% variation in stress or strain range could produce a factor of 10 or 100 in the number of cyclics to failure. Thus the desired level of accuracy has to be balanced by the cost and effort required to obtain the accuracy and the need for the accuracy.

5.11.1 Yield Surface Plasticity

Yield surface models for plasticity were developed for strain-rate-independent response; thus it is limited to applications below about one-half the melting temperature (on an absolute temperature scale). All time effects, such as strain-rate dependence, creep, stress relaxation, and strain recovery, are excluded by definition.

The method is very efficient for numerical simulations. The yield surface approach has been included in several commercial finite-element codes. The equation permits a forward-time-marching integration procedure where the current plastic strain increment is added to the previous accumulated plastic strain to get the current value of the total plastic strain. Since the plastic straining is path dependent, the proper definition of the loading history is essential. Also, the number of increments or steps to get to a certain load level will affect the results since the accuracy of the numerical procedure improves with smaller increments. A very desirable feature is the simplicity of the input data for a plasticity analysis. Nonlinear material input data usually consist of a piecewise linear (or some other relatively simple) representation of the stress–strain curve.

Any uniaxial model, such as the Ramberg–Osgood model, can be used for multiaxial calculations by substituting effective stress and strain for the uniaxial stress and strain. The initial yield surface is well defined by the von Mises yield surface, and the radius of the circle is determined from the initial yield stress. The model has been widely accepted for predicting the initial yield surface.

An experimental program to determine the initial yield surface (or any yield surface) is not easy to execute because it is not possible to observe the elastic limit until the limit has been exceeded. The plastic deformation hardens the material so that the material in the second reading does not correlate with the first reading.

The isotropic and kinematic models of hardening are simple to implement but are not very realistic for complex loading histories. Experimental evidence shows that the yield surface changes size and shape extensively during nonproportional multiaxial deformation. Further, the evolution of the yield surface during deformation is not easy to quantify experimentally. Experimental evidence also shows that the plastic strain increment vector is not normal to the yield surface during nonproportional loading, as proposed by some models. These models should be limited to proportional loading applications.

5.11.2 Creep Modeling

The definition of creep usually implies that the stress is limited to values less than the yield stress or the critical resolved shear stress required to produce slip. Temperatures are generally greater than one-half the absolute melting temperature. The creep rate is a strong function of temperature, as discussed in Chapter 3.

The method is widely used for numerical simulations and is also included in several commercial finite-element codes. The methods of time and strain hardening permit a forward-time-marching integration procedure. The hereditary integral that arises in viscoelasticity is not nearly as efficient numerically. Since most creep models predict large changes in creep strain rate, even for constant stress and temperature conditions, efficient implementation demands an adaptive time-stepping procedure. Even so, a typical creep analysis will usually require far more computer time than a linear elastic analysis or even a plasticity analysis of the same structure.

The constants in the constant stress creep models can be determined from the creep data and used with time- or strain-hardening rules. The parameters in many of the creep models are functions of temperature, thus the method may require several sets of creep data, one set for each temperature. Also, fitting most creep models to data requires a nonlinear curve-fitting procedure. Often, this is not a trivial task and requires considerable engineering judgment. The viscoelasticity approach requires solving an integral equation to determine the material parameters.

The formulations of the time- and strain-hardening rule are theoretically incorrect from two points of view: (1) by assuming that the creep rate is the same as the partial derivative of the creep function with respect to time (equation 5.7.2), which excludes the effect of the stress and temperature changes although the approach was developed to model time-varying stress and temperature histories; and (2) because several of the simple creep models involve time as a variable in the creep-rate equation, and time as an absolute variable does not satisfy the principle of objectivity in a rate equation that must be integrated.

Finally, the models are not applicable to the combined effects of plasticity and creep. Summing plastic and creep strain excludes the interaction between plasticity and creep. For example, it was shown that the amount of primary creep in a creep test is a function of the initial loading rate. It is also difficult to separate plastic and creep strain experimentally.

REFERENCES

Bathe, K. (1982). *Finite Element Procedures in Engineering Analysis,* Prentice Hall, Englewood Cliffs, NJ.

Crisfield, M. (1991). *Nonlinear Finite Element Analysis of Solids and Structures, Vol. 1: Essentials,* Wiley, Chichester, West Sussex, England.

Dieter, G. (1986). *Mechanical Metallurgy,* 3rd ed., McGraw-Hill, New York.

Drucker, D. C. (1956). On uniqueness in the theory of plasticity, *Quarterly of Applied Mathematics,* Vol. 14, pp. 35–42.

Drucker, D. C. (1950). Some implications of work hardening and ideal plasticity, *Quarterly of Applied Mathematics,* Vol. 7, no. 4, pp. 411–418.

Drucker, D. C. (1951). Proceedings of the First U.S. National Congress of Applied Mechanics, American Society of Mechanical Engineers, p. 487.

Hinton, E., ed. (1992). *NAFEMS Introduction to Nonlinear Finite Element Analysis,* NAFEMS, Glasgow.

Hodge, P. (1957). Discussion of the Prager hardening law, *Journal of Applied Mechanics,* Vol. 23, pp. 482–484.

Kraus, H. (1980). *Creep Analysis,* Wiley, New York.

Liu, K. C., and W. L. Greenstreet (1976). Experimental studies to examine the behavior of metal alloys used in nuclear reactors, in *Constitutive Equations in Viscoplasticity,* ASME Applied Mechanics Division Publication 20, ASME, New York.

Lode, W. (1926). Versuche ueber einfluss den einfluss der mittleren hauptspannung auf das fliessen der metalle eisen kupfer und nickel, *Z. Physik,* Vol. 36, pp. 913–939.

Mair, W. M., and H. D. Pugh (1964). Effect of prestrain on the yield surface in copper, *Journal of Mechanical Engineering Science,* Vol. 6, No. 2.

Maistkowski, J., and W. Szczepinski (1965). An experimental study of the yield of prestrained brass, *International Journal of Solids and Structures,* Vol. 1, pp. 189–194.

McDowell, D. L. (1983). Transient nonproportional cyclic plasticity, *Report 107, UILU-Eng-83-4003,* University of Illinois at Urbana–Champaign.

McDowell, D. L. (1985a). A two surface model for transient nonproportional cyclic plasticity, *Transactions of ASME, Journal of Applied Mechanics,* p. 10.

McDowell, D. L. (1985b). Significance of nonproportional loading tests for characterization of cyclic response of metals, Spring Conference on Experimental Mechanics, Las Vegas, pp. 229–236.

McDowell D. L. (1985c). An experimental study of the structure of constitutive equations for nonproportional cyclic plasticity, *Journal of Engineering Materials and Technology,* Vol. 107, pp. 307–315.

McDowell, D. L. (1987). Simple experimentally motivated cyclic plasticity model, *Journal of Applied Mechanics,* Vol. 113, No. 3, pp. 387–397.

Owen, D., and E. Hinton (1980). *Finite Elements in Plasticity: Theory and Practice,* Pineridge Press, Swansea, Wales.

Phillips, A., and J. L. Tang (1972). The effect of loading path on the yield surface at high temperature, *International Journal of Solids and Structures,* Vol. 8, pp. 463–474.

Prager, W. (1955). *Proceedings of the Institute of Mechanical Engineers,* Vol. 169, p. 41.

Prager, W. (1956). *Journal of Applied Mechanics,* Vol. 23, p. 93.

Prandtl, L. (1924). *Proceedings of the First International Conference on Applied Mechanics,* Delft, The Netherlands, p. 43.

Reuss, A. (1930). *Mathematik und Mechanik,* Vol. 10, p. 266.

Taylor, G. I., and H. Quinney (1931). The plastic distortion of metals, *Philosophical Transactions of the Royal Society,* London, Vol. A230, pp. 323–363.

Tresca, H. (1864). *Comptes Rendus de l'Academie des Sciences, Paris,* Vol. 59, p. 754.

Tresca, H. (1868). *Memories de l'Sav. Academie des Sciences, Paris,* Vol. 18, p. 733.

Von Mises, R. (1928). *Zeitschrift fur Angewandte Mathematik und Mechanik,* Vol. 8, p. 161.

Von Mises, R. (1913). Mechanics of solids in the plastically deformable state, *NASA Technical Memorandum 88488,* 1986. (translation of Mechanik der festen Koerper im plastisch-deformablem Zustrand, Nachrichten von der Königlichen Gesellschaft der Wissenschaften, pp. 582–592).

Westergaard, H. M. (1920). On the resistance of ductile materials to combined stresses, *Journal of the Franklin Institute,* Vol. 189, pp. 627–640.

Williams, J. F., and N. L. Svensson (1970). Effect of tensile prestrain on the yield locus of 11OO°F aluminum, *Journal of Strain Analysis,* Vol. 5, No. 2, p. 130.

Zienkiewicz, 0., and R. Taylor (1991). *The Finite Element Method, Vol. 2, Solid and Fluid Mechanics, Dynamics and Non-linearity,* 4th ed., McGraw-Hill, London.

Ziegler, H. (1959). *Quarterly of Applied Mathematics,* Vol. 17, No. 1, pp. 55–65.

PROBLEMS

5.1 Determine the direction cosines of the projection of the principal stress axes on the pi plane.

5.2 Verify that radial distance $\sqrt{x_1^2 + x_2^2}$ in Figure 5.1.3 is equal to the magnitude of the deviatoric stress vector.

5.3 Verify that the von Mises yield criterion in plane stress is given by equation 5.1.15. Show that equations 5.1.14 and 5.1.15 can be represented by Figure 5.1.7 as a function of the two nonzero principal stresses in a two-dimensional orthogonal coordinate system.

5.4 Verify equations 5.1.16 and 5.1.17 for yield under tension–torsion loading, and plot the resulting equations as a function of shear and normal stress to verify the representation shown in Figure 5.1.8.

5.5 What is the maximum difference between the Tresca and von Mises yield criteria in a three-dimensional state of loading as shown in Figure 5.1.6?

5.6 It is generally known that the Tresca and von Mises yield criteria agree in tension and disagree in pure shear, as shown in Figure 5.1.8. Show that if the Tresca and von Mises criteria agree in pure shear, they will disagree in tension and determine the magnitude of the difference. Sketch the resulting yield surfaces in the pi plane.

5.7 A closed-ended thin-walled pressure vessel of radius r and wall thickness t is subjected to an internal pressure p. Estimate the pressure required to produce yield and identify the pending slip planes.

5.8 A composite shaft is made of a 1.0-in.-diameter core of 2040-0 aluminum alloy (E = 10,000 ksi and σ_y = 22 ksi) and a 4130 case-hardened steel (E = 30,000 ksi and σ_y = 190 ksi) with an outside diameter of 1.10 in. A 5-ft section of shaft is subjected to a torque T and an axial force of 500 lb. What is the maximum torque that can be applied before yielding starts? What are the potential slip planes?

5.9 A thin-walled sphere of radius r and thickness t is subjected to internal pressure p. Find the pressure that will initiate plastic flow according to the von Mises yield criterion. What are the slip planes?

5.10 An alloy is subjected to a pure three-dimensional state of stress; that is, $t_{11} = t_{22} = t_{33} = 0$ and $t_{12} = t_{23} = t_{31} = \tau$. What is the maximum value of shear stress if the yield stress in tension is 55 ksi? Use both the von Mises and Tresca yield criteria.

5.11 The plane strain elastic solution for a thick-walled tube with an inside radius a and outside radius b loaded with an internal pressure is given by

$$\sigma_r = -p\left(\frac{b^2 - r^2}{b^2 - a^2}\right)\frac{a^2}{r^r}, \qquad \sigma_\theta = p\left(\frac{b^2 + r^2}{b^2 - a^2}\right)\frac{a^2}{r^r}, \qquad \sigma_z = \frac{pa^2}{b^2 - a^2}$$

What pressure produces yielding according the Tresca and von Mises criteria if the yield stress in tension is σ_y?

5.12 Sketch the Tresca yield surface in the (σ_1, σ_2) plane for the case when $\sigma_3 \neq 0$. Assume that the yield stress in tension is σ_y and assume that σ_3 is a positive value between 0 and σ_y.

5.13 Determine specific representations for the tangent and plastic moduli from the tensile or cyclic response curves predicted by the Ramberg–Osgood equation.

5.14 Sketch the terms defining plastic work in equation 5.3.4 for a tensile response curve. Show that the argument used in proving convexity is reasonable, namely, $(\sigma_{ij} - \sigma_{ij}^*)\, d\varepsilon_{ij}^P >> d\sigma_{ij}\, d\varepsilon_{ij}^P$.

5.15 Use the Tresca yield surface as the potential function in equation 5.2.13 and derive the associated flow law. Discuss the physical consequences of the resulting equations.

5.16 Show that the strain invariant given in equation 5.2.21 reduces to the uniaxial strain under uniaxial loading. What is the reduced form of the invariant for the loading in Example Problem 5.1?

5.17 Calculate the response of the material in Example Problem 5.1 to the same loading loop but traced in reverse order. Sketch the yield surface at the end of each load step.

5.18 Determine the direction of the plastic strain increment vectors for each of the following principal stress vectors:

$$\sigma = [\sigma_y, 0, 0], \qquad \sigma = [0, \sigma_y, \sigma_y], \qquad \sigma = [\sigma_y, -\sigma_y, 0]$$

where σ_y is the yield stress in tension. Determine the size of the yield surface and sketch the plastic strain vectors and the yield surface in the pi plane.

5.19 Determine the principal stresses for the following principal plastic strain increments:

(a) $d\varepsilon_3^P = d\varepsilon_1^P$, $d\varepsilon_2^P = -\frac{1}{2}d\varepsilon_1^P$, and a mean stress of $-\frac{2}{3}\sigma_y$ where σ_y is the yield stress in tension.

(b) $d\varepsilon_3^P = 0.8d\varepsilon_1^P$, $d\varepsilon_2^P = -2d\varepsilon_1^P$, and a mean stress of $\frac{2}{3}Y$.

5.20 A thin sheet of aluminum is subjected to a biaxial stress history that produces the following linear paths in principal stress space, $(\sigma_1; \sigma_2)$:

A: (0; 0) to (10,000; 10,000) psi
B: (10,000; 10,000) to (10,000; 0) psi
C: (10,000; 0) to (30,000; 0) psi
D: (30,000; 0) to (0; 0) psi

Use both the Prager and Ziegler kinematic hardening rules and assume that the material obeys the Tresca yield criterion. The uniaxial stress–strain curve has a yield stress in tension of 8000 psi with an elastic modulus of 10,000 ksi and a constant tangent modulus of 1000 ksi. Determine the plastic strain and yield surface at the end of each load step. Report your answer for the yield surface in a sketch.

5.21 A material is loaded from (0; 0) to (200; −400) MPa and unloaded on the same path in two-dimensional principal stress space $(\sigma_1; \sigma_2)$. Estimate the plastic strain tensor at the end of the stress cycle using kinematic and isotropic hardening rules. The yield stress in tension is 100 MPa and the plastic tangent modulus is $1/E_p = J_2/10^{10}$ MPa^{-1}. Assume that the material follows the von Mises yield criterion. Sketch the yield surface at the end of the load step.

5.22 A metal has an initial yield stress in tension of40,000 psi. The material is loaded to 60,000 psi in the σ_2 direction. Assuming 50% kinematic and 59% isotropic hardening, sketch the yield surface in the pi plane before and after loading and give the key values of stress.

5.23 Constants for the Bailey–Norton Law (equation 3.5.3) for a material at a particular temperature are $A = 2.5 \times 10^{-5}$, $\sigma_0 = 10^5$ MPa, $m = 2$, and $n = 0.5$, when time is in hours. The material is subjected to a uniaxial stress of $2 \times 1O^5$ MPa for 20 h and then a stress of 4×10^5 MPa for 10 h. Using the time-hardening rule, what is the uniaxial creep strain after 30 h? Using the strain-hardening rule, what is the creep strain after 30 h?

5.24 What are the components of the creep strain rate for the material in problem 5.23 when the stress components are $\sigma_{xx} = 1 \times 10^5$, $\sigma_{zz} = 3 \times 10^5$, and $\tau_{xy} = 2 \times 10^5$, assuming **(a)** strain hardening and the accumulated effective creep strain is 0.01 and **(b)** time hardening and the accumulated creep time is 5 h.

5.25 What is a possible physical reason for creep–fatigue interaction? What deformation mechanisms could possibly interact under variable loading conditions? What types of experiments could be conducted to detect creep–fatigue interaction?

PART THREE

STATE VARIABLE APPROACH

The state variable approach developed from a synthesis of modeling, physical metallurgy, and experimental results of many researchers and practitioners in the last two decades. The development was motivated by three general conclusions about the state of classical modeling.

First, the yield surface approach to plasticity is not totally adequate. It is excellent for predicting the initial yield of metals, but the subsequent yield surfaces that evolve during a general multiaxial deformation history are not easy to model or quantify experimentally. The plastic strain increment vector is not always normal to the yield surface; nor is it parallel to the deviatoric stress vector. Fundamentally, the approach excludes all time and rate effects that occur at temperatures above one-half the melting temperature.

Second, the addition of a creep strain to account for time effects is not fully adequate. The plastic and creep strain are not independent. The coupling can be significant even in the traditional constant-stress creep test, where it is known that the initial loading rate affects the subsequent creep response. Time is not a good variable for constitutive modeling because it is not objective. Elapsed time, which is objective, is not computationally efficient when the model is formulated as a superposition integral.

Third, the results of physical metallurgy are largely ignored. An enormous body of fundamental knowledge about metals has been developed during the last half-century. Information about the physical basis of deformation and the associated variables should be useful for modeling. Knowledge of the active deformation mechanism under different loading conditions will provide insight about the macroscopic response and clues for the choice of variables. A priori knowledge of the deformation mechanisms can guide experimental programs and the selection of materials for specific applications.

CHAPTER 6

Foundation of State Variable Modeling

In this chapter we present the philosophy of the state variable approach to constitutive modeling and show how to use the results from physical metallurgy in the development of models. The results from classical plasticity, creep, and physical metallurgy are reviewed to establish a foundation for the state variable approach. Three basic forms for the inelastic flow law are introduced (exponential, power law, and hyperbolic sine) and the uniaxial exponential model is developed as an example for the tensile, cyclic, and creep response of metals. Chapter 7 contains an extension to the multiaxial and thermomechanical response of metals and several three-dimensional constitutive models.

6.1 GENERAL PHILOSOPHY OF STATE VARIABLES

This section is used to establish many of the key ideas of the state variable approach. It begins with a discussion of the variables that are used in state variable modeling and the structure of the equations that form the philosophy of the approach. The results from physical metallurgy and the classic approaches are examined to identify some specific choices for the state variables and the effect of these choices on the mathematical structure of the equations.

Characterizing the properties of dislocations and their reactions with grain boundaries, precipitates, and other dislocations could take many variables. The precise characterization of each dislocation and interaction with the microstructure appears impossible and impractical since there are typically 10^6 to 10^{12} cm of dislocation lines per cubic centimeter in a plastically deformed metal. Further, an identical microstructure is not repeatable in two samples of the same alloy. Thus it is necessary to classify and limit variables for use in modeling. Two types of internal state variables have been suggested by Swearengen and Holbrook (1985). Explicit internal state variables are used to define the local details of the microstructure, such as grain size, precipitate orientation, and dislocation properties. A large number of explicit internal state variables are required to fully characterize a microstructure. Alternatively, implicit internal state variables are defined as macroscopic averages of events associated with the development of the microstructure. They characterize average properties associated with dislocation propagation, and their interaction with precipitates, grain boundaries, and other dislocations. In this work, implicit variables are used, and the term *state variable* means implicit internal state variable. This is the primary difference between a materials and a mechanics approach to modeling the microstructure. The material scientist uses explicit internal state variables to model events in the microstructure to develop new materials and processes, whereas the objective of this work is to develop material models for use in finite element simulations of structural response.

In addition to the state variables, there are external variables that are macroscopically defined or observed, and controlled, measured, or calculated in an experimental program. These variables must be correlated with the internal state variables of the material during deformation. Typical external variables include stress, pressure, strain, strain rate, inelastic strain rate, work, strain energy density, and temperature.

Plastic and creep strain variables are not used. Instead, a single, unified, inelastic strain variable is used to characterize all the inelastic effects. Recall the kinematic equation used to determine the total strain tensor, $\boldsymbol{\varepsilon}$, is the sum of the elastic $\boldsymbol{\varepsilon}^{\mathrm{E}}$, inelastic, $\boldsymbol{\varepsilon}^{\mathrm{I}}$, and thermal strain tensors, $\boldsymbol{\varepsilon}^{\mathrm{TH}}$. Mathematically, the kinematic equation 5.8.2 is replaced by

$$\varepsilon_{ij} = \varepsilon_{ij}^{\mathrm{E}} + \varepsilon_{ij}^{\mathrm{I}} + \varepsilon_{ij}^{\mathrm{TH}} \tag{6.1.1}$$

The inelastic strain is the sum of the plastic and creep strain components, $\varepsilon_{ij}^{\mathrm{I}} = \varepsilon_{ij}^{\mathrm{P}} + \varepsilon_{ij}^{\mathrm{C}}$. Equation 6.1.1 avoids many of the issues associated with defining and characterizing plasticity separately and developing a rule for creep–fatigue interaction. Experimentally, the inelastic strain can be determined uniquely from the total measured strain since the elastic and thermal strains are almost entirely unaffected by inelastic deformation.

Time is avoided as a variable in rate equations. Recall the results from time hardening. Use of time did not produce realistic results in cyclic loading because the principle of objectivity was not satisfied. However, when elapsed time is used and objectivity is satisfied, the resulting mathematical structure for the flow equation, for example, is a hereditary integral, which is not desirable for computation or the evaluation of the material parameters. Time-dependent variables will be characterized by their rates, such as the inelastic strain rate or state variable evolution rate. The inelastic strain rate, for example, can be integrated to obtain the accumulated inelastic strain at any point in deformation history without ever using time explicitly as a state variable.

One objective of state variable constitutive modeling is to develop a correlation between the external variables and the internal state variables. The existence of this correlation permits the inelastic strain rate of a metal at any point in the deformation history to be determined from the current values of the external and internal variables. The inelastic strain rate at a later point in the history is determined by later values of the mechanical and thermal loads, and the later values of the state variables. Thus the state variables must be updated continuously to account for the changes in the microstructure. The relationship between the current microstructure and previous deformation history; that is, the "history dependence" is included in the model through the evolution of the state variables. The entire previous history of the deformation is characterized by the current state of the microstructure. This is not fundamentally different from yield surface plasticity, where the coordinates of the center and yield surface size embody the deformation history and are, in fact, macroscopic state variables.

Mathematically, the state variable approach can be characterized by three types of equations: a kinematic equation, a kinetic equation, and a set of state variable evolution equations. The kinematic equation is used with the integral of the inelastic strain rate; that is,

$$\varepsilon_{ij} = \varepsilon_{ij}^{E} + \int_0^t \dot{\varepsilon}_{ij}^{I}\, d\tau + \varepsilon_{ij}^{TH} \tag{6.1.2}$$

where t is some time in the deformation history. The total strain and inelastic strain can be large or small, but the elastic and thermal strains are assumed to be small, so a linear elastic stress–strain law can be used (equation 4.5.8 or 4.5.10 depending on symmetry). The kinetic equation, or inelastic flow law, is written formally as

$$\dot{\varepsilon}_{ij}^{I} = \dot{\varepsilon}_{ij}^{I}(\sigma_{kl}, T, \zeta_k, \ldots) \tag{6.1.3}$$

where $\dot{\varepsilon}_{ij}^{I}$ is the inelastic strain rate tensor that must be integrated to obtain the accumulated inelastic strain. The stress, σ_{kl}, temperature T, other external variables, and internal state variables ζ_k are expected to be present. The state variables are used as macroscopic measures of the characteristics of the microstructure. The evolution equations to update the current values of the state variables are formally given as

$$\dot{\zeta}_k = \dot{\zeta}_k(\sigma_{ij}, \dot{\varepsilon}_{ij}^{I}, T, \zeta_i, \ldots) \tag{6.1.4}$$

and must also be integrated to determine their current values. At this point in the development the exact definition or number of state variables is undefined and needs to be established from the principles of physical metallurgy. Further, the exact list of independent external and internal state variables in equations 6.1.3 and 6.1.4 are undefined and must be established. Equations 6.1.2 to 6.1.4 establish the essential mathematical structure for all state variable models.

The next step is to identify some potential state variables. The state variables must be selected to describe the critical microstructural effects in anticipation that the predicted macroscopic inelastic response will be accurate to a broad class of loading conditions and temperatures. Thus let us review the critical metallurgical and experimental observations that can be used to define the state variables and external variables. These important concepts are:

1. Inelastic deformation occurs at low temperature due to slip. The slip occurs on the slip planes closest to the planes with the maximum shear stress (Section 1.2.1).
2. Inelastic deformation occurs at high temperature due to planar slip and dislocation climb, a diffusion-controlled mechanism (Section 1.2.3).
3. The accumulated inelastic strain does not produce a significant change in the volume. Recall from Figure 4.4. 1 that the volumetric strain during tensile testing of several alloys was two orders of magnitude smaller than the tensile strain.
4. The accumulated inelastic strain and strain rate appear to be independent of hydrostatic stress (Figure 4.4.2) except for some alloys under extreme hydrostatic loading conditions, as shown in Figure 4.4.3.
5. Back stress (a resistance to slip) results from the interaction of dislocations with other dislocations, grain boundaries, precipitates, and other barriers. The net stress producing slip or inelastic strain is the overstress or the difference between the shear stress and back stress (Section 1.3.1). The back stress is in the direction opposite to the local shear stress in uniaxial loading and is therefore expected to be orientation dependent in three-dimensional loading.
6. The initial hardness of a metal is determined from the resistance to slip that is created by obstacles such as grain boundaries, precipitates, and interstitial atoms in the initial microstructure. These effects are isotropic in polycrystalline metals and are modeled by the initial value of the drag stress (Sections 1.5 and 2.2).
7. Hardening is an increase in the resistance to deformation. It results primarily from the interaction of dislocations with other dislocations, precipitates, and grain boundaries.
8. Dynamic recovery is the annihilation of dislocations during deformation that reduces the effective rate of hardening. It was shown in Section 1.4.2 and Figure 2.4.4 that temperature and strain-rate dependence result from the balance between the hardening and recovery rates.
9. Recovery is achieved by the formation of cells and subgrains. Cells and subgrains are orientation independent and monotonically approach a saturated state under uniform loading conditions. This effect is shown in the micrographs in Figures 1.1.1 and 1.3.8.
10. Creep results primarily from the temperature-dependent mechanisms of diffusion, dislocation glide, and climb (Section 3.2). Creep in ten-

sion and compression are different (Figure 3.3.4), due to the effect of lattice expansion or compression on the diffusion rate.

11. Static recovery results from the interaction stress between the dislocations themselves. As discussed in Section 1.4.1, higher temperatures increase dislocation mobility and promote diffusion. Static recovery lowers the effective steady-state hardening during creep. The rates of dynamic and static thermal recovery are much different.

The net effect of hardening and softening can be separated mathematically into two groups by their orientation-dependent or orientation-independent characteristics. This establishes the need for a minimum of two types of hardening state variables. A back-stress tensor is introduced to model the hardening and recovery effects associated with dislocation pileups. In addition, a scalar drag stress is used to model the development of hardening or softening associated with isotropic effects such as point defects, precipitates, cells and subgrains and grains. Other scalar state variables are necessary to model the net effect of hardening and static thermal recovery during creep. These variables will be introduced as necessary in the development of the different models.

The back-stress effect has been incorporated in state variable models by Chaboche (1977), Lee and Zaverl (1978, 1980a, b), Miller (1976), Walker (1980), Ramaswamy et al. (1984, 1990), and Stouffer et al. (1990). However, use of a back-stress type of variable can be observed in the kinematic hardening models of Prager and Ziegler in Section 5.3.3. In all cases, the kinetic equation for the inelastic strain rate (equation 6.1.3) has the form

$$\dot{\varepsilon}^{\mathrm{I}}_{ij} = \lambda(\sigma_{kl}, T, \zeta_k, \ldots)(S_{ij} - \Omega_{ij}) \tag{6.1.5}$$

where the deviatoric strain is used to preserve the condition of no inelastic volume changes. This equation is similar to equation 5.3.9 for kinematic hardening except that inelastic strain rate is used in place of the plastic strain increment, and the meaning for back stress and the kinematic variable α_{ij} are different. The function λ depends on the other macroscopic and state variables, similar to equation 6.1.3. Equation 6.1.5 includes the effects of back stress by permitting reverse inelastic flow after unloading, as shown in Figures 1.3.3 to 1.3.5, when the overstress, defined as the difference between the applied stress and the back stress, becomes negative. Equation 6.1.5 satisfies inelastic incompressibility if the back stress is a second-order deviatoric tensor since both the stress and inelastic strain rate are second-order deviatoric tensors and λ is a scalar. The tensorial property of the back stress also allows for orientation dependence that is consistent with the dislocation pileups discussed above. If the applied stress direction is changed, the direction of shear stress in the crystallographic slip system will also change, and slip would occur in another slip plane in a different direction, producing different dislocation interactions. Thus the components of the back stress tensor will also change. Finally, the initial value of the back stress is usually assumed to be zero because dislocation interactions do not develop until after slip has started.

The drag stress variable, Z, is introduced to model the initial hardness and the subsequent hardening or softening that develops during cycling. The initial value of drag stress, $Z(0) = Z_0$, is used to characterize the initial resistance to slip that is present in the undeformed microstructure. The evolution of the drag stress is associated with the interaction of dislocations with point (orientation-independent) defects and the growth of cells and subgrains. Since the interaction of dislocations with precipitates, interstitial atoms, cells, and subgrains is largely orientation independent, the drag stress is assumed to be a scalar. The drag stress is assumed to increase monotonically (for hardening) or decrease (for softening) to a saturated state in constant-amplitude cycling. Bodner (1987) uses a scalar variable similar to drag stress to model both isotropic and directional hardening as a function of the accumulated inelastic work, whereas Walker (1981), Miller (1976), Robinson (1978, 1982), and others use drag stress to model isotropic hardening or softening as a function of the accumulated inelastic strain. The drag stress is roughly equivalent to isotropic hardening in the classical plasticity.

The representation for plastic flow with kinematic hardening in the yield surface theory (equation 5.3.9) is similar to equation 6.1.5. Recall that inelastic strain rate and deviatoric stress tensors are parallel without kinematic hardening or back stress. However, when the latter are present the inelastic strain rate and deviatoric stress are not parallel. This result is consistent with the experimental effects that were discussed in Section 5.3.5.

The relationship between deviatoric stress and inelastic strain rate was investigated by Lindholm et al. (1984) for nonproportional cyclic loading. Tension/torsion samples were loaded in 90° out-of-phase control, as discussed in Section 5.3.5. In 90° out-of-phase loading, the principal strain directions and the maximum shear strain directions sweep continuously through all the material planes. The experiment by Lindholm et al. (1984) for Hastelloy X is shown in Figure 6.1.1. The angle between the inelastic strain rate vector and the deviatoric stress vector also varies continuously. This experiment also confirms another important macroscopic mechanical characteristic that results from the back stress. Equation 6.1.5 permits the variation in angle between inelastic strain rate and stress. The effect will be discussed in Chapter 7 when the multiaxial response is presented.

6.2 INELASTIC FLOW EQUATION

The section begins with a method to model loading and unloading without a yield surface. The section also contains the general methods that are used to develop the state variable flow equations by combining the results of yield

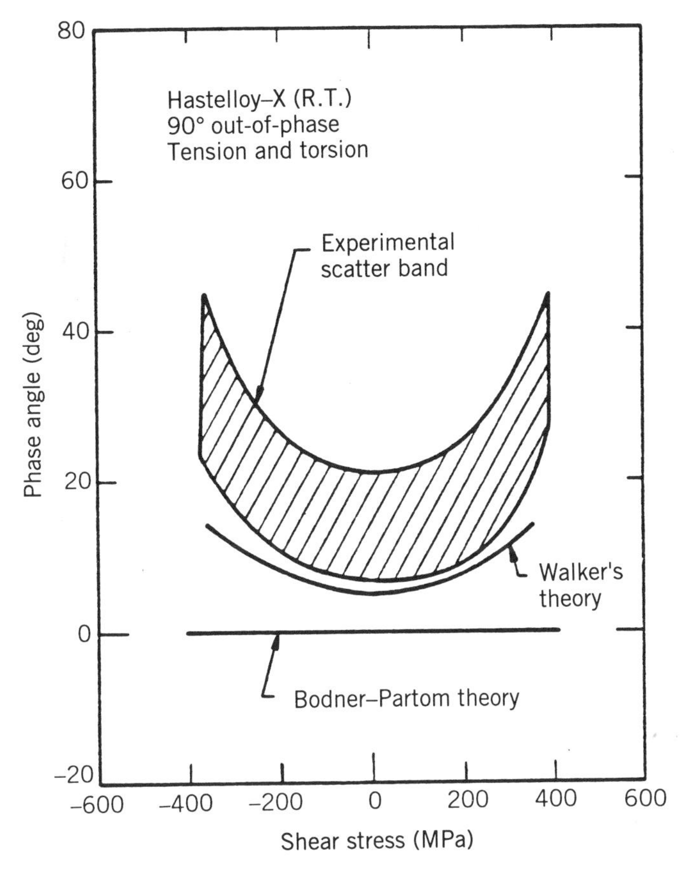

Figure 6.1.1 Angle between the shear stress and inelastic strain rate vectors for Hastelloy X during an axial-torsion experiment. (After Lindholm et al., 1984.)

surface plasticity with metallurgical observations of the dislocation velocity. This analysis gives the four basic types of flow equations for the inelastic strain rate.

Before developing an inelastic flow equation it is useful to show how to accomplish loading and unloading without the use of a yield surface. Consider the time derivative of kinematic equation 6.1.1 for the special case of uniaxial isothermal response. In this situation the total strain rate can be written as

$$\dot{\varepsilon} = \frac{\dot{\sigma}}{E} + \dot{\varepsilon}^{\mathrm{I}}(\sigma, \zeta_k) \tag{6.2.1}$$

where the inelastic strain rate is assumed to be determined from the current values of the stress and state variables. Assume that the material is loaded in tension at the constant strain rate $\dot{\varepsilon} = \dot{\varepsilon}_0$ to point A, where the stress strain curve is nearly flat and then unloaded. This hypothetical test is shown in Figure 6.2.1. At point A the inelastic strain rate is given by

$$\dot{\varepsilon}^{\mathrm{I}}(\sigma_A, \zeta_k^A) \approx \dot{\varepsilon}_0$$

since the stress rate is nearly zero. The inelastic strain rate at point A depends on the stress and state variables at point A. Immediately after unloading at point B the inelastic strain rate is given by

$$\dot{\varepsilon}^{\mathrm{I}}(\sigma_B, \zeta_k^B) = \dot{\varepsilon}^{\mathrm{I}}(\sigma_A, \zeta_k^A) = \dot{\varepsilon}_0$$

which is the same as point A because the stress and state variables do not change the instant the strain rate changes. The stress rate at point B can now be determined by writing equation 6.2.1 as

$$-\dot{\varepsilon}_0 = \frac{\dot{\sigma}_B}{E} + \dot{\varepsilon}_0$$

which gives $\dot{\sigma}_B = -2E\dot{\varepsilon}_0$. Thus the stress rate is initially $2E$ faster than the inelastic strain rate and "linear elastic" unloading is achieved when $\dot{\varepsilon}^{\mathrm{I}} = 0$. This method of determining the unloading response is also numerically more efficient than the loading criteria used with yield surface plasticity in Section 5.4.

A general representation for the mathematical form of the parameter λ can be determined by repeating the steps leading to equation 5.2.22 for the inelastic strain rate as given in equation 6.1.5. In this case

$$\dot{\varepsilon}_{ij}^{\mathrm{I}} = \sqrt{D_2}\,\frac{S_{ij} - \Omega_{ij}}{\sqrt{K_2}} \tag{6.2.2}$$

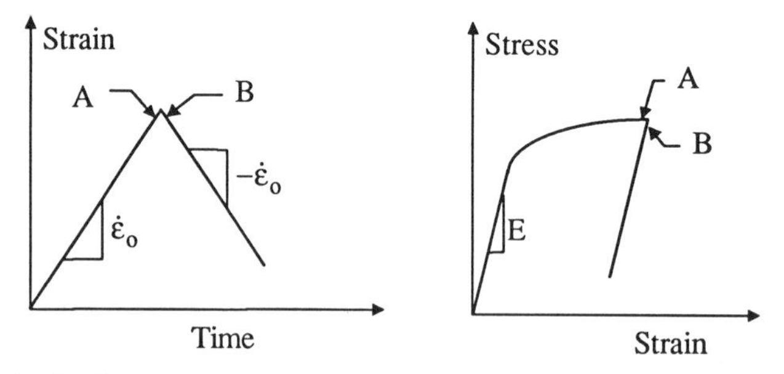

Figure 6.2.1 Strain control loading history and resulting stress–strain response of a tensile specimen.

where D_2 and K_2 are the second invariants of the inelastic strain rate and overstress tensors, that is

$$D_2 = \tfrac{1}{2}\dot{\varepsilon}^{\mathrm{I}}_{ij}\dot{\varepsilon}^{\mathrm{I}}_{ij} \qquad \text{and} \qquad K_2 = \tfrac{1}{2}(S_{ij} - \Omega_{ij})(S_{ij} - \Omega_{ij}) \tag{6.2.3}$$

Equation 6.2.2 is the state variable analog to the Prandtl–Reuss equation 5.5.11. The parameter $\sqrt{D_2}$ is proportional to the effective inelastic strain rate in equation 5.2.21, $\dot{\varepsilon}^{\mathrm{I}}_e = 2\sqrt{D_2/3}$. In this case the components of the inelastic strain rate are proportional to the components of the overstress tensor rather than the deviatoric stress. Equation 6.2.2 also embodies the ideas of kinematic hardening and is consistent with the Bauschinger effect.

The next step is to establish a representation $\sqrt{D_2}$. The back stress is used to model dislocation pileups. The net stress producing slip is the difference between the applied stress and back stress. It is primarily responsible for the strain hardening observed in tensile curves and the Bauschinger effect that occurs on unloading and reloading in the opposite direction. Since the back stress develops with plastic straining, the initial value of the back stress is assumed to be zero, $\Omega_{ij}(0) = 0$. The drag stress is used to model the isotropic hardening associated with the formation of a dislocation microstructure. The initial hardness, which is assumed to be isotropic, is characterized by the initial value of the drag stress, that is, $Z(0) = Z_0$. Since precipitates and solid solution atoms are part of the initial microstructure their presence contributes to the initial hardness of the alloy. The drag stress enters equations for the inelastic strain rate as the inverse ratio with the overstress. The combined effect of back stress and drag stress has the general form

$$D_2 = D_2\left(\frac{|S_{ij} - \Omega_{ij}|}{Z}\right) \text{ or } D_2 = D_2\left(\frac{\sqrt{K_2}}{Z}\right) \tag{6.2.4}$$

in almost all the state variable constitutive equations. Finally, note that the two effects, back stress and drag stress, are present simultaneously. The drag stress evolves from an initial value to a final value with the development of a microstructure. Thus the initial value of the drag stress is used to model experimental results with relatively small amounts of accumulated inelastic strain or inelastic work. The drag stress, Z, is generally assumed to be near its initial value for a small strain tensile test or the first few cycles of a fatigue test. The drag stress also remains almost constant for materials with an initial microstructure that is very hard. In this case the additional hardening due to dislocation interactions is small compared to the initial hardness. These properties will be used to establish constants in the state variable equations.

Several specific representations for D_2 have been proposed in the literature. In most cases the specific mathematical representations are based on metallurgical equations for the deformation rate as a function of stress and temperature. These equations fall into one of three categories of functions: power law, exponential, and hyperbolic sine, similar to equations 3.4.1, 3.4.3, 3.4.5, and 3.4.6 for steady-state creep. For example, one of the first representations was due to Bodner and Partom (1972) for the special case when the back stress is zero in equation 6.2.3. They observed from the results of Gilman (1959) that the dislocation velocity as a function of stress is represented by two fundamental mathematical forms:

$$v = A \exp\left(-\frac{B}{\tau}\right) \qquad \text{and} \qquad v = C\tau^n \tag{6.2.5}$$

where v is the dislocation velocity on the slip plane, τ the local shear stress, A, B, and C material parameters, and n an exponent. The exponential equation is best for slip at high velocities, correlates well with data, and has a relatively firm theoretical foundation in physical metallurgy. The power law equation is more representative of experimental data for dislocation movements at low velocities. Bodner proposed a representation that embodied both functions by assuming that

$$D_2 = D_0^2 \exp\left[-\left(\frac{Z^2}{3J_2}\right)^n\right] \tag{6.2.6}$$

where Z is a single scalar state variable used to model both isotropic and kinematic hardening mechanisms and J_2 is the second invarient of the deviatoric stress tensor. The constant D_0 is the limiting value of the inelastic strain rate and the exponent n is used to couple power law properties to the exponential function. The general form of the Bodner equation (Section 7.7) is obtained by substituting equation 6.2.6 into equation 6.2.2 for the special case of no back stress:

$$\dot{\varepsilon}_{ij}^{\mathrm{I}} = D_0 \exp\left[-\frac{1}{2}\left(\frac{Z^2}{3J_2}\right)^n\right]\frac{S_{ij}}{\sqrt{J_2}} \tag{6.2.7}$$

The Bodner equation was later modified to include back stress by Ramaswamy and Stouffer (1990), that is,

$$\dot{\varepsilon}_{ij}^{\mathrm{I}} = D_0 \exp\left[-\frac{1}{2}\left(\frac{Z^2}{3T^2K_2}\right)^n\right]\frac{S_{ij} - \Omega_{ij}}{\sqrt{K_2}} \tag{6.2.8}$$

where absolute temperature (kelvin) is included for extension to thermal cycling in Chapter 7. Other representations were established by similar arguments. Walker (1981) and Kreig, Swearengen, and Rohde (1978) proposed separately a power law representation (Section 7.9) that has the form

$$\dot{\varepsilon}^{\mathrm{I}}_{ij} = D_0 \frac{|S_{kl} - \Omega_{kl}|^n}{Z} \frac{(\frac{3}{2}S_{ij} - \Omega_{ij})}{|S_{kl} - \Omega_{kl}|} \tag{6.2.9}$$

which follows directly from equation 6.2.5. The quantity $|S_{kl} - \Omega_{kl}|$ in equation 6.2.9 is given by

$$|S_{kl} - \Omega_{kl}| = \sqrt{\tfrac{2}{3}(\tfrac{3}{2}S_{ij} - \Omega_{ij})(\tfrac{3}{2}S_{ij} - \Omega_{ij})} \tag{6.2.10}$$

Miller (1976) proposed a representation (Section 7.8) using the hyperbolic sine function that is a physically based model for the steady-state creep rate, $\dot{\varepsilon}_{\mathrm{SS}}$, as a function of the constant creep stress, σ_{SS}, in equation 3.4.3. Introducing the back stress and drag stress in the format described in equation 6.2.4, Miller obtained a flow equation in the form

$$\dot{\varepsilon}^{\mathrm{I}}_{ij} = \frac{3}{2} D_0 \left[\sinh\left(\frac{\sqrt{K_2}}{Z}\right)^{3/2}\right]^n \frac{S_{ij} - \Omega_{ij}}{\sqrt{K_2}} \tag{6.2.11}$$

The results in equations 6.2.7, 6.2.8, 6.2.9 and 6.2.11 are representative of the basic types of state variable flow equations that are currently in use. The role of the state variables in the three representations with back stress and drag stress are equivalent. The hardening parameter used by Bodner is not exactly the drag stress but a single scalar parameter that attempts to include both the scalar and vector hardening effects.

An important difference between the state variable flow equations and the yield surface approach to plasticity is that the inelastic strain rate is present for all values of stress. The computed strain rates for small stresses are very small, and the accumulated inelastic strain is very small compared to the elastic strain. The idea of inelastic strain at all values of stress appears to be consistent with experimental observations. Recall the effect of initial loading rate on creep that is shown in Figure 3.6.4. The higher loading rates must have developed more dislocation mobility that produces more primary creep strain. This conclusion follows since the accumulated creep strain resulting from dislocation glide and dislocation creep will increase with increasing dislocation mobility and dislocation density.

The next step is to reduce the three-dimensional models to uniaxial representations for use with tensile and fatigue test data. Let us define ε, σ, and Ω as the uniaxial values of the strain, stress, and back stress, respectively. Since these variables are all deviatoric tensors, uniaxial values must be written as tensor components. The back-stress tensor can be defined like the deviatoric stress tensor for uniaxial loading or like α in equation 5.3.13. Defining back stress like deviatoric stress, $K_2 = (\sigma - \Omega)^2/3$, the three-dimensional flow equations become:

$$\text{Bodner–Partom:}\quad \dot{\varepsilon}^{\mathrm{I}} = \frac{2}{\sqrt{3}} D_0 \exp\left[-\frac{1}{2}\left(\frac{Z}{|\sigma|}\right)^{2n}\right]\frac{\sigma}{|\sigma|} \tag{6.2.12}$$

$$\text{Ramaswamy–Stouffer:}\quad \dot{\varepsilon}^{\mathrm{I}} = \frac{2}{\sqrt{3}} D_0 \exp\left[-\frac{1}{2}\left(\frac{Z}{T|\sigma-\Omega|}\right)^{2n}\right]\frac{\sigma-\Omega}{|\sigma-\Omega|} \tag{6.2.13}$$

$$\text{Miller:}\quad \dot{\varepsilon}^{\mathrm{I}} = D_0\left[\sinh\left(\frac{|\sigma-\Omega|}{Z}\right)^{3/2}\right]^{n}\frac{\sigma-\Omega}{|\sigma-\Omega|} \tag{6.2.14}$$

Using the definition in equation 5.3.13 for back stress, the power law model becomes

$$\dot{\varepsilon}^{\mathrm{I}} = D_0\frac{|\sigma-\Omega|^{n}}{Z}\frac{\sigma-\Omega}{|\sigma-\Omega|} \tag{6.2.15}$$

In the Bodner and Ramaswamy–Stouffer equations the parameter D_0 is a scale factor that corresponds to the maximum inelastic strain rate. The parameter $D_0 = 1\ \mathrm{sec}^{-1}$ in the power law models, or is a function of temperature when nonisothermal cycling is considered. The exponent n controls the strain-rate sensitivity in all four models. An example of the effect of n on the strain-rate sensitivity is shown in Figure 6.2.2.

Rather than giving the complete running development of all four models, the ideas behind the state variable approach are given in detail, and the Ramaswamy–Stouffer model is developed as an example. Concepts from other authors and examples of equations are introduced when appropriate. A summary of the Bodner, Miller, Kreig–Swearengen–Rohde, and Walker models is given in Part C of Chapter 7. Finally, only isothermal response is considered in the remaining sections of this chapter; thermal history effects are discussed in Chapter 7. Thus the temperature dependence will be supressed for the remaining sections of this chapter.

6.3 BACK-STRESS EVOLUTION EQUATION

Recall that back stress was introduced to characterize the strain hardening and Bauschinger effects that arise from dislocation pileups. The drag stress is introduced to characterize the initial resistance to slip from precipitates, grains, and point defects in the initial microstructure and cyclic hardening that arises from dislocation structures such as cells and subgrains in fatigue. It is generally assumed that the drag stress does not evolve very much from the initial value during small strain tensile response. Thus it can be seen from equation 6.2.13 that if the stress and inelastic strain rate are constant in a tensile test the back stress is also constant (isothermal response). It is assumed

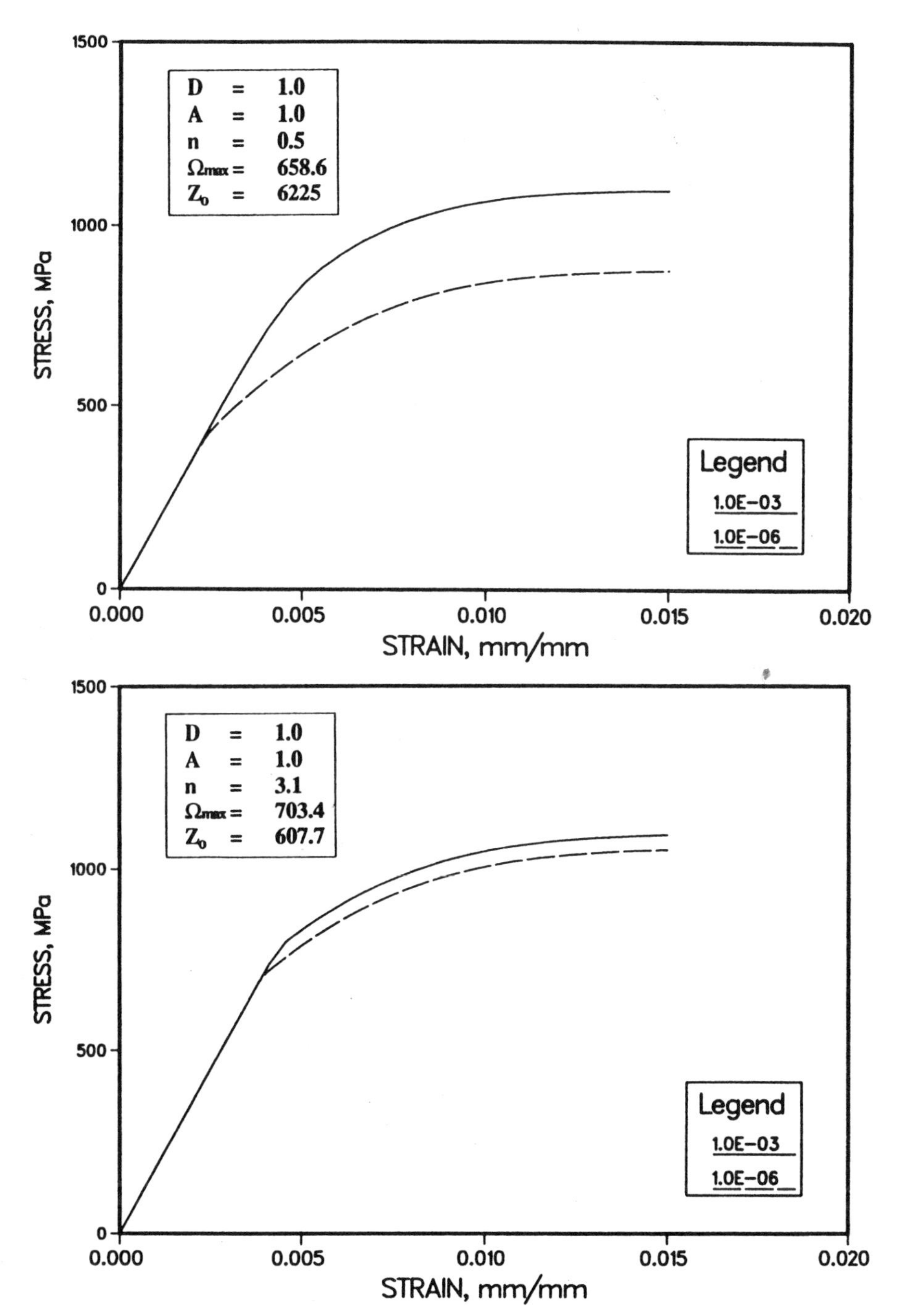

Figure 6.2.2 Effect of n on the strain-rate sensitivity of the exponential model for (a) rate sensitive and (b) rate insensitive materials. (Courtesy of M. Y. Sheh.)

that this constant value for the back stress at saturation of the tensile curve, at time t_s, is the maximum value of the back stress. The initial value of the back stress is taken to be zero. The initial value of the back-stress rate is unknown, but the back-stress rate at saturation must be zero since the back stress is constant. The initial and saturation values of the mechanical variables and state variables are summarized in Figure 6.3.1. In this section specific representations for the back-stress evolution equation are presented. The methods to evaluate the back-stress and flow equation parameters are discussed in Section 6.4.

The primary function of the back-stress equation is to characterize the effects of strain hardening and dynamic recovery. Recall that recovery is a reduction of the dislocation density and stored energy, which results in a reduction in the shear stress required to produce slip. Dynamic recovery occurs simultaneous with deformation and reduces the effect of hardening. The balance between the rate of hardening and recovery gives the strain-rate effect shown in Figure 2.4.4. Probably the simplest model for the back-stress rate that includes the simultaneous effect of strain hardening and dynamic recovery is

$$\dot{\Omega} = f_1 \dot{\varepsilon}^{\mathrm{I}} - g_1 \Omega |\dot{\varepsilon}^{\mathrm{I}}| \tag{6.3.1}$$

where f_1 and g_1 are constants. The first term, $f_1 \dot{\varepsilon}^{\mathrm{I}}$, simulates the growth in back stress or strain hardening as proportional to the inelastic strain rate. This follows since the dislocation density increases with increasing amounts of inelastic strain up to an equilibrium microstructure. The second term, $-g_1 \Omega |\dot{\varepsilon}^{\mathrm{I}}|$, simulates the rate of dynamic recovery and it is assumed to depend on the current value of the back stress and the inelastic strain rate. Thus initially there is no recovery, and the rate of recovery increases with increasing

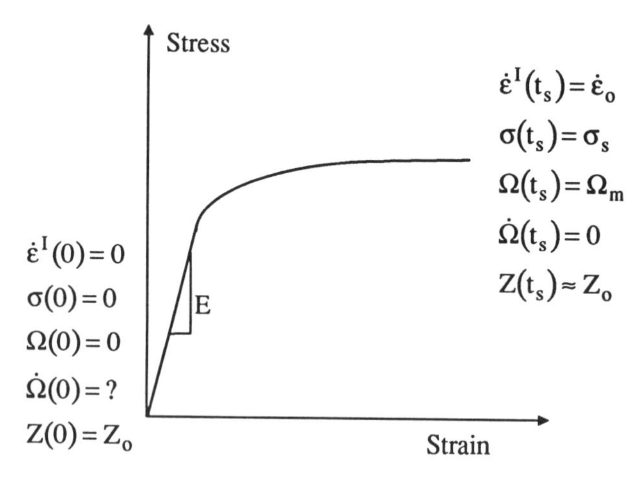

Figure 6.3.1 Initial and final values for several state variables in a tensile test.

dislocation density. Since dynamic recovery is simultaneous with deformation, the rate of dynamic recovery is assumed to depend directly on the rate of inelastic straining. The rate of dynamic recovery is zero when the inelastic strain rate vanishes. The absolute value is used with the inelastic strain rate in the recovery term to provide the correct sign for the dynamic recovery in both tension and compression.

The back-stress evolution equation 6.3.1 must provide a balance between hardening and recovery as the tensile curve approaches saturation (becomes flat). Referring to any of the flow equations shows that when the inelastic strain rate and stress are constant, the back stress must be constant since the drag stress is constant. In this situation

$$\dot{\Omega} = 0 = f_1 \dot{\varepsilon}^{\mathrm{I}} - g_1 \Omega_m |\dot{\varepsilon}^{\mathrm{I}}| \tag{6.3.2}$$

and the back stress is at a maximum value that is independent of strain rate. Thus $f_1 = g_1 \Omega_m$ and the back-stress evolution can be written as

$$\dot{\Omega} = f_1 \dot{\varepsilon}^{\mathrm{I}} - f_1 \frac{\Omega}{\Omega_m} |\dot{\varepsilon}^{\mathrm{I}}| \tag{6.3.3}$$

The parameter Ω_m would not expected to be a constant for all strain rates from very slow to very fast. This result is a consequence of the simple mathematical structure for equation 6.3.3. The formulation will be shown to be valid for relatively fast strain rates typical of tensile testing but will have to be modified for the back stress that is developed in dislocation glide and dislocation creep.

Equation 6.3.3 is a fairly common representation for the back-stress evolution (see, e.g., Walker, 1981, or Krieg, Swearengen, and Rohde, 1978). In general, most back-stress models fall into the form

$$\dot{\Omega} = h(\ldots)\dot{\varepsilon}^{\mathrm{I}} - r(\dot{\varepsilon}^{\mathrm{I}}, \Omega, \ldots) \tag{6.3.4}$$

where the hardening term is often taken as proportional to the inelastic strain rate and a function $h(\ldots)$ as shown in equation 6.3.4, and the dynamic recovery function, r, is usually a function of the inelastic strain rate and back stress. These functions frequently depend on temperature and are coupled to other state variables. In fact, it is fairly common to substitute the flow equation into the back-stress equation for the inelastic strain rate and deduce a structure for the recovery term when $\Omega = 0$. Various back-stress models are presented later.

Equation 6.3.3 is a first try for the back-stress evolution equation. It can be implemented with equation 6.2.13, for example, to determine the stress–strain response of an alloy. Let us assume that the constants have been determined for René 80 at 982° C. The experimental and model results are shown in Figure 6.3.2. The plot shows that the strain-rate sensitivity is correlated rather well but that the shape of the curves, especially for the slower strain rates, is not correct. Since the calculated response is below the experimental response, the amount of hardening, or the value of the back stress, should be larger early in the calculated response. A review of equation 6.3.3 shows that the initial value of the back-stress rate is zero, $\dot{\Omega}(0) = 0$, and it remains zero until inelastic straining begins.

The back-stress rate can be determined from the back-stress history by using the experimental data and a flow equation. Solving equation 6.2.13, the back stress can be determined from the data as

$$\Omega(t) = \sigma(t) - Z_0 \left[-2 \ln \frac{\sqrt{3}\, \dot{\varepsilon}^{\mathrm{I}}(t)}{2D_0} \right]^{-1/2n} \tag{6.3.5}$$

The back-stress rate history was then determined for Rene’ 80 and plotted in Figure 6.3.3a. Thus the tensile data tend to show that the initial value of the back-stress rate should not be zero. This effect has also been observed in creep as shown in Figure 3.6.4. Recall that the creep response depends on

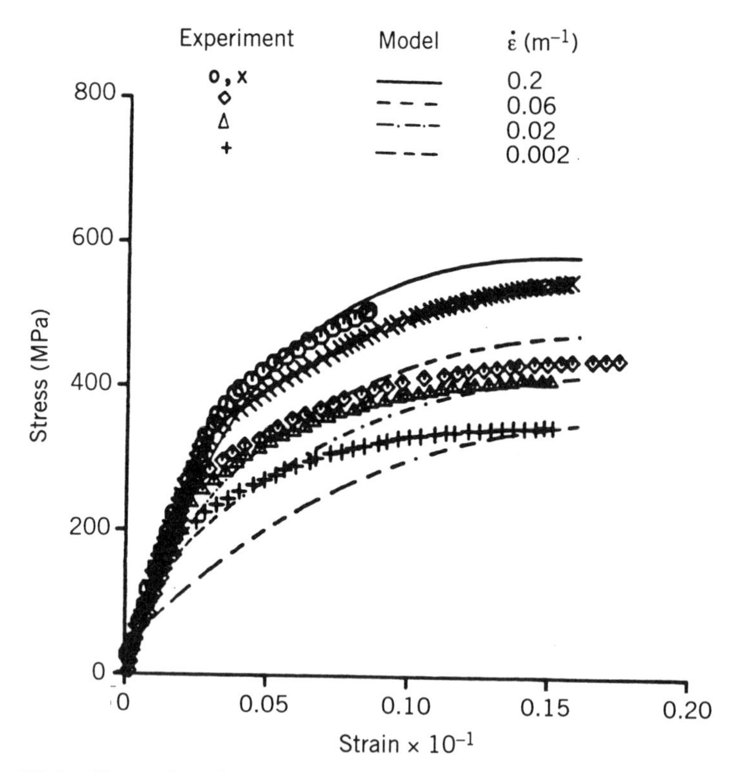

Figure 6.3.2 Comparison between the calculated and experimental tensile response of René 80 at 982°C without stress rate in the back-stress equation. (After Ramaswamy, 1986.)

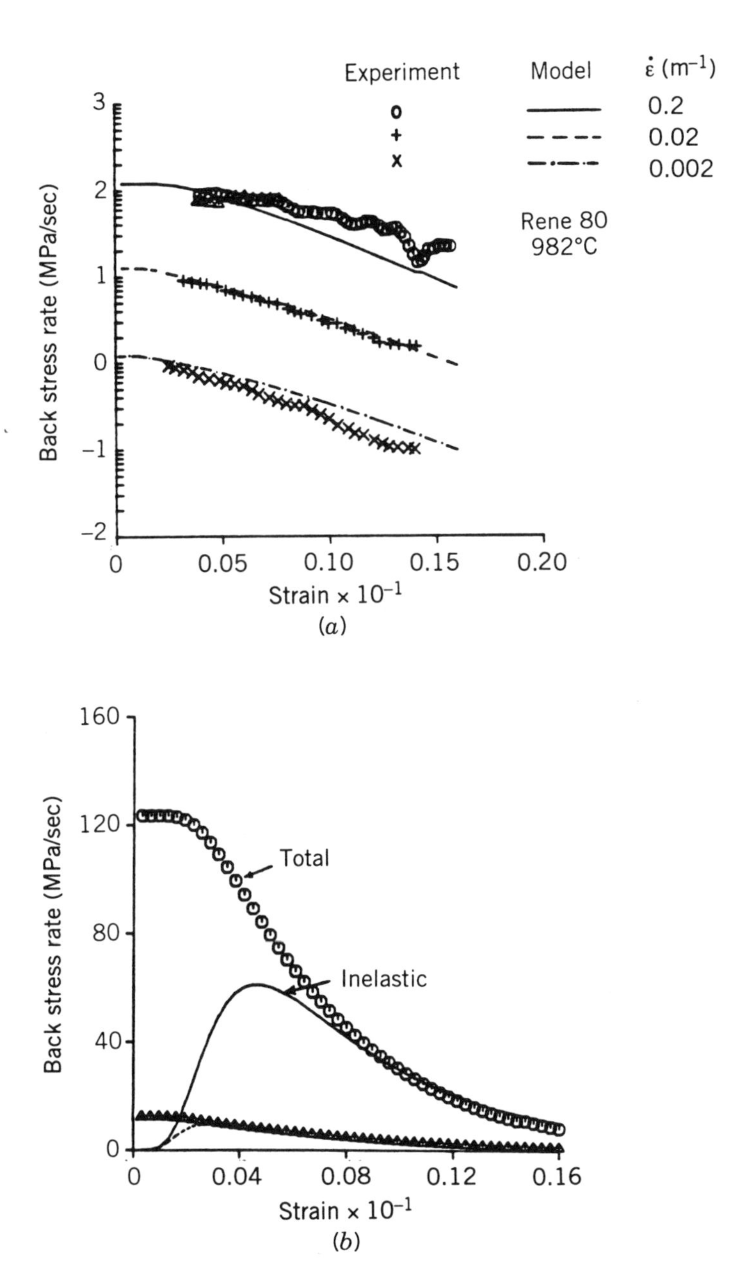

Figure 6.3.3 (*a*) Back-stress rate history, and (*b*) back stress history for René 80 at 982°C that was determined from the tensile response data. Solid lines show the correlation with the back-stress equation with stress rate included. (After Ramaswamy, Stouffer, and Laflen, Journal of Engineering Materials and Technology, V112, pg. 282, 1990. Copyright © The American Society of Mechanical Engineers, 345 East 47th St., New York, NY 10017. Used with permission.)

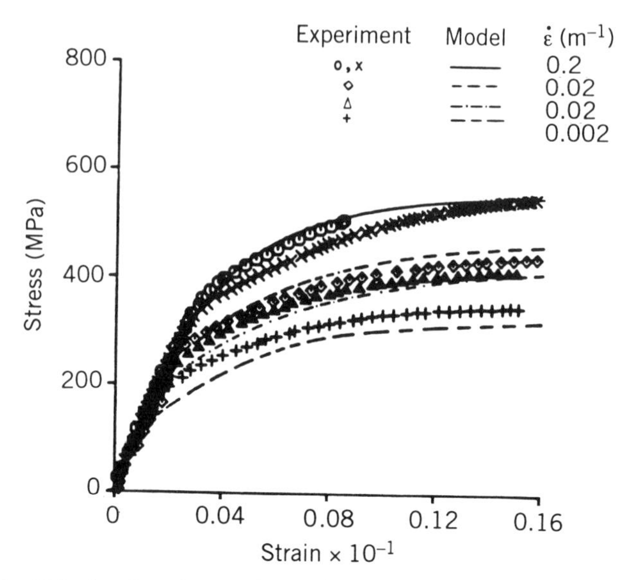

Figure 6.3.4 Calculated and experimental tensile response curves for René 80 at 982°C using stress rate in the back-stress equation. (After Ramaswamy, Stouffer, and Laflen, Journal of Engineering Materials and Technology, V112, pg. 282, 1990. Copyright © The American Society of Mechanical Engineers, 345 East 47th St., New York, NY 10017. Used with permission.)

the initial loading rate. The back-stress evolution equation can be modified to include the initial loading rate by assuming that the initial back-stress rate is proportional to the stress rate; that is,

$$\dot{\Omega} = f_1\, \dot{\varepsilon}^{\mathrm{I}} - f_1 \frac{\Omega}{\Omega_m} |\dot{\varepsilon}^{\mathrm{I}}| + f_2\, \dot{\sigma} \tag{6.3.6}$$

where f_2 is a material parameter. The stress-rate term will dominate early in the loading history and approach zero as the tensile response approaches saturation. The stress-rate term also implies that the back stress will be reduced from the maximum value on elastic unloading. This is reasonable since a dislocation pileup would be expected to relax at least some upon removal of the shear stress. Figure 6.3.4 shows that the stress-rate term is successful in increasing the initial rate of hardening and that the shape of the tensile curves are better. The correlation between the back-stress history and the model with stress rate in the back-stress rate equation is shown in Figure 6.3.3; and the tensile response in the strain rate independent regime is shown in Figure 6.3.5.

Consider next a parametric study of the model characterized by equations 6.2.13 and 6.3.6 to show the effect of the two constants in the back-stress equation. This type of study helps to establish methods to determine the

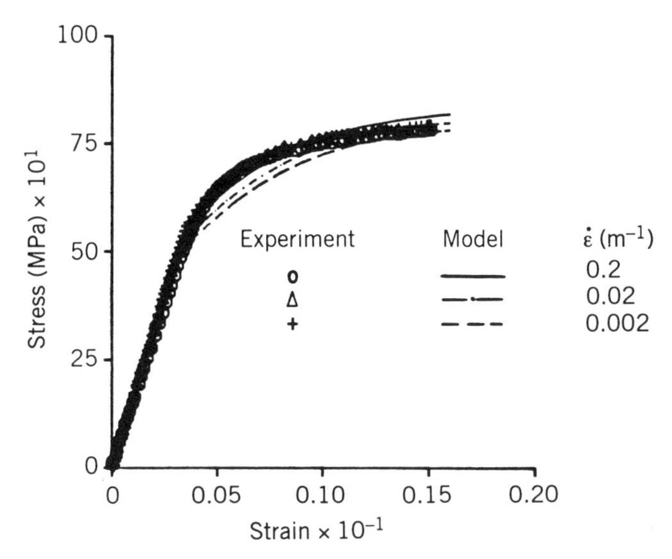

Figure 6.3.5 Comparison of calculated and experimental response of René 80 at 538°C. (After Ramaswamy, Stouffer, and Laflen, Journal of Engineering Materials and Technology, V112, pg. 282, 1990. Copyright © The American Society of Mechanical Engineers, 345 East 47th St., New York, NY 10017. Used with permission.)

material constants. The parameter f_2 controls the tensile yield stress as shown in Figure 6.3.6 and can produce a stress overshoot, a sharp yield, or a smooth yield stress. The stress overshoot, typical of low-carbon steel, is loading-rate sensitive, due to the effect of the dislocation atmosphere. The parameter f_2 also allows the initial loading rate to influence the amount of primary creep as shown in Figure 6.3.7. The parameter f_1 controls the amount of strain required to reach saturation, as demonstrated in Figure 6.3.8.

The use of stress rate in an evolution equation is relatively unique. Equation 6.3.6 satisfies the principle of objectivity for the linear strain formulation and Freed, Caboche, and Walker (1991) showed that equation 6.3.6 satisfies the principles of thermodynamics. Further, it is possible to redefine the parameters to eliminate the stress-rate term in equation 6.3.6. The back stress evolution can be rewritten after integrating as

$$\Omega = \Omega^{\mathrm{I}} + f_2 \sigma$$

where $\dot{\Omega}^{\mathrm{I}}$ is given by the remaining terms of equation 6.3.6. Substituting this result into equation 6.2.13 gives

$$\dot{\varepsilon}^{\mathrm{I}} = \frac{2}{\sqrt{3}} D_0 \exp\left[-\frac{1}{2}\left(\frac{\hat{Z}}{|\sigma - \Omega^{\mathrm{I}}|}\right)^{2n}\right] \frac{\sigma - \Omega^{\mathrm{I}}}{|\sigma - \Omega^{\mathrm{I}}|}$$

where

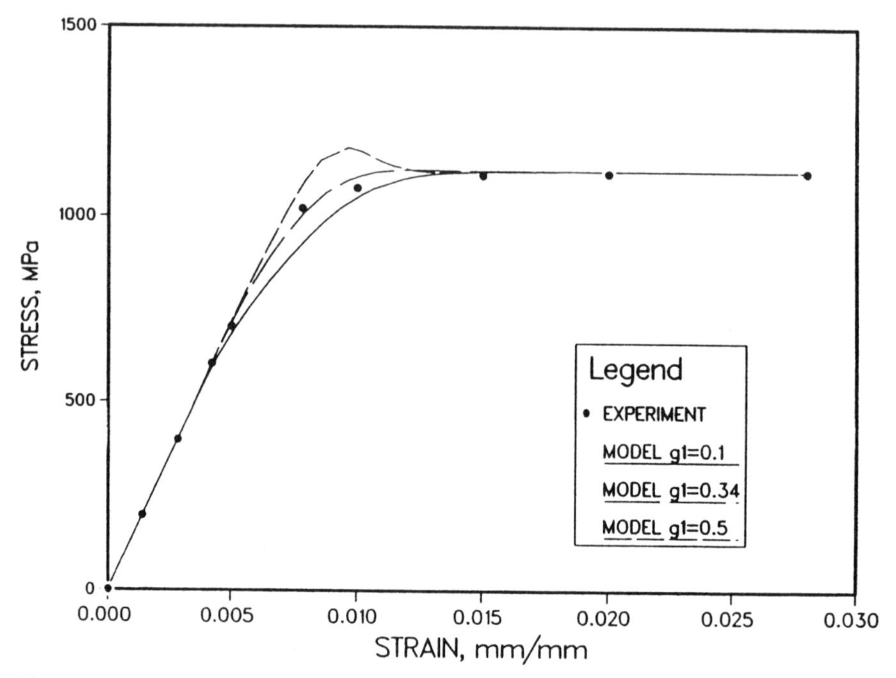

Figure 6.3.6 Effect of f_2 (g_1 in the legend) on the shape of the tensile response curve. (Courtesy of M. Y. Sheh.)

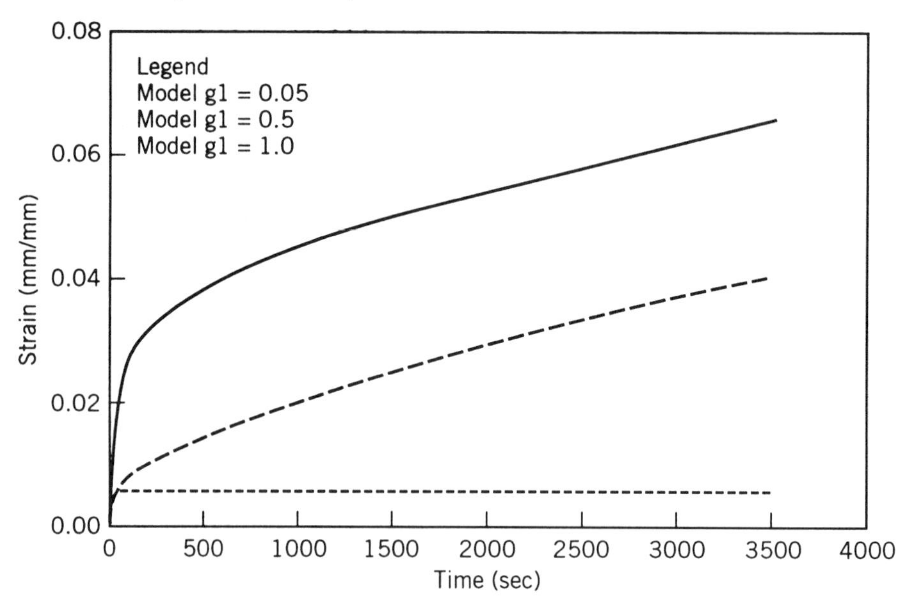

Figure 6.3.7 Effect of f_2 (g_1 in the legend) on the amount of primary creep. (Courtesy of M. Y. Sheh.)

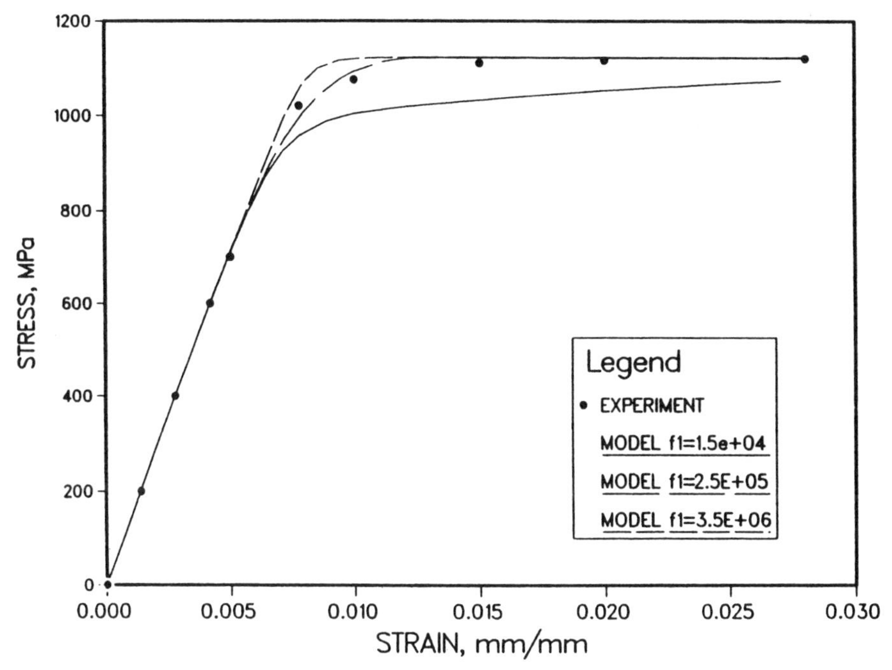

Figure 6.3.8 Effect of the parameter f_1 on the strain to saturation. (Courtesy of M. Y. Sheh.)

$$\hat{Z} = \frac{Z}{1 - f_2} \qquad \text{and} \qquad \dot{\Omega}^{\mathrm{I}} = \frac{1}{1 - f_2}\left(f_1\,\dot{\varepsilon}^{\mathrm{I}} - f_1\,\frac{\Omega}{\Omega_m}\,|\dot{\varepsilon}^{\mathrm{I}}|\right)$$

This result shows that the stress-rate term is not necessary. However, it is easier to determine the constitutive parameters with the stress-rate term present because it is easier to understand the effects of each parameter.

Other examples of the back stress evolution equation are shown to indicate some of the differences that may arise. Equation 6.3.6 has been extended by Sheh and Stouffer (1990) to include more material parameters for use in the single-crystal alloy René N4. This result,

$$\dot{\Omega} = f_1(\dot{\varepsilon}^{\mathrm{I}})^{n_1} - f_1\,\frac{\Omega}{\Omega_m}\,|\dot{\varepsilon}^{\mathrm{I}}|^{n_1} + f_2\left(\frac{\sigma}{\sigma_0}\right)^{n_2}\dot{\sigma} \tag{6.3.7}$$

still satisfies the saturation condition in equation 6.3.2. Miller (1976), for example, developed his back stress model from equation 6.3.4 where r is a function of the back stress and temperature. Then at saturation $r = h\,\dot{\varepsilon}^{\mathrm{I}}$, where the inelastic strain rate is evaluated at saturation using the flow equation. This gives

$$\dot{\Omega} = f_1 \dot{\varepsilon}^{\mathrm{I}} - f_1[\sinh(A_1|\Omega|]^n \frac{\Omega}{|\Omega|} \tag{6.3.8}$$

for constant temperatures after using a relationship between the drag stress and back stress at saturation, The quantities f_1 and A_1 are material constants. Notice that equation 6.3.8 permits recovery to continue after plastic deformation has been terminated, and the recovery will not stop until the back stress has decayed to zero.

6.4 EVALUATION OF PARAMETERS AND VERIFICATION

One issue that has prevented the use of state variable models by many people is the ease in determining the material parameters. This task is frequently complicated by the lack of experimental data. Methods will be developed to determine the constants in equations 6.2.13 and 6.3.6 as an example. These equations will then be verified for several different materials and loading conditions. The methods presented in this section can be applied to other constitutive equations, but the details of the results will depend on the mathematical characteristics of each model.

The section is divided into three subsections. Sections 6.4.1 and 6.4.2 show how to determine constants for the flow and back stress equations. The general approach is designed to determine the flow equation constants and verify the results before determining the back-stress parameters. The parameters for the drag stress and creep equations are then determined after the flow and back-stress parameters are verified. This process can be described by the following flow diagram in Figure 6.4.1. The drag stress and creep equation parameters

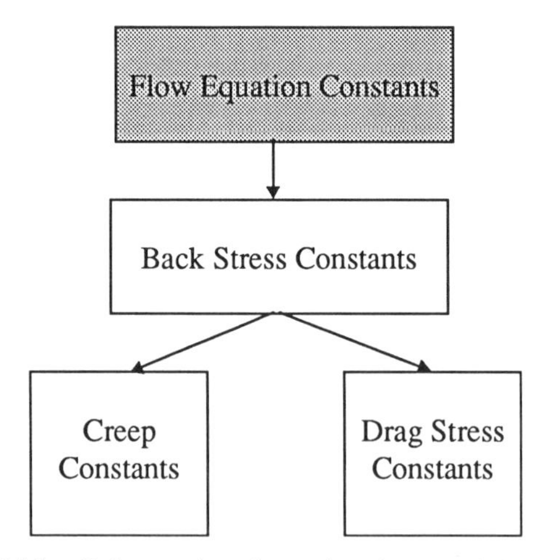

Figure 6.4.1 Order used to determine the material parameters.

can be determined in any order once the flow and back stress parameters are known.

Section 6.4.3 contains four relationships between the parameters. The equations were determined from data for the saturation of a tensile response curve, the Bauschinger effect, stress relaxation, and strain recovery. The results in this section are new and have not been fully tested, and should be used with caution. However, these four algebraic relationships reduce the number of independent parameters that must be determined and significantly simplify the task of finding parameters. Finally, the parameters for several materials are given in Appendices 6.1 to 6.3.

6.4.1 Flow Equation Parameters

It is often necessary to determine the inelastic strain rate as part of the process to find the material constants. If digital values of the stress, strain, and time variables are available from a computer-acquired record of the test, there is a large continuous data file. The inelastic strain can be determined from equation 6.1.1 using the stress and strain histories, and elastic modulus. The inelastic strain rate is then determined from the slope of a sliding spline fit of typically five or seven time-inelastic strain pairs. This process is satisfactory if there are at least 50 smooth data pairs. However, for fewer than 50 data pairs or for noisy data sets from data digitized by hand, it is often necessary to correlate the data with an equation that can be differentiated to give the inelastic strain rate. In this case the strain axis can be replaced by time ($\varepsilon = \dot{\varepsilon}_0 t$) and the knee of the stress–strain curve can be fitted with a simple function that can be differentiated. The inelastic strain rate is then given by $\dot{\varepsilon}^{\mathrm{I}} = \dot{\varepsilon}_0 - \dot{\sigma}/E$. Exponential and power law equations can be used to fit the tensile curve. This exercise provides a clue for the selection of the most accurate flow law.

In the Bodner and Ramaswamy equations the parameter D_0 is a somewhat arbitrary scale factor that correlates with the maximum inelastic strain rate. Typically, it is a factor 10^4 sec^{-1} greater than the maximum expected strain rate in a particular application. For example, $D_0 = 1$ sec^{-1} for creep, $D_0 = 10^4$ sec^{-1} for tensile and fatigue response for strain rates up to 1.0 sec^{-1}, and $D_0 = 10^6$ sec^{-1} or greater for wave propagation and high-rate loading applications.

Since n controls the strain-rate sensitivity, let us begin with the parameters in the flow equation and experimental tensile response data. At tensile data saturation (the strain at which the stress–strain curve becomes flat) the inelastic strain rate is equal to the total strain rate, $\dot{\varepsilon}_0$, which is constant in a constant strain rate tensile test. At saturation the back stress is at the maximum value, Ω_m (see equation 6.3.2) and the drag stress is near the initial value Z_0 and remains almost constant. Further, assume that the temperature is constant so that the specific dependence on temperature can be dropped. If there is any chance of extension to thermal cycling, the temperature should be retained. The presence of temperature does not complicate the material parameter evaluation procedure. Thus for saturated, isothermal, tensile loading equation 6.2.13 becomes

$$\dot{\varepsilon}_0 = \frac{2}{\sqrt{3}} D_0 \exp\left[-\frac{1}{2}\left(\frac{Z_0}{|\sigma_s - \Omega_m|}\right)^{2n}\right] \tag{6.4.1}$$

where the last term, $\sigma - \Omega/|\sigma - \Omega|$, in Equation 6.2.13 is equal to $+1$. Using this result, equation 6.4. 1 can be inverted and written in the form

$$X^{1/2n} = \left(-2 \ln \frac{\sqrt{3}\dot{\varepsilon}_0}{2D_0}\right)^{1/2n} = \frac{Z_0}{\sigma_s - \Omega_m} \tag{6.4.2}$$

which can be used to find n, Ω_m, and Z_0 assuming that D_0 is assigned a value as described above.

The maximum value of the back stress can vary widely in different types of materials. The value during tensile loading must be positive (opposite to the tensile stress) and less than the saturated tensile stress; that is, $0 \leq \Omega_m \leq \sigma_s$. In an extensively warm worked material that has developed subgrains and raised the yield strength considerably, the back stress is quite small when compared to the drag stress (Miller, 1976). In undeformed large-grain materials the maximum value of the back stress can easily be greater than one-half of the saturated tensile stress. Milligan and Antolovich (1989) determined values for the saturated back stress as large as $0.85\sigma_s$ in PWA 1480, a single-crystal alloy with large precipitates. Large values of back stress are also verified by cyclic loading if reverse plastic flow begins before the applied stress reaches zero on unloading. To summarize, the following assumptions can be used if no other data are available:

$$\Omega_m \approx \begin{cases} 0.0 \text{ to } 0.3\sigma_s & \text{for small-grain materials} \\ 0.5 \text{ to } 0.7\sigma_s & \text{for large-grain materials} \end{cases} \tag{6.4.3}$$

It may be difficult to determine Ω_m from tensile data only; therefore, it may be necessary to use an approximation. If cyclic data are available, the methods of Section 6.4.3 can be used to determine Ω_m.

The evaluation method for the constants n, Ω_m, and Z_0 depends on whether the response is strain-rate dependent or strain rate independent. In general, values of n greater than or equal to 3, $n \geq 3.0$, correspond to strain rate independent response for equation 6.2.13. For a material with almost no rate sensitivity, the value of n would be 3.0, and $D_0 = 10^4$ sec^{-1} is selected as discussed above. The corresponding value $X^{1/2n}$ in equation 6.4.2 is 1.62 and 1.77 for $\dot{\varepsilon}_0 = 1.0$ sec^{-1} and $\dot{\varepsilon}_0 = 0.001$ sec^{-1}, respectively. Using an average value of 1.70 in equation 6.4.2 gives

$$Z_0 = 1.70(\sigma_s - \Omega_m) \tag{6.4.4}$$

Equations 6.4.3 and 6.4.4 can be used to estimate the two parameters Z_0 and Ω_m for experimental data from a tensile test. One advantage of the state variable approach is that the equations appear to be relatively robust, so reasonable results can be obtained with only estimates of some of the parameters. In this case the remaining parameters that are determined from data are not the true parameters but are compensated to match the data.

The values of the parameters for strain rate sensitive material response are determined by the following procedure. In this case it is necessary to determine the exponent n to characterize the strain rate sensitive response. Let us assume that a family of strain rate sensitive tensile curves are available and there are i data pairs of the strain rate and saturated stress values from the $i \geq 3$ tensile tests. In this case there are at least three simultaneous equations for the three parameters n, Ω_m, and Z_0. The inverted flow equation 6.4.2 can be rewritten in the form

$$\ln X = 2n \ln Z_0 - 2n \ln(\sigma_s - \Omega_m) \tag{6.4.5}$$

Equation 6.4.5 is a linear function of $\ln(\sigma_s - \Omega_m)$ and $\ln X$. The three material parameters Ω_m, Z_0, and n can be determined by first assuming a value for Ω_m or making an estimate from equation 6.4.3. This value is then used in equation 6.4.5 to determine the values for Z_0 and n by the method of least squares and the correlation coefficient of the fit between the data and line. The value of Ω_m is then iterated until the value of the correlation coefficient is close to 1.00. If the value of the correlation coefficient does not approach 1.00, a different flow equation should be investigated. Thus the method can also be used to suggest the best flow equation for a particular material. The three parameters Ω_m, Z_0, and n should be close to the true values for the material under consideration. At this point it is strongly recommended that equation 6.4.2 be used to calculate the values of saturated stresses and compare these values to the experimental data.

6.4.2 Back Stress Parameters

If only tensile data are available, it is reasonable to assume that the maximum value of the back stress can be evaluated, and the values of Z_0 and n have been determined as described above. In this case a back-stress history can be determined from the tensile data by using the inverted flow equation, 6.3.5. The back-stress history can then be used to determine the back-stress rate as a function of time. The back-stress rate can be obtained by numerical differentiation of the back-stress history. This is reasonable for a computer-acquired data file with hundreds of data points, but it not easy for hand-digitized data.

In this case fit the back stress curve with an equation and differentiate to find the back stress rate history.

The constant f_2 in the back stress equation 6.3.6 can be determined from the initial value of the back stress rate during the initial "elastic" loading phase of the tensile history. Since the initial back stress rate is proportional to the stress rate or strain rate,

$$\dot{\Omega}(0) = f_2\,\dot{\sigma}(0) = f_2\,E\dot{\varepsilon}(0) \tag{6.4.6}$$

the initial value of the back stress rate in Figure 6.3.3a can be correlated to the strain rate as shown in Figure 6.4.2. The parameter f_2 is determined from the slope of the line.

The parameter f_1 can be established by arranging the back stress equation 6.3.6 in the form

$$\frac{\dot{\Omega} - f_2\,\dot{\sigma}}{\dot{\varepsilon}^{\mathrm{I}}} = f_1\left(1 - \frac{\Omega}{\Omega_m}\right) \tag{6.4.7}$$

The parameter f_1 was determined by the method of least squares using the data in Figure 6.4.3. The constants $E = 125$ GPa, $D = 1\ \mathrm{sec}^{-1}$, $n = 0.242$, $Z_0 = 51{,}000$ MPa, $f_1 = 28{,}800$ MPa, $f_2 = 0.3005$, and $\Omega_m = 283$ MPa were

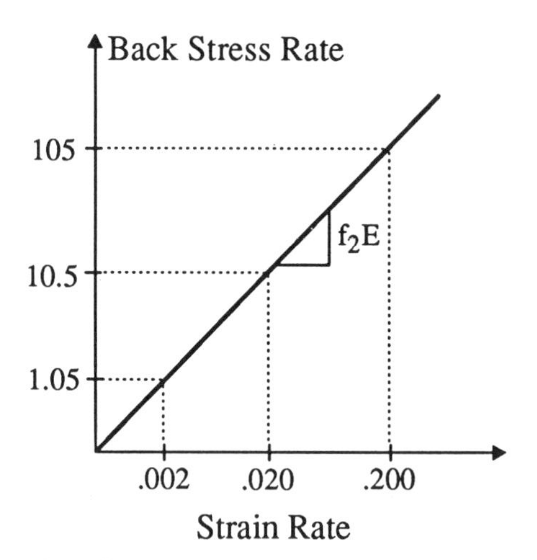

Figure 6.4.2 Correlation of the initial back-stress rate in Figure 6.3.3 with the strain rate for René 80 at 982°C for the elastic modulus in Figure 4.5.1. (After Ramaswamy, 1986.)

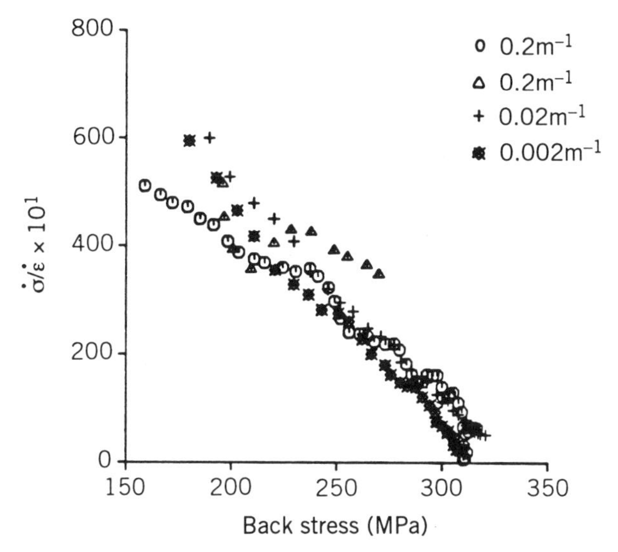

Figure 6.4.3 Evaluation of the back stress parameter f_1 for René 80 at 982°C for four strain rates. (After Ramaswamy, 1986.)

used to calculate the model response in Figures 6.3.3 and 6.3.4. The value $D = 1\ \text{sec}^{-1}$ was used rather than $D = 10^4\ \text{sec}^{-1}$ because the temperature was rather high. The results shown in Figure 6.3.5 are for the tensile response of René 80 at 538°C. The constants for René 80 at four temperatures are summarized in Appendix 6.1.

6.4.3 Relationships Between Parameters

This section contains the development of four new relationships between the material parameters for the Bauschinger effect, saturation of the tensile response, stress relaxation, and strain recovery. Since fatigue data are relatively common, the Bauschinger effect can be used as a criteria to establish a relationship between the material parameters. To begin, notice that the value of the overstress required to produce yield in flow equation 6.2.13 is a constant for constant values of n, D_0 and Z_0. Thus the value of the overstress producing yield in tension, σ_{yt}, must be the same value required to produce yield in compression, σ_{yc}, as shown in Figure 6.4.4; thus

$$|\sigma_{yc} - \Omega_{yc}| = |\sigma_{yt} - \Omega_{yt}| \tag{6.4.8}$$

On initial loading the back stress at yield is equal to $f_2\sigma_{yt}$ from equation 6.3.6. Further on unloading from a saturated state in a tensile test at point A the back stress at yield in compression is equal to $\Omega_m - f_2|\Delta\sigma|$, where $|\Delta\sigma|$ is the magnitude of the stress drop. Thus equation 6.4.8 becomes

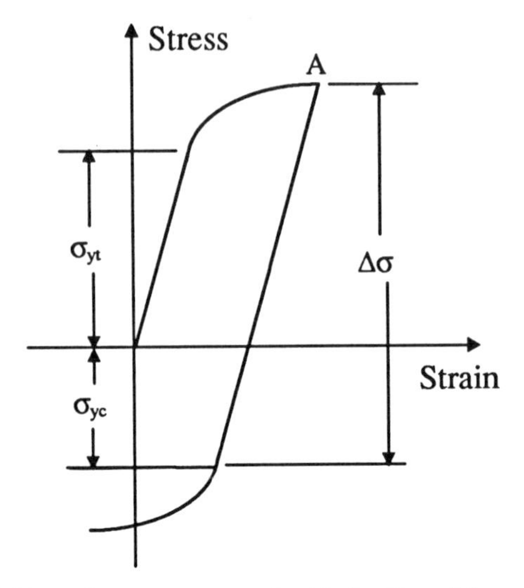

Figure 6.4.4 Definition of parameters used with the Bauschinger effect to determine a relationship between the constants.

$$|\sigma_{yc} - \Omega_m + f_2\,|\Delta\sigma|| = \sigma_{yt}(1 - f_2) \tag{6.4.9}$$

This is an equation between Ω_m and f_2, using the Bauschinger effect from the first loop of a fatigue test since equation 6.4.9 is valid only when $Z = Z_0$.

Next, the back stress equation can be manipulated to develop a representation for f_1, the parameter that defines the inelastic strain to reach saturation. The back-stress equation 6.3.6 can be rearranged and written in integral form for a tensile test. This procedure gives

$$\int_0^{\Omega} \frac{d\Omega}{1 - \Omega/\Omega_m} = \int_0^{\sigma} \frac{f_2\,d\sigma}{1 - \Omega/\Omega_m} + \int_0^{\varepsilon^{\mathrm{I}}} f_1\,d\varepsilon^{\mathrm{I}} \tag{6.4.10}$$

Let us next define the parameter A by

$$-\Omega_m \ln A = \int_0^{\sigma} \frac{f_2\,d\sigma}{1 - \Omega/\Omega_m}$$

Then integrating equation 6.4.10, the accumulated change in back stress (first term of equation 6.4.10) can be written in the form

$$1 - \frac{\Omega}{\Omega_m} = A \exp\left(-\frac{f_1\,\varepsilon^{\mathrm{I}}}{\Omega_m}\right) \tag{6.4.11}$$

As the tensile curve approaches saturation, the back stress approaches the maximum value and the exponential term must approach zero. If saturation is assumed to occur when

$$\exp\left(-\frac{f_1\,\varepsilon_S^{\mathrm{I}}}{\Omega_m}\right) = 0.01 \tag{6.4.12}$$

the parameter f_1 can be estimated from

$$f_1 = \frac{4.6\Omega_m}{\varepsilon_S^{\mathrm{I}}} \tag{6.4.13}$$

where $\varepsilon_S^{\mathrm{I}}$ is the accumulated inelastic strain at saturation, the value of strain for which the stress strain curve becomes flat. This value can be estimated from a tensile curve.

Stress relaxation can also be used to establish another relationship between the material parameters. When a material is loaded into the inelastic region and the strain is then held constant, the stress will decay and asymptotically approach a stable value. In Figure 6.4.5, point *A* denotes the start of stress relaxation from a fully saturated tensile state and point *B* corresponds to the fully relaxed stress state. Since the total strain is constant during stress relaxation, the sum of changes in elastic and inelastic strain must vanish; that is,

$$0 = \frac{\Delta\sigma_{AB}}{E} + \Delta\varepsilon_{AB}^{\mathrm{I}} \tag{6.4.14}$$

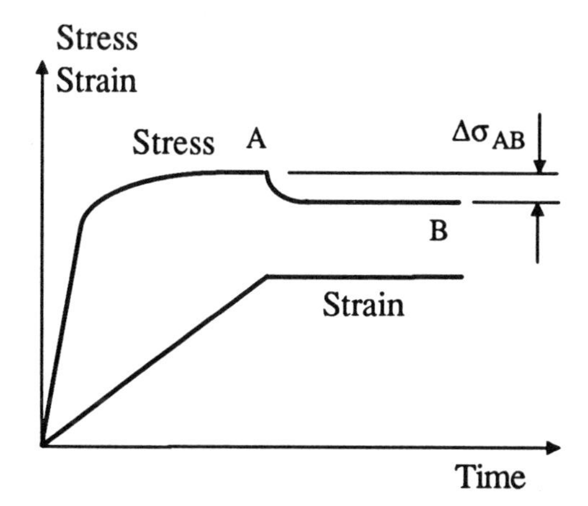

Figure 6.4.5 Stress-time and strain-time histories during stress relaxation and the values of stress used for determining the material constants.

where $\Delta\sigma_{AB}$ and $\Delta\varepsilon^{\mathrm{I}}_{AB}$ represent the incremental changes in the stress and inelastic strain during relaxation from point A to point B. It then follows that

$$\Delta\varepsilon^{\mathrm{I}}_{AB} = -\frac{\Delta\sigma_{AB}}{E} \tag{6.4.15}$$

It is expected that the stress relaxation stops when the overstress drops to the critical value for slip. Referring to the development of equation 6.4.9, the overstress at point B can be written as

$$|\sigma_B - \Omega_B| = \sigma_{yt}(1 - f_2)$$

as long as the drag stress is close to Z_0. Solving for the back stress at point B gives

$$\Omega_B = \sigma_B - \sigma_{yt}(1 - f_2) \tag{6.4.16}$$

since $\sigma_B > \Omega_B$ for this application. Next consider back stress equation 6.3.6 as an incremental equation applied to the stress relaxation between points A and B in one increment, that is,

$$\Delta\Omega_{AB} = \Omega_B - \Omega_A = f_1\,\Delta\varepsilon^{\mathrm{I}}_{AB} - f_1\frac{\Omega}{\Omega_m}|\Delta\varepsilon^{\mathrm{I}}_{AB}| + f_2\,\Delta\sigma_{AB} \tag{6.4.17}$$

Then using equations 6.4.13 and 6.4.15, the average value of the back stress in equation 6.4.17 gives

$$\Omega_B - \Omega_m = -\frac{4.6\Omega_m}{\varepsilon^{\mathrm{I}}_S}\left(\frac{\Delta\sigma_{AB}}{E} + \frac{\Omega_B + \Omega_m}{2\Omega_m}\left|\frac{\Delta\sigma_{AB}}{E}\right|\right) + f_2\,\Delta\sigma_{AB} \tag{6.4.18}$$

noting that the back stress at point A is Ω_m. Equation 6.4.18 can be used with equation 6.4.16 to obtain

$$f_2 = \frac{C_1[\Omega_m - \sigma_B + \sigma_{yt}]}{|\Delta\sigma_{AB}| + C_1\sigma_{yt}} \quad \text{where} \quad C_1 = 1 + \frac{2.3|\Delta\sigma_{AB}|}{E\,\varepsilon^{\mathrm{I}}_S} \tag{6.4.19}$$

Equation 6.4.19 is the second equation between Ω_m and f_2 that can be used with equation 6.4.9 to determine Ω_m and f_2. All other values in equation 6.4.19 can be determined from tensile and stress relaxation tests.

As a final example of the relationships that can be developed between material parameters, consider high temperature and strain recovery as shown in Figures 1.3.3 and 1.3.5. Assume that the initial tensile curve is fully saturated so that the back stress at point C in Figure 6.4.6 can be estimated from equation 6.3.6 as

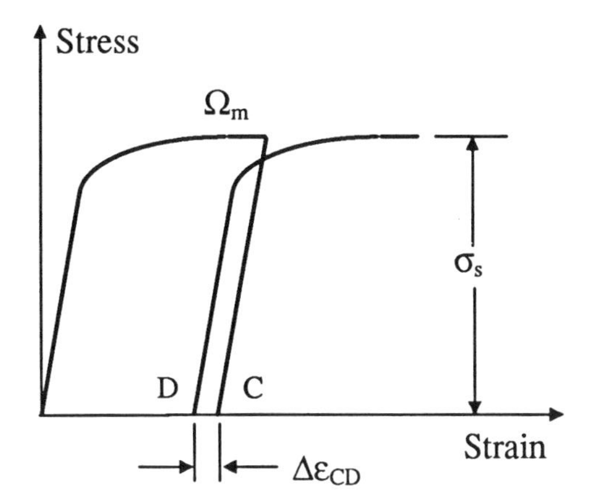

Figure 6.4.6 Diagram used for the evaluation of the material constants from a strain recovery test.

$$\Omega_C = \Omega_m - f_2\,\sigma_S \tag{6.4.20}$$

noting that there is only elastic unloading. Next observe that the inelastic strain rate corresponding to point D in Figure 1.3.5 is near zero and the accumulated inelastic strain during recovery is nearly complete. Thus point D corresponds to a nearly fully recovered strain state. Since the value of overstress required to initiate inelastic flow is approximately the same as the value at which inelastic deformation stops, the back stress at point D can be determined from the initial yield $|\sigma_y - f_2\,\sigma_y|$ for $Z = Z_0$ as

$$|\sigma_D - \Omega_D| = \Omega_D = \sigma_y(1 - f_2) \tag{6.4.21}$$

since $\sigma_D = 0$ and $\Omega_D > 0$. The change in the back stress during strain recovery can be estimated from the incremental back-stress equation 6.4.17 as

$$\Omega_D - \Omega_C = -f_1|\Delta\varepsilon^{\mathrm{I}}_{CD}|\left(1 + \frac{\Omega_C + \Omega_D}{2\Omega_m}\right) \tag{6.4.22}$$

The average value of the back stress was used again in equation 6.4.22. Combining equations 6.4.20, 6.4.21, and 6.4.22 and using equation 6.4.13 gives

$$\Omega_m = C_2 f_2 + C_3 \tag{6.4.23}$$

where the parameters C_2 and C_3,

$$C_2 = (\sigma_S - \sigma_y)\frac{\varepsilon_S^{\mathrm{I}} - 2.3|\Delta\varepsilon_{CD}^{\mathrm{I}}|}{\varepsilon_S^{\mathrm{I}} - 4.6|\Delta\varepsilon_{CD}^{\mathrm{I}}|} \quad \text{and} \quad C_3 = \sigma_y \frac{\varepsilon_S^{\mathrm{I}} + 2.3|\Delta\varepsilon_{CD}^{\mathrm{I}}|}{\varepsilon_S^{\mathrm{I}} - 4.6|\Delta\varepsilon_{CD}^{\mathrm{I}}|}$$

can be entirely determined from the recovery response data. Equation 6.4.23 is a third equation between Ω_m and f_2. Thus the two parameters can be determined from stress relaxation or recovery data used in combination with tensile data.

Let us consider use of these relationships in finding parameters for strain rate insensitive tensile materials. Recall that $D_0 = 10^4\ \text{sec}^{-1}$ and $n = 3$; then equations 6.4.4 for strain rate independent response, 6.4.9 from the Bauschinger effect, 6.4.13 from saturated tensile response, and 6.4.19 for stress relaxation can be used to determine the four parameters Z_0, Ω_m, f_1, and f_2 if the experimental data are available. Strain recovery (equation 6.4.23) is a high-temperature effect that would probably not occur with low-temperature strain rate independent response. If all these data are not available, it is necessary to make some assumptions about the response. These assumptions, of course, depend on the particular material under study. However, there are two assumptions that appear to have some generality. First, if pure kinematic hardening is assumed, $\Delta\sigma \approx 2\sigma_{yt}$ and $\sigma_{yc} \approx \sigma_S - 2\sigma_{yt}$ in equation 6.4.9. Second, the maximum value of the back stress can be estimated from equation 6.4.3. Thus the parameter estimation process is much simpler.

The flow equation parameters (Z_0, Ω_m, and n) for strain rate sensitive materials still must be determined from Equation 6.4.6 and a set of strain rate dependent tensile curves. The value of Ω_m can be used with the strain to saturation to estimate the value of f_1 from equation 6.4.13. If stress relaxation or strain recovery or cyclic data is available then equation 6.4.9 or 6.4.19 or 6.4.23, respectively, can be used to determine f_2. If this data is not available then it is reasonable to use equation 6.4.9 with the assumption $\Delta\sigma \approx 2\sigma_{yt}$.

These two cases show that the state variable parameters for tensile response can be uniquely defined if enough data are available. As more experience is developed with the use of these equations, it will be possible to develop better rules of thumb to estimate parameters when data are not available.

Finally, approximating one or two parameters and using data to determine the remaining parameters will produce a reasonable correlation with experimental results even though the parameters are not optimum. Prediction of results other than those used to find the parameters depends on the accuracy of the assumed parameters; however, the equations appear to be reasonably robust, so predicted results are usually reasonable. This is one reason why the parameters are not unique and there are not tables of parameters for state variable models. The parameters in the appendix for state variable models represent parameters that have been used successfully, but they are probably not the only parameters that could be used for that particular material at the temperature specified.

6.5 CREEP AND CREEP APPLICATIONS

Consider next creep at higher stresses when dislocation glide and dislocation creep are present. In this case back stress, the resistance to glide due to dislocation pileups at barriers, is present but its maximum value is generally less than the maximum back stress, Ω_m. Physically there are two reasons for the lower maximum back stress value. First, the creep response is driven by thermal vibrations and assisted by stress; therefore, the applied stresses and interaction stresses are less that those associated with slip in tensile tests. Second, climb and static thermal recovery are present, reducing the number of dislocations in pileups over long periods of time.

The effect of back stress has been included in a power law model by Cadek (1987). The temperature compensated steady-state creep rate, $\dot{\varepsilon}'_{ss}$, is determined by

$$\dot{\varepsilon}'_{ss} = C\left(\frac{\sigma_{\mathrm{SS}} - \Omega}{G}\right)^{\beta}$$

where the value of the exponent β depends on temperature and the overstress, $\sigma_{\mathrm{SS}} - \Omega$, is normalized by the shear modulus, G. The correlation between the creep rate and stress for the power law model is shown in Figure 6.5.1 for Fe–21Cr–37Ni alloy.

Let us begin the application to creep by modifying the maximum value of the back stress in equation 6.3.6 to account for the difference between the dislocation creep and dislocation glide, and slip. The steady-state creep rate, $\dot{\varepsilon}^{\mathrm{C}}_{\mathrm{ss}}$, and associated creep stress, σ_{ss}, can be used in equation 6.3.5 to determine the corresponding steady-state back stress, Ω_{ss}, as

$$\Omega_{\mathrm{SS}} = \sigma_{\mathrm{SS}} - Z_0\left[-2\ \ln\left(\frac{\sqrt{3}\dot{\varepsilon}^{\mathrm{C}}_{\mathrm{SS}}}{2D_0}\right)\right]^{-1/2n} \tag{6.5.1}$$

The steady-state value of the back stress has been determined for several materials at different temperatures. An example of the steady state stress and saturated back stress from creep and tensile tests, respectively, for René 80 at four temperatures is shown in Figure 6.5.2. It can be seen that the linear relationship between the back stress and stress from the creep tests is independent of temperature (except at 760°C) during creep, but that the maximum value of the back stress does depend on temperature. This result suggests that it is necessary to modify the model to give the correct saturation value for the back stress and develop an evolution equation for the steady-state (or equilibrium) value of back stress during creep.

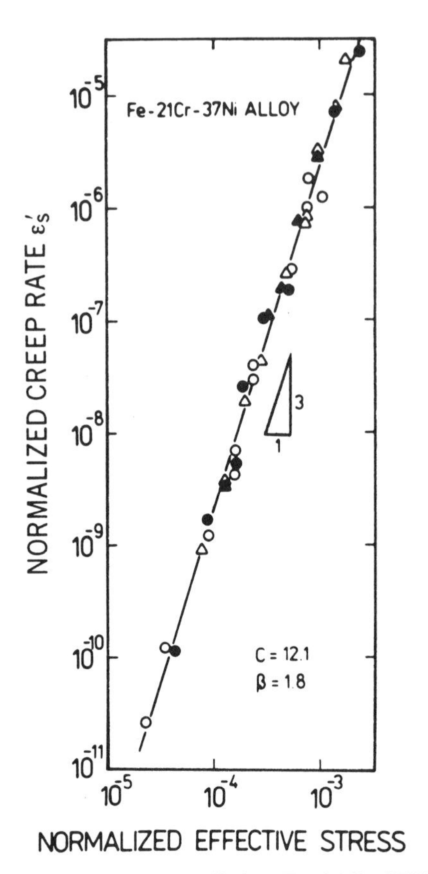

Figure 6.5.1 Power law creep model applied to Fe–21Cr–37Ni: ▲ 873°K; △ 923°K; ● 993°K; ○1023°K. (From J. Cadek, Back stress concept in power law creep of metals, *Materials Science Engineering,* Vol. 94, 1987, pp. 79–92.)

Using the René 80 data as an example (see Figure 6.5.2), the equilibrium value of the back stress, ω_e, for both tensile and creep response is a linear function of the creep stress in the form

$$\omega_e = \begin{cases} \Omega_m(T) & \text{for } \sigma > \sigma_y \\ b_1(T)\sigma + b_2(T) & \text{for } \sigma_2 < \sigma < \sigma_y \\ b_3 & \text{for } \sigma < \sigma_2 \end{cases} \tag{6.5.2}$$

where σ_y, the yield stress, is the transition stress between dislocation glide and slip, and b_3 at σ_2 is an arbitrary low stress cutoff to prevent numerical defaults. The constants in the function $\omega_e = b_1(T)\sigma + b_2(T)$ are picked to characterize the inclined lines in Figure 6.5.2 as a function of temperature.

The equilibrium value of the back stress, $\omega_e = \Omega_m$, in equation 6.3.6 must be modified to reflect the difference between tensile and creep response. Re-

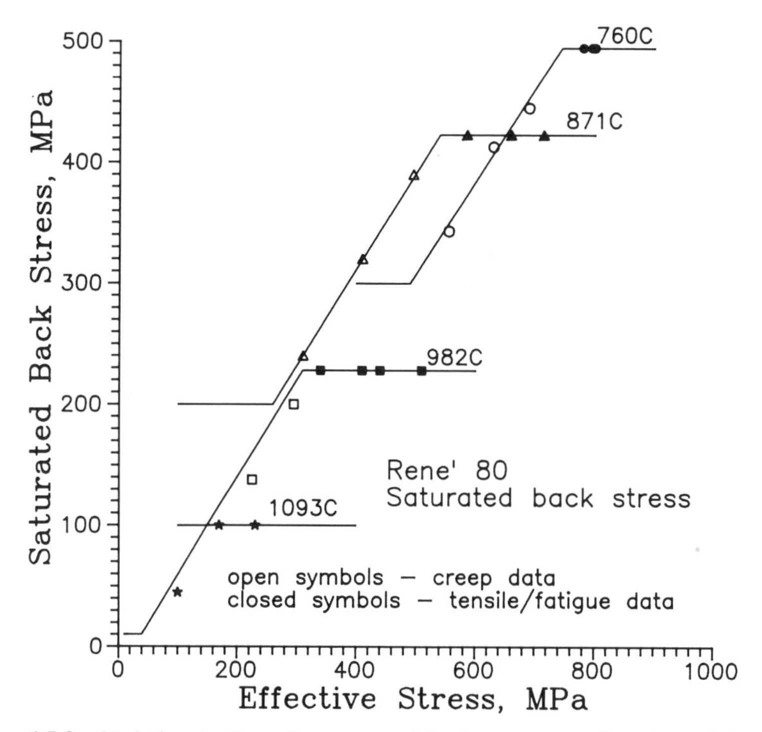

Figure 6.5.2 Variation in the value-saturated back stress as a function of the creep stress and temperature for René 80. (After Bhattachar and Stouffer, Journal of Engineering Materials and Technology, V115, pg. 353, 1993. Copyright © The American Society of Mechanical Engineers, 345 East 47th St., New York, NY 10017. Used with permission.)

place Ω_m in equation 6.3.6 with an equilibrium variable ω_e that depends on temperature. The equilibrium back stress $\omega_e(T)$ is assumed to have the initial value $\omega_0(T) = \Omega_m(T)$ and evolve into the appropriate equilibrium value, ω_e, depending on the current value of the stress. This implies that the back stress model (equation 6.3.6) must be revised to

$$\dot{\Omega} = f_1\, \dot{\varepsilon}^{\mathrm{I}} - f_1\, \frac{\Omega}{\omega}|\dot{\varepsilon}^{\mathrm{I}}| + f_2\, \dot{\sigma} \tag{6.5.3}$$

Since the recovery process is controlled by stress, let us assume that the equilibrium back stress $\omega(T)$ decays exponentially into ω_e; that is, let

$$\dot{\omega} = f\left(\frac{\sigma}{\sigma_0}\right)(\omega_e - \omega) \tag{6.5.4}$$

where f is an undetermined function of the normalized stress. The constants are determined so that during tensile and cyclic loading the stress changes at a rate that is much faster than the decay that is predicted by equation 6.5.4. Thus equation 6.5.4 has little effect during tensile and cyclic loading. During creep over long periods of time the stress never gets above the yield stress and $\omega(T)$ decays to the appropriate value of ω_e before the material achieves the steady-state creep rate.

A specific representation for the function f can be determined from the creep response data. To begin, notice that equation 6.5.4 can be integrated for a creep test to give

$$-\ln \frac{\omega - \omega_e}{\Omega_m - \omega_e} = f\left(\frac{\sigma}{\sigma_0}\right) t \tag{6.5.5}$$

since the creep stress is constant. The value of $\omega - \omega_e$ approaches zero at the onset of steady-state creep. Letting t_{ss} denote the time to the beginning of steady-state creep response or the minimum creep rate, the function f becomes

$$f\left(\frac{\sigma_{ss}}{\sigma_0}\right) t_{ss} = -\ln(0.01) = 4.61 \tag{6.5.6}$$

for a correlation within 1%. The stress σ_{ss} is the constant value creep stress. The specific representation can be determined from a plot of the creep stress against the inverse of the time to steady-state creep. For most metals a power law representation has proven quite accurate; thus as an example, let

$$f\left(\frac{\sigma_{ss}}{\sigma_0}\right) \equiv A\left(\frac{\sigma_{ss}}{\sigma_0}\right)^p = \frac{4.61}{t_{ss}} \tag{6.5.7}$$

The stress is normalized to avoid fractional units. The values for the creep coefficient A and creep exponent p can be evaluated using a regression analysis with the creep data. A comparison between the experiment and model is shown in Figure 6.5.3 for René 80 at 982°C using the parameters in Appendix 6.1. Recall from Section 3.5 that it is important to control both the creep stress and initial loading rate when creep testing since the loading rate can alter subsequent creep response significantly. Combining the above results gives

$$\dot{\omega} = A\left(\frac{\sigma}{\sigma_0}\right)^p (\omega_e - \omega) \tag{6.5.8}$$

as one possible representation for the reduced value of the back stress during creep. This equation is limited to use for creep by dislocation glide or dislocation creep when back stress is present.

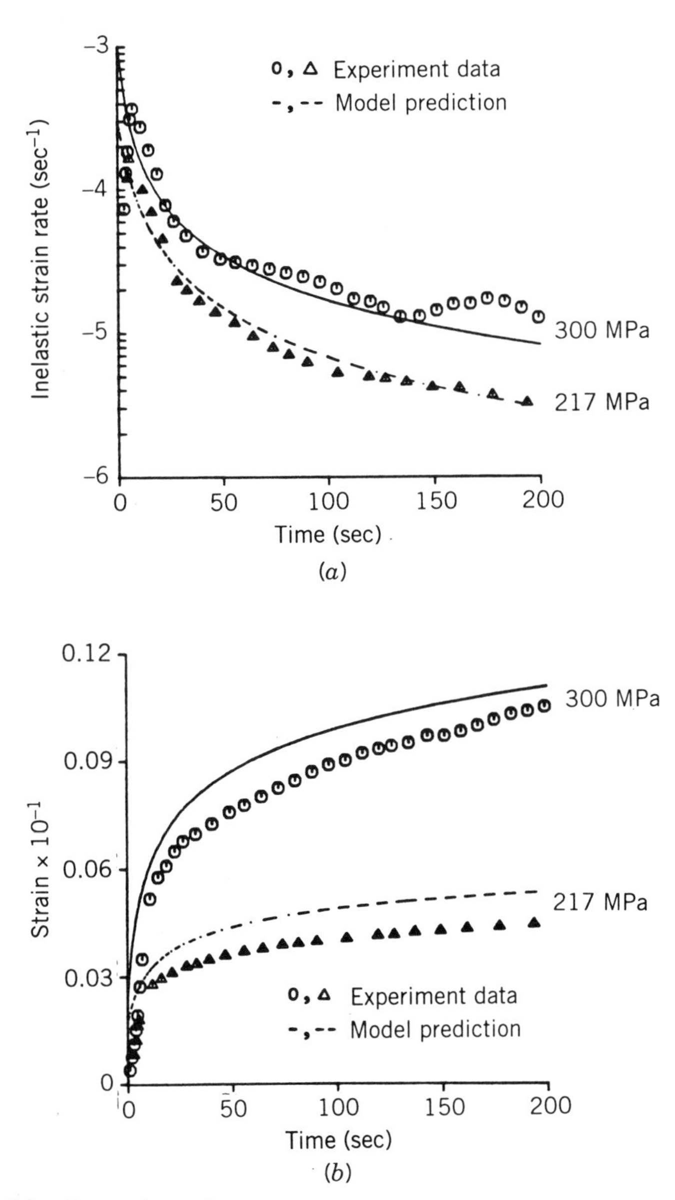

Figure 6.5.3 Comparison of the calculated and experimental creep strain and creep strain rate for René 80 at 982°C. (After Ramaswamy, 1986.)

6.6 DRAG STRESS AND FATIGUE CALCULATIONS

In this section the equations will be extended to include the cyclic hardening and softening observed in metals. The section will include a review of the physical basis for hardening and softening, development of the drag stress evolution equations, evaluation of the drag stress parameters, and a parameter sensitivity study.

Before proceeding to develop the drag stress equations, it is a useful check of a model to calculate the cyclic response using the parameters determined primarily from the tensile response. The calculations should be reasonably accurate for the early response before the development of dislocation cells and subgrains. Two examples of the early cyclic response are shown in Figures 6.6.1 and 6.6.2. Figure 6.6.1 shows the cyclic response with mean stress relaxation, and Figure 6.6.2 shows the effect with stress relaxation in compression. In both cases the correlation between the experimental response and model were adequate, indicating that it is appropriate to proceed with development of the drag stress evolution equations.

Recall from the discussions in Chapter 2 that a metal can be cyclically stable, harden, or soften during cyclic loading. It was shown that initially hard materials soften, initially soft materials harden, and materials that are initially partially hardened do not harden or soften very much. Hardening results from an increase in dislocation density and the interaction between dislocations,

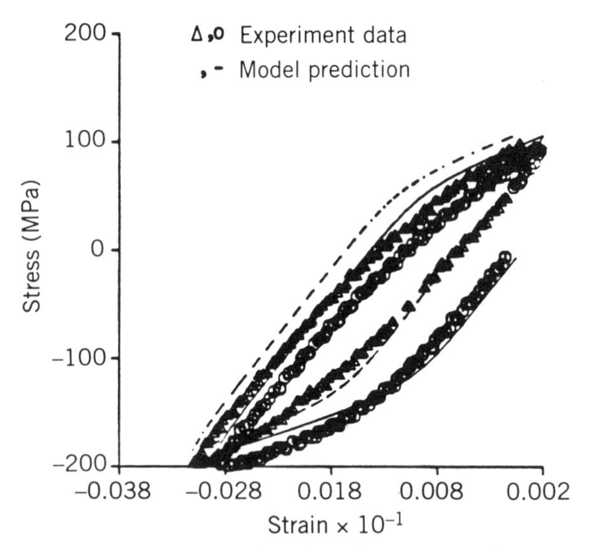

Figure 6.6.1 Comparison of the predicted and experimental response for a cyclic test with compressive mean strain. The material is René 80 at 982°C with a strain rate of 0.005 min^{-1}. (After Ramaswamy, Stouffer, and Laflen, Journal of Engineering Materials and Technology, V112, pg. 285, 1990. Copyright © The American Society of Mechanical Engineers, 345 East 47th St., New York, NY 10017. Used with permission.)

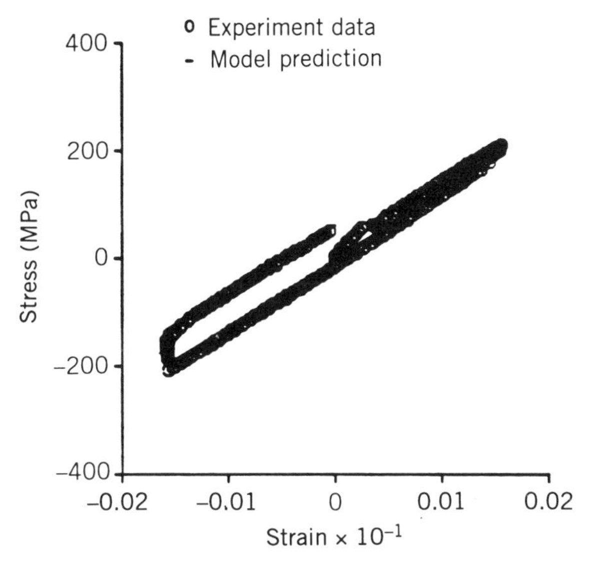

Figure 6.6.2 Comparison of the predicted and experimental response for the first cycle of a test with 12 sec of stress relaxation in compression. The material is René 80 at 982°C with a strain rate of 0.005 min^{-1}. (After Ramaswamy, Stouffer, and Laflen, Journal of Engineering Materials and Technology, V112, pg. 285, 1990. Copyright © The American Society of Mechanical Engineers, 345 East 47th St., New York, NY 10017. Used with permission.)

making the material more resistant to slip. Softening can result from the shearing of precipitates. Cells develop from cross slip at low temperatures and subgrains develop from climb at high temperatures, and a stable microstructure is approached that is characteristic of the material, temperature, and strain rate. This microstructure represents a balance between hardening and dynamic recovery. The cyclically hardened state is stable at low temperature. Exposure to high temperatures after loading allows static thermal recovery to proceed. This is annealing or the first step in recrystallization.

The drag stress evolution equation is developed to model the rate of change of hardness from the development of the dislocation microstructure. The drag stress is a scalar variable. The hardness is assumed to increase or decrease monotonically from the initial state, $Z(O) = Z_0$, to the final state, Z_1, during uniform cyclic loading. Two types of variables are typically used to describe the growth in dislocation density: the accumulated inelastic strain, $|\varepsilon^I| = \int|\dot{\varepsilon}^I|$, and the accumulated inelastic work, $W^I = \int|\sigma\, d\varepsilon^I|$. Since the cell or subgrain microstructure grows monotonically and approaches an equilibrium state one of two exponential growth laws are commonly used:

$$\text{Bodner and Partom (1975):} \quad \dot{Z} = m(Z_1 - Z)\dot{W}^I - R_2 \tag{6.6.1}$$

$$\text{Walker (1981):} \quad \dot{Z} = m(Z_1 - Z)|\dot{\varepsilon}^I| - R_2 \tag{6.6.2}$$

where R_2 is a different static thermal recovery term in each equation. Cyclic hardening is modeled if $Z_1 > Z_0$, cyclic softening is modeled by $Z_1 < Z_0$, and for cyclically stable materials $Z = Z_0$. The parameter m controls the rate of hardening or softening. Figures 6.6.3 and 6.6.4 show the effect of Z_1 and m on the response of equation 6.6.1 when static thermal recovery, R_2, is neglected. The use of inelastic work or inelastic strain in the drag stress is essentially the same if the response is strain-rate independent. However, the inelastic work or area inside the hysteresis loop is a more sensitive variable for strain-rate-sensitive materials because it depends on changes both in stress and inelastic strain rate. Finally, if static thermal recovery is neglected in equations 6.6.1 and 6.6.2, they can be integrated to obtain

$$Z = Z_1 - (Z_1 - Z_0)\exp(-mW^{\mathrm{I}}) \tag{6.6.3}$$

where $|\varepsilon^{\mathrm{I}}|$ can be substituted for W^{I} to obtain a representation for equation 6.6.2 which is similar to Walker's formulation.

The next step is to develop a method to evaluate the material parameters. First recall that the parameters in the flow equation 6.2.13 and the back-stress equation 6.3.5 have been evaluated. Flow equation 6.2.13 can be inverted and rewritten as

$$Z = X^{1/2n}(\sigma - \Omega) \tag{6.6.4}$$

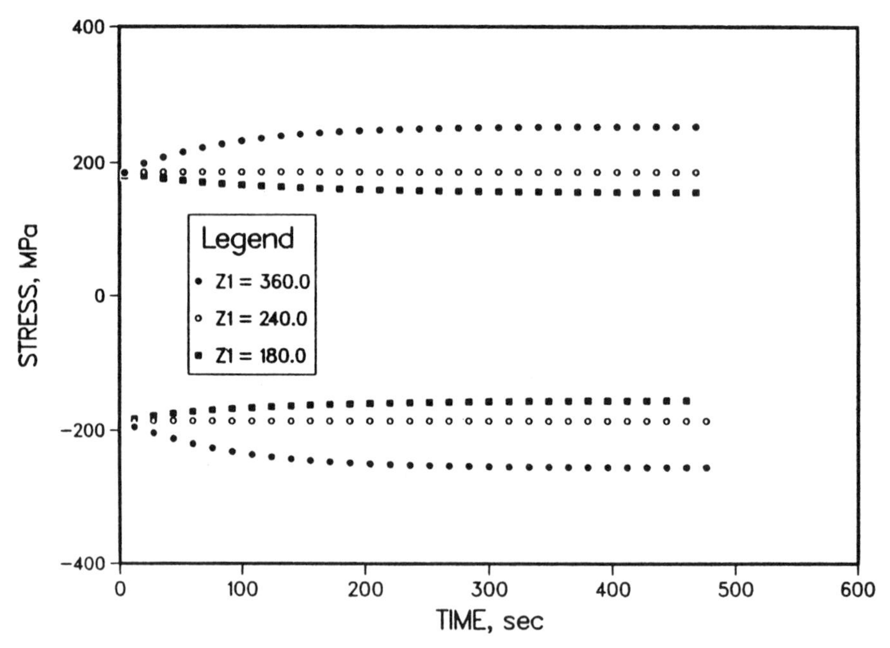

Figure 6.6.3 The parameter Z_1 allows for hardening, softening, or stable cyclic response. (Courtesy of M. Y. Sheh.)

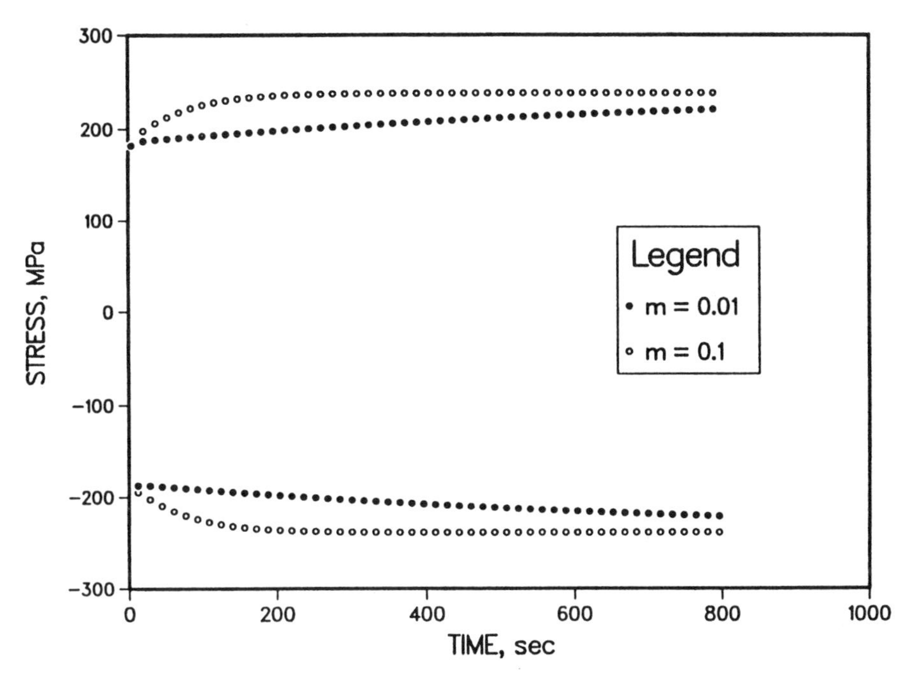

Figure 6.6.4 Example of how the parameter m controls the rate of hardening. (Courtesy of M. Y. Sheh.)

for the tensile loading part of the cycle. The parameter $X^{1/2n}$ is the mapped inelastic strain rate defined in equation 6.4.2. Since all four variables in equation 6.6.4 change continuously during loading, this result is not generally useful. However, consider a fatigue test with wide-open stable loops or a tensile curve after cycling the material until a stable hysteresis loop is obtained. In this case the drag stress has the value Z_1 and the response at tensile saturation of the cyclically stable material is given by

$$Z_1 = X_s^{1/2n} (\sigma_s - \Omega_m) \tag{6.6.5}$$

where the stress and strain at saturation are observed experimentally and the back stress is at the maximum value from a tensile test. Thus Z_1 can be determined directly from a special test. If these data are not available, a cyclically stable hysteresis loop can be extrapolated to the strain required to reach saturation in a tensile test and the corresponding stress can be used as an estimate of the saturation stress in equation 6.6.5. This was done for a cyclically softening material, as shown in Figure 6.6.5.

The parameter m that controls the rate of hardening or softening can be estimated by estimating the total accumulated inelastic strain or work required to produce the cyclically stable response. As the material response approaches the cyclically stable response the exponential function in equation 6.6.3 must

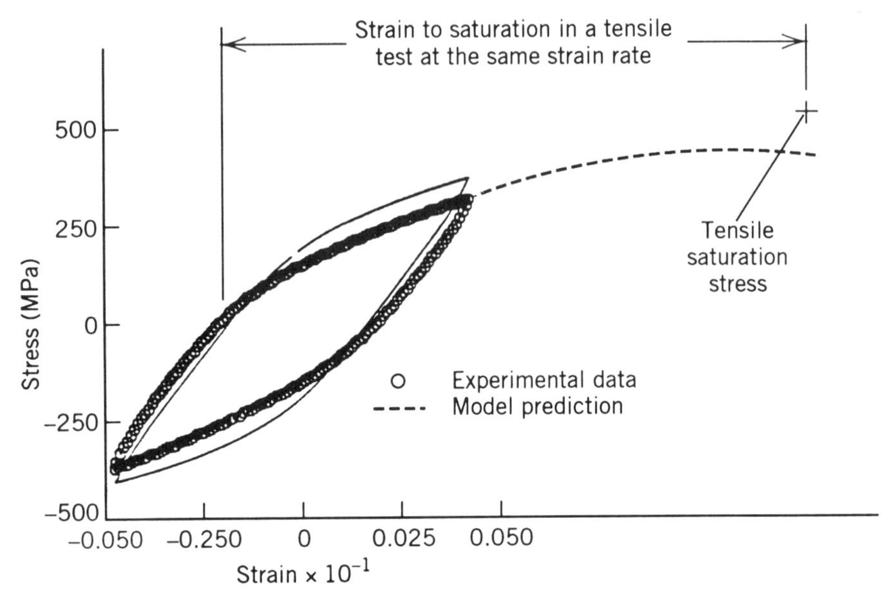

Figure 6.6.5 Extrapolation of a stable fatigue loop to find the parameter Z_1 for René 80 at 982°C and a strain rate of 0.2 min^{-1}. René 80 at 982°C was found to be cyclically stable at 0.002 min^{-1} and cyclically soften at 0.2 min^{-1}. The model does not match the data exactly because an average value was used for Z_1 rather the making Z_1 a function of strain rate.

approach zero. Assuming that a cyclically stable response occurs when the drag stress is within 1% of saturation, equation 6.6.3 becomes

$$\frac{Z - Z_1}{Z_0 - Z_1} = 0.01 = \exp(-mW_s^{\mathrm{I}}) \tag{6.6.6}$$

where W_s^{I} is the accumulated inelastic work required to reach the cyclically stable state. The value of m can then be estimated from the accumulated inelastic work at saturation. Once again $|\varepsilon^{\mathrm{I}}|$ can be substituted for W^{I} to obtain the constants for Walker's equation.

Before verifying the drag stress equation, it is helpful to make a modification for numerical efficiency. Direct application of the equations for a cyclic load history requires integration over the entire history. For example, if it is desired to calculate the response of the 10,000th loop in a load history it is necessary to integrate over the first 9999 loops to establish the correct values of the state variables at the beginning of the 10,000th loop. However for uniform cyclic loads an extrapolation procedure can be used to eliminate some of the integration time. In general, it is necessary to integrate through periods of significant transient response such as in the beginning of the load history or during variable-amplitude cycling. However, during periods of relatively

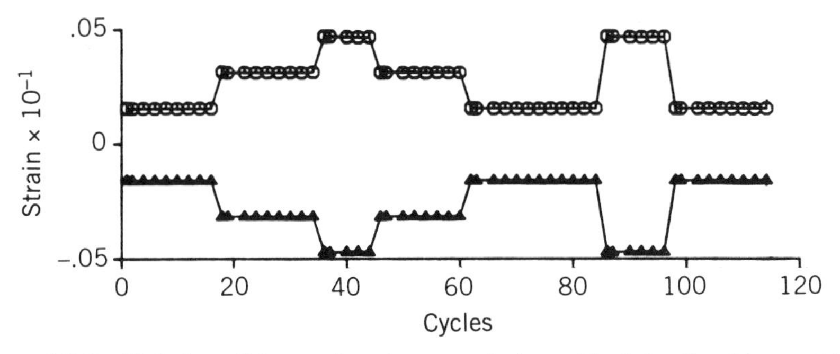

Figure 6.6.6 Variation of the total strain range during a block loading history. (After Ramaswamy, 1986.)

constant-amplitude cycling or when the response is slowly changing, the variables can be extrapolated, For example, the drag stress $Z(t_2)$ at some time t_2 can be estimated from the response at an earlier time t_1 by

$$Z(t_2) = Z(t_1) + n_{\text{cyc}}\, \Delta Z \tag{6.6.7}$$

where n_{cyc} is the number of cycles between t_1 and t_2, and ΔZ in the change in the drag stress in one cycle at time t_1. This technique was used to estimate the drag stress for the variable amplitude loading shown in Figure 6.6.6. The corresponding changes in drag stress shown in Figure 6.6.7 were determined by integrating when there was a change in the strain amplitude and extrap-

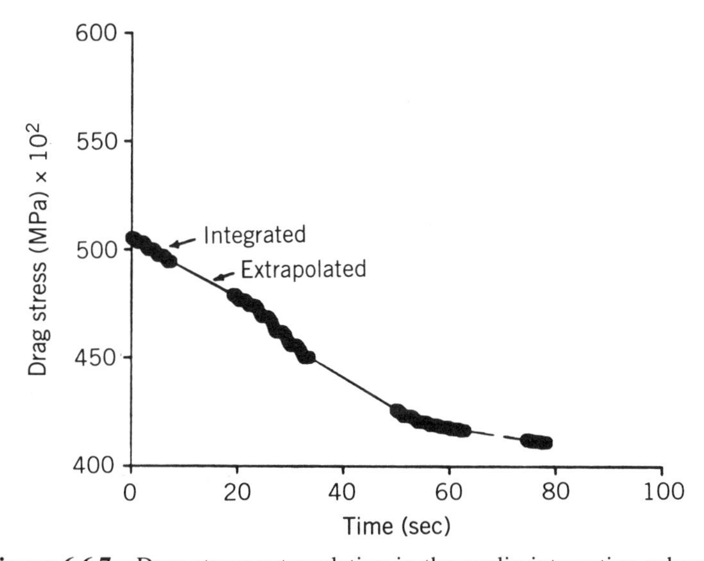

Figure 6.6.7 Drag stress extrapolation in the cyclic integration scheme.

olating during periods of constant-amplitude cycling. Figure 6.6.8 shows the 78th cycle of the same test reported in Figure 6.6.1 after the strain amplitude was increased and decreased. Figure 6.6.9 is for 30th cycle of the test shown in Figure 6.6.2. Caution should be exercised when using equation 6.6.7 because it is a linear growth law, whereas the drag stress is an exponential growth law. Therefore, n_{cyc} should not be too large compared to the total number of cycles to saturation.

6.7 NUMERICAL IMPLEMENTATION OF STATE VARIABLE EQUATIONS

The numerical implementation of state variable equations for a stand alone code involves three major elements: development of the flow diagram, integration of the rate equations, and selection of the time step. In many applications the trapezoidal rule has proven adequate for integration of state variable rate equations because it is second order accurate. The trapezoidal rule (discussed in Section 5.10.1) will be used as a starting point. Another method, an asymptotic integration rule, is presented in Section 9.3. In this section we focus on the flow diagram for state variable models and present a method to pick the appropriate time step for the trapezoidal rule.

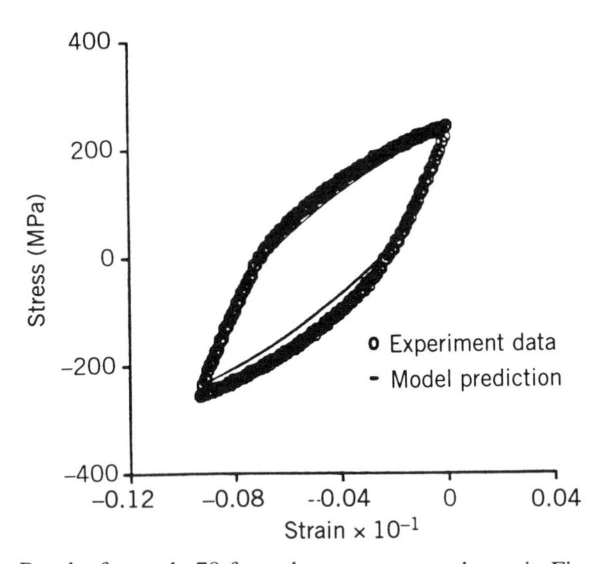

Figure 6.6.8 Results for cycle 78 from the same test as shown in Figure 6.6.1 using the parameters in Appendix 6.1. (After Ramaswamy, Stouffer, and Laflen, Journal of Engineering Materials and Technology, V112, pg. 285, 1990. Copyright © The American Society of Mechanical Engineers, 345 East 47th St., New York, NY 10017. Used with permission.)

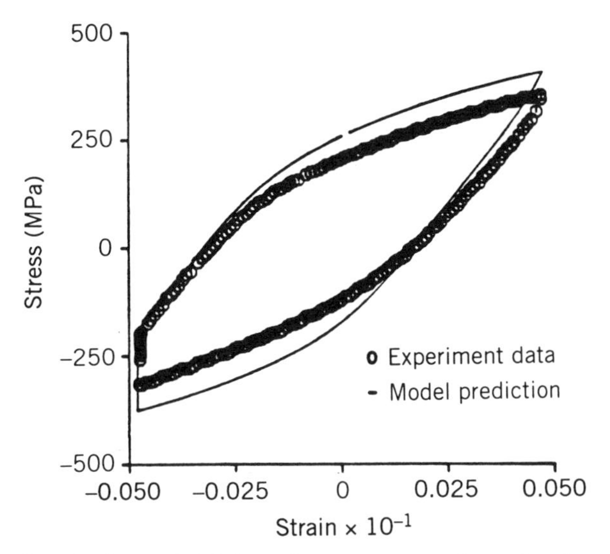

Figure 6.6.9 Results for cycle 30 from the same test as shown in Figure 6.6.2. (After Ramaswamy, Stouffer, and Laflen, Journal of Engineering Materials and Technology, V112, pg. 285, 1990. Copyright © The American Society of Mechanical Engineers, 345 East 47th St., New York, NY 10017. Used with permission.)

A typical flow diagram for a stand alone state variable code is shown in Figure 6.7.1. The load history defines the input history as a function of time and specifies if the values in the load history are stress or strain. (The same code is used for both stress and strain histories.) There can be several load cases in one file to model situations like a mix of creep and fatigue histories. The time loop controls the termination time for the current load case, and it is used to determine the current time increment, Δt, and the stress or strain at time $t + \Delta t$. The iteration loop is entered knowing the input variable (stress or strain) at time $t + \Delta t$ and all other variables at time t. The first step in the iteration loop is to estimate the stress at $t + \Delta t$ if the input data is strain. For the case of uniaxial loading an initial estimate of the stress can be determined from the total strain at $t + \Delta t$ as

$$\sigma(t + \Delta t) = E[\varepsilon(t + \Delta t) - \varepsilon^{\mathrm{I}}(t + \Delta t)] \simeq E[\varepsilon(t + \Delta t) - \varepsilon^{\mathrm{I}}(t) - \dot{\varepsilon}^{\mathrm{I}}(t)\,\Delta t] \tag{6.7.1}$$

after using a forward Euler estimate of the current inelastic strain increment. Equation 6.7. 1 is used the first time through the iteration loop. A trapezoidal rule is used to estimate the current inelastic strain increment during subsequent iterations when a estimate of the inelastic strain rate, $\dot{\varepsilon}^{\mathrm{I}}(t + \Delta t)$, is available; that is,

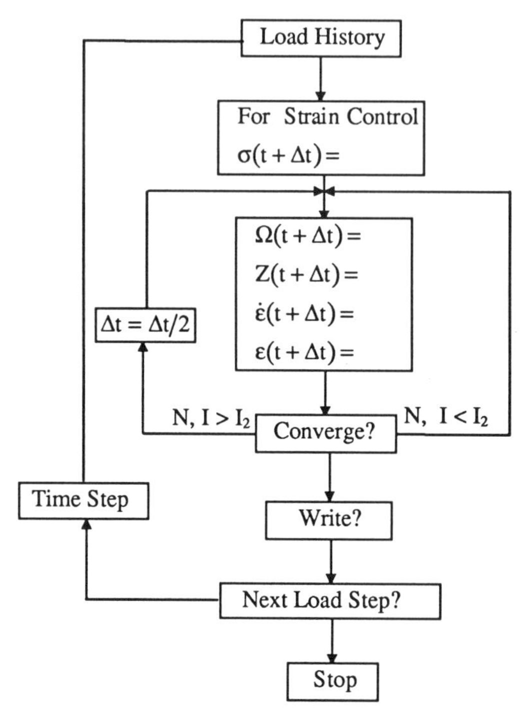

Figure 6.7.1 Typical flow diagram for the implementation of a state variable constitutive model in a stand-alone computer code.

$$\sigma(t + \Delta t) = E\left[\varepsilon(t + \Delta t) - \varepsilon^{\mathrm{I}}(t) - (\dot{\varepsilon}^{\mathrm{I}}(t) + \dot{\varepsilon}^{\mathrm{I}}(t + \Delta t)) \frac{\Delta t}{2} \right] \quad (6.7.2)$$

The back stress at $t + \Delta t$ is determined by dividing equation 6.3.6 into two terms, and defining an inelastic back-stress rate as

$$\dot{\Omega}^{\mathrm{I}}(t + \Delta t) = f_1 \dot{\varepsilon}^{\mathrm{I}}(t + \Delta t) - \frac{f_1}{\Omega_m} \Omega(t + \Delta t) |\dot{\varepsilon}^{\mathrm{I}}(t + \Delta t)| \quad (6.7.3)$$

which can be integrated with a trapezoidal rule (equation 5.9.2) and added to equation 6.7.3 to get

$$\Omega(t + \Delta t) = \Omega^{\mathrm{I}}(t + \Delta t) + f_2 \sigma(t + \Delta t) \quad (6.7.4)$$

The first time through the iteration loop $\Omega(t)$ can be used in place of $\Omega(t + \Delta t)$ in equation 6.7.3, or a forward Euler estimate can be used for $\Omega(t + \Delta t)$. The other state variables and inelastic strain are determined at $t + \Delta t$ in the same way. Convergence is assumed to occur when difference in the stress, inelastic strain, and inelastic strain rate are within a specified limit on two successive iterations, that is,

$$
\begin{aligned}
|\Delta\sigma_k - \Delta\sigma_{k-1}| &< \delta_\sigma \\
|\Delta\varepsilon_k^{\mathrm{I}} - \Delta\varepsilon_{k-1}^{\mathrm{I}}| &< \delta_{\varepsilon^{\mathrm{I}}} \\
|\Delta\dot{\varepsilon}_k^{\mathrm{I}} - \Delta\dot{\varepsilon}_{k-1}^{\mathrm{I}}| &< \delta_{\dot{\varepsilon}^{\mathrm{I}}}
\end{aligned}
\tag{6.7.5}
$$

The iteration loop is controlled by two special rules. First, if the total number of iterations reaches a specified value, the algorithm reduces the time step by 50% and returns to the beginning of the time-step loop. This process is repeated until convergence is achieved. Second, no iteration is required during elastic response and when the inelastic strain rate is constant, such as during steady-state creep or when the tensile curve is flat. That is, if

$$
\begin{aligned}
\dot{\varepsilon}^{\mathrm{I}} &< 10^{-8} \qquad \text{no iteration is required} \\
\frac{\dot{\varepsilon}_k^{\mathrm{I}} - \dot{\varepsilon}_{k-1}^{\mathrm{I}}}{\dot{\varepsilon}_k^{\mathrm{I}} + \dot{\varepsilon}_{k-1}^{\mathrm{I}}} &< 10^{-3} \qquad \text{no iteration is required}
\end{aligned}
$$

A loading history with wide variations in loads can produce large excursions in stress and inelastic strain rate. In this case a dynamic time-step scheme is very helpful to reduce CPU costs. The dynamic time-step scheme is based on three separate control criteria:

1. Maximum stress increment during elastic loading and unloading
2. Maximum strain increment during steady-state creep or saturated tensile response
3. Maximum strain rate increment during transients in the loading or response histories.

The time step for the current time increment is chosen as the minimum value from the criteria above. These values are determined from the response properties in the previous time step.

The stress increment is used during elastic loading and unloading to prevent an overestimate of the stress during the transition to inelastic strain. Calculation for the time step is given by

$$
\Delta t_k = \Delta t_{k-1} \frac{\Delta\sigma_{\max}}{\Delta\sigma_{k-1}} \tag{6.7.6}
$$

where Δt_{k-1} is the previous time step. The stress increment $\Delta\sigma_{k-1}$ is the previous stress increment or the maximum of the previous stress increments in a finite element calculation. The parameter $\Delta\sigma_{\max}$ is the maximum desired stress increment in a loading history. Typical values for $\Delta\sigma_{\max}$ are from 1 to 15 MPa. The initial time step can be determined from equation 6.7.6.

The maximum inelastic strain increment controls the time step when the stress and inelastic strain rate are not changing. The situation occurs during steady-state creep and saturated tensile loading. The time-step calculation is similar to equation 6.7.6, and is written as

$$\Delta t_k = \Delta t_{k-1} \frac{\Delta \varepsilon^{\mathrm{I}}_{\max}}{\Delta \varepsilon^{\mathrm{I}}_{k-1}} \tag{6.7.7}$$

where $\Delta\varepsilon^{\mathrm{I}}_{\max}$ is the maximum value of the desired inelastic strain increment and $\Delta\varepsilon^{\mathrm{I}}_{k-1}$ is the previous change in the inelastic strain increment. The values of $\Delta\varepsilon^{\mathrm{I}}_{max}$ range from 10^{-4} to 10^{-6}, depending on the material. This criterion is very important during creep when the creep rates are very small.

The maximum inelastic strain-rate criterion controls the time step when the inelastic strain rate is changing rapidly. This occurs, for example, in the knee of the tensile curve and the criterion is very important for elastic–perfectly plastic response. Too large a time step in these situations will require many iterations to converge. Time-step control is established by limiting the value of the second derivative, $\ddot{\varepsilon}^{\mathrm{I}}$. To begin, note that the inelastic strain increment can be written as

$$\Delta \varepsilon^{\mathrm{I}}_k = \frac{\Delta t_k}{2} (\dot{\varepsilon}^{\mathrm{I}}_k + \dot{\varepsilon}^{\mathrm{I}}_{k+1}) \tag{6.7.8}$$

The value of $\dot{\varepsilon}^{\mathrm{I}}_{k+1}$ can be approximated by the first two terms of a Taylor series expansion about the current time, $\dot{\varepsilon}^{\mathrm{I}}_{k+1} = \dot{\varepsilon}^{\mathrm{I}}_k + \Delta t_k \ddot{\varepsilon}^{\mathrm{I}}_k$. Inserting this result into equation 6.7.8 gives

$$\Delta \varepsilon^{\mathrm{I}} = \frac{\Delta t_k}{2} (\dot{\varepsilon}^{\mathrm{I}}_k + \dot{\varepsilon}^{\mathrm{I}}_k + \Delta t_k \ddot{\varepsilon}^{\mathrm{I}}_k) = \Delta t \dot{\varepsilon}^{\mathrm{I}}_k + \frac{\Delta t_k^2}{2} \ddot{\varepsilon}^{\mathrm{I}}_k \tag{6.7.9}$$

The last term in equation 6.7.9 is an estimate of the contribution of $\ddot{\varepsilon}^{\mathrm{I}}$ to the current inelastic strain step. Thus the time increment Δt_k can be chosen to limit the influence of $\ddot{\varepsilon}^{\mathrm{I}}$. Defining e to be the value of the last term and solving for Δt_k gives

$$\Delta t_k = \sqrt{\frac{2e}{\ddot{\varepsilon}^{\mathrm{I}}}} \tag{6.7.10}$$

But $\ddot{\varepsilon}^{\mathrm{I}}$ can be estimated by the backward difference $\ddot{\varepsilon}^{\mathrm{I}} = (\dot{\varepsilon}^{\mathrm{I}}_k - \dot{\varepsilon}^{\mathrm{I}}_{k-1})/\Delta t_{k-1}$, so equation 6.7.10 can be rewritten as

$$\Delta t_k = \sqrt{\frac{2e\ \Delta t_{k-1}}{\dot{\varepsilon}_k^{\mathrm{I}} - \dot{\varepsilon}_{k-1}^{\mathrm{I}}}} \tag{6.7.11}$$

The parameter e is the maximum value by which the forward Euler estimation can be in error. A typical value for e is around 0.01.

To prevent the time step from changing too rapidly, two other constraints are imposed: $\Delta t_k < a_1\ \Delta t_{k-1}$ and $\Delta t_k > a_2\ \Delta t_{k-1}$, where a_1 and a_2 are scale parameters. This prevents the subsequent time increment from becoming too large or too small too fast. The values $a_1 = 1.5$ and $a_2 = 0.5$ have been used successfully in many applications. In addition, an absolute maximum value for the time step is imposed; it is required that $\Delta t_k < \Delta t_{\max}$. The load case is also allowed to overwrite a time increment to prevent missing key values in a load history: for example, the peak values of a fatigue loop. This procedure is not unique and many variations are possible.

6.8 SUMMARY

It is generally accepted that at least two state variables are required in constitutive modeling. The back stress tensor results from the pileup of dislocations against barriers. The stress to produce slip, the overstress, is the difference between the shear stress on the slip plane in the slip direction and the back stress. The effect of the back stress is observed by the Bauschinger effect and anelastic recovery. The drag stress, a scalar variable, is used to model hardening resulting from the interaction of dislocations with precipitates, interstitial atoms, grains, cells, subgrains, and other isotropic barriers. The initial value of the drag stress is used to define the hardness, or resistance to slip, in the initial microstructure.

The flow equations, or representation for the inelastic strain rate, are based on the mathematical structure of the deformation mechanisms for slip and/or creep. The models of Bodner, Miller, Walker, Kreig et al., and Ramaswamy and Stouffer show the basic representations used in state variable modeling. Back-stress equations generally consist of a hardening term and a dynamic recovery term that are in equilibrium during saturated-state tensile response and steady-state creep. The drag stress evolution equation also generally consists of cyclic hardening and static thermal recovery terms.

There is no general consensus on which representation is best; however, the power law formulation using equations 6.2.14 and 6.3.3 has been reported to have difficulty modeling both strain-rate-dependent and strain-rate-independent tensile response (Ramaswamy, 1986). The importance of having two state variables to model tensile strain hardening and cyclic hardening separately can be demonstrated by considering an early form of the Bodner (1972) model, which is represented by equations 6.2.12 and 6.6.3 when creep effects are neglected. In this application the material parameters were picked to model the tensile response, and Figure 6.8.1 shows the agreement between the experimental and model results. However, application to cyclic response in Figure 6.8.2 was not particularly successful because the hardening parameter Z monotonically approaches the saturated value Z_1 and remains constant. This gives the sharp corners in the hysteresis loop and does not account for hardening and recovery during cycling. A more recent form of the Bodner model is presented in Section 7.7 overcomes this difficulty. However, this simple model is useful when only strain hardening or cyclic hardening needs to be modeled.

Let us next review the uniaxial models presented in this chapter for computational efficiency, ease in determining the material parameters, and accuracy. Numerical implementation of state variable constitutive equations is not particularly easy. State variable models lead to differential equations that are highly nonlinear and mathematically very stiff. The mathematical definition of "stiff" implies that the linear differential equations have widely separated eigenvalues or time constants. This results because the response of the metal can change very rapidly in a short period of time. For example, the inelastic

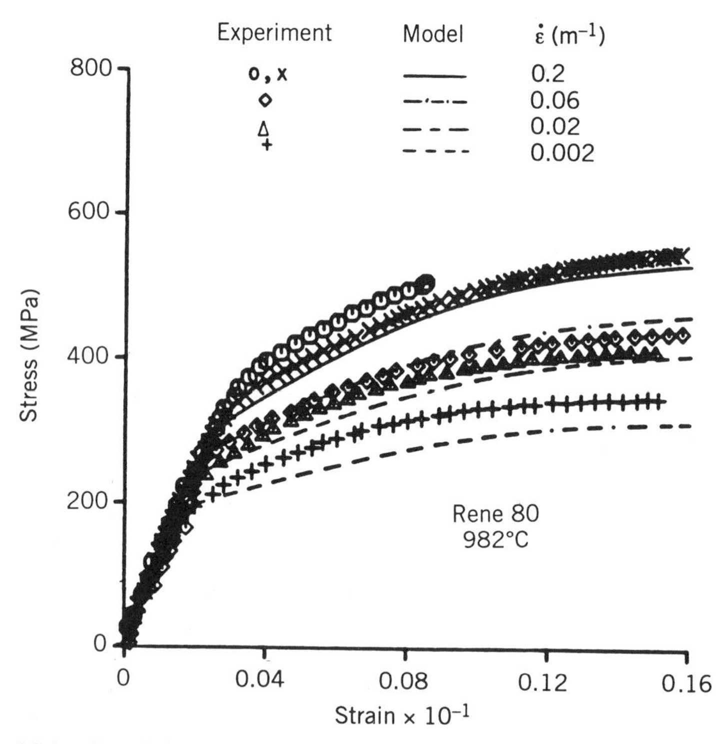

Figure 6.8.1 Correlation between the experimental response of René 80 at 982°C and an early form of the Bodner model. (After Ramaswamy, 1986.)

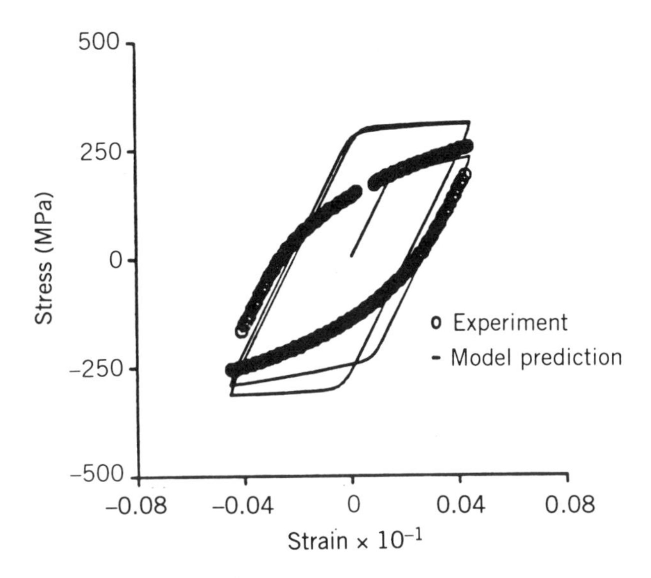

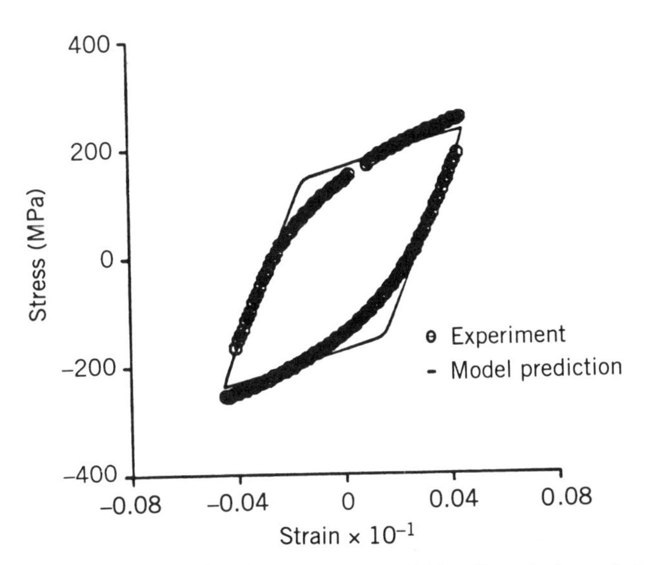

Figure 6.8.2 Disadvantage of only one state variable. Correlation of the cycle response using the constants determined from the tensile response in Figure 6.8.1. (After Ramaswamy, 1986.)

strain rate can change by many orders of magnitude in the beginning of a creep test or at yield in a tensile test. This requires small time steps in the integration algorithm at certain times in the loading history. However, an efficient integration algorithm must also permit large time steps during other loading periods. The integration rules presented in Section 6.7 show some simple methods to integrate stiff constitutive equations relatively efficiently. Additional integration methods are given in Chapter 9. There are also library subroutines for stiff equations that can be used efficiently.

One of the primary factors that has slowed the use of state variable equations is the difficulty in determining the material parameters. Systematic methods of determining parameters in Sections 6.4.1, 6.4.2, 6.5, and 6.6 are effective but somewhat laborious. The inverted flow equation and differentiation for the inelastic strain rate and back stress rate can be programmed for use with data files. The equations between parameters in Section 6.4.3, although not a totally proven method, should make evaluation of the parameters much easier.

Accuracy is subjective. The desired accuracy clearly depends on the application and associated costs. State variable equations have been developed in an effort to obtain additional accuracy for structural analysis and life prediction. For example, the strain-life equations involve the stress to an exponent. Thus a small error in stress could dramatically change the expected life. Thus accuracy can be very important.

State variable constitutive equations have been proven rather accurate for many materials for both strain rate dependent and strain rate independent response. The accuracy clearly depends on having a good reliable experimental program or data set to determine the parameters. This means that all variables, including the initial loading rate in creep, must be controlled or measured. Uniaxial low cycle fatigue response with stress relaxation and mean strain shifts can be modeled by updating the state variables. Static thermal recovery and pressure dependent plastic flow have not been included in the equations above, but static thermal recovery is involved in other models, as discussed in Chapter 7.

REFERENCES

Abuelfoutouh, N. (1983). A thermodynamically consistent constitutive model for inelastic flow of materials, Ph.D. dissertation, University of Cincinnati.

Bhattachar, V. S., and D. C. Stouffer (1993a). Constitutive equations for the thermomechanical response of René 80, Part 1: Development from isothermal data, *Journal of Engineering Materials and Technology,* Vol. 115, pp. 351–357.

Bhattachar, V. S., and D. C. Stouffer (1993b). Constitutive equations for the thermomechanical response of René 80, Part 1: Effects of temperature history, *Journal of Engineering Materials and Technology,* Vol. 115, pp. 358–364.

Bodner, S. R., and Y. Partom (1972). A large deformation elastic plastic analysis of a thick walled spherical shell, *Journal of Applied Mechanics,* Vol. 39, pp. 751–757.

Bodner, S.R., and Y. Partom (1975). Constitutive equations for elastic-viscoplastic strain hardening materials, *Journal of Applied Mechanics,* Vol. 24, p. 283.

Bodner, S.R. (1987). Review of a unified elastic–viscoplastic theory, in *Unified Constitutive Equations for Creep and Plasticity*, A. K. Miller, ed., Elsevier Applied Science, Barking, Essex, England.

Cadek, J. (1987). The back stress concept in power law creep of metals, *Materials Science and Engineering*, Vol. 94, pp. 79–92.

Chaboche, J. L. (1977). *Bulletin de l'Academie des Sciences Techniques*, Vol. 25, No. 1, p. 33.

Chaboche, J. L. (1986). Time dependent constitutive theories for cyclic plasticity, *International Journal of Plasticity,* Vol. 2, No. 2, pp. 149–188.

Chaboche, J. L., and G. Rousselier (1983). On plastic and viscoplastic constitutive equations, Part I: Rules developed with internal variable concept; Part II: Application of the internal variable concept to 316 stainless steel, *Journal of Pressure Vessel and Piping Technology,* Vol. 105, pp. 153–170.

Chan, K. S., and R. A. Page (1988). Inelastic deformation and dislocation structure of a nickel alloy: Effects of deformation and thermal histories, *Metallurgical Transactions,* Vol. 19A, pp. 2477–2486.

Chan, K. S., et al. (1986). Constitutive modeling for isotropic materials, *NASA CR 179522.*

Dafalios, Y. F., and E. P. Papov (1976). Plastic internal variables formation of cyclic plasticity, *Journal of Applied Mechanics,* Vol. 43, pp. 645–651.

Eftis, J., M. S. Ardel-Kader, and D. L. Jones (1989). Comparisons between the modified Chaboche and Bodner–Partom viscoplastic theories at high temperature, *International Journal of Plasticity,* Vol. 5, pp. 1–27.

Eftis, J., and D. L. Jones (1981). Evaluation and development of constitutive relations for inelastic behavior, *Final Report AFOSR-80-0096 and AFOSR-81-0241.*

Freed, A. D., J. L. Chaboche, and K. P. Walker (1991). On the thermodynamics of stress rate in the evolution of back stress in viscoplasticity, *NASA TM 103794.*

Gilman, J. J. (1969). *Micromechanics of Flow in Solids,* McGraw-Hill, New York.

James, G. H., P. K. Imbrie, P. S. Hill, D. H. Allen, and W. E. Haisler (1987). An experimental comparison of several current viscoplastic constitutive models at elevated temperature, *Journal of Engineering Materials and Technology,* Vol. 109, p. 130.

Jones, W. B., R. W. Rohde, and J. C. Swearengen (1982). Deformation modeling and the strain transient dip test, *ASTM STP 765,* pp. 102–118.

Krempl, E. (1979). An experimental study of room temperature rate-sensitivity, creep and relaxation of AISI type 304 stainless steel, *Journal of the Mechanics and Physics of Solids,* Vol. 27, pp. 363–375.

Krempl, E., and V. V. Kallianpur (1984). Some critical uniaxial experiments for viscoplasticity at room temperature, *Journal of the Mechanics and Physics of Solids,* Vol. 32, pp. 302–314.

Krieg, R. D., J. C. Swearengen, and R. W. Rohde (1978). A physically based internal variable model for rate dependent plasticity, in *Inelastic Behavior of Pressure Vessel and Piping Components,* ASME/CSME, PVP-PB-028, p. 15.

Kujawski, D., V. V. Kallianpur, and E. Krempl (1980). An experimental study of uniaxial creep, cyclic creep and relaxation of AISI type 304 stainless steel at room temperature, *Journal of the Mechanics and Physics of Solids,* Vol. 28, pp. 129–148.

Kuruppu, M. D., J. F. Williams, N. Bridgeford, R. Jones, and D. C. Stouffer (1992). Constitutive modeling of the elastic plastic behavior of 7050-T7459 aluminum alloy, *Journal of Strain Analysis,* Vol. 27, No. 2, pp. 85–92.

Lee, D., and F. Zaverl (1978). A generalized strain rate dependent constitutive equation for anisotropic metals, *Acta Metallurgica,* Vol. 26, pp. 1771–1780.

Lee, D., and F. Zaverl (1980a). Further development of generalized constitutive relations for metal deformation, *Metallurgical Transactions,* Vol. 11A, pp. 983–991.

Lee, D., and F. Zaverl (1980b). Development of constitutive equations for nuclear reactor core materials," *Journal of Nuclear Materials,* Vol. 88, pp. 104–110.

Lehmann, T. (1989). Some thermodynamic considerations on inelastic deformations including damage, *Mechanica,* Vol. 79, pp. 1–24.

Lemaitre, J., and J. L. Chaboche (1990). Published in English as *Mechanics of Solid Materials,* Cambridge University Press, Cambridge.

Lindholm, U. S., K. S. Chan, and B. H. Thacker (1993). On determining material constants for unified thermo-visco-plastic constitutive equations, in *Material Parameter Estimation for Modern Constitutive Equations,* MD Vol. 43, AMD Vol. 168, ASME, New York, pp. 183–193.

Lindholm, U. S., K. S. Chan, S. R. Bodner, R. M. Weber, K. P. Walker, and B. N. Cassenti (1984). Constitutive modeling for isotropic materials, *NASA CR 174718.*

McDowell, D., R. K. Payne, D. Stahl and S. D. Antolovich (1984). Effects of nonproportional loading histories on type 304 stainless steel, Spring Meeting of Société Française de Métallurgie, Paris, p. 73.

Metzer, A. M. (1982). Steady and transient creep behavior based on unified constitutive equations, *Journal of Engineering Materials and Technology,* Vol. 104, pp. 18–25.

Metzer, A. M. and S. R. Bodner (1979). Analytical formulation of a rate and temperature dependent stress strain relation, *Journal of Engineering Materials and Technology,* Vol. 101, pp. 254–257.

Miller, A. K. (1976). An inelastic constitutive equation for monotonic, cyclic and creep deformation: Part 1: Equations development and analytical procedures, Part 2: Application to 304 stainless steel, *Journal of Engineering Materials and Technology,* Vol. 98H, pp. 97–113.

Miller, A. K. (1978). An inelastic constitutive model for monotonic, cyclic, and creep deformation; Part I: Equations, development, and analytical procedures; Part II: Application to type 304 stainless steel, *Journal of Engineering Materials and Technology,* Vol. 98, No. 2, pp. 97–113.

Miller, A. K. (1987). The MATMOD equations, in *Unified Constitutive Equations for Creep and Plasticity,* A. K. Miller, ed., Elsevier Applied Science, Barking, Essex, England.

Milligan, W. W., and S. D. Antolovich (1989). Deformation modeling and constitutive modeling of anisotropic superalloys, *NASA CR 4215.*

Murakami, S., M. Kawai, Y. Ohmi (1989). Effects of amplitude history on the multiaxial cyclic behavior of 319 stainless steel, *Journal of Engineering Materials and Technology,*, Vol. 111, pp. 278–285.

Murakami, S., M. Kawai, Y. Yamada (199O). Creep after cyclic plasticity at elevated temperature, *Journal of Engineering Materials and Technology,* Vol. 112, pp. 346–358.

Neu, R. W. (1993). Nonisothermal material parameters for the Bodner model, in *Material Parameter Estimation for Modern Constitutive Equations,* MD Vol. 43, AMD Vol. 168, ASME, New York, pp. 211–226.

Ramaswamy, V. G. (1986). A constitutive model for the inelastic multiaxial cyclic response of a nickel base superalloy René 8O, *NASA CR 3998.*

Ramaswamy, V. G., D. C. Stouffer, and J. H. Laflen (1990). A unified constitutive model for the inelastic uniaxial response of René 80 at temperatures between 538°C and 982°C, *Journal of Engineering Materials and Technology,* Vol. 112.

Ramaswamy, V. G., R. H. Van Stone, L. T. Dame, and J. H. Laflen (1984). Constitutive modeling for isotropic materials, *NASA Conference Publication 2339.*

Rice, J. R. (1975). Continuum mechanics and thermodynamics of plasticity in relation to microscale deformations, in *Constitutive Equations in Plasticity,* A. S. Argon, ed., MIT Press, Cambridge, MA.

Robinson, D. N. (1978). A unified creep plasticity model for structural metals at high temperature, *ORNL TM-5969.*

Robinson, D. N. and R. W. Swindeman (1982). Unified creep plasticity constitutive equations for 2 1/4 Cr-1 Mo steel at elevated temperature, *ORNL TM-8444.*

Sheh, M. Y., and D. C. Stouffer (1990). A crystallographic model for the tensile and fatigue response of René N4 at 982°C, *Journal of Applied Mechanics,* Vol. 57, pp. 25–31.

Stouffer, D. C. (1981). A constitutive representation for IN 100, *AFWAL-TR-81-4039,* Materials Laboratory, Wright Patterson AFB, OH.

Stouffer, D. C., and S. R. Bodner (1882). A relationship between theory and experiment for a state variable constitutive equation, in *Mechanical Testing for Deformation Model Development, ASTM STP 765,* ASTP, Philadelphia, pp. 239–250.

Stouffer, D. C., V. G. Ramaswamy, J. H. Laflen, R. H. Van Stone, and R. Williams (1990). A constitutive model for the inelastic multiaxial response of René 80 at 871C and 982C, *Journal of Engineering Materials and Technology,* Vol. 112, pp. 241–246.

Stouffer, D. C., M. Y. Sheh, M. R. Williams, and M. D. Kuruppu (1990). Development, implementation, and verification of a constitutive model for AL7050-T7451 aluminum alloy, *First Annual Report for the Commonwealth of Australia.*

Swearengen, J. C., and J. C. Holbrook (1985). Internal variable models for rate dependent plasticity: analysis of theory and experiment, *Research Mechanica,* Vol. 13, pp. 93–128.

Tanka, T. G. (1983). A unified numerical method for integrating stiff time-dependent constitutive equations for elastic/viscoplastic deformation of metals and alloys, Ph.D. dissertation, Department of Material Science, Stanford University.

Walker, K. P. (1981). Research and development program for nonlinear structural modeling with advanced time–temperature dependent constitutive relationships, *NASA CR 165533.*

PROBLEMS

6.1 Show that $K_2 = (\sigma - \Omega)^2/3$ for uniaxial loading and verify that the three-dimensional plastic flow models given in equations 6.2.7, 6.2.8, 6.2.9, and 6.2.11 can be reduced to the uniaxial models given in equations 6.2.12 to 6.2.15.

6.2 Determine a relationship between the parameters in equations 6.2.13 and 6.3.6 for the case of isothermal stress relaxation from an elastic state as shown in Figure 6.6.2.

6.3 Repeat Problem 6.2 except for the case of stress relaxation in compression from a fully open fatigue loop. How does you result compare to Equation 6.4.13?

6.4 Appendix 2.3 contains data for three tensile curves and one fatigue loop with stress relaxation in compression for IN 100 at 1350°F. Determine the parameters for equations 6.2.13 and 6.3.6 and make plots to compare the model and experimental tensile results. Run a computer program to determine the predicted shape of the fatigue loop and compare the result to the data. Discuss your model results.

6.5 The hardening parameter Z for use with Bodner–Partom equation 6.2.12 was given by Bodner as equation 6.6.1 with $R_2 = 0$ when the model is used for tensile loading. Assume that $D_0 = 10^4\ \text{sec}^{-1}$ and use the methods in Section 6.4 to develop equations for the material parameters n, m, Z, and Z_1 for strain-rate-dependent materials. Assume that there is tensile data and fatigue data with stress relaxation available.

6.6 Use the results of Problem 6.5 to determine the constants for the IN 100 data in Appendix 2.3. Program the Bodner–Partom equations and plot the model and experimental results, and compare. Use the computer program to determine the predicted shape of the fatigue loop as saturation. Discuss your results.

6.7 An alternaltve form of the Bodner–Partom equation (Lindholm, Chan, and Thacker, 1993) involves both isotropic, Z^{I} and directional, Z^D, hardening to model the Bauschinger effect. The uniaxial form of the model is

$$\sigma = \left(-2 \ln \frac{2D_2}{\sqrt{3}\,|\dot{\varepsilon}^{\mathrm{I}}|}\right)^{1/2n} (Z^{\mathrm{I}} + Z^{\mathrm{D}})\ \mathrm{sgn}(\dot{\varepsilon}^{\mathrm{I}})$$

where

$$\dot{Z}^{\mathrm{I}} = m_1(Z_1 - Z^{\mathrm{I}})\dot{W}^{\mathrm{I}} - A_1 Z_1 \left(\frac{Z^{\mathrm{I}} - Z_0}{Z_1}\right)^{r_1}$$

$$\dot{Z}^{\mathrm{D}} = \left[m_2 \left(Z_3 \frac{\sigma}{|\sigma|} - Z^{\mathrm{D}} \right) \dot{W}^{\mathrm{I}} - A_2 Z_1 \left(\frac{Z^{\mathrm{D}}}{Z_1}\right)^{r_2} \frac{Z^{\mathrm{D}}}{|Z^{\mathrm{D}}|} \right] \frac{\sigma}{|\sigma|}$$

and where the material parameters can be identified by the presence of subscripts. Determine the parameters for the model for the data given in Appendix 2.3, and compare the model results to the experimental results.

6.8 There is a family of yield surface constitutive models (Lemaitre and Chaboche, 1990) with nonlinear kinematic and isotropic hardening that are very similar to the state variable models. A law that deals with this family has two internal variables, $\chi(t)$ and $R(t)$, and six parameters: k, K, n, C, g, b and Q. Two of the parameters, $K(\varepsilon_e)$ and $g(\varepsilon_e)$, are frequently taken as a function of the accumulated inelastic strain, ε_e.The model is formulated as the following set of differential equations:

$$\dot{\varepsilon}_{ij} = \frac{3}{2}\dot{\varepsilon}_e \frac{S_{ij} - \Omega_{ij}}{K_2}$$

$$\dot{\varepsilon}_e = \left\langle \frac{K_2 - R - k}{K(\varepsilon_e)} \right\rangle^n$$

$$\dot{\Omega}_{ij} = \frac{2}{3} C\dot{\varepsilon}_{ij} - g(\varepsilon_e)\Omega_{ij}\dot{\varepsilon}_e$$

$$\dot{R} = b[Q - R]\dot{\varepsilon}_e$$

where $\Omega(0) = R(0) = 0$, and $\langle x \rangle = x$ for $x > 0$ and $\langle x \rangle = 0$ for $x < 0$. Note that the last equation can be integrated. Reduce the model to a system of equations for uniaxial response. Determine the parameters for the uniaxial model for the data given in Appendix 2.3, and compare the model results to the experimental response.

CHAPTER 7

Multiaxial and Thermomechanical Modeling

In isothermal uniaxial cycling there is a characteristic amount of hardening or softening that depends on the material properties. The hardening or softening results from the development of a microstructure (cells or subgrains) that depends on the temperature. However, the hardening that results from multiaxial cycling, or even combined mechanical and thermal cycling, is different from that observed in uniaxial, isothermal cycling. Further, the combined hardening in multiaxial thermal cycling is not the simple composition of thermal and multiaxial cycling effects due to the interaction of the active deformation mechanisms. A simple model for extra hardening in multiaxial loading was first introduced by Bodner (1987), who developed a scalar parameter to detect the presence of nonproportional loading and activate an extra hardening term in the drag stress equation.

This chapter consists of three major parts: a discussion of extra hardening in multiaxial loading, a discussion of thermomechanical loading, and a review of the other models introduced in Chapter 6. The chapter begins with a discussion of the extra hardening properties that are observed in thermal and multiaxial cycling. In the next sections we show how to extend a uniaxial model to a multiaxial loading and add terms for extra hardening. Part B contains an application to thermomechanical cycling. The methods and limitations of interpolation and extrapolation are presented first, and then thermal history effects are examined. Part C contains a review of the Bodner model with extra hardening, the development of the Miller model from the concepts of steady-state creep, and the power law models of Walker and Kreig et al. The models presented are relatively new and have not been fully tested and verified with comprehensive experimental programs; however, it is hoped that this chapter will provide some useful tools for analysis and design, as well as a description of how materials behave in real operating environments.

7.1 RELATIONSHIP BETWEEN MULTIAXIAL AND THERMAL CYCLING

Multiaxial material loading and the prediction of the subsequent response poses major challenges both experimentally and theoretically. The development of multiaxial extensometers has been carried out in individual laboratories, and reliable commercial units have just recently become available. The response under multiaxial loading is very different from uniaxial response, and the results vary significantly with temperature. The accurate prediction of multiaxial response is important for analysis of structural deformation and life modeling. Many structures in service are in multiaxial load environments.

Recall from Section 5.3.5 that there are two general types of multiaxial loading: proportional and nonproportional. In proportional loading the ratios of the components of the control variable, stress or strain, are constant throughout the loading history, as demonstrated in equation 5.3.22. In nonproportional loading the values of the stress or strain components are varied independently. The degree of nonproportionality in a two-dimensional test such as a tension torsion test can easily be controlled experimentally by defining a frequency and phase shift as shown in equation 5.3.23. If $\beta = 0$, the axial and shear strain are in-phase or proportional, and if $\beta = \pi/2$, the axial and shear strains are 90° out-of-phase and the test is nonproportional. The condition of $\beta = \pi/2$ is referred to as a circular cycle

Examples of the two effects of nonproportional loading were demonstrated in Section 5.3. It was shown that the directions of the deviatoric stress and plastic strain increment (or strain rate) vectors are not parallel in nonproportional loading. As a result, models like the Prandtl–Reuss equation 5.2.17 or Bodner–Partom equation 6.2.7, where the inelastic strain rate and deviatoric stress are parallel, will not predict the correct response. Models with back stress (equations 6.2.8, 6.2.9, and 6.2.11) or kinematic hardening (equation 5.5.11) will predict a variation in orientation between deviatoric stress and inelastic strain-rate vectors.

There is also extra hardening observed in nonproportional loading that is not present in proportional loading. The definitions of blocks of a tension–torsion strain-controlled history are shown in Figure 7.1.1*a–c*. The nonproportional strain histories were constructed to develop an increasing degree of nonproportionality, starting from proportional loading. The corresponding stress histories are shown in Figure 7.1.1*d–f* and the magnitude of the extra hardening is plotted using the effective stress or von Mises stress vector $\boldsymbol{\sigma} = \sigma \mathbf{n}_1 + \sqrt{3}\tau \mathbf{n}_2$ in Figure 7.1.2. The angle between the deviatoric stress and inelastic strain from this experiment was given in Figure 5.3.10. The results show that the amount of hardening increases as the degree of nonproportionality, β, increases. This effect has been widely documented in the literature for other materials and is not special for type 304 stainless steel. The result indicates that general nonproportional response cannot be predicted from observations of uniaxial or proportional testing alone.

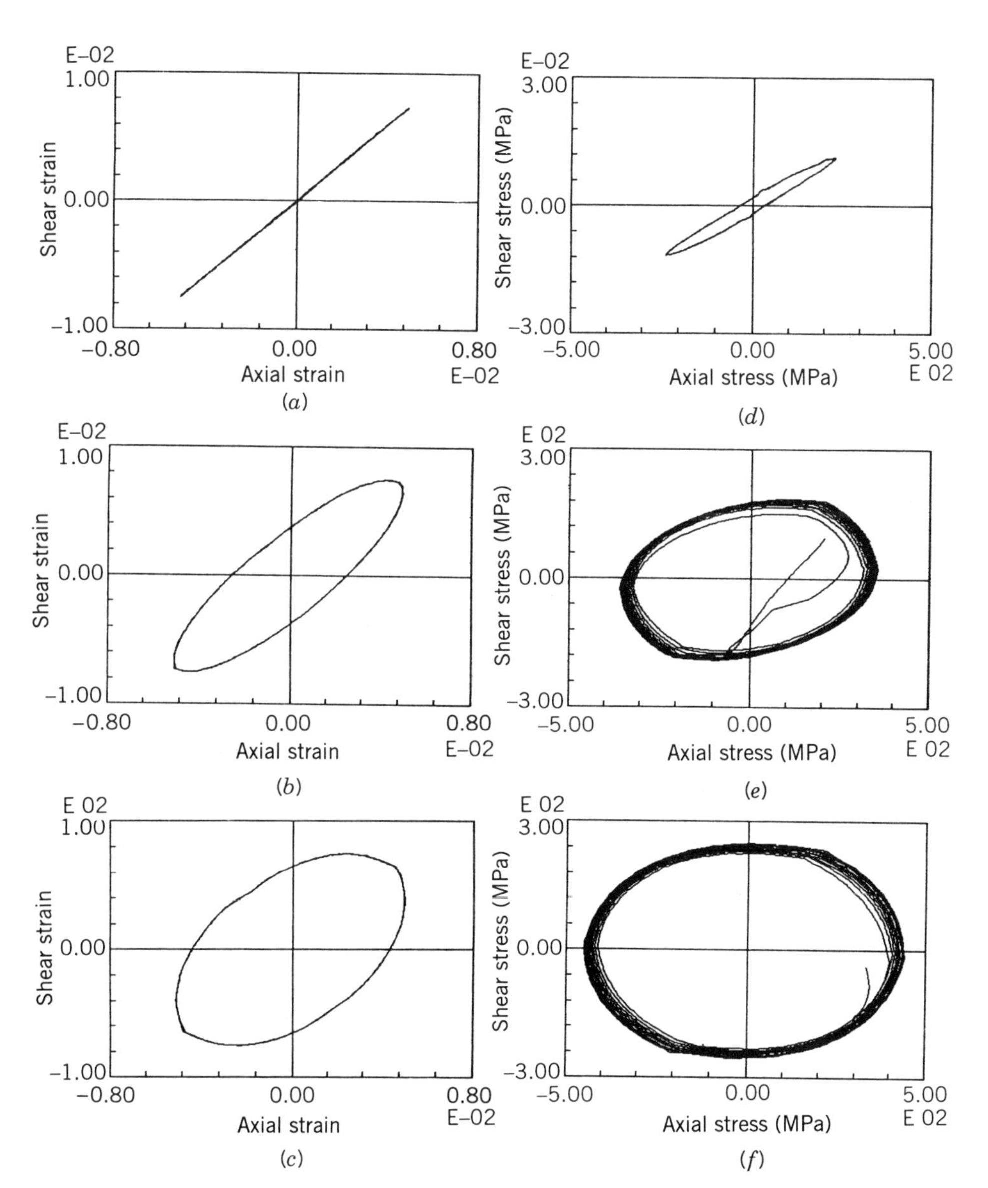

Figure 7.1.1 The load histories are defined by equation 5.3.23 with $\epsilon_a = 0.007$ and $\gamma_a = 0.0075$ held constant for (*a*) 16 cycles with $\beta = 0$ followed by (*b*) 25 cycles with $\beta = 30°$ followed by (*c*) 25 cycles with $\beta = 60°$. The stress response of 304 stainless steel at room temperature is shown in (*d*), (*e*), and (*f*), respectively, (After McDowell, 1983b.)

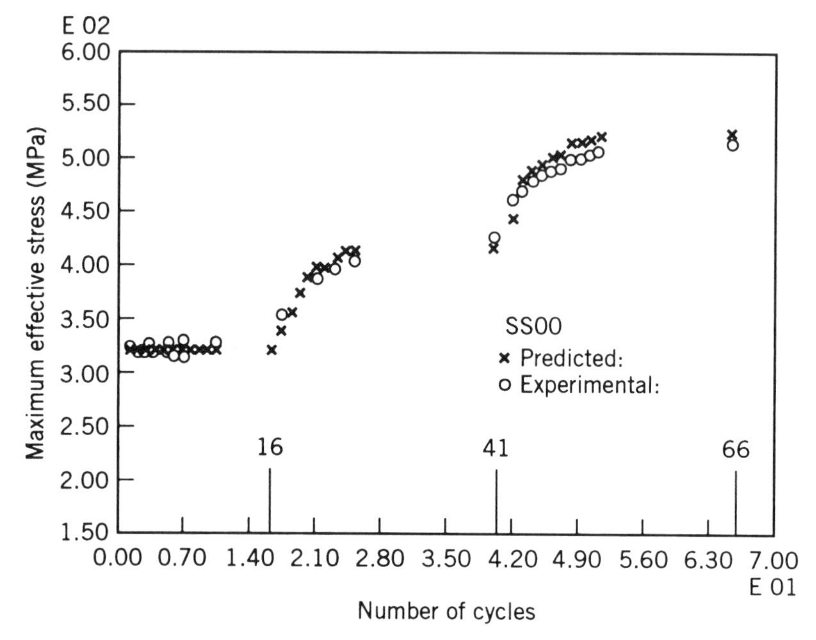

Figure 7.1.2 Value of the effective stress resulting from the cyclic loading defined in Figure 7.1.1. (After McDowell, 1983b.)

The reason for the extra hardening in nonproportional loading can be deduced by considering the deformation mechanism. Recall that at room temperature inelastic deformation occurs by planar slip. In proportional loading (tension, torsion, or combined tension torsion) there is a constant ratio of axial strain to shear strain, and only the magnitude of the strain components change. As a result, the direction of the maximum shear stress is constant, and slip occurs on the same planes throughout the test. In nonproportional loading the direction of the maximum shear stress is changing during the load history and slip is occurring on different planes at different times. Thus the number of dislocation intersections are increased, more dislocations are pinned, the dislocation density increases, and extra hardening is observed. The 90° out-of-phase test develops the maximum amount of extra hardening in a tension–torsion test because the direction of the maximum shear stress rotates through 360° and slip occurs on all the possible slip planes.

The amount of extra hardening is also related to the magnitude of the cyclic strain range. A comprehensive set of experiments that show the effect of strain range and temperature on the hardening was presented by Murakami, Kawai, and Ohmi (1989). Consider first the response at room temperature in Figures 7.1.3 and 7.1.4. Figure 7.1.3 shows the hardening resulting from in-phase and 90° out-of-phase testing of type 316 stainless steel for increasing strain range. The total amount of hardening increased with increasing strain range, and

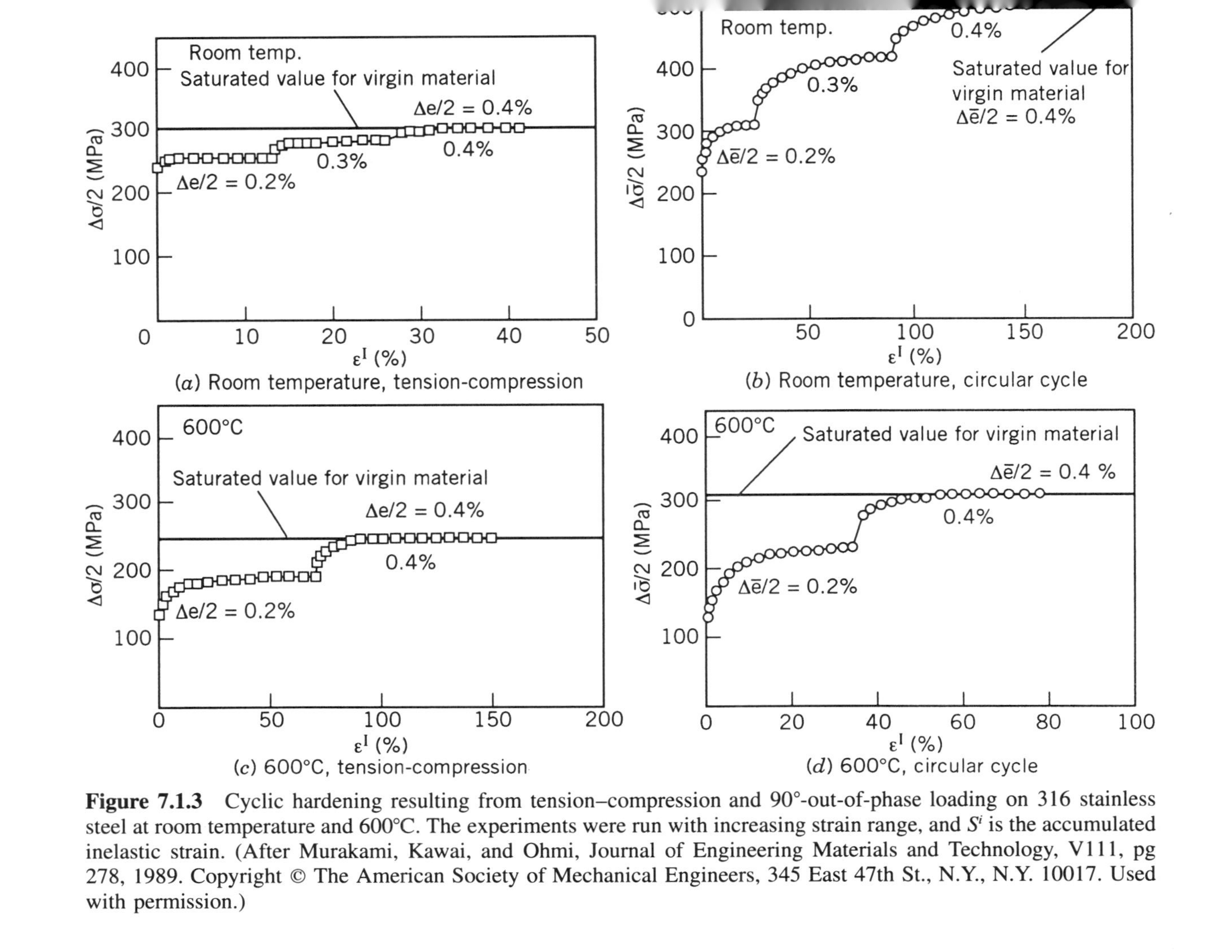

Figure 7.1.3 Cyclic hardening resulting from tension–compression and 90°-out-of-phase loading on 316 stainless steel at room temperature and 600°C. The experiments were run with increasing strain range, and S^i is the accumulated inelastic strain. (After Murakami, Kawai, and Ohmi, Journal of Engineering Materials and Technology, V111, pg 278, 1989. Copyright © The American Society of Mechanical Engineers, 345 East 47th St., N.Y., N.Y. 10017. Used with permission.)

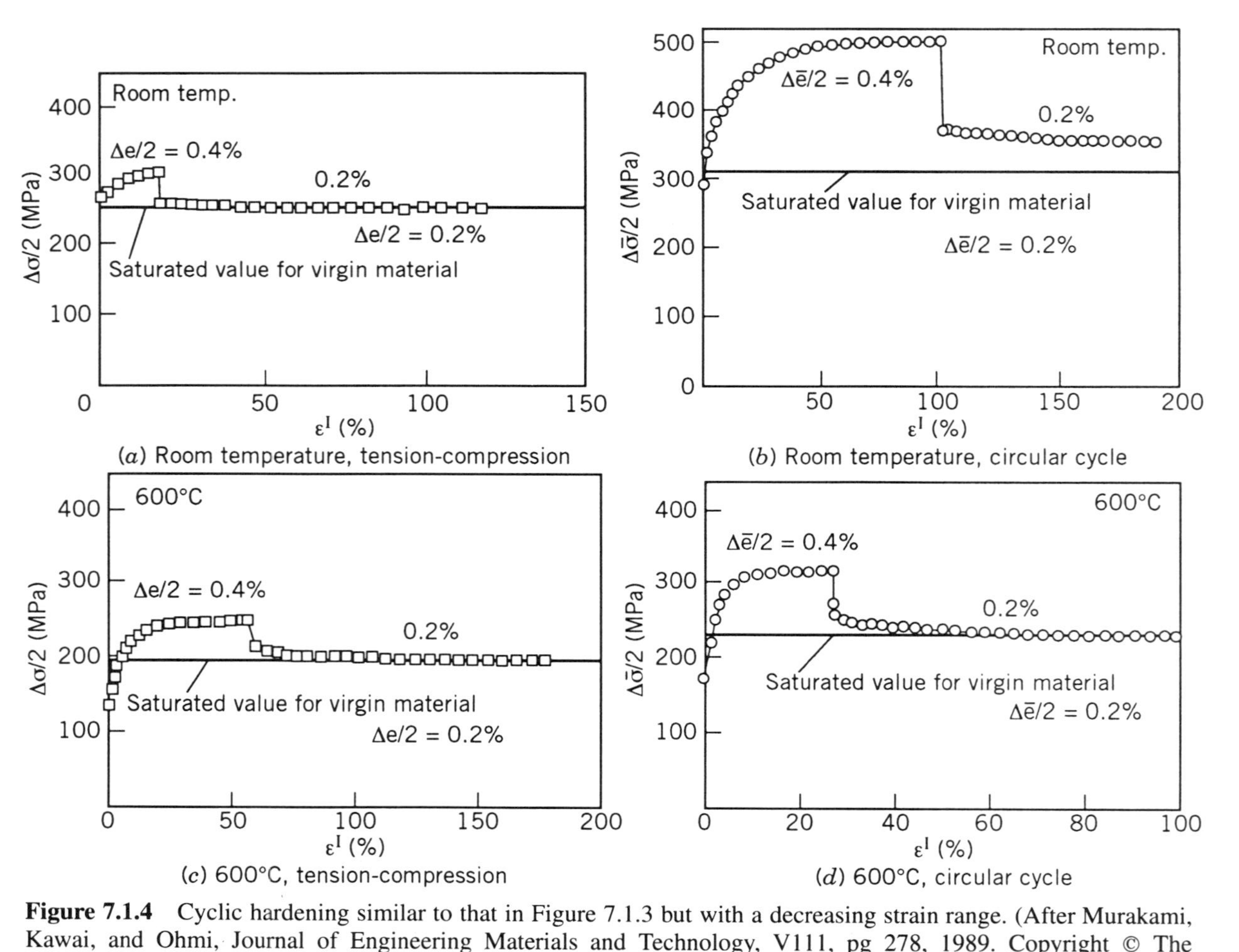

Figure 7.1.4 Cyclic hardening similar to that in Figure 7.1.3 but with a decreasing strain range. (After Murakami, Kawai, and Ohmi, Journal of Engineering Materials and Technology, V111, pg 278, 1989. Copyright © The American Society of Mechanical Engineers, 345 East 47th St., N.Y., N.Y. 10017. Used with permission.)

there was much more hardening in the 90° out-of-phase test than in the tension–compression test. This result suggests that the amount of extra hardening depends on the strain range or inelastic work and that there is a characteristic microstructure and dislocation density at saturation for each strain range. The amount of hardening in Figure 7.1.3*a* and *b* was independent of the loading history since the saturation at 0.4% strain is the same as the response of the virgin material. To further investigate the effect of strain-range history in nonproportional hardening, Murakami at al. conducted a second set of tests at the same strain ranges, except that the tests were started with the largest strain range. In the proportional test at room temperature in Figure 7.1.4, the material saturated at the stress, corresponding to the 0.2% strain range of a virgin material. This shows that there is no strain-range sequence effect in proportional loading. However, in nonproportional cycling the results were different. At room temperature the saturated value for the 0.2% strain range cycling after 0.4% strain range cycling was above that observed for the 0.2% strain range cycling of the virgin material. This shows that there are different characteristic microstructures at different strain ranges in nonproportional cycling at room temperature. It is not clear from these experiments, but it appears that the response of the nonproportional test at room temperature would eventually saturate at the 0.2% virgin material strain range with continued cycling. This suggests that a material will gradually evolve into the microstructure that is characteristic of its current loading, and that extra hardening from previous cycling will eventually be eliminated.

The discussion above is valid only for temperatures below about one-half of the melting temperature. At elevated temperatures when recovery is more predominate, the strain-range sequence effect above is reduced or even eliminated. Figures 7.1.3 and 7.1.4 also show the hardening resulting from in-phase and 90° out-of-phase testing at 600°C. At both temperatures there was more hardening in the circular (nonproportional) tests than in the uniaxial tests. However, the same strain-range history at 600°C and room-temperature did not produce the same results. The amount of hardening shown in Figures 7.1.3*a* and *c* in the tension–compression tests at room temperature is less than at 600°C, but the saturation levels were lower at 600°C. However, the amount of extra hardening was reduced in the test at 600°C compared to the room-temperature test for nonproportional hardening, Figures 7.1.3*b* and *d*. When the strain range was reduce from 0.4% to 0.2% as shown in Figures 7.1.4*b* and *d*, the stress was initially above the 0.2% saturation stress for the virgin material, but it relaxed rapidly to the 0.2% saturation stress in a few cycles. Thus the sequence effect was effectively eliminated by the presence of stress relaxation.

The response at temperatures above one-half the melting temperature is also complicated by the interaction of slip and climb. In proportional tests at high temperature, deformation occurs by slip and climb simultaneously. As a result of the climb, dislocations move normal to the slip plane and interact with dislocations on other slip planes. Thus the dislocation network has more intersections, resulting in more kinks and jogs, and there is more hardening relative to the initial yield stress in high-temperature deformation than in low-temperature deformation. The stress levels are lower at high temperature than at room temperature, due to dynamic recovery. These effects are seen in Figures 7.1.3*a* and *c* and 7.1.4*a* and *c* for proportional loading by comparing the response at 600°C with the response at room temperature.

The amount of hardening in nonproportional loading compared to that in the proportional loading is shown in Figures 7.1.3 (or Figure 7.1.4). The increase in hardening at room temperature (Figure 7.1.3*b* compared to Figure 7.1.3*a*) is larger than at high temperature (Figure 7.1.3*d* compared to Figure 7.1.3*c*) because the dislocation intersections arising from climb in high temperature proportional loading offset some of the effect of the extra hardening in the circular cycle. This shows the amount of extra hardening is reduced with increasing hardening. However, the total amount of hardening (stress level of the curve) at room temperature is higher than that at high temperature in the nonproportional tests in Figures 7.1.3 and 7.1.4 due to the increased action of dynamic recovery at high temperature.

The observations above are supported by another set of experiments by Murakami, Kawai, and Ohmi (1989). Figure 7.1.5 shows the effect of a temperature sequence on proportional and nonproportional cycling at a constant strain range of 0.30%. Consider first the proportional tests. Observe that the virgin material saturation stress is almost the same for the 200 and 600°C cycling. In the 600–200–600°C sequence there is no sequence effect; the saturation stress due to second cycling at 600°C was the same as the first cycling at 600°C. However, the saturated stress for the 200°C cycling was above the 200°C saturation stress for the virgin material. The 200–600–200°C sequence was different. These experiments show that the saturation stress from the initial cycling at 200°C was less than the 200°C cycling after cycling at 600°C. Thus cycling at 200°C does not affect subsequent cycling, but cycling at 600°C does affect the response at 200°C. Recalling the discussion above, the cycling at 600°C with climb and slip produces a dislocation microstructure that is harder than pure slip. Thus cycling at a high temperature will affect the subsequent saturation stress at low temperature, but there is no effect if the temperature order is reversed. This implies that prior slip does not influence the microstructure resulting from slip and climb.

Figure 7.1.5*c* and *d* shows that in nonproportional cycling temperature, sequencing does not have an effect. The experiments show that prior cycling at 200 or 600°C did not change the subsequent saturation stress. This indicates that the extra hardening produced by the nonproportional cycling controls the final saturation stress level, and that the difference in the two levels at 200 and 600°C results from the stronger effect of dynamic recovery at 600°C.

The results of these experiments show the close correlation between multiaxial fatigue and thermomechanical fatigue. In both cases there can be extra hardening that is not observed in uniaxial, isothermal loading alone. These experiments provide a useful insight for constitutive modeling.

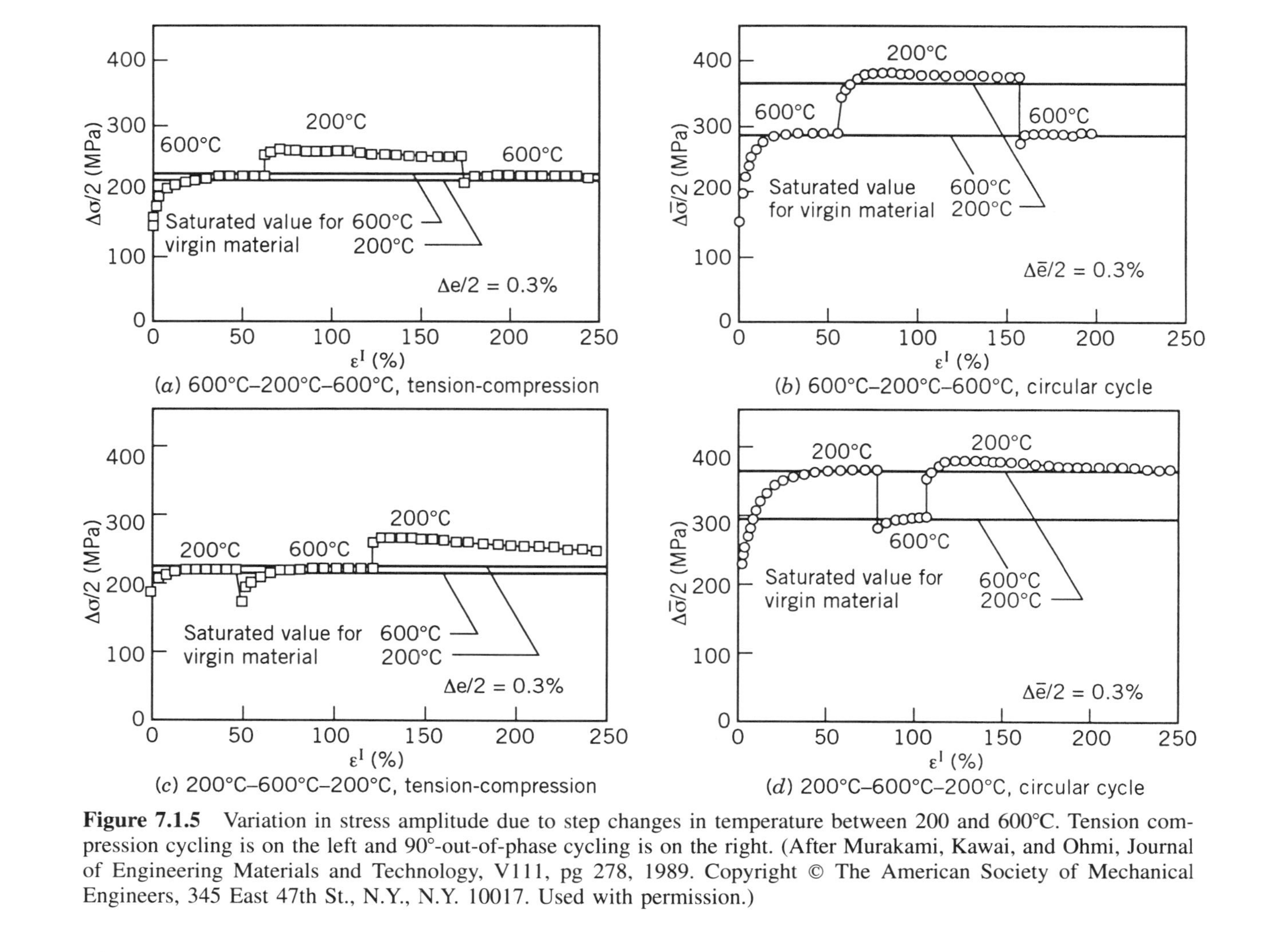

Figure 7.1.5 Variation in stress amplitude due to step changes in temperature between 200 and 600°C. Tension compression cycling is on the left and 90°-out-of-phase cycling is on the right. (After Murakami, Kawai, and Ohmi, Journal of Engineering Materials and Technology, V111, pg 278, 1989. Copyright © The American Society of Mechanical Engineers, 345 East 47th St., N.Y., N.Y. 10017. Used with permission.)

A. MULTIAXIAL LOADING

7.2 EXTENSION TO MULTIAXIAL LOADING

Consider first the extension to multiaxial fatigue by examining the tensor structure of the equations developed in Chapter 6 for uniaxial response. First note that kinematic equation 6.6.1 (relating the total strain to the elastic, inelastic, and thermal strains) and the inelastic flow equation 6.2.8 are both three-dimensional tensor equations. The drag stress equation 6.6.5 is a scalar hardening equation, so it can be used for multiaxial loading in its current form. The back-stress equation 6.3.6, and the creep equations 6.5.2 and 6.5.7 were formulated for uniaxial response, so they are not applicable for multiaxial loading.

The first step is to develop a three-dimensional tensor representation for the back-stress and creep equations with the assumption that no additional parameters are necessary for the extension to multiaxial loading. This will not apply when extra hardening is present, but from the previous experiments the resulting formulation should be valid for proportional loading and for high-temperature nonproportional cycling when the extra hardening is minimal. A modification for extra hardening that can be added later for low-temperature nonproportional cycling is presented in Section 7.3.

As part of the foregoing approach, the multiaxial model should reduce to the uniaxial model. It is also reasonable to assume that the inelastic response should be isotropic for polycrystalline metals. This is consistent with the observation that the elastic response of polycrystalline materials is isotropic even though the individual grains exhibit anisotropic elastic response. Finally, it is useful to note from the flow equation 6.2.8 that the inelastic strain rate $\dot{\varepsilon}^{\mathrm{I}}_{ij}$, deviatoric stress S_{ij}, and back stress Ω_{ij} must all be deviatoric tensors to guarantee that there is no inelastic volume change (i.e., $\dot{\varepsilon}^{\mathrm{I}}_{ii} = 0$). Thus it is reasonable to expect that Ω_{ij} is a second-order deviatotic isotropic tensor.

Since the three-dimensional back-stress equation must reduce to the uniaxial equation, the two equations should have the same functional relationships between the variables. Therefore, replace the material parameters in equation 6.3.5 by tensor material parameters and the scalar variables by deviatoric tensor variables, to obtain

$$\dot{\Omega}_{ij} = f_{ijkl}\dot{\varepsilon}^{\mathrm{I}}_{kl} - g_{ijkl}\Omega_{kl}\dot{\varepsilon}^{\mathrm{I}}_{e} + h_{ijkl}\dot{S}_{kl} \tag{7.2.1}$$

The parameters f_{ijkl}, g_{ijkl}, and h_{ijkl} are the tensor equivalents of f_1, f_1/Ω_s, and f_2, respectively, and $\dot{\varepsilon}^{\mathrm{I}}_{e}$ is the effective inelastic strain rate defined by equation 5.2.21. The parameters f_{ijkl}, g_{ijkl}, and h_{ijkl} are fourth-order isotropic tensors; that is, the tensor components must have the same value in all orthogonal coordinate systems at a point. Using the expansion for isotropic tensors given in equation 4.5.4, the first term of equation 7.2.1 can be written as

$$f_{ijkl}\dot{\varepsilon}^{\mathrm{I}}_{kl} = [a\delta_{ij}\delta_{kl} + b(\delta_{ik}\delta_{jl} + \delta_{il}\delta_{jk})]\dot{\varepsilon}^{\mathrm{I}}_{kl} \tag{7.2.2}$$

where a and b are two independent scalar material parameters. Expanding the right-hand side of equation 7.2.2 and noting that $\dot{\varepsilon}^{\mathrm{I}}_{ij}$ is a symmetric deviatoric tensor gives

$$f_{ijkl}\dot{\varepsilon}^{\mathrm{I}}_{kl} = a\delta_{ij}\dot{\varepsilon}^{\mathrm{I}}_{kk} + b(\dot{\varepsilon}^{\mathrm{I}}_{ij} + \dot{\varepsilon}^{\mathrm{I}}_{ji}) = 2b\dot{\varepsilon}^{\mathrm{I}}_{ij} \equiv k_1\dot{\varepsilon}^{\mathrm{I}}_{ij}$$

after defining k_1 as a new scalar material parameter. Repeating these steps for the second and third terms in equation 7.2.2 gives

$$\dot{\Omega}_{ij} = k_1\dot{\varepsilon}^{\mathrm{I}}_{ij} - k_2\Omega_{ij}\dot{\varepsilon}^{\mathrm{I}}_{e} + k_3\dot{S}_{ij} \tag{7.2.3}$$

The material parameters k_1, k_2, and k_3 can be related to the parameters in the uniaxial back-stress equation by comparing $\dot{\Omega}_{11}$ from equation 7.2.3 to equation, 6.3.6. Using the same definitions for the back-stress and deviatoric stress tensors ($\Omega_{11} = 2\Omega/3$, where Ω is the uniaxial back stress in equation 6.2.13) and defining the inelastic-strain-rate tensor similar to that given in equation 5.3.13, the expansion for $\dot{\Omega}_{11}$ becomes

$$\dot{\Omega}_{11} = \tfrac{2}{3}\,\dot{\Omega} = k_1\dot{\varepsilon}^{\mathrm{I}} - \tfrac{2}{3}\,k_2\Omega|\dot{\varepsilon}^{\mathrm{I}}| + k_3\,\tfrac{2}{3}\,\dot{\sigma}$$

and comparing to equation 6.5.3,

$$\dot{\Omega} = f_1\dot{\varepsilon}^{\mathrm{I}} - f_1\,\frac{\Omega}{\omega}\,|\dot{\varepsilon}^{\mathrm{I}}| - f_2\dot{\sigma} \tag{6.5.3}$$

gives

$$k_1 = \tfrac{2}{3}f_1, \qquad k_2 = f_1\,\frac{1}{\omega}, \qquad k_3 = f_2 \tag{7.2.4}$$

Combining equations 7.2.3 and 7.2.4 gives a representation for the multiaxial back stress as

$$\dot{\Omega}_{ij} = \tfrac{2}{3}f_1\dot{\varepsilon}^{\mathrm{I}}_{ij} - f_1\,\frac{\Omega_{ij}}{\omega}\,\dot{\varepsilon}^{\mathrm{I}}_{e} + f_2\dot{S}_{ij} \tag{7.2.5}$$

Since the parameters in equation 7.2.5 are determined from the uniaxial response, this result is expected to be valid for proportional loading and approximately correct for high-temperature nonproportional loading.

The first step in verifying these equations is to use the representation for torsion. For pure torsion the overstress and inelastic strain rate invariants are $K_2 = (\sigma_{12} - \Omega_{12})^2$ and $\dot{\varepsilon}_e^{\mathrm{I}} = 2\dot{\varepsilon}_{12}^{\mathrm{I}}/\sqrt{3}$. The equations for pure torsion can then be written as

$$\dot{\varepsilon}_{12}^{\mathrm{I}} = D_0 \exp\left[-\frac{1}{2}\left(\frac{Z}{\sqrt{3}\,|\sigma_{12} - \Omega_{12}|}\right)^{2n}\right]\frac{\sigma_{12} - \Omega_{12}}{|\sigma_{12} - \Omega_{12}|} \tag{7.2.6}$$

$$\dot{\Omega}_{12} = \tfrac{2}{3} f_1 \dot{\varepsilon}_{12}^{\mathrm{I}} - \frac{2}{\sqrt{3}} f_1 \frac{\Omega_{12}}{\omega} |\dot{\varepsilon}_{12}^{\mathrm{I}}| + f_2 \dot{\sigma}_{12} \tag{7.2.7}$$

$$\dot{Z} = 2m(Z_1 - Z)\sigma_{12}\dot{\varepsilon}_{12}^{\mathrm{I}} \tag{7.2.8}$$

Figure 7.2.1 shows the experimental data and predicted results using equations 7.2.6 to 7.2.8 for René 80 at 871°C in pure torsion. The fatigue response was calculated using the constants determined from uniaxial experiments and the extrapolation technique characterized by equation 6.6.7. Application to tension–torsion cyclic loading was investigated as shown in Figure 7.2.2 for the strain-range history shown in Figure 6.6.6. The equations are similar to equations 7.2.6 to 7.2.8 except that both axial and shear stress are present in the invariants K_2 and $\dot{\varepsilon}_e^{\mathrm{I}}$ (see Problem 7.2). The strain-controlled loading in Figure 7.2.2 was defined by $\varepsilon_{12} = C\varepsilon_{11}$, where C is constant throughout the test.

A representation for multiaxial creep response can be developed by using the same assumptions that were used for back stress. Unfortunately, these assumptions are not totally valid because uniaxial creep in tension and compression is different. Thus the assumption of isotropy is not totally correct, and the traditional formulation should be used with caution. The classical assumptions for extending uniaxial creep models to multiaxial response are:

1. Multiaxial creep of polycrystalline metals is isotropic.
2. The inelastic volume is approximately constant, so the stress, strain, and back stress are all deviatoric tensors.
3. The uniaxial and multiaxial representations should have the same functional relationships between variables, so the multiaxial model reduces to the uniaxial model.

All of the existing equations satisfy these assumptions except equations 6.5.4 and 6.5.8. These assumptions can be incorporated by replacing the uniaxial stress in the creep equations by the von Mises effective stress (equation 5.2.21), which is a simple function of the second invariant of the deviatoric stress tensor. This procedure gives

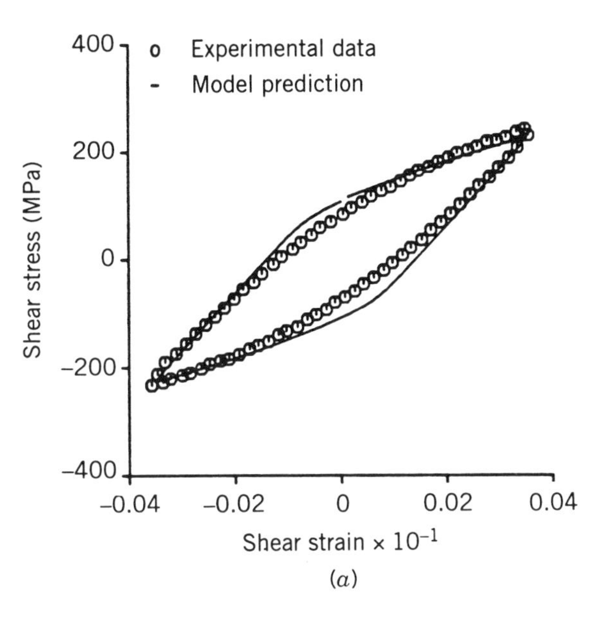

(*a*)

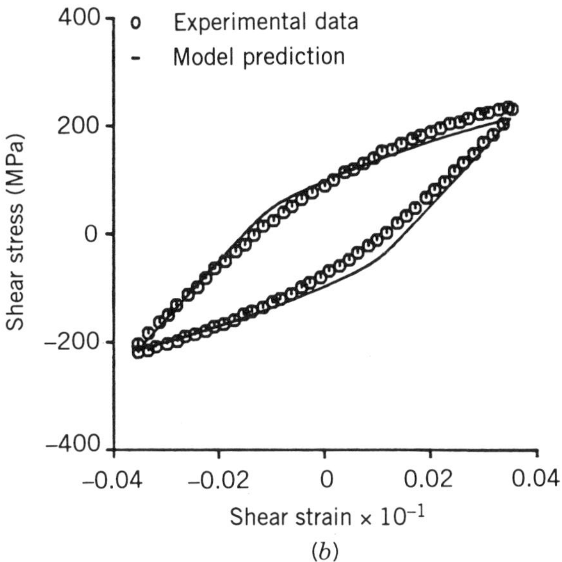

(*b*)

Figure 7.2.1 Predicted and experimental cyclic shear response of René 80 at 871°C at a strain rate of 0.002 min^{-1} for (*a*) cycle 31 and (*b*) cycle 54. (After Ramaswamy et al., 1985.)

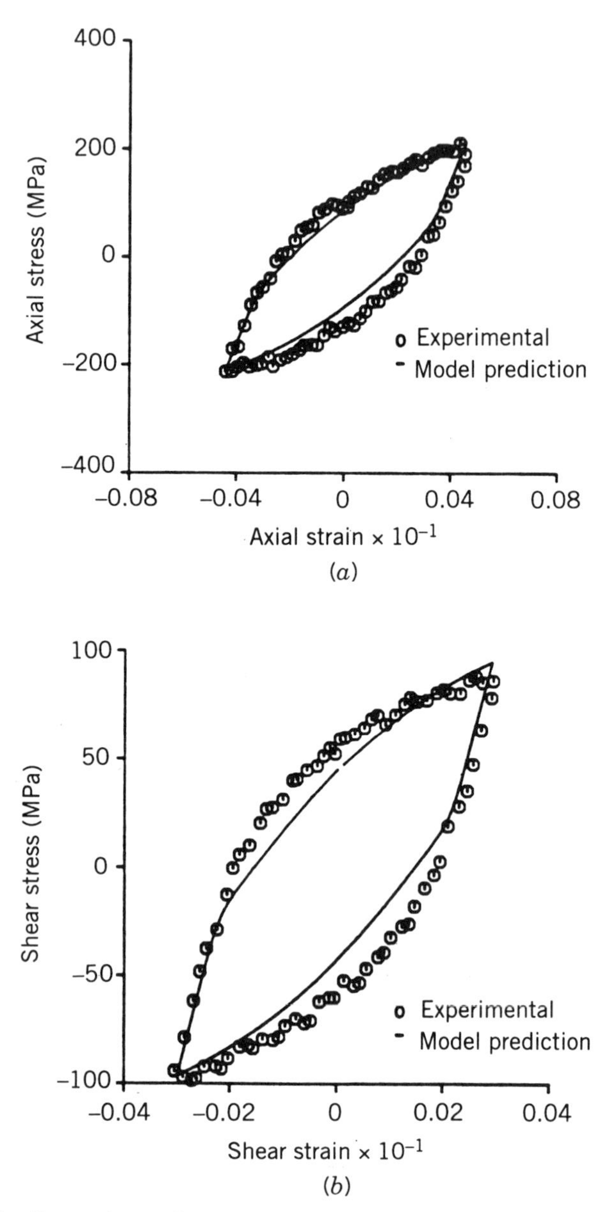

Figure 7.2.2 Comparison of predicted and experimental response of René 80 for cycle 39 for an in-phase tension torsion test at 982°C for (*a*) axial and (*b*) shear response. (After Stouffer et al., Journal of Engineering Materials and Technology, V112, pg 241, 1990. Copyright © The American Society of Mechanical Engineers, 345 East 47th St., N.Y., N.Y. 10017. Used with permission.)

$$\omega_e = \begin{cases} \Omega_m & \text{for } \sqrt{3J_2} > \sigma_y \\ b_{1T}\sqrt{3J_2} + b_{2T} & \text{for } \sigma_2 < \sqrt{3J_2} < \sigma_y \\ b_3 & \text{for } \sqrt{3J_2} < \sigma_2 \end{cases} \tag{7.2.9}$$

and

$$\dot{\omega} = A \left(\frac{\sqrt{3J_2}}{\sigma_0} \right)^p (\omega_e - \omega), \qquad \sigma_m > 0 \tag{7.2.10}$$

as a representation that uses the uniaxial material parameters.

The use of effective stress for predicting creep under combined states of stress was investigated by Murakami, Kawai, and Yamada (1990). Their results shown in Figure 7.2.3 for type 316 stainless steel at 600°C verify that the same values of von Mises stress produces the same creep strain curves. These observations make it easy to extend most uniaxial creep models to multiaxial creep as long as the stresses are positive.

As a final application of the multiaxial model described above, let us investigate the use of the parameters from uniaxial tests to predict the response of high-temperature nonproportional loading. Nonproportional testing was investigated for René 80 at 871 and 982°C. The axial and shear loading paths for the tests are given in Figure 7.2.4. The axial stress–shear stress response curves for the model and experiments are given in Figure 7.2.5 for two dif-

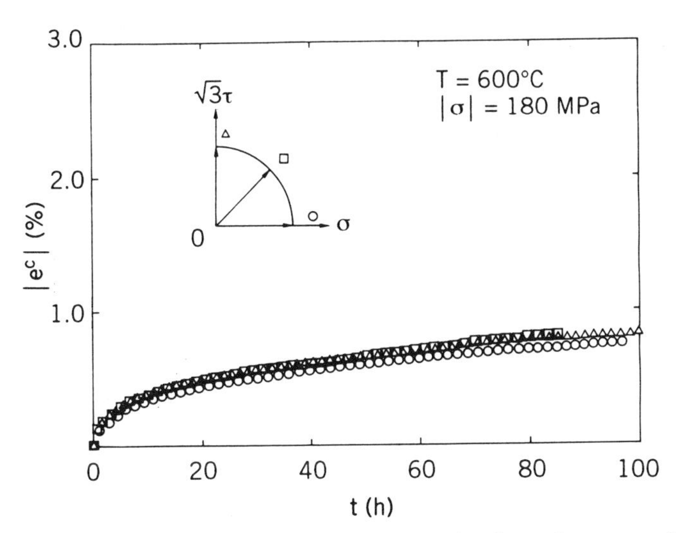

Figure 7.2.3 Creep under combined states of stress that have the same value of von Mises effective stress. (After Murakami, Kawai, and Yamada, Journal of Engineering Materials and Technology, V112, pg 349, 1990. Copyright © The American Society of Mechanical Engineers, 345 East 47th St., N.Y., N.Y. 10017. Used with permission.)

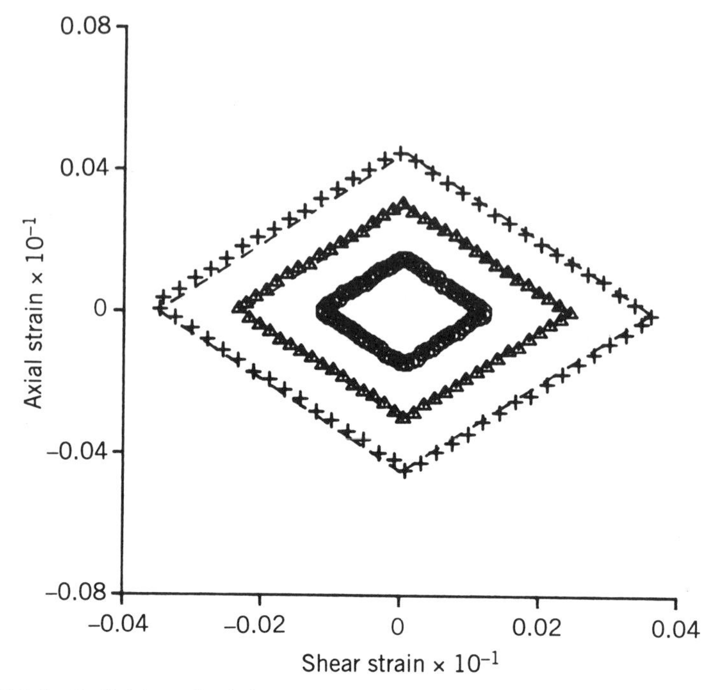

Figure 7.2.4 Definition of axial and shear strain paths for the nonproportional testing on René 80 at 871 and 982°C. (After Stouffer et al., Journal of Engineering Materials and Technology, V112, pg 241, 1990. Copyright © The American Society of Mechanical Engineers, 345 East 47th St., N.Y., N.Y. 10017. Used with permission.)

ferent strain ranges. The corresponding axial stress–strain and shear stress–strain hysteresis loops are given in Figure 7.2.6. The response was again calculated using the material parameters from the uniaxial experiments and the extrapolation technique. The material exhibited a little more hardening than was predicted by the model. This was expected from the experiments on type 316 stainless steel in Section 7.1. Comparing the orientation of the curves in Figures 7.2.4 and 7.2.5, the rotation of the stress response was correctly predicted. The angle between the stress and inelastic strain rate vectors was calculated for 90° out-of-phase loading. The results in Figure 7.2.7 for René 80 at 982°C are similar to that in Figure 6.1.1 for Hastelloy X at room temperature. To summarize, these results show that tensor properties of the model are correct but that the scalar hardening (drag stress) needs to be modified for nonproportional loading.

7.3 MODIFICATIONS FOR EXTRA HARDENING

Recall that the extra hardening observed in nonproportional cycling is different from proportional cycling because the slip on several intersecting planes

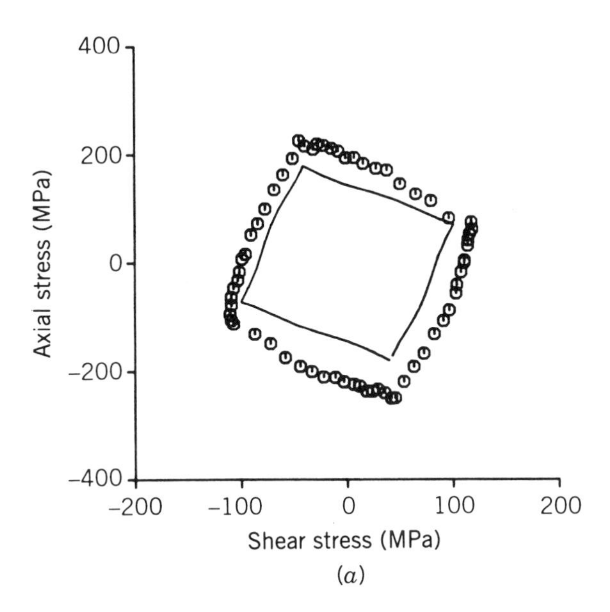

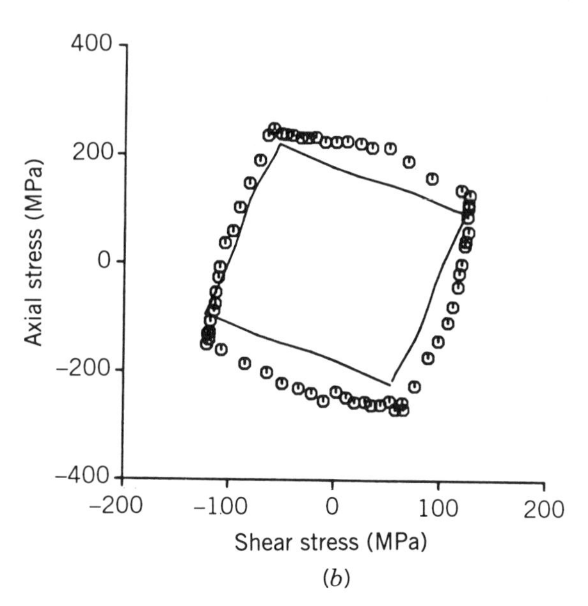

Figure 7.2.5 Experimental and predicted stress response for the nonproportional strain histories defined in Figure 7.2.4 for (*a*) cycle 11 and (*b*) cycle 18. (After Ramaswamy et al., 1985.)

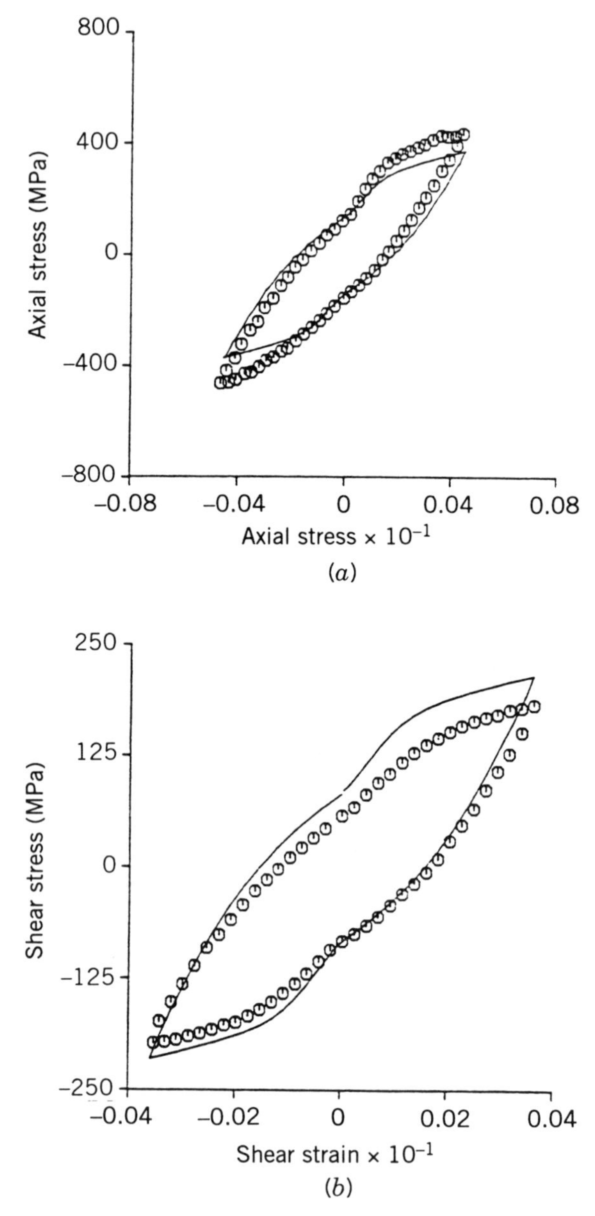

Figure 7.2.6 Experimental (circles) and predicted (line) stress–strain curves for the nonproportional loading of René 80 at 982°C. (After Ramaswamy et al., 1985.)

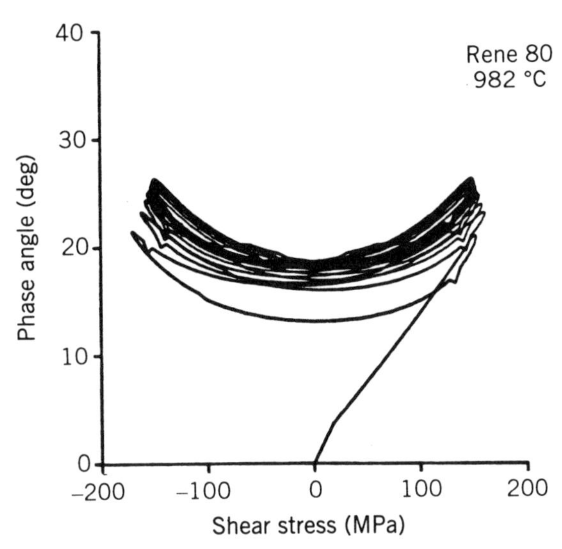

Figure 7.2.7 Predicted angle between the inelastic strain rate and deviatoric stress vectors for a 90°-out-of-phase tension torsion loading using the constants for René 80. (After Stouffer et al., Journal of Engineering Materials and Technology, V112, pg 241, 1990. Copyright © The American Society of Mechanical Engineers, 345 East 47th St., N.Y., N.Y. 10017. Used with permission.)

produces more dislocation intersections than in a single slip loading condition. Bodner (1987) (with Lindholm et al., 1984, and Chan et al., 1990) developed a method to model extra hardening by formulating a variable to sense the presence of nonproportional loading and adding terms to the drag stress equation to account for the increase in the hardening. He proposed that the extra hardening increment grows in nonproportional cycling from zero to a maximum value at saturation, and that it is fully reversible if proportional cycling is resumed after nonproportional cycling. The reversibility assumption is based on the notion that the extra jogs and kinks would eventually be absorbed by the cell or subgrain walls, and the current microstructure would evolve toward the microstructure that is characteristic of proportional loading.

The variable used to detect this type of loading is based on the observation that if the strain and strain-rate vectors are parallel during a loading history, the orientation of the strain vector will remain constant. This is proportional loading, as shown in Figure 7.3.1. Conversely, if the strain and strain rate vectors are not parallel, the components of the strain vector will change relative to each other and that nonproportional loading results. Thus a variable to detect nonproportional loading is the angle θ, between the strain and strain-rate vectors which can be determined from the dot product; that is,

$$\cos\theta = \frac{\varepsilon_{kl}\dot{\varepsilon}_{kl}}{\sqrt{\varepsilon_{pq}\varepsilon_{pq}}\sqrt{\dot{\varepsilon}_{mn}\dot{\varepsilon}_{mn}}} \tag{7.3.1}$$

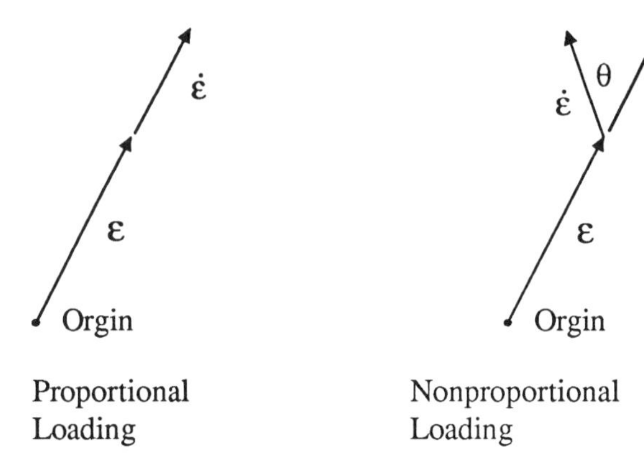

Figure 7.3.1 Schematic diagram of how the relative direction of the strain and strain rate vectors can produce proportional and nonproportional loading.

Stress can be substituted for strain in equation 7.3.1 for stress-controlled loading histories.

Consider next a modification for extra hardening. Since the drag stress is used to model cyclic hardening and softening, the drag stress is the appropriate equation to include an extra hardening term. Since Z_1 is used to define the final hardened state, the drag stress equation 6.6.3 can be modified for nonproportional loading as

$$Z = (1 + \alpha)Z_1 - [(1 + \alpha)Z_1 - Z_0] \exp(-mW^I) \tag{7.3.2}$$

where α is a material parameter that defines the amount of extra hardening. A simple representation for α is

$$\alpha = \alpha_m(1 - |\cos \theta|) \tag{7.3.3}$$

where α_m is a scale factor that defines the maximum amount of extra hardening from 90° out-of-phase loading. The parameter α_m is usually assumed to vary with strain range and temperature for most materials. A multiaxial cyclic stress–strain curve similar to the uniaxial curves discussed in Section 2.5 could be established for 90° out-of-phase cycling to determine α_m as a function of strain range. The parameter α_m is the scale factor that maps the uniaxial cyclic stress–strain curve onto the 90° out-of-phase curve as shown in Figure 7.3.2. Recall that the parameter m in equation 7.3.2 controls the rate of hardening. If m is a constant, the same rate of hardening is assumed to occur in proportional and nonproportional loading. Equation 7.3.2 will also predict softening for proportional cycling following nonproportional cycling.

The equations have been implemented to improve the accuracy of the prediction of the nonproportional loading shown in Figures 7.2.4 to 7.2.6. The

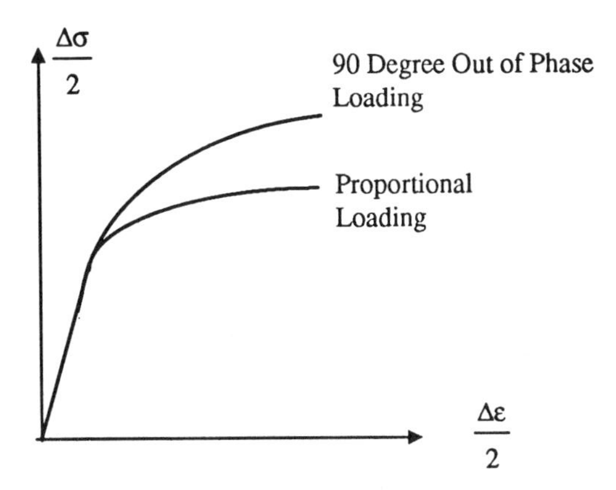

Figure 7.3.2 Diagram showing proportional and nonproportional cyclic stress–strain curves.

parameter α_m was assumed to be a constant in this exercise, $\alpha_m = 1.0$; the other parameters are given in Appendix 6.1. Figure 7.3.3 shows the same data with the extra hardening equations operational. In summary, at low temperature extra hardening occurs in nonproportional cycling from the extensive dislocation network created by continuously changing the slip plane. At high temperature the apparent amount of extra hardening in nonproportional loading is reduced due to the additional effect of climb and recovery.

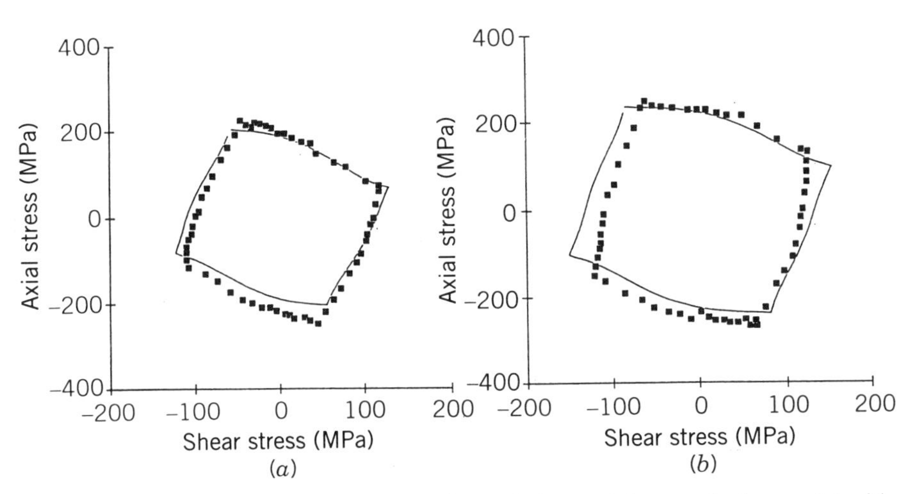

Figure 7.3.3 Comparison between experimental data and the constitutive model with extra hardening for the experimental results in Figure 7.2.5: (*a*) cycle 11; (*b*) cycle 18. (Courtesy of G. C. Salemme, 1993.)

B. THERMOMECHANICAL LOADING

7.4 INTRODUCTION TO MODELING THERMOMECHANICAL EFFECTS

It was shown in Section 7.1 that thermomechanical cycling can produce extra hardening due to the interaction of planar slip at low temperature and slip and climb at high temperature. In particular, there is extra hardening at low temperature after high-temperature deformation because the resistance to slip is increased by a dislocation network that was created by the prior climb and slip at high temperature. We begin Part B with a discussion of thermomechanical experiments on René 80. Section 7.5 contains an analysis of the equations for thermal cycling but without the thermal history effects, and Section 7.6 contains a model for extra hardening due to thermomechanical cycling.

The elasticity equations with thermal expansion and temperature-dependent material parameters are given in equation 4.5.7. Figure 4.5.1 shows how the elastic modulus, Poisson ratio and coefficient of expansion vary with temperature for René 80. Notice that errors can result for extrapolation to temperatures outside the data range; thus caution must be exercised when extrapolating.

Thermomechanical experiments involve superposition of the elastic, inelastic, and thermal strains. Frequently, thermomechanical experiments are conducted by separating the thermal strain from the mechanical strain, defined as the sum of the elastic and inelastic strain, that is,

$$\varepsilon^{M} = \varepsilon^{E} + \varepsilon^{I} = \varepsilon - \varepsilon^{TH} \tag{7.4.1}$$

This separation is achieved experimentally by first thermally cycling the specimen and recording the thermal strain over the temperature range of the test. Then the mechanical strain used to control the experiment is the difference between the desired total strain history and the recorded thermal strain. All experimental results reported in this study are with the thermal strain removed. The mechanical and temperature histories for four typical experiments are shown in Figure 7.4.1. The out-of-phase test is easier to control than the in-phase test, with the highest loads and temperature applied simultaneously. Heating can be achieved rapidly by supplying adequate power, but cooling with air is usually rather slow. As a result, most thermomechanical load cycles are 5 minutes or longer.

Figure 7.4.2 shows the result of two out-of-phase thermomechanical experiments on René 80. The temperature ranges are between 760 and 982°C, and 649 and 1093°C. Climb is present at the higher temperatures. The response changes significantly between the first and second cycles, with almost no change after the second cycle. Comparing the first cycle to the later cycles

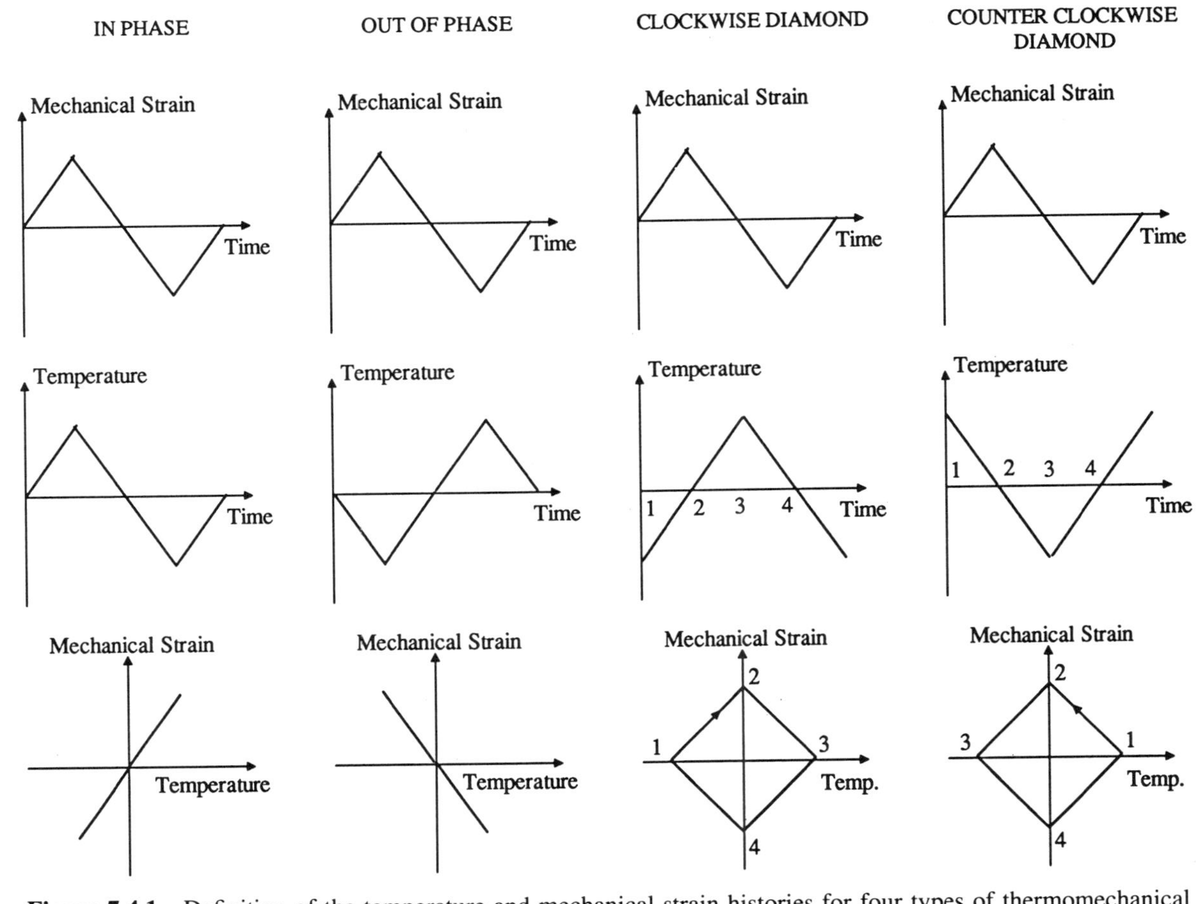

Figure 7.4.1 Definition of the temperature and mechanical strain histories for four types of thermomechanical cyclic experiments.

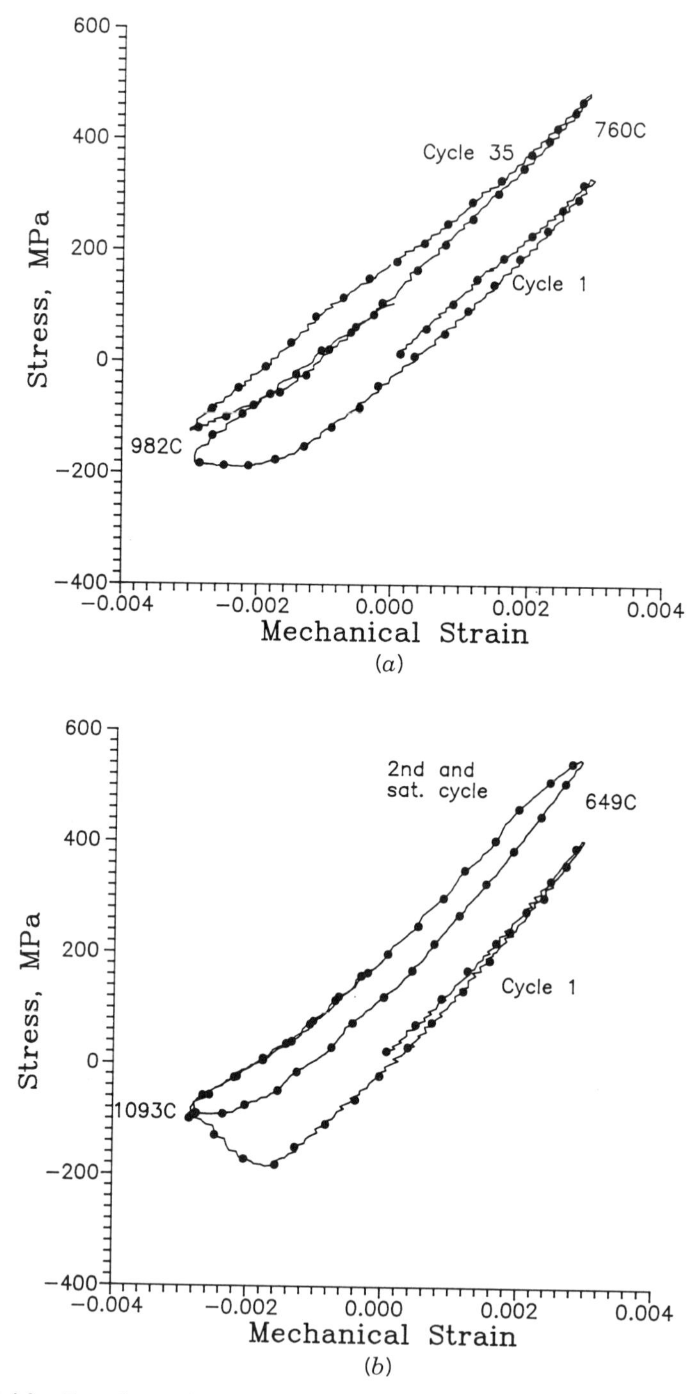

Figure 7.4.2 Experimental response of René 80 to two out-of-phase thermomechanical load histories at a strain rate of 0.002 min^{-1}. (After Ramaswamy, 1986.)

in both figures, it appears that extra hardening is present. The peak stresses at the lowest temperatures are greater in the later cycles than in the first cycle, indicating that high-temperature deformation affected subsequent low-temperature deformation similar to Figure 7.1.5. The shape of the curves on the compression side of the hysteresis loop also shows that there is additional hardening in the later cycles that is not present in the first cycle.

7.5 METHODS OF INTERPOLATION AND EXTRAPOLATION

This section is a step toward establishing a model with the extra hardening that results in thermomechanical cycling. However, it is advantageous first to extract as much information as possible from isothermal data. In particular, it should be possible to model the first hysteresis loop from a thermomechanical load history using the results from isothermal tests alone. Thermal history effects will not occur until after the first low temperature–high temperature–low temperature sequence. Extra hardening is not expected to be present for thermal cycling over a temperature range that does not introduce a change in the deformation mechanism.

To begin, observe that the inelastic flow equation is very similar to the model of Weertman (1956, 1960, 1963) for dislocation creep when dislocation climb plays a major role. For intermediate to high stress levels and for temperatures above one-half of the melting temperature, the steady-state creep rate can be described by Arrhenius equation 3.4.1, with the stress functions given in equations 3.4.3 to 3.4.5. Using the exponential representation leads to the inelastic flow equation

$$\dot{\varepsilon}^{\mathrm{I}}_{ij} = D_0 \exp\left[-\left(\frac{Z'^2}{3T^2K_2}\right)^n\right] \frac{S_{ij} - \Omega_{ij}}{\sqrt{K_2}} \tag{7.5.1}$$

where formally

$$Z' = \frac{ZQ}{k}$$

recalling that Q is the activation energy, T the absolute temperature, and k is Boltzmann's constant. Unfortunately, Q can change significantly with temperature and deformation mechanism; thus explicit dependence on the activation energy is dropped to avoid another unknown function. This result can be reduced for uniaxial loading in the form

$$\dot{\varepsilon}^{\mathrm{I}} = \frac{2}{\sqrt{3}} D_0 \exp\left[-\left(\frac{Z'}{T|\sigma - \Omega|}\right)^{2n}\right] \frac{\sigma - \Omega}{|\sigma - \Omega|} \tag{7.5.2}$$

The revised flow equation now includes a physically motivated basis for temperature (noting that T is absolute temperature). Elimination of the $\frac{1}{2}$ factor and addition of the absolute temperature required revision of the isothermal material parameters determined earlier. This time-consuming step can be eliminated in the future by using equation 7.5.1 or 7.5.2 with the absolute temperature to determine the isothermal material parameters. The prime notation is used to emphasize that Z' and Z represent different numerical values even though the same evolution equation is used.

Next note that equation 7.2.10 for the saturated value of the back stress will not be correct for thermal cycling. Consider, for example, a combined thermal and mechanical cycle where the temperature is changed from T_0 to T_1 immediately after the test is started, before the onset of any inelastic response. At the start of the calculations ω_e is initialized to $\Omega_m(T_0)$ and held constant throughout the calculation. As a result, the wrong value of ω_e will be used since the test temperature is really T_1. Recall during the inelastic deformation that the back-stress saturation equation 7.2.10 was developed for constant stress and temperature creep, and the creep constants were determined by integrating the saturation rate equation. Since the thermal rates are usually rather slow in thermomechanical cyclic applications, it is reasonable to use an integrated form of the back-stress saturation equation 6.5.5 in the form

$$\omega = \omega_e + (\Omega_\mathrm{m} - \omega_e) \exp \left[-A \left(\frac{\sqrt{3J_2}}{\sigma_0} \right)^p t_r \right] \tag{7.5.3}$$

In this representation the parameter t_r represents the time duration of high-temperature creep response since the beginning of the load history and hence is an elapsed time. Equation 7.5.3 contains Ω_m, which is updated to its value at the current temperature. (Recall that Ω_m is determined from isothermal tensile test data at stresses above the yield stress.) Equation 7.5.3 can be used to model the evolution of ω for time-varying stress and temperature loads by continuously updating the values of Ω_m and ω_e by interpolation. When diffusion is not important, ω is equal to Ω_m. If the effect of diffusion is present, ω tracks ω_e and the elapsed time t_r is important. In René 80, for example, diffusion becomes important above about 700°C. The term t_r is therefore the time exposure to temperatures above 700°C when diffusion is important. The rate of recovery of ω to ω_e depends on the stress and parameters A and p.

There are two additional comments about the ideas in equation 7.5.3. First, the back-stress and drag-stress evolution equations, equations 7.2.5 and 6.6.3, are not subject to these modifications. The integrated form of the drag stress involves Z_0' and Z_1', which are updated by interpolation during temperature changes, and the initial value of the back stress is zero, so the equation is not subject to the foregoing inconsistency. Second, time is not always a good variable for constitutive modeling. This comes about because time must be replaced by elapsed time to satisfy the Principle of Objectivity (Section 5.10). However elapsed time, which arises in a hereditary (superposition) integral, is not efficient to integrate numerically. The time variable in equation 7.5.3 is elapsed time and is not integrated. The time advances only when the temperature is above 700°C.

Consider next use of the methods of interpolation and extrapolation over a wide temperature range when more than one deformation mechanism is present. As an example, the experimental program on René 80 consisted of a number of isothermal tests at 538, 760, 871, and 982°C. The new model parameters were determined for each isothermal temperature as described in Chapter 6, and the results were verified. The thermomechanical test program included temperatures up to 1093°C, so it was desirable to use extrapolation to predict the isothermal cyclic response at 1093°C. The values of the drag stress parameters, Z_0' and Z_1', were assumed to be linear with temperature at the higher temperatures for René 80, as shown in Figure 7.5.1. Unfortunately, linear extrapolation to 1093°C was not successful since the extrapolated response overpredicted the hardness observed. The parameters at 1093°C were then determined from additional tests and it was found that the correct values for Z_0' and Z_1' at 1093°C (1366 K) were below the extrapolated values, as shown in Figure 7.5.1. A comparison between the model and experimental results is shown in Figure 7.5.2 for both set of constants.

Consider next the interpolation of parameters between the isothermal test temperatures. It was found that three parameters, Z_0' and Z_1' from the drag

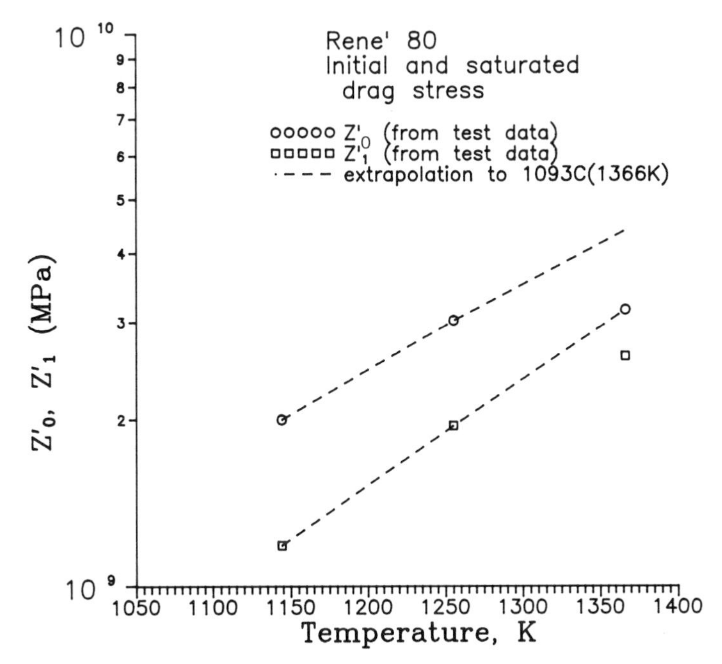

Figure 7.5.1 The lines show an attempt at extrapolation of the parameters Z_0 and Z_1 to determine to response at 1093°C (1366°K). The data points at 1366°K were determined from a test at 1366°K. (After Bhattachar, 1991).

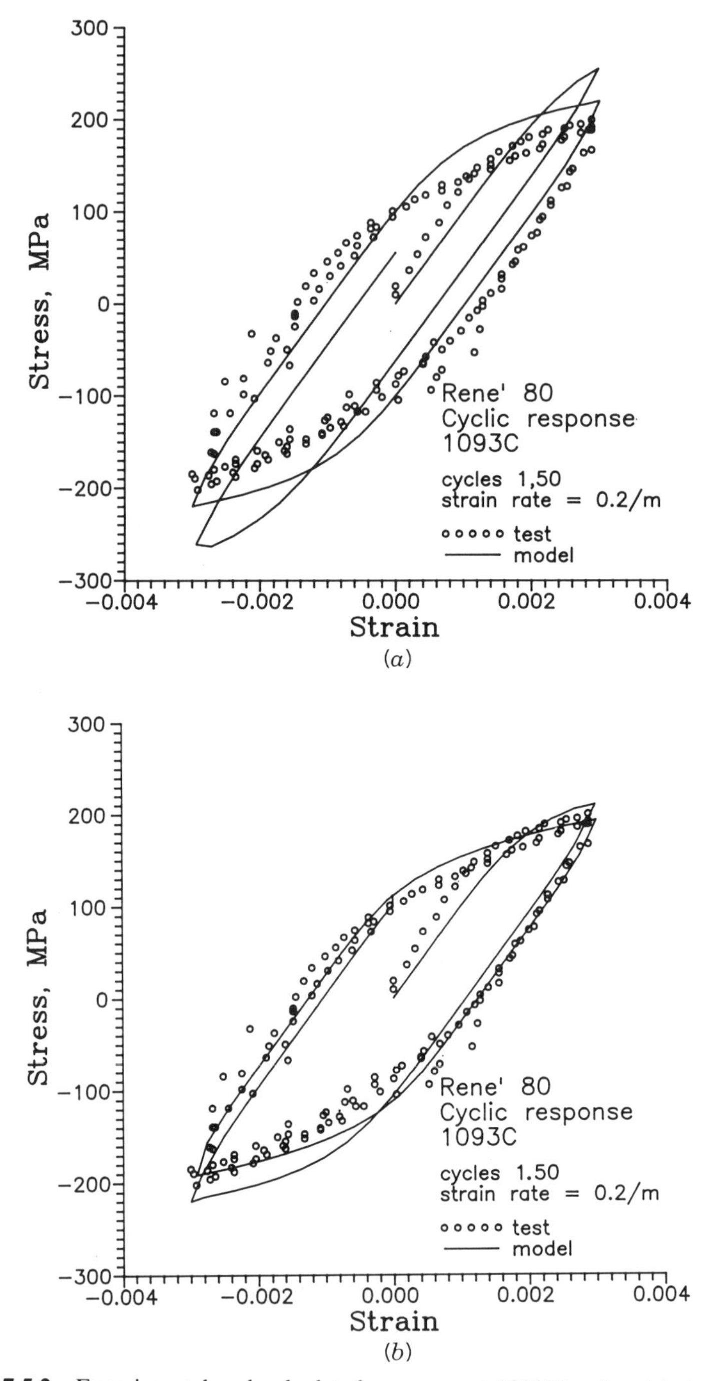

Figure 7.5.2 Experimental and calculated response at 1093°C using (*a*) the extrapolated values of Z_0 and Z_1 and (*b*) the values determined from experimental data at 1093°C. (After Bhattachar, 1991).

stress equation and n from the flow equation, exhibited very large variations over the entire temperature range (538 to 1093°C) so that interpolation is difficult. The values of these parameters are shown in Figure 7.5.3 as a function of temperature. The increase in drag stress starts at about 700°C (973°K), the temperature at which climb (diffusion) becomes important. Below 700°C deformation is mostly by planar slip, and above 700°C deformation occurs by slip and climb. The increase in drag stress corresponds to the extra hardening resulting from the interaction between slip and climb. The decrease in n at about 700°C correlates to the change from strain rate independent response to strain rate sensitive response. Recall that the strain rate sensitivity results from the balance between strain hardening and dynamic recovery. Above 700°C recovery becomes important and strain rate sensitivity occurs.

The shapes of the curves in Figure 7.5.3 were determined numerically by trial and error, so that the stress–strain curves between the isothermal temperatures are ordered with temperature between the isothermal test temperatures as shown in Figure 7.5.4. The shapes of the curves in Figure 7.5.3 are not consistent with any common nonlinear interpolation rule between the isothermal test temperatures. The values for Z_0, Z_1, and n in Figure 7.5.3 were used to calculate the response to constant strain rate loading with a slow ramp temperature change, as shown in Figure 7.5.5. An example of the calculated response to the temperature ramp history is shown in Figure 7.5.6. The erratic response during the temperature ramp comes from the steepness of the curves in Figure 7.5.3. It is interesting to observe that the response of a similar alloy to a ramp temperature change is also erratic. Figure 7.5.7 shows the nonisothermal straining of B1900+Hf between 538 and 760°C. The data suggest that transition between planar slip and slip with climb may not be smooth and uniform.

As a step toward developing an equation for thermomechanical fatigue, let us calculate the response of the first cycle without thermal history effects to determine if the previous modifications are correct. Figure 7.5.8 shows a comparison of the first cycle from two out-of-phase experiments cycled between 649 and 1039°C. The second experiment was run with a 120-sec hold in compression at 1039°C.

To summarize the results of the section, recall that:

1. Linear or higher-order interpolation between the isothermal test data was not possible for René 80, due to the change of deformation mechanism at about 700°C. Trial-and-error estimation of the parameters Z_0', Z_1', and n was necessary to obtain ordered tensile response curves between the test temperatures.
2. Extrapolation to 1039°C from the lower temperatures was not possible for René 80, due to the sharp increase in recovery. Data were necessary at 1039°C to obtain the correct values of Z_0' and Z_1'.

The parameters for the model are given in Appendix 7.1.

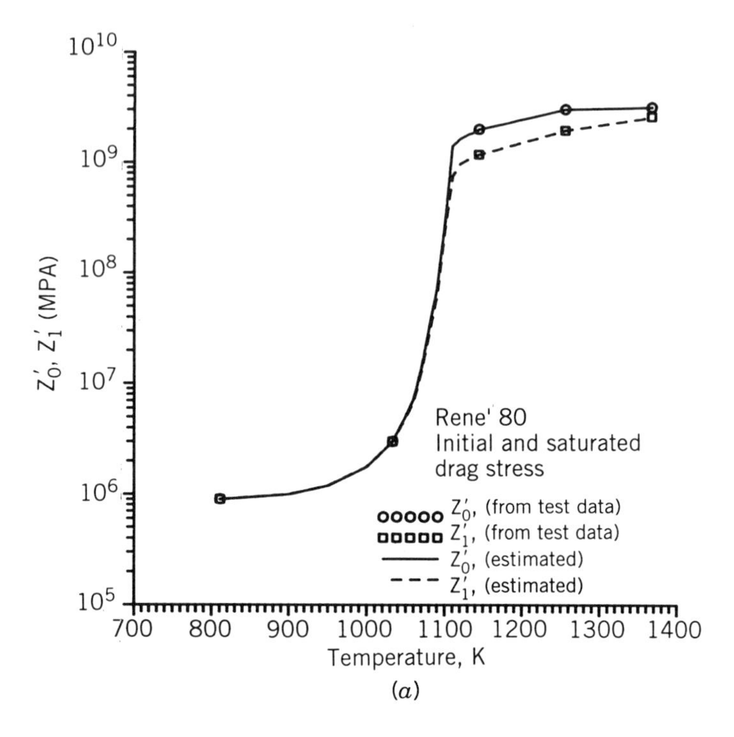

(*a*)

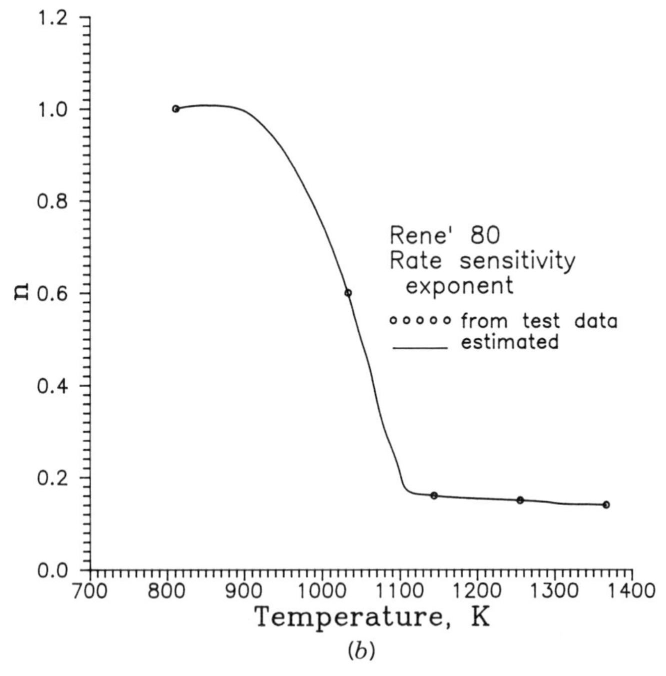

(*b*)

Figure 7.5.3 Variation of Z_0, Z_1, and n with absolute temperature. The steep slopes made direct interpolation impossible. (After Bhattachar, 1991).

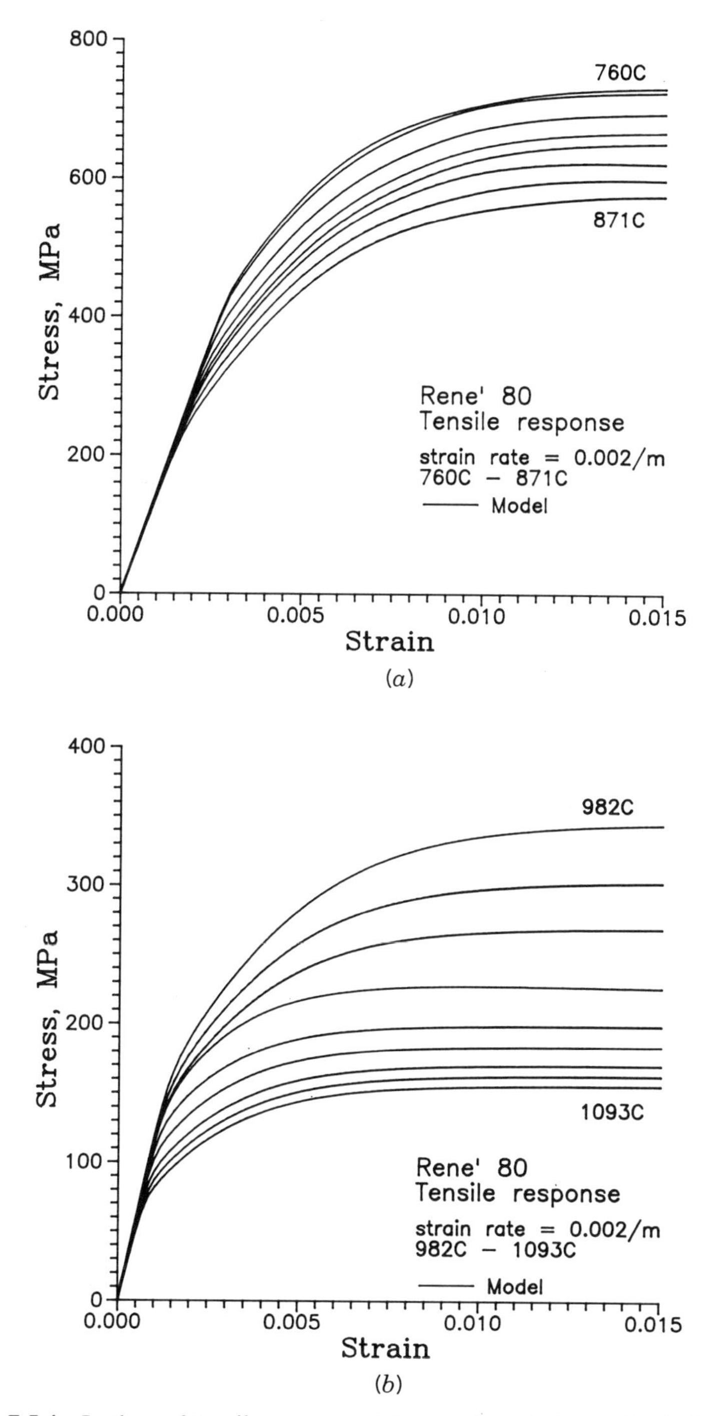

Figure 7.5.4 Isothermal tensile response at temperatures between the test temperatures with the parameters Z_0, Z_1, and n adjusted by trial and error to give a smooth, ordered response with temperature (After Bhattachar, 1991.)

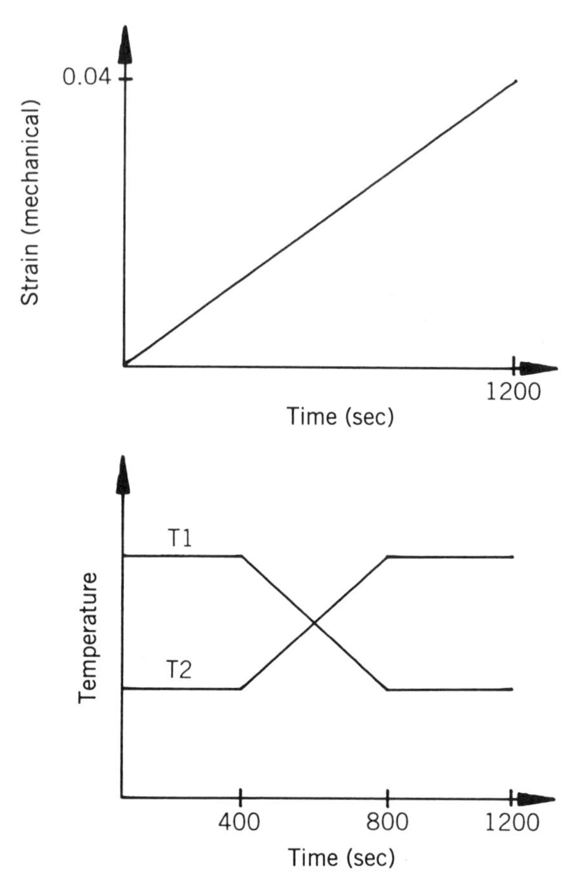

Figure 7.5.5 Typical temperature and mechanical strain histories for the nonisothermal (ramp) calculations in Figure 7.5.6. (After Bhattachar, 1991.)

7.6 TEMPERATURE HISTORY EFFECTS

To deduce the effect of temperature history on the response it is reasonable to compare the results of interpolation to the first few thermomechanical cycles. This was done as shown in Figure 7.6.1. The first $1\frac{1}{2}$ cycles of the model match the experimental response well. The next $\frac{1}{4}$ cycle shows that the material response is harder than the model response. The predicted response at 1093°C is about 40 MPa above the experimental result in the figure, and the shape of the curve on loading into compression in the second cycle is not correct. Thus the inelastic deformation at 649°C affected the subsequent compressive deformation at 1039°C in the second cycle.

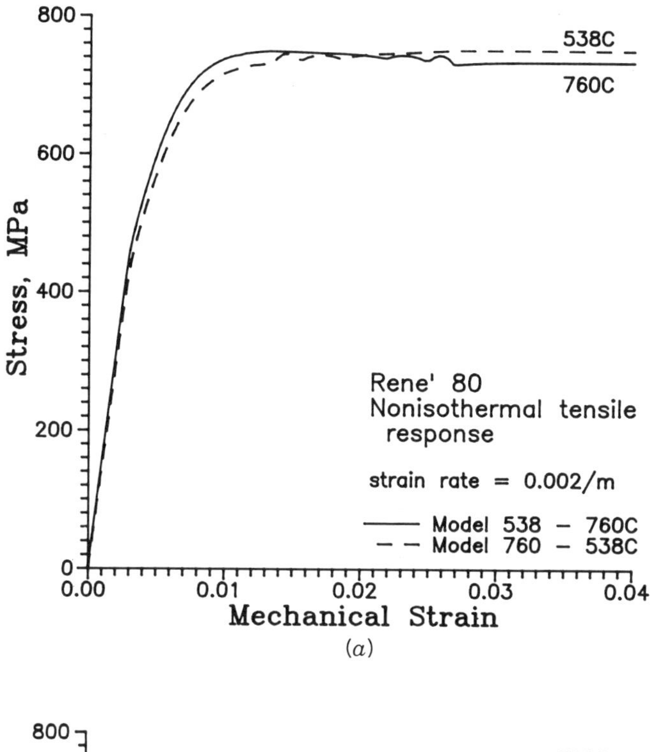

(*a*)

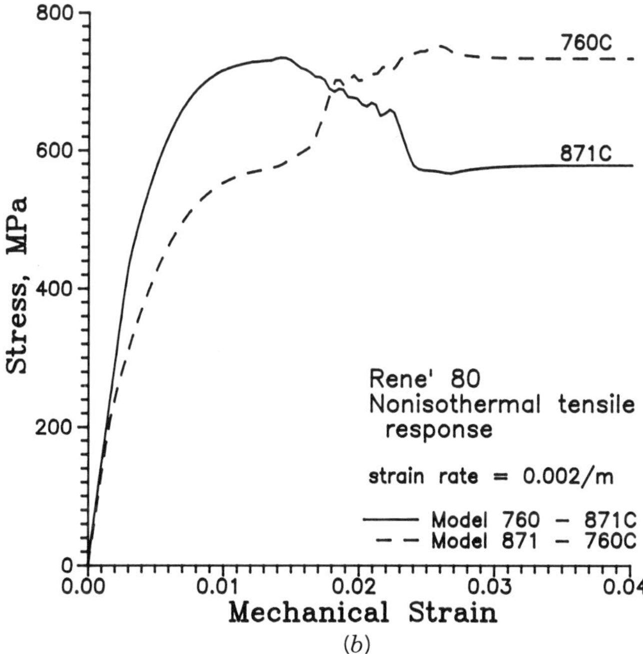

(*b*)

Figure 7.5.6 Calculated nonisothermal (ramp) tensile response using the adjusted valued for Z_0, Z_1, and n. (After Bhattachar, 1991.)

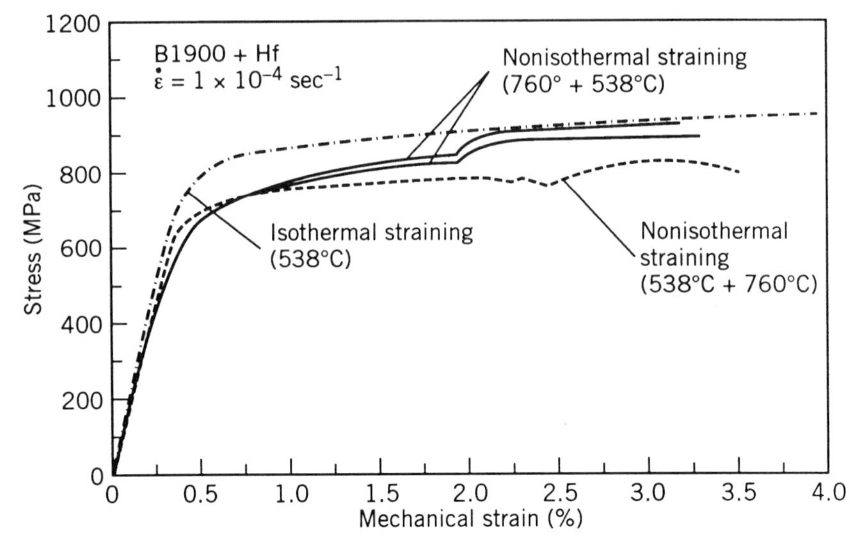

Figure 7.5.7 Instabilities in the nonisothermal tensile response of B1900+Hf for out-of-phase loading between 538 and 760°C. (From Chan and Page, 1988; reproduced with permission of The Minerals, Metals & Materials Society.)

The increase in hardness at 1093°C may have resulted from an increase in dislocation density that developed during the inelastic deformation at 649°C. In particular, it is reasonable to expect that isothermal deformation at 1093 and 649°C will produce dislocation microstructures that are characteristic of the test temperature and strain rate. At 649°C deformation occurs mostly by planar slip. At 1093°C deformation occurs by slip and climb; however, the extremely high temperature will also accelerate the rate of recovery. Evidence of this is seen in the first cycle by the reduction in stress at 1093°C. Thus it is assumed that deformation at 649°C increases the dislocation density at 1093°C. Unfortunately, microscopy was not available to confirm this hypothesis.

A model for the extra hardening is based on three assumptions. First, it is assumed that the deformation at 649°C increases the hardness at 1093°C. Second, temperature variations must be accompanied by inelastic deformation to produce extra hardening. (*Note:* Static thermal recovery can produce softening at high temperature after inelastic deformation, but this is not the situation under consideration.) Third, isothermal deformation after thermomechanical loading can eventually eliminate the extra hardening as the material approaches the dislocation microstructure that is characteristic of the isothermal deformation.

The model for extra hardening depends on two variables. The variable β detects and characterizes the relative amount of interaction between the residual microstructure and the current mode of deformation. The variable ΔZ

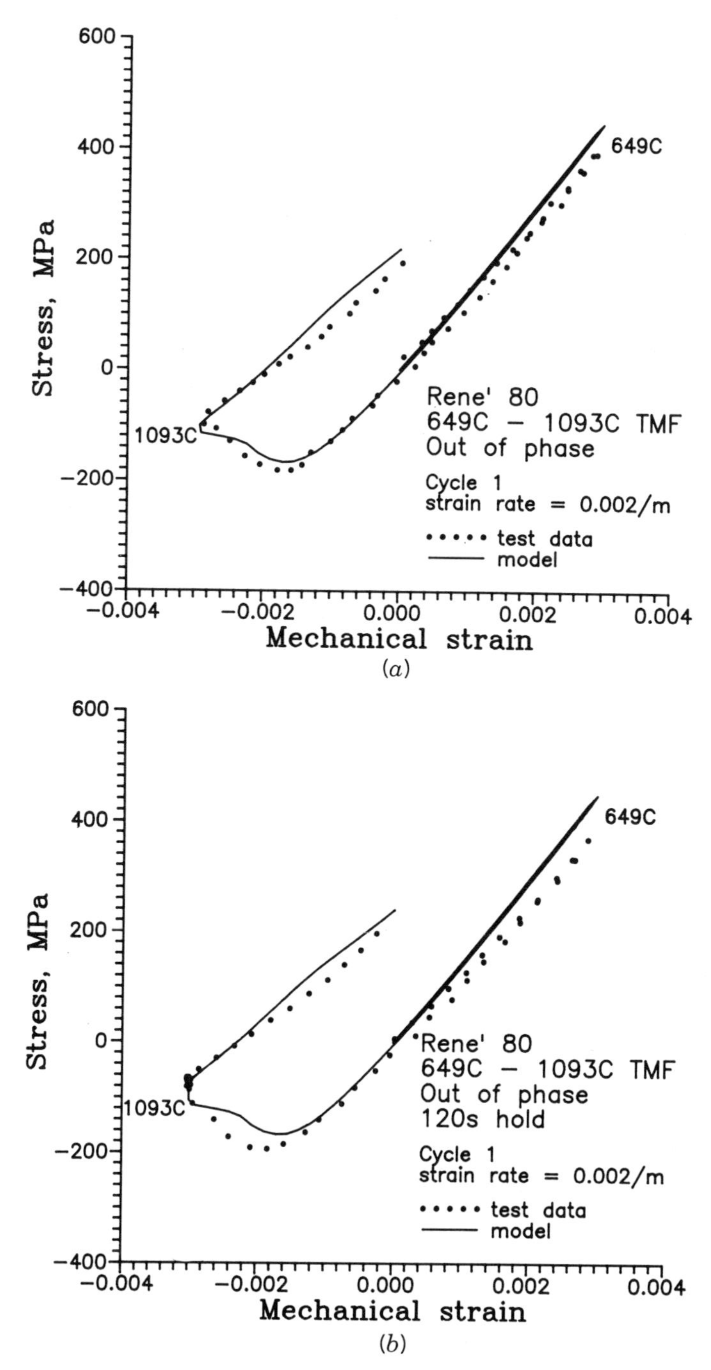

Figure 7.5.8 Experimental and calculated results for the first cycle of response of René 80: (*a*) with thermal cycling between 649 and 1093°C; (*b*) has a 120-sec hold in compression. (After Bhattachar, 1991.)

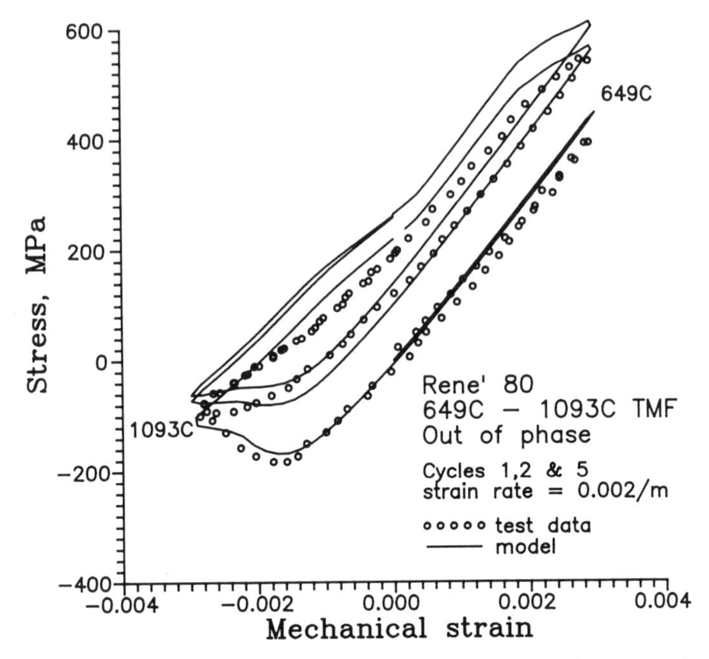

Figure 7.6.1 Comparison between experimental and calculated response in a 649 to 1093°C test using the isothermal parameters. (After Bhattachar and Stouffer, Journal of Engineering Materials and Technology, V115, pg 351, 1993. Copyright © The American Society of Mechanical Engineers, 345 East 47th St., N.Y., N.Y. 10017. Used with permission.)

is the amount of extra hardening resulting from this interaction. Since the extra hardening is not expected to depend on the direction of the applied load, it is an isotropic or scalar variable. Since the extra hardening occurs during cyclic loading, the drag stress is modified by the addition of the extra hardening variable

$$Z' = Z_1' - (Z_1' - Z_0')\exp(-mW^{\mathrm{I}}) + \Delta Z \tag{7.6.1}$$

where ΔZ is defined to vanish for isothermal deformation so that it will not affect the value of Z' for isothermal calculations.

The interaction variable β is designed to sense the presence of a dislocation structure characteristic of low temperature, planar slip, deformation. The initial value of β is assumed to be zero. It remains zero if deformation occurs at high temperature ($T > 700$°C for René 80) when diffusion is important, but its value is not zero if inelastic work is done at low temperature. The growth of β is assumed to depend on the rate of inelastic working, $\dot{W}^{\mathrm{I}}$. It is also assumed that there is a limiting value β_1 for β. These properties can be characterized by the exponential growth law

$$\dot{\beta} = m_2(\beta_1 - \beta)\dot{W}^{\mathrm{I}} \tag{7.6.2}$$

where m_2 controls the rate of growth of β. The parameters β_1 and m_2 are functions of temperature, with $\beta_1 = 0$ for $T > 700°\mathrm{C}$ see Figure 7.6.2. The initial is $\beta = 0$ for $W^{\mathrm{I}} = 0$.

The evolution of ΔZ is the amount of extra hardening, and it is assumed to depend on the inelastic working rate, similar to the drag stress Z'. Since extra hardening occurs only when interactions occur, the growth of ΔZ should depend on β. A review of the data indicates that ΔZ should have a limiting value, Z_2', that represents the maximum amount of extra hardening that can occur from a deformation mechanism operating on a dislocation microstructure that was established at a different temperature. An representation similar to drag stress is

$$\Delta Z = Z_2'[1 - \exp(-\beta W^{\mathrm{I}})] \tag{7.6.3}$$

where W^{I} is the accumulated inelastic work. The function Z_2' has two distinct values. At temperatures greater than a critical temperature T_c ($T_c \approx 700°\mathrm{C}$ for René 80) Z_2' must be equal to the amount of extra hardening, and below the critical temperature the value is zero, $Z_2' = 0$. Therefore the value of ΔZ in equation 7.6.3 changes during thermomechanical cycling. Consider a typical thermomechanical cycle starting at low temperature:

	Using equation 7.6.2.	Using equation 7.6.3
Low temperature:	$\beta(0) = 0$, $\dot{\beta} > 0, \beta \Rightarrow \beta_1$	$Z_2' = 0$ $\Delta Z = 0$
High temperature:	$\beta \neq 0, \beta_1 = 0$, $\dot{\beta} < 0, \beta \Rightarrow 0$	$Z_2' \neq 0$ $\Delta Z = Z_2'$
Low temperature:	$\beta(0) = 0$, $\dot{\beta} > 0, \beta \Rightarrow \beta_1$	$Z_2' = 0$ $\Delta Z = 0$

Starting the cycle at high temperature will not produce any hardening because $\beta = \beta_1 = 0$, then it follows that $Z_2' = 0$ and $\Delta Z = 0$.

The values of Z_2', β_1, and m_2 are determined from the thermomechanical data shown in Figure 7.6.1. Since the second cycle in Figure 7.6.1 is essentially the saturated cycle, and since the stresses in the first and second cycles at 1093°C are very close to saturation at 100 MPa, the value of the drag stress becomes $Z' = Z_1' + Z_2'$. Inverting the flow equation 7.5.3 and using the saturation values for the stress, back stress, and inelastic strain rate variables gives

$$T\left[-\ln\left(\frac{\sqrt{3}\dot{\varepsilon}_0}{2D_0}\right)\right]^{1/2n} = \frac{Z_1' + Z_2'}{\sigma_S - \Omega_m} \tag{7.6.4}$$

Since the values of all the variables in equation 7.6.4 are known, the value of Z_2' can be determined. Unfortunately, no direct measures for the variables β_1 and m_2 could be found. Since β_1 represents the saturated values of β in the low-temperature regime, and since $\beta_1 = 0$ during high-temperature deformation, it was assumed to be a stiff exponential function typical of a step function at $T = 700°C$, as shown in Figure 7.6.2. The parameter m_2 controls the change of β. Numerical exercises showed that $m_2 = 1.0$ for $T < 700°C$, and $m_2 = 0.5$ for $T > 700°C$ were acceptable values for the remaining parameter. The values of the model parameters are summarized in Appendix 7.1.

A comparison between the experimental and calculated response is shown in Figures 7.6.3 and 7.6.4 for cycling between 649 and 1093°C with a strain rate of 0.002 min^{-1}. These results are for two out-of-phase experiments, one with a 120-sec hold in compression at 1093°C. The model was also applied to a block isothermal test at five temperatures, as shown in Figure 7.6.5. There was no extra hardening in the experiment because the specimen was not cycled through a low temperature–high temperature–low temperature sequence. The model was next used to predict the data of Cook, Kim, and McKnight (1988) for one out-of-phase, one in-phase, and two clockwise diamond histories. A comparison between the model and data is shown in Figures 7.6.6 and 7.6.7. The test temperature ranges were smaller than the other tests at 649 to 871°C and 871 to 982°C, but a strain rate of 0.01 min^{-1} was higher.

The study of thermal history effects has brought out an important point about thermomechanical cycling. If the strain range and temperature levels activate only one deformation mechanism, the isothermal values of the constitutive parameter can be expected to model the response satisfactorily by interpolation. However, if more than one deformation mechanism is activated, temperature history effects may be very important. In this case extrapolation and interpolation will not be successful over the entire temperature range.

C. SUMMARY OF OTHER MODELING OPTIONS

This part of the manuscript contains a summary of the models of Bodner, Miller, Walker, and Kreig et al. that were introduced in Chapter 6. The models are presented using the previous notation for convenience of the reader. Parameters for all four models for a number of materials are given in Appendices 7.2 to 7.4.

Before proceding it is important to note there are two ways to develop a back stress tensor from uniaxial loading. Recall that the back stress is a

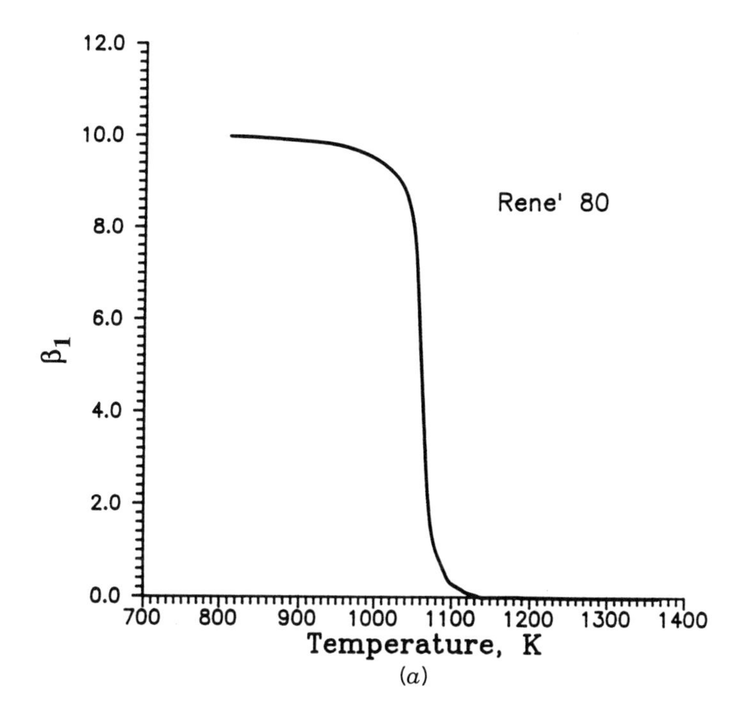

(*a*)

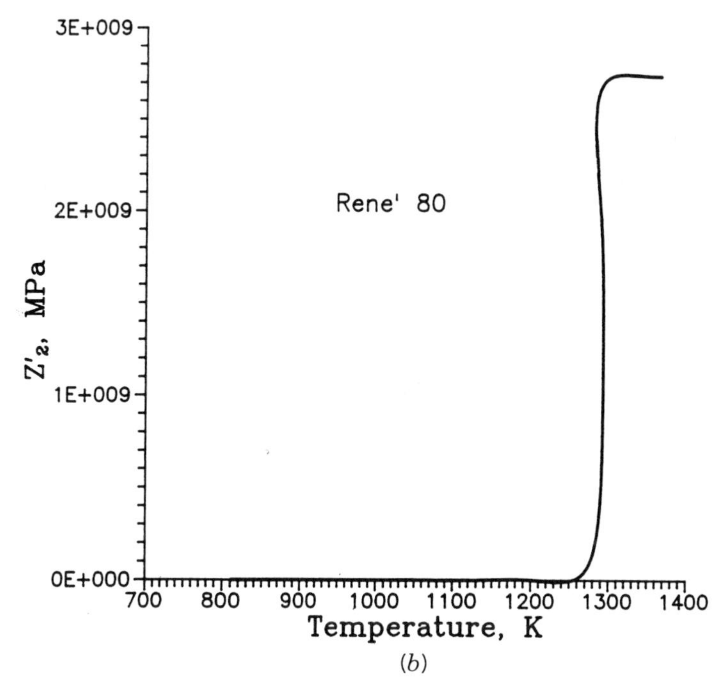

(*b*)

Figure 7.6.2 Parameters β_1, Z_2', and m as a function of temperature. The parameter β_1 was assumed to be a stiff exponential function with the asymptotic values shown. (After Bhattachar, 1991.)

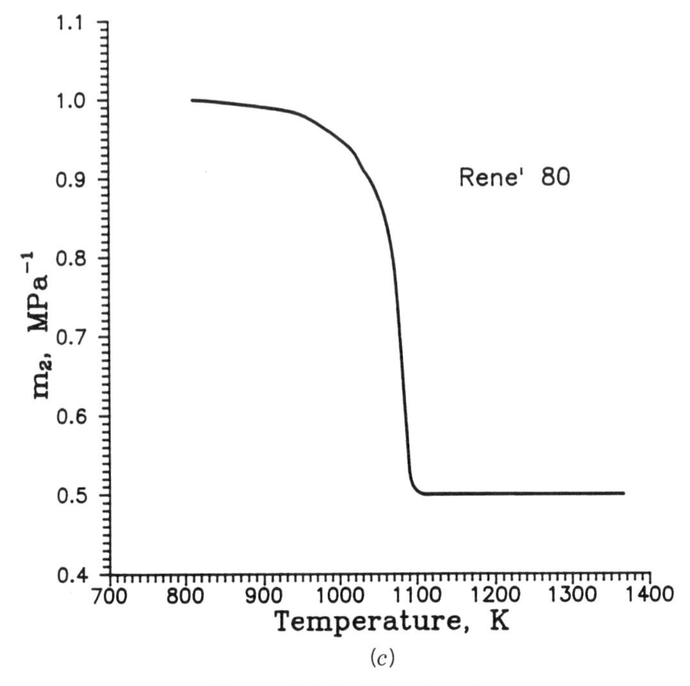

(c)

Figure 7.6.2 (*Continued*)

deviatoric tensor, $\Omega_{11} + \Omega_{22} + \Omega_{33} = 0$. Then defining Ω as the back stress in a uniaxial test in the $\hat{e}_1$ coordinate direction, the tensor components can be written as

$$\Omega_{11} = \Omega \qquad \text{and} \qquad \Omega_{22} = \Omega_{33} = -\tfrac{1}{2}\,\Omega$$

However, some authors define back stress similar to deviatoric stress, that is,

$$\Omega_{11} = \tfrac{2}{3}\,\Omega \qquad \text{and} \qquad \Omega_{22} = \Omega_{33} = -\tfrac{1}{3}\,\Omega$$

Converting between one and three dimensions will produce different numerical factors with the foregoing definitions. Both definitions are acceptable, but the equations must be consistent.

7.7 BODNER'S MODEL

There was a relatively complete review of the model by Bodner (1987), and Neu (1993) published a paper on parameter estimation for nonisothermal re-

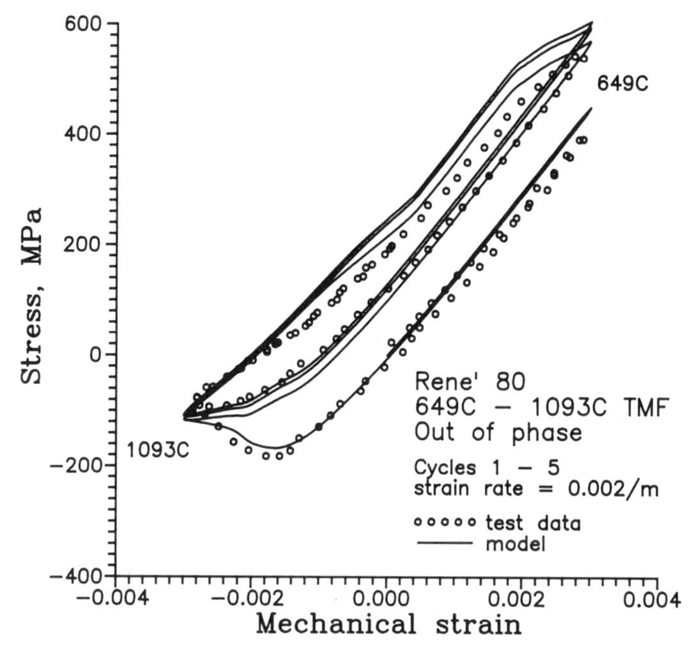

Figure 7.6.3 Experimental and calculated response for the first five cycles of a 649 to 1093°C thermomechanical test. (After Bhattachar and Stouffer, Journal of Engineering Materials and Technology, V115, pg 351, 1993. Copyright © The American Society of Mechanical Engineers, 345 East 47th St., N.Y., N.Y. 10017. Used with permission.)

sponse using the Bodner model. The flow equation in the Bodner model was given earlier as

$$\dot{\varepsilon}^{\mathrm{I}}_{ij} = D_0 \exp\left[-\frac{1}{2}\left(\frac{Z^2}{3J_2}\right)^n\right]\frac{S_{ij}}{\sqrt{J_2}} \tag{6.2.7}$$

The absence of a back stress state variable is the major difference between Bodner's model and the others discussed in Chapter 6. The scalar hardening variable Z replaces both the back stress and drag stress variables used in other models; and it is developed to model both the strain hardening observed in the tensile response and the cyclic hardening observed in fatigue. To model both isotropic and kinematic hardening, the variable Z is taken to consist of sum of isotropic, Z^{I}, and directional, Z^{D}, components; that is,

$$Z = Z^{\mathrm{I}} + Z^{\mathrm{D}} \tag{7.7.1}$$

The variable Z is not exactly the same as the drag stress defined earlier but is a variable that characterizes the resistance to slip or the hardening.

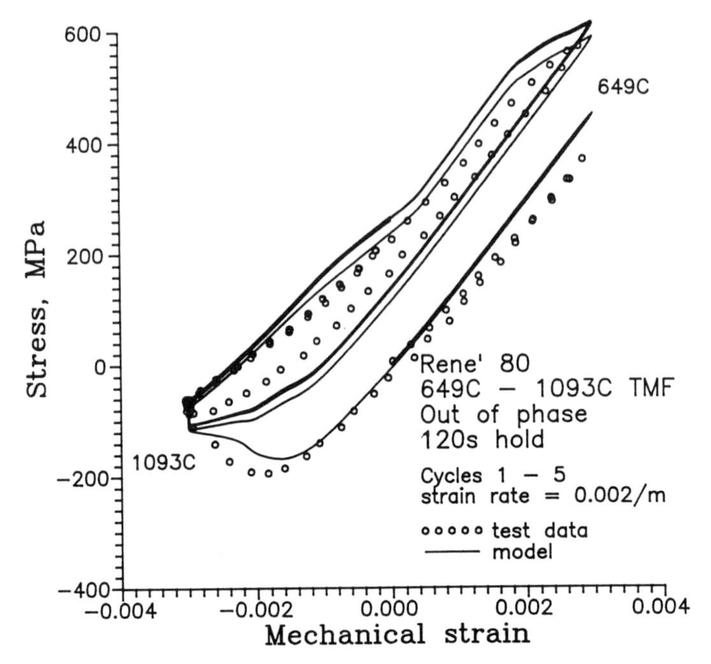

Figure 7.6.4 Experimental and calculated response for the first five cycles of a 649 to 1093°C thermomechanical test with a 120-sec hold in compression. (After Bhattachar and Stouffer, Journal of Engineering Materials and Technology, V115, pg 351, 1993. Copyright © The American Society of Mechanical Engineers, 345 East 47th St., N.Y., N.Y. 10017. Used with permission.)

The evolution equation for the isotropic hardening component is similar to equation 6.6.1, but with a thermal recovery term included it becomes

$$\dot{Z}^{\mathrm{I}} = m_1(Z_1 - Z^{\mathrm{I}})|\dot{W}^{\mathrm{I}}| - A_1 Z_1 \left(\frac{Z^{\mathrm{I}} - Z_2}{Z_1}\right)^{r_1} \qquad (7.7.2)$$

with the initial condition $Z^{\mathrm{I}}(0) = Z_0$. The parameter Z_1 defines the limiting (saturation) value for Z^{I} and m_1 controls the hardening rate. The thermal recovery gives the minimum value for Z^{I} as Z_2, and A_1 and r_1 are temperature-dependent material parameters. The material parameters in the first term are defined from the strain hardening observed in the tensile response, and the parameters in the recovery term are determined from the creep response. During steady-state creep $\dot{Z}^{\mathrm{I}} = 0$; thus a relationship exits between the parameters. Methods for determining constants have also been reported by Metzer (1979, 1982), Stouffer (1981, 1982), and Chan (1986).

The directional hardening is represented by a second-order tensor β_{ij} and a scalar effective value Z^{D}, that is part of the total hardening Z. The variable Z^{D} is defined as

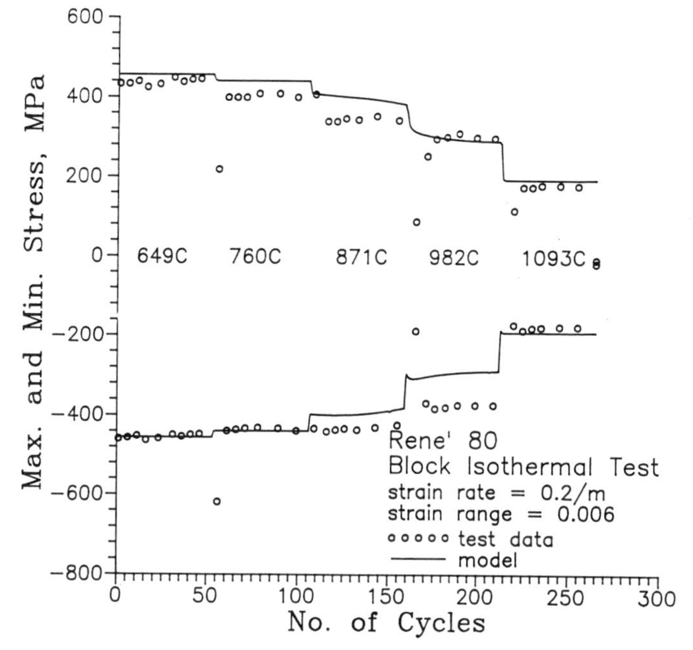

Figure 7.6.5 Predicted and experimental values of the maximum and minimum stresses in a block isothermal test. (After Bhattachar and Stouffer, Journal of Engineering Materials and Technology, V115, pg 351, 1993. Copyright © The American Society of Mechanical Engineers, 345 East 47th St., N.Y., N.Y. 10017. Used with permission.)

$$Z^{\mathrm{D}} = \beta_{ij}u_{ij} = \beta_{ij}\frac{\sigma_{ij}}{|\sigma_{ij}|} \tag{7.7.3}$$

where u_{ij} is the sign of each stress component and β_{ij} is the magnitude of the directional hardening parameter for each stress component. In uniaxial cycling $Z^{\mathrm{D}} = \pm\beta_{11}$ for tension and compression loading, respectively. The hardening rate in each direction is assumed to be a function of the inelastic rate of working in that direction, and the rate of recovery is assumed to be determined from the current hardness, β_{ij}, in each direction.

The function representation for $\dot{\beta}_{ij}$ is similar to the isotropic equation and is given by

$$\dot{\beta}_{ij} = m_2(Z_3u_{ij} - \beta_{ij})\dot{W}^{\mathrm{I}} - A_2Z_1\left(\frac{|\beta_{ij}|}{Z_1}\right)^{r_2}\frac{\beta_{ij}}{|\beta_{ij}|} \tag{7.7.4}$$

where Z_3 represents the maximum possible value of β_{ij}. Similar to equation 7.7.2, the first term defines the rate of directional hardening and the second

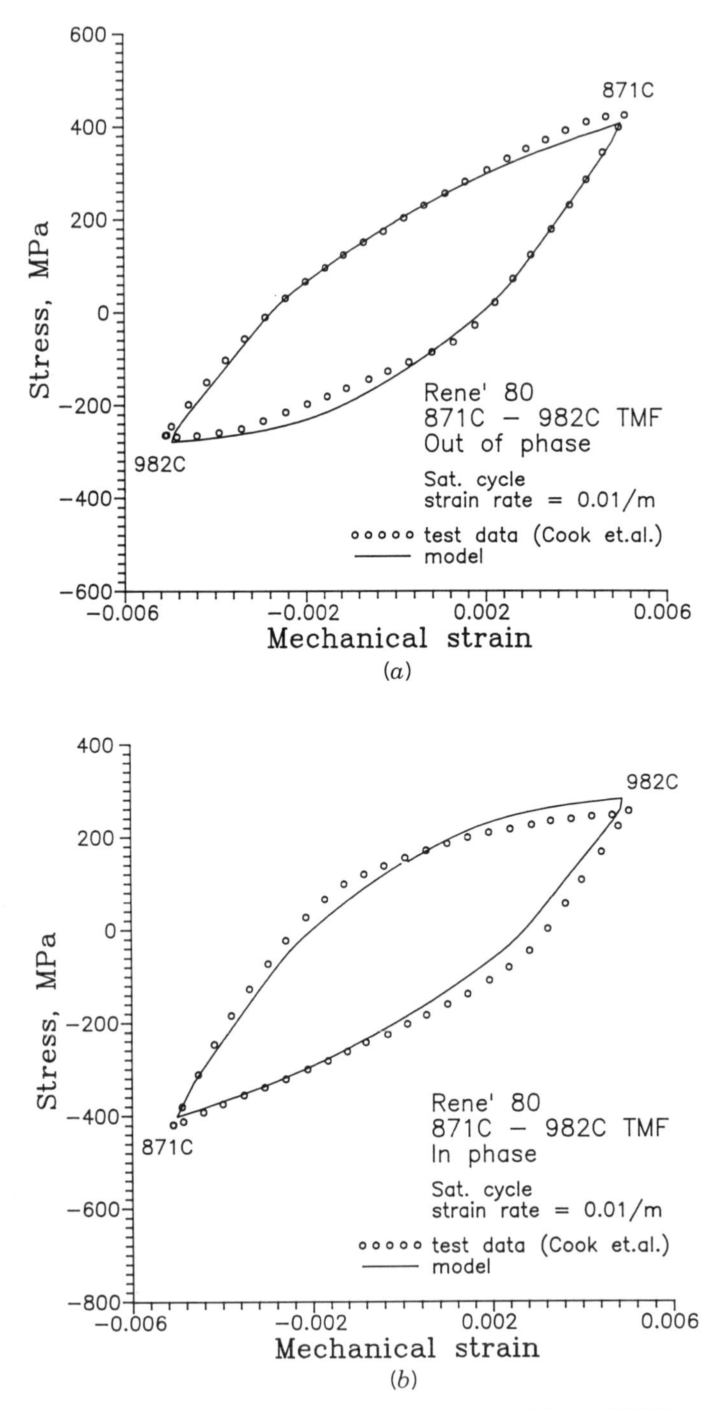

Figure 7.6.6 Experimental and calculated response for 871 to 982°C out-of-phase and in-phase tests. (Data after Cook, Kim, and McKnight, 1988; model of Bhattachar, 1991).

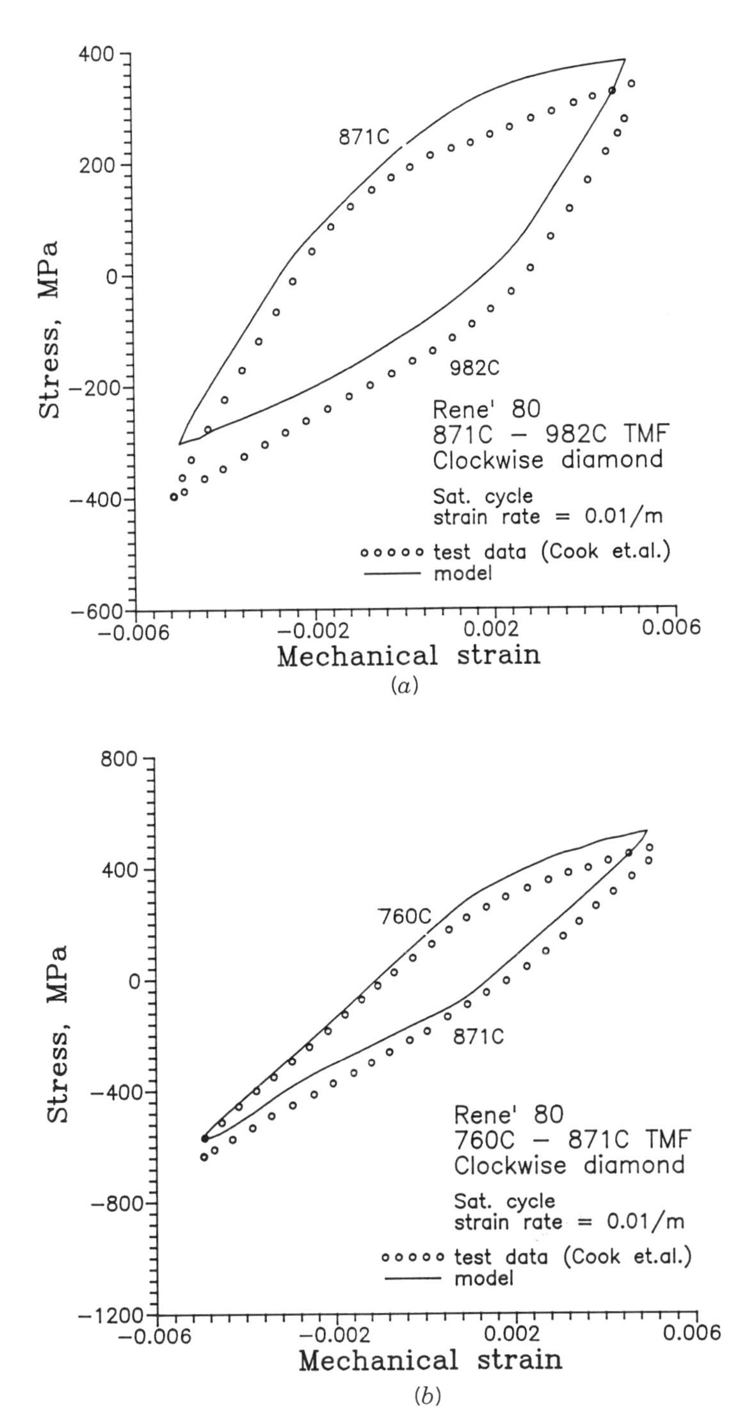

Figure 7.6.7 Predicted and experimental response for two clockwise diamond tests at 871 to 982°C and 760 to 871°C. (Data after Cook, Kim, and McKnight, 1988; model of Bhattachar, 1991).

the rate of recovery. The parameter Z_1 is used to nondimensionalize $|\beta_{ij}|$, and m_2, r_2, and A_2 are temperature-dependent parameters. One feature of the model is that $Z = Z^I + Z^D$ behaves like kinematic hardening in cyclic loading. Neu (1993) reported that Z_3 should be less than Z_0 and Z_2 to prevent $Z = Z^I + Z^D$ from becoming negative. "However, this limitation can make it difficult, or even impossible, to predict the rate sensitivity over a large range of high temperatures and low stresses." Cyclic hardening and softening is controlled by Z^I. Cyclic hardening occurs when $Z_1 > Z_0$ and softening occurs with $Z_1 < Z_0$. In some cases. the shape of the cyclic loop is "too square", as shown in Section 6.8, so Bodner suggests using m_2' in place of m_2 in equation 7.7.4, where

$$m_2' = \frac{m_2}{2}\,[1 + \exp(-m_3 Z^D)] \qquad (7.7.5)$$

where m_3 is a material parameter that can be determined from cyclic data.

As reported earlier, nonproportional loading can produce extra hardening. Bodner introduced the method in Section 7.3 to model the extra hardening in nonproportional loading. Since the drag stress equation in Sections 6.6 and 7.3 is the same as the first term of equation 7.7.2, the results from Section 7.3 can be used with the Bodner model. Bodner applied his model to Hastelloy X at room temperature under nonproportional loading. A plot of the model predictions for the extra hardening shown in Figure 7.7.1 is in excellent agreement with the experimental data. However, the angle between the deviatoric stress and inelastic strain rate is not correct, as shown in Figure 6.1.1. The parameters for the Bodner–Partom model are given in Appendix 7.2 for several materials.

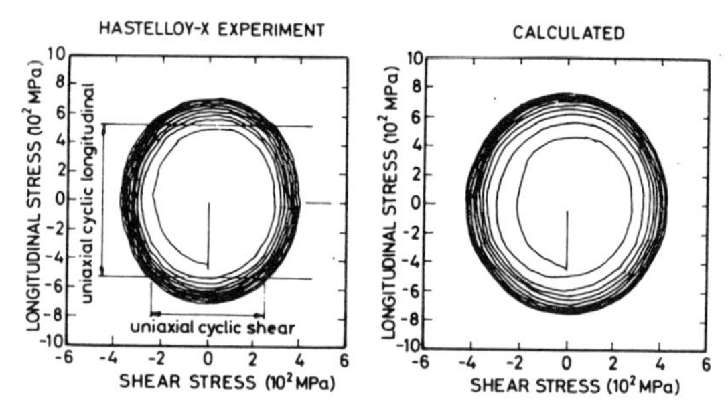

Figure 7.7.1 Comparison of experimental and calculated results from the Bodner model for a 90°-out-of-phase tension–torsion test on Hastelloy X at room temperature. (From Bodner, 1987; Unified Constitutive Equations for Plastic Deformation and Creep of Engineering Alloys, A. K. Miller ed., reprinted by permission of Chapman & Hall.)

7.8 MILLER'S MODEL

Miller (1976) developed a representation for a flow law using the hyperbolic sine function that is a physically based model for steady-state creep. The model is creative in the way that temperature is included in the flow and evolution equations. Only two parameters are temperature dependent. The remaining parameters are all temperature independent. This section contains a review of the original Miller model to show the underlying ideas and method of development. There are several extensions in the literature (Miller, 1987) that are useful for special situations such as solution-hardened materials and recrystallization. The development is focused on the uniaxial model; however, the multiaxial representation is given at the end of this section.

The steady-state creep rate, $\dot{\varepsilon}_{SS}$, can be written as a function of the constant creep stress, σ_{SS}, for many stainless steel, aluminum, and ferrous alloys. Redefining the constants in equation 3.4.5, the steady-state creep rate is given by

$$\dot{\varepsilon}_{SS} = D_0\,[\sinh(A\sigma_{SS})]^n \tag{7.8.1}$$

where D_0 is a function of temperature, and A and n are constants. Miller (1987) later observed that steady-state creep data is log linear when the creep stress is normalized by the elastic modulus; that is, σ_{SS} is replaced by σ_{SS}/E in equation 7.8.1. The stress dependence in equation 7.8.1 is identical to equation 3.4.5, except that the final form of the temperature dependence will be extended to cover a broader temperature range. Back stress and drag stress can be introduced in the format described in equation 6.2.4; thus a general representation for uniaxial loading can be obtained as

$$\dot{\varepsilon}^{\mathrm{I}} = D_0\left\{\sinh\left[Af\left(\frac{\sigma - \Omega}{Z}\right)\right]\right\}^n \operatorname{sgn}(\sigma - \Omega) \tag{7.8.2}$$

The function f is defined so that equation 7.8.2 will reduce to equation 7.8.1 during steady-state creep; therefore, the function f must satisfy

$$f\left(\frac{\sigma_{SS} - \Omega_{SS}}{Z_{SS}}\right) = \sigma_{SS} \tag{7.8.3}$$

Miller conducted experiments on warm-worked stainless steel that produced considerable strain hardening due to the development of subgrains. In this case the drag stress is very large compared to the back stress. In the absence of back stress, equation 7.8.2 can be inverted and rearranged to obtain a representation of the form $\sigma = k(Z)$, where the inverse flow function k defines the state of the material as a function of the drag stress, inelastic strain rate, and temperature. Miller found from his experiments that a simple relationship exists in steady-state loading, namely

$$Z_{SS} = k_1 \sigma_{SS}^{1/3} \tag{7.8.4}$$

where k_1 is a constant defining the steady-state material properties. Dividing both sides of equation 7.8.4 by σ_{SS} and rearranging gives

$$\left(k_1 \frac{\sigma_{SS}}{Z_{SS}} \right)^{1.5} = \sigma_{SS} \tag{7.8.5}$$

The back stress can be reintroduced since its value is relatively small, and redefining the constant as $k_1 = A^{-2/3}$ gives

$$\frac{1}{A} \left(\frac{\sigma_{SS} - \Omega_{SS}}{Z_{SS}} \right)^{1.5} = \sigma_{SS} \tag{7.8.6}$$

a definition for the function f in equation 7.8.3. Thus the flow law can be rewritten as

$$\dot{\varepsilon}^{\mathrm{I}} = D_0 \left\{ \sinh \left[\left(\frac{|\sigma - \Omega|}{Z} \right)^{3/2} \right] \right\}^n \operatorname{sgn}(\sigma - \Omega) \tag{7.8.7}$$

where the absolute value and sgn functions are used to provide the correct signs for loading in compression. In a later formulation of the flow law, Miller defines $Z = \sqrt{F}$, where F is the friction stress resisting slip. The square-root factor is introduced to give the stress–strain curve a parabolic shape when the drag stress dominates the response. A parabolic response curve occurs when hardening is controlled primarily by dislocation multiplication. This is observed in pure metals such as aluminum and copper.

The flow law in equation 7.8.7 has the property that only D_0 is a function of temperature. The temperature dependence is established by noting that the activation energy (discussed in Section 3.4) is approximately constant above a critical temperature, T_{cr}, that depends on the deformation mechanism. This observation can be used with equation 3.4.1 to establish the temperature dependence. In particular, D_0 in the flow law can be written as

$$D_0 = D_0' \exp\left(\frac{-Q}{kT} \right) \qquad \text{for } T > T_{\mathrm{cr}} \tag{7.8.8a}$$

with temperature on the Kelvin scale and D_0' a material constant. The critical temperature is typically 50 to 60% of the melting temperature for creep by lattice diffusion. The constant value of activation energy above the critical temperature is denoted as Q. Below T_{cr}, Miller noted that the activation energy, Q_d, decreases nearly linearly from Q to zero at $T = 0$, that is,

$$Q_d = \frac{T}{T_{cr}} Q \qquad \text{for } T < T_{cr}$$

Using equation 3.4.1 without the stress function $f(\sigma)$, the value of Q_d can be determined as

$$\frac{Q_d}{R} = \left. \frac{d(\ln \dot{\varepsilon}^{\mathrm{I}})}{d(1/T)} \right|_{\sigma} = \left. \frac{d(\ln D_0)}{d(1/T)} \right|_{\sigma} = \frac{TQ}{T_{cr}}$$

and integrating and solving for D_0 gives

$$D_0 = D_0' \exp\left(\frac{-Q}{kT_{cr}}\right)\left(\ln \frac{T_{cr}}{T} + 1\right) \qquad \text{for } T < T_{cr} \tag{7.8.8b}$$

The general representations for the back-stress and drag stress evolution equations are similar to equation 6.3.3 except that the recovery terms include temperature; that is,

$$\dot{\Omega} = f_1 \dot{\varepsilon}^{\mathrm{I}} - f(\Omega, T) \qquad \text{and} \qquad \dot{Z} = g_1 \dot{\varepsilon}^{\mathrm{I}} - g(Z, T)$$

In steady-state conditions, $\dot{\Omega} = \dot{Z} = 0$; then the inelastic strain rate in equation 7.8.7 can be used with the properties of creep to define the recovery functions. For example, the drag stress becomes

$$g(Z_{SS}, T) = g_1 \dot{\varepsilon}_{SS}^{\mathrm{I}} = g_1 D_0[\sinh(A\sigma_{SS})]^n \tag{7.8.9}$$

after using equation 7.8.1. However, the steady-state stress, σ_{SS}, can be expressed as a function of the drag stress through equation 7.8.4, and recalling that $k_1 = A^{-2/3}$, equation 7.8.9 becomes

$$g(Z_{SS}, T) = g_1 D_0[\sin(A^3 Z_{SS}^3)]^n \tag{7.8.10}$$

Equation 7.8.10 gives the functional form for $g(Z,T)$, which can be used for any value of drag stress; thus the drag stress evolution equation becomes

$$\dot{Z} = g_1 \dot{\varepsilon}^{\mathrm{I}} - g_1 D_0[\sinh(A_2 Z^3)]^n \tag{7.8.11}$$

after defining $A^3 = A_2$. The process can be repeated to establish a back-stress evolution equation. The steady-state back stress can be obtained by substituting equation 7.8.4 into equation 7.8.6 and simplifying to obtain $\Omega_{SS} = (1 - k_2)\sigma_{SS}$. The back stress equation is then given by

$$\dot{\Omega} = f_1\dot{\varepsilon}^{\mathrm{I}} - f_1 D_0[\sinh(A_1|\Omega|)]^n \operatorname{sgn} \Omega \tag{7.8.12}$$

where the first term defines the strain hardening and dynamic recovery is modeled by the second term. The temperature effects are included through the function D_0 in both the drag stress and back stress equations. Finally, equations 7.8.11 and 7.8.12 can be used to establish a relationship between the drag stress and back stress at saturation in a creep or tensile test. Since $\dot{\Omega} = \dot{Z} = 0$ at saturation, it follows that the arguments of the hyperbolic functions must be equal; thus

$$\Omega_{SS} = \frac{A_2}{A_1} Z_{SS}^3 \tag{7.8.13}$$

The last step in the development of the model is to include cyclic hardening in the drag stress. The equations above were established from the monotonic tensile and creep properties of the material. The drag stress parameter g_1 is therefore an isotropic hardening parameter that must be modified for the Bauschinger effect. Recall, the dislocation mechanisms that occur in cyclic and monotonic loading are different. The back stress is motivated by dislocation pileups associated with tensile loading whereas the drag stress corresponds to the development of dislocation microstructure during fatigue. During cyclic loading the magnitude of back stress, $|\Omega|$, will oscillate between zero and some value up to it's maximum value. The drag stress (or hardness) will grow until it reaches the average value of the hardness created by the oscillating back stress. This suggests that the parameter g_1 should be replaced by

$$g_1 = g_1' \left(c_2 + |\Omega| - \frac{A_2}{A_1} Z^3 \right) \tag{7.8.14}$$

where g' and c_2 are constants. During steady state monotonic loading $g_1 = g_1' c_2$ from equation 7.8.13, and the monotonic properties are preserved. During cyclic loading the drag stress will increase (in the absence of thermal recovery) until the average value reaches a saturation value that corresponds to $g_1 = 0$. The final form of the drag stress is obtained by using equation 7.8.14 in equation 7.8.11 to get

$$\dot{Z} = g_1'\dot{\varepsilon}^{\mathrm{I}} \left(c_2 + |\Omega| - \frac{A_2}{A_1} Z^3 \right) - g_1' c_2 D_0[\sinh(A_2 Z^3)]^n \tag{7.8.15}$$

Observe that the temperature-dependent recovery terms in equation 7.8.15 facilitates the modeling of an annealing process. If the parabolic stress–strain shape is required, set $Z = \sqrt{F}$ in equation 7.8.7 and replace equation 7.8.15 by

$$\dot{F} = g_1'\dot{\varepsilon}^{\mathrm{I}}\left(c_2 + |\Omega| - \frac{A_2}{A_1}F^{1.5}\right) - g_1'c_2D_0[\sinh(A_2F^{1.5})]^n \quad (7.8.16)$$

The Miller flow equation is given in three dimensions as

$$\dot{\varepsilon}_{ij}^{\mathrm{I}} = \frac{3}{2}D_0\left[\sinh\left(\frac{\sqrt{K_2}}{Z}\right)^{3/2}\right]^n \frac{S_{ij} - \Omega_{ij}}{\sqrt{K_2}} \quad (6.2.11)$$

The back-stress and drag stress equations are written as

$$\dot{\Omega}_{ij} = f_1\dot{\varepsilon}_{ij}^{\mathrm{I}} - f_1D_0[\sinh(A_1\Omega_e)]^n \frac{\Omega_{ij}}{\Omega_e} \quad (7.8.17)$$

and

$$\dot{Z} = g_1'\dot{\varepsilon}_e^{\mathrm{I}}\left(c_2 + \Omega - \frac{A_2}{A_1}Z^3\right) - g_1'c_2D_0[\sinh(A_2Z^3)]^n \quad (7.8.18)$$

where the subscript e is defined in equations 5.2.21 and 5.2.22 for stresslike and strainlike variables, respectively.

The Miller equations are interesting because of the underlying relationship to physical metallurgy and methods used to derive the flow and evolution equations. This methodology can be applied to other specific representations to obtain models for special situations. A major advantage of the Miller model is that temperature effects are included through the function D_0 and the initial value of the drag stress, $Z_0(T)$. All other parameters are independent of temperature. One of the criticisms is the difficulty in numerical integration of equations because they are very stiff, This difficulty appears to be overcome in a paper by Miller and Tanaka (1988) and an asymptotic integration method by Chula and Walker (1989) that is presented in Section 9.3. Systematic methods for determining the material parameters are given by Miller (1976) and Adebanjo (1993). An example of the cyclic response characteristics of the model is given in Figure 7.8.1 using the parameters in Appendix 7.3.

7.9 POWER LAW MODELS

Walker (1981) and Kreig, Swearengen, and Rohde (1978) proposed separately a power law representation for the flow law of the form shown in equations 6.2.9 and 6.2.15, that is,

$$\dot{\varepsilon}_{ij}^{\mathrm{I}} = D_0 \frac{|S_{kl} - \Omega_{kl}|^n}{Z} \frac{(\frac{3}{2}S_{ij} - \Omega_{ij})}{|S_{kl} - \Omega_{kl}]} \quad (6.2.9)$$

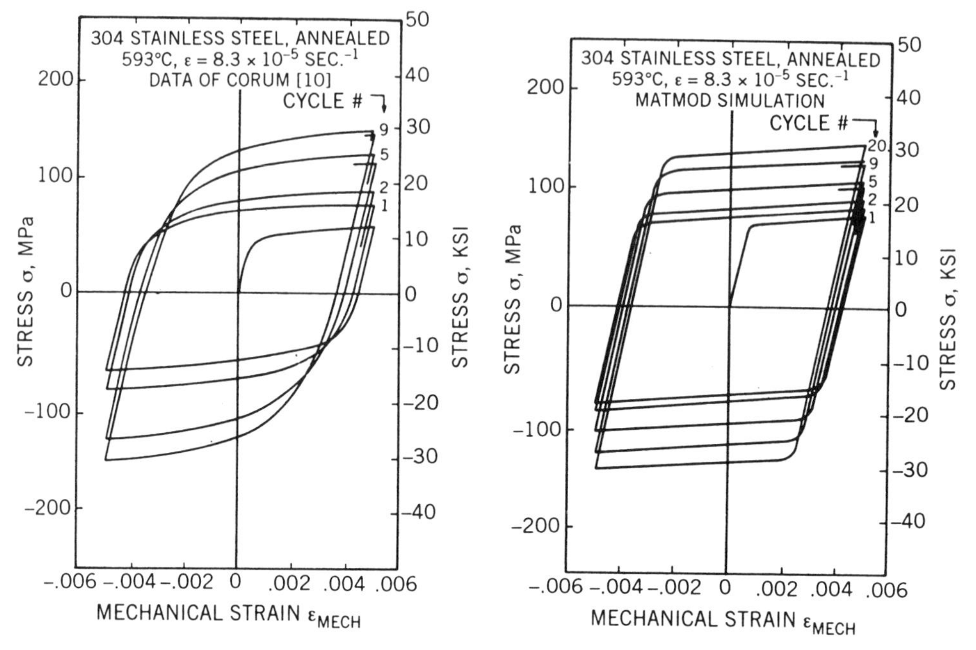

Figure 7.8.1 (*a*) Experimental fatigue response showing cyclic hardening; (*b*) simulated response of the data using the Miller model. The model parameters are given in Appendix 7.3. (After Miller, Journal of Engineering Materials and Technology, V98, pg 109, 1976. Copyright © The American Society of Mechanical Engineers, 345 East 47th St., N.Y., N.Y. 10017. Used with permission.)

where $|S_{kl} - \Omega_{kl}|$ is defined by equation 6.2.10. The major differences in the models are in the evolution equations for the back stress and drag stress. This section includes a summary of the development of the Kreig et al. model and a summary of the Walker model since it has several similar features.

The back-stress and drag stress evolution equations have hardening and recovery terms as similar to those discussed previously. The hardening rates in both equations is assumed to be linear in the strain rate, and are taken to be of the following form in three dimensions:

$$\dot{\Omega}_{ij} = f_1 \dot{\varepsilon}^{\mathrm{I}}_{ij} - f \frac{\Omega_{ij}}{|\Omega_{ij}|} \quad \text{and} \quad \dot{Z} = g_1 \dot{\varepsilon}^{\mathrm{I}}_{e} - g \tag{7.9.1}$$

where f_1 and g_1 are constants, and f and g are the back stress and drag stress recovery functions, respectively. The back stress recovery is assumed to be parallel to the current direction of inelastic straining (or slip) since it is associated with the hardening due to dislocation pileups. The definition of the recovery functions is the unique feature of the two models.

The drag stress is used to model cyclic hardening characteristic of the development of dislocation cells and subgrains. The recovery is assumed to depend on the current state of the dislocation network, which is modeled by the current value of drag stress, Z. However, it is expected that the current dislocation state can approach the fully annealed Z_A; thus the recovery is assumed to be a function of $(Z - Z_A)$. The rate of recovery is also a strong function of the temperature. Thus, using the Arrhenius equation, Kreig et al. assumed a drag stress recovery function in the form

$$g = g_2 \frac{(Z - Z_A)^m}{T} \exp\left(\frac{-Q}{kT}\right) \tag{7.9.2}$$

where the constants g_2, Z_A, and m can be determined from annealing data. The activation energy is a constant for temperatures above about $0.6T_m$. It can be determined from creep data provided that diffusion is the underlying mechanism (see Section 3.4). The constant g_1 controls the rate of hardening from the initial state Z_0 (not necessarily Z_A) to the fully hardened state Z_1.

Recall that back stress results from dislocation pileups at barriers such as subgrain and cell walls. The recovery of back stress by climb is controlled by the stress concentration at the head of the pileup. Kreig et al. developed a model from the kinetics of the process as given by Friedel (1967); therefore, the function f is assumed to be

$$f = f_2\Omega_e^2 \left[\exp\left(\frac{f_3\Omega_e^2}{kT}\right) - 1\right] \tag{7.9.3}$$

where f_2 and f_3 are material constants, and Ω_e, the effective back stress, is defined the same as the effective stress. The evolution equations for the Kreig et al. model can be summarized as

$$\dot{\Omega}_{ij} = f_1\dot{\varepsilon}_{ij}^{\mathrm{I}} - f_2\Omega_e\Omega_{ij}\left[\exp\left(\frac{f_3\Omega_e^2}{kT}\right) - 1\right]\Omega_{ij} \tag{7.9.4}$$

and

$$\dot{Z} = g_1\dot{\varepsilon}_e^{\mathrm{I}} - g_2 \frac{(Z - Z_A)^m}{T} \exp\left(\frac{-Q}{kT}\right) \tag{7.9.5}$$

with equation 6.2.9 as the flow equation. The parameters in the flow and back-stress equations are related by the condition of steady-state creep or tensile response, $\dot{\Omega} = 0$.

In 1987, Kreig, Swearengen, and Jones extended the model to include finite deformation (see reference for details) and added a unique approach to isotropic hardening. They observed that hardening results from reducing the mean free distance λ that a dislocation can propagate before being stopped. This concept is characterized by assuming that flow stress is a function of the grain or subgrain size. The Hall–Petch equation (equation 1.5.1, Figures

1.5.1 and 1.5.2) gives the flow stress as the inverse root of the grain size, $\sigma \propto 1/\sqrt{d}$. But the relation breaks down by the formation of cells or subgrains (Problem 1.13) and does not include strain or strain-rate variables. Kreig et al. investigated the variation of the cell size with strain and found a good correlation with data using an exponential function

$$\lambda = A_0 + B\exp(-\beta\varepsilon^{\mathrm{I}})$$

Since the materials experience both hardening and dynamic recovery during deformation (the change in cell size), the above representation can be differentiated to obtain

$$\dot{\lambda} = \beta(\lambda - A_0)\dot{\varepsilon}^{\mathrm{I}} \tag{7.9.6}$$

The kinetic relationship for the inelastic strain rate was then assumed to be

$$\varepsilon_{ij}^{\mathrm{I}} = A_2\lambda^m \exp\left(\frac{-Q}{kT}\right)(\sinh A_1|S_{ij} - \Omega_{ij})^n \frac{S_{ij} - \Omega_{ij}}{|S_{ij} - \Omega_{ij}|} \tag{7.9.7}$$

The traditional drag stress has been replaced by the mean free slip distance, which is represented by temperature and hyperbolic sine functions similar to the Miller formulation. The back-stress rate is given by equation 7.9.4. Equation 7.9.6 should be viewed as a preliminary result because the cell or subgrain size is generally accepted to be a function of the deformation history.

The evolution equations for the Walker model are summarized by Chula and Walker (1989). The back-stress evolution equation is given by

$$\dot{\Omega}_{ij} = n_2\dot{\varepsilon}_{ij}^{\mathrm{I}} - (\Omega_{ij} - \Omega_{ij}^0)\dot{G} + \dot{\Omega}_{ij}^0 \tag{7.9.8}$$

and

$$\dot{G} = n_3\dot{\varepsilon}_e^{\mathrm{I}} + n_6(\tfrac{2}{3}\Omega_{ij}\Omega_{ij})^{(m-1)/2} \tag{7.9.9}$$

where Ω_{ij}^0 and $\dot{\Omega}_{ij}^0$ are the initial values of the back stress and back-stress rate. The drag stress is based on accumulated inelastic strain rather than the accumulated inelastic work. In the absence of thermal recovery, the integrated drag stress is

$$Z = Z_1 - Z_2\exp(-n_7\varepsilon_e^{\mathrm{I}}) \tag{7.9.10}$$

where n, m, Z_1, Z_2, n_1, n_2, n_3, n_6, and n_7 are material constants.

The power law models have been used widely for structural simulations in finite-element codes involving many materials. An example of the predictive capability of a very complicated loading is shown in Figures 7.9.1 and 7.9.2.

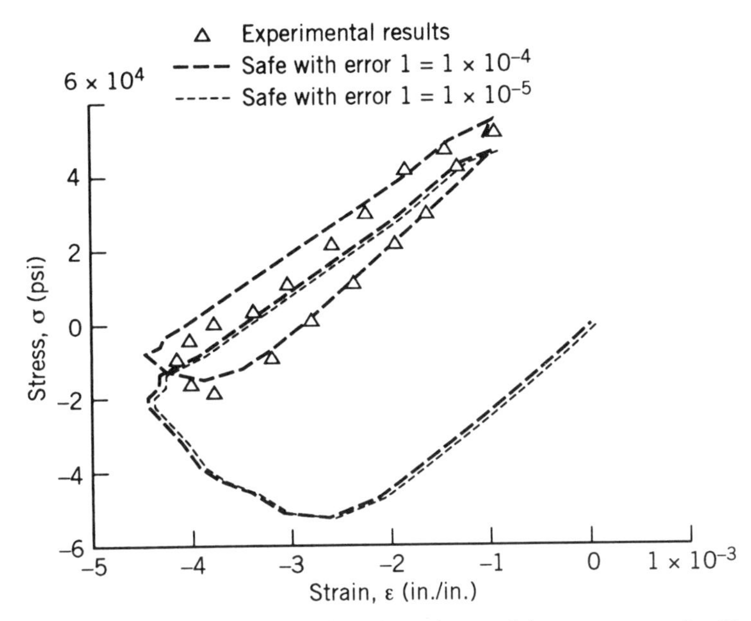

Figure 7.9.1 Thermomechanical fatigue loop from a laboratory test of a Hastelloy X combustor linear loaded to simulate an engine mission cycle (takeoff, cruise, landing, taxi). Prediction of the Walker model in a three-dimensional solid finite-element model of Louver 5 consisting of 546 elements and 1247 nodes. The model was integrated with the asymptotic integration algorithm in Section 9.3. The model parameters are given in Appendix 7.4. (After Chula and Walker, 1989.)

An earlier form of the Kreig et al. back stress equation was used to produce the results in Figure 7.9.2. The earlier back stress equation and parameters for both models ares given in Appendices 7.4 and 7.5. It is clear that state variable models can produce some rather impressive results once the material parameters have been determined.

7.10 SUMMARY

The key property in multiaxial cycling is the classification of proportional or nonproportional loading. In proportional loading the slip planes remain constant throughout the loading history and the hardening response is similar to uniaxial cycling. In nonproportional loading the slip planes change during the loading history and there is extra hardening at low temperature due to the development of a more extensive dislocation microstructure created by intersecting slip planes. Increasing temperatures increase the effect of climb, and the amount of extra hardening is reduced due to the effect of climb on slip. Thus the parameters determined from uniaxial tests should be successful in

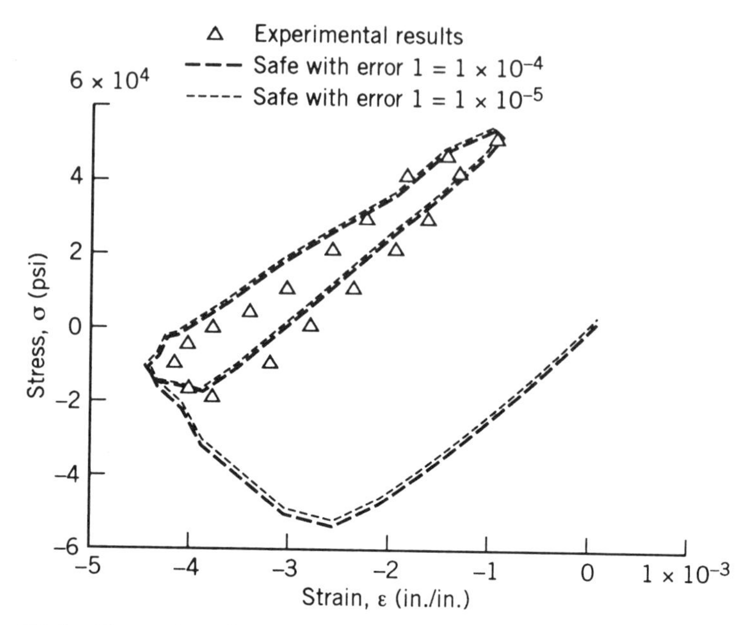

Figure 7.9.2 Comparison between experimental data of Figure 7.9.1 and finite-element simulation using the Kreig, Swearengen, and Rohde model. The model parameters are given in Appendix 7.5. (After Chula and Walker, 1989.)

predicting the response of proportional loading and reasonably applicable for nonproportional high-temperature loading, but not low-temperature nonproportional loading. Many of the constitutive models in the literature do not reflect the difference between proportional and nonproportional hardening.

Thermomechanical cycling is similar to multiaxial cycling. The methods of interpolation and extrapolation are expected to be successful for low- and high-temperature cycling if the deformation mechanism is constant, either slip or slip and climb. However, if the temperature oscillates between low-temperature slip and high-temperature slip with climb, there will be extra hardening due to the effect of climb on the low-temperature slip. Thus interpolation and extrapolation are not expected to be accurate when thermomechanical history effects are present (see, e.g., Cassenti, 1985).

REFERENCES

Adebanjo, R. O. (1993). Determination of the material constants for the MATMOD-ReX constitutive model, in *Material Parameter Estimation for Modern Constitutive Equations*, MD Vol. 43, AMD Vol. 168, ASME, New York, p. 1.

Anand, L. (1982). Constitutive equations for the rate dependent deformation of metals at at elevated temperature, *Journal of Engineering Materials and Technology*, Vol. 104, p. 12.

Anand, L. (1985). Constitutive equations for hot working metals, *International Journal of Plasticity*, Vol. 1, pp. 213–231.

Bhattachar, V. S. (1991). A unified constitutive model for the thermomechanical fatigue response of a nickel base superalloy René 80, Ph.D. disseration, Department of Aerospace Engineering and Engineering Mechanics, University of Cincinnati.

Bhattachar, V. S., and D. C. Stouffer (1993a). Constitutive equations for the thermomechanical response of René 80, Part 1: Development from isothermal data, *Journal of Engineering Materials and Technology*, Vol. 115, pp. 351–357.

Bhattachar, V. S., and D. C. Stouffer (1993b). Constitutive equations for the thermomechanical response of René 80, Part 2: Effects of temperature history, *Journal of Engineering Materials and Technology*, Vol. 115, pp. 358–364.

Bodner, S. R. (1987). Review of a unified elastic–viscoplastic theory, in *Unified Constitutive Equations for Creep and Plasticity*, A. K. Miller, ed., Elsevier Applied Science, Barking, Essex, England.

Cassenti, B. N. (1993). Research and development program for the development of advanced time-temperature dependent constitutive equations, Vol 1: Theoretical discussion, NASA CR-182132.

Chan, K. S., U. S. Lindholm, S. R. Bodner, S. R. Hill, R. W. Weber and T. G. Meyer (1986). Constitutive modeling for isotropic materials, NASA CR-179522.

Chan, K. S., and R. A. Page (1988). Inelastic deformation and dislocation structure of a nickel alloy: effects of deformation and thermal histories, *Metallurgical Transactions*, Vol. 19A, pp. 247–248.

Chan, K. S., U. S. Lindholm, S. R. Bodner, and A. Nagy (1990). High temperature deformation of the B1900+Hf alloy under multiaxial loading: theory and experiment, *Journal of Engineering Materials and Technology*, Vol. 112, pp. 7–13.

Chula, A., and K. P. Walker (1989). A new uniformly valid asympotic integration algorithm for elastic–plastic creep and unified viscoplastic theories including continuum damage, *NASA TM 102344*, NASA Lewis Research Center, Cleveland OH.

Cook, T. S., K. S. Kim, and R. L. McKnight (1988). Thermal mechanical fatigue of cast René 80, in *Low Cycle Fatigue*, ASTM STP 942, pp. 692–708.

Friedel, J. (1967). *Dislocations*, Addison-Wesley, Reading, MA.

Helling, D. E., and A. K. Miller (1987). The incorporation of yield surface distortion into a unified constitutive model, Part 1: Equation development; Part 2: Predictive capablities, *Acta Mechanica*, Vol. 69, pp. 9–53.

Henshall, G. A., D. E. Helling and A. K. Miller (1995). Improvements in the MATMOD equations for modeling solute effects and yield surface distortion, in *Unified Constititutive Laws of Plastic Deformation* A. S. Krausz, ed., Academic Press, New York.

Henshall, G. A., and A. K. Miller (1990). Simplifications and improvements in unified constitutive equations for creep and plasticity, Part 1: Equations development; Part 2: Behavior and capability of the model, *Acta Metallurgica Materials*, Vol. 38, No. 11, pp. 2101–2128.

Jones, R. M. (1975). *Mechanics of Composite Materials* McGraw-Hill, New York.

Krieg, R. D., J. C. Swearengen, and W. B. Jones (1987). A physically based internal variable model for rate dependent plasticity, in *Unified Constitutive Equations for Creep and Plasticity* A. K. Miller, ed., Elsevier Applied Science, Barking, Essex, England.

Krieg, R. D., J. C. Swearengen, and R. W. Rohde (1978). A physically based internal variable model for rate dependent plasticity, in *Inelastic Behavior of Pressure Vessel and Piping Components*, ASME PVP-PB-028, T. Y. Chang and E. Krempl, eds., ASME, New York.

Lindholm, U. S., K. S. Chan, S. R. Bodner, R. M. Weber, K. P. Walker, and B. N. Cassinti (1984). Constitutive modeling of isotropic materials, *NASA CR 174718.*

Metzer, A. M. and S. R. Bodner (1979). Analytical formulation of a rate and temperature dependent stress strain relation, *Journal of Engineering Materials and Technology*, Vol. 101, pp. 254–257.

Metzer, A. M. (1982). Steady state and transient creep based on unified constitutive equations, *Journal of Engineering Materials and Technology*, Vol. 104, pp. 18–25.

McDowell, D. S. (1983a). Transient nonproportional cyclic plasticity, *Report 107, UILU-Eng-83-4003*, University of Illinois at Urbana–Champaign.

McDowell, D. S. (1983b). On the path dependence of transient hardening and softening to stable states under complex biaxial loading, *Proceedings of the NSF Supported International Conference on Constitutive Laws for Engineering Materials*, C. S. Desai and R. H. Gallagher, eds., Tucson, AZ.

McDowell, D. F. (1983c) On the path dependence of transient hardening and softening to stable states under complex cyclic loading, in *Proceedings of the International Conference on Constitutive Laws for Engineering Materials*, C. S. Desai and R. H. Gallager, eds., Tucson, AZ.

McDowell, D. L., D. F. Socie, and H. S. Lamba (1983). Multiaxial nonproportional cyclic deformation, *ASTM STP 770*, pp. 500–518.

McDowell, D. L., R. K. Payne, D. Stahl, and S. D. Antolovich (1984). Effects of nonproportional loading histories on type 304 stainless steel, Spring Meeting of Société Française de Métallurgie, Paris.

McKnight, R. W., J. H. Laflen, and G. T. Spamer (1982). Turbine blade tip durability analysis, *NASA CR 165268.*

Miller, A. K. (1976). An inelastic constitutive equation for monotonic, cyclic and creep deformation, Part 1: Equations development and analytical procedures; Part 2: Application to 304 stainless steel, *Journal of Engineering Materials and Technology*, Vol. 98H, pp. 97–113.

Miller, A. K. (1987). The MATMOD equations, in *Unified Constitutive Equations for Creep and Plasticity*, A. K. Miller, ed., Elsevier Applied Science, Barking, Essex, England.

Miller, A. K., and T. G. Tanka, (1988). A new method for integrating unified constitutive equations under complex histories, *Journal of Engineering Materials and Technology*, Vol. 110, pp. 205–211.

Murakami S., M. Kawai, and Y. Ohmi (1989). Effects of amplitude history on multiaxial cyclic behavior of 316 stainless steel, *Journal of Engineering Materials and Technology*, Vol. 111, pp. 278–285.

Murakami, S., M. Kawai, and Y. Yamada (1990). Creep after cyclic plasticity under multiaxial conditions for 316 type stainless steel at elevated temperature, *Journal of Engineering Materials and Technology*, Vol. 112, pp. 346–352.

Murakami, S., M. Kawai, K. Aoki, and Y. Ohmi (1989). Temperature dependence of multiaxial nonproportional cyclic behavior of 316 stainless steel, *Journal of Engineering Materials and Technology*, Vol. 111, pp. 32–39.

Neu, R. W. (1993). Nonisothermal material parameters for the Bodner–Partom model, in *Material Parameter Estimation for Modern Constitutive Equations*, MD Vol. 43, AMD Vol. 168, ASME, New York, p. 211.

Ramaswamy, V. G. (1986). A constitutive model for the inelastic multiaxial cyclic response of a nickel base superalloy René 80, *NASA CR 3998*.

Ramaswamy, V. G., R. H. Van Stone, L. T. Dame, and J. H. Laflen (1985). Constitutive modeling of isotropic materials, *NASA CR 175004*.

Robinson, D. N. (1978). A unified creep plasticity model for structural metals at high temperature, ORNL TM-5969.

Robinson, D. N. and R. W. Swindeman (1982). Unified creep plasticity constitutive equations for 2 1/4 Cr-1 Mo steel at elevated temperature, ORNL TM-8444.

Salemme, G. C., (1993). A nonproportional loading activated extra hardening correction for an inelastic constitutive model, M.S. thesis, Department Aerospace Engineering and Engineering Mechanics, University of Cincinnati.

Stouffer, D. C., V. G. Ramaswamy, J. H. Laflen, R. H, Van Stone, and R. Williams (1990). A constitutive model for the inelastic multiaxial response of René 80 at 871C and 982C, *Journal of Engineering Materials and Technology*, Vol. 112, pp. 241–246.

Stouffer, D. C. (1981). A constitutive representation for IN100, AFWAL TR-81-4039, Wright Patterson AFB, Ohio.

Stouffer, D. C. and S. R. Bodner (1982). A relationship between theory and experiment for a state variable constitutive equation, *Mechanical Testing for Deformation Model Development*, ASTM STP 756, pp. 239–250.

Walker, K. P. (1981). Research and Development Program for Nonlinear Structural Modeling with Advanced Time-Temperature Dependent Constitutive Relationships, *NASA CR 165533*.

Weertman, J. (1956). *Mechanics and Physics of Solids*, Vol. 4, p. 230.

Weertman, J. (1960). *Transactions of AIME*, Vol. 218, p. 207.

Weertman, J. (1963). *Transactions of AIME*, Vol. 227, p. 1475.

PROBLEMS

7.1 For review, define briefly each of the following terms:

dynamic recovery	screw dislocation
static recovery	partial dislocation
critical resolved shear stress	back stress
drag stress	Nabarro creep
slip line	interstitial solid solution
dislocation cell	dislocation subgrain
cyclic softening	dislocation glide

7.2 Write out explicit representations of equations 6.2.8, 6.6.3, and 7.2.5 for combined tension torsion and combined biaxial tension. What value of C for $\varepsilon_{12} = C\varepsilon_{11}$ was used to define the proportional experiments in Figure 2.2.2?

7.3 Show that equation 7.3.4 at saturation for a uniaxial stress in the $\mathbf{e}_2$ direction gives $\Omega_{22} = \Omega_m - \Omega_{22}^0$.

7.4 The following was given by Chula and Walker (1989) for a unified state variable equation for the viscoplastic response of metals:

$$\text{Inelastic strain rate:} \quad \dot{\varepsilon}_{ij}^{\mathrm{I}} = \left(\frac{\sqrt{K}}{Z}\right)^n \frac{S_{ij} - \frac{2}{3}\Omega_{ij}}{\sqrt{K}} \tag{a}$$

$$K = \tfrac{3}{2}(S_{ij} - \tfrac{2}{3}\Omega_{ij})(S_{ij} - \tfrac{2}{3}\Omega_{ij}) \tag{b}$$

$$\text{Back stress rate:} \quad \dot{\Omega}_{ij} = n_2\dot{\varepsilon}_{ij}^{\mathrm{I}} - (\Omega_{ij} - \Omega_{ij}^0)\dot{G} + \dot{\Omega}_{ij}^0 \tag{c}$$

$$\dot{G} = n_3\dot{\varepsilon}_e^{\mathrm{I}} + n_6(\tfrac{2}{3}\Omega_{ij}\Omega_{ij})^{(m-1)/2} \tag{d}$$

$$\dot{\varepsilon}_e^{\mathrm{I}} = \sqrt{\tfrac{2}{3}\dot{\varepsilon}_{ij}^{\mathrm{I}}\dot{\varepsilon}_{ij}^{\mathrm{I}}} \tag{e}$$

$$\text{Drag stress:} \quad Z = Z_1 - Z_2\exp(-n_7\varepsilon_e^{\mathrm{I}}) \tag{f}$$

where n, m, n_1, n_2, n_3, n_6, n_7, Z_1, and Z_2 are material constants, and S_{ij} and Ω_{ij} are deviatoric tensors. (**a**) Let ε^{I}, σ, and Ω denote the values of the inelastic strain, stress, and back stress observed in uniaxial loading. Determine the values of $\varepsilon_{ij}^{\mathrm{I}}$, S_{ij}, and Ω_{ij} as a function of the uniaxial variables so that equation (a) satisfies incompressibility for uniaxial loading. (**b**) Evaluate equations (b) and (e) for K and $\dot{\varepsilon}_e^{\mathrm{I}}$ as a function of the uniaxial variables. (**c**) Rewrite equations (a), (c), and (d) as a function of the uniaxial variables. (**d**) What relationship must exist between n_2, n_3, and n_6 if the stress saturates during tensile loading at the constant strain rate $\dot{\varepsilon}_0$. (**e**) Develop a relationship between the exponent n and the other material parameters in equation (a) at tensile saturation as described in part (**d**).

7.5 A constitutive equation proposed by Anand (1982) for hot-working metals was developed using one state variable S to model hardening, dynamic recovery, and static thermal recovery. The uniaxial form of the model is given by:

$$\text{Inelastic flow:} \quad \dot{\varepsilon}^{\mathrm{I}} = A\exp\left(\frac{-Q}{RT}\right)\left[\sinh\left(\xi\frac{|\sigma|}{S}\right)\right]^{1/m} \times \operatorname{sgn}\sigma \tag{a}$$

$$\text{Hardening rate:} \quad \dot{S} = h_0\left|1 - \frac{S}{S^*}\right|^n |\dot{\varepsilon}^{\mathrm{I}}|\operatorname{sgn}\left(1 - \frac{S}{S_{\mathrm{sat}}}\right) - B\mu\exp\left(\frac{-Q_1}{RT}\right)\left(\frac{S}{\mu}\right)^p \tag{b}$$

$$\text{Hardening saturation:} \quad S_{\mathrm{sat}} = |S|\left[\frac{|\dot{\varepsilon}^{\mathrm{I}}|}{C}\exp\left(\frac{Q_2}{RT}\right)\right]^n \tag{c}$$

where T is the absolute temperature, and A, B, C, R, Q, Q_1, Q_2, μ, ξ, h_0, m, n, and p are material parameters. The first term of equation (b) models hardening and dynamic recovery, and the second term models static thermal recovery that is present at very high temperatures. (**a**) Verify that the model saturates for tensile loading at constant strain rate. (**b**) Establish a three-dimensional tensor representation for the Anand model that is isotropic, satisfies incompressibility, and reduces to equations (a), (b), and (c) for uniaxial loading. (**c**) What is the specific form of the equations for pure shear loading? (**d**) Determine the parameters for the data given in Appendix 2.3 and compare the model predictions to experimental data.

7.6 The isothermal constitutive model of Robinsion (1982) with two state variables, back stress and drag stress, can be written as

$$2\mu\dot{\varepsilon}_{ij} = \begin{cases} F^n \dfrac{S_{ij} - \Omega_{ij}}{\sqrt{K_2}} & \text{for } F > 0 \text{ and } S_{ij}(S_{ij} - \Omega_{ij}) > 0 \\ 0 & \text{for } F < 0 \text{ or } F > 0 \text{ and } S_{ij}(S_{ij} - \Omega_{ij}) \leq 0 \end{cases}$$

$$\dot{\Omega}_{ij} = \begin{cases} \dfrac{H}{G^\beta}\dot{\varepsilon}_{ij} - RG^{m-\beta}\dfrac{\Omega_{ij}}{\sqrt{I_2}} & \text{for } G > G_0 \text{ and } S_{ij}\Omega_{ij} > 0 \\ \dfrac{H}{G_0^\beta}\dot{\varepsilon}_{ij} - RG_0^{m-\beta}\dfrac{\Omega_{ij}}{\sqrt{I_2}} & \text{for } G \leq G_0 \text{ and } S_{ij}\Omega_{ij} \leq 0 \end{cases}$$

where $I_2 = \Omega_{ij}\Omega_{ij}/2$ and the functions F and G are defined by

$$F = \frac{K_2}{Z^2} - 1 \qquad \text{and} \qquad G = \frac{I_2}{Z^2}$$

and μ, β, n, m, R, and H are material parameters. Values for the parameters for $2\text{–}\frac{1}{4}$Cr–1Mo steel at 538°C are $\mu = 3.6 \times 10^7$, $n = 4.0$, $\beta = 0.75$, $m = 3.87$, $R = 8.97 \times 10^{-8}$, and $H = 9.92 \times 10^3$. (a) Verify that the model saturates for tensile loading at constant strain rate. (b) What is the specific form of the equations for pure shear loading? (c) Program the equations and plot the tension–torsion response for a proportional load history $\sigma = 3\tau$ and nonproportional load history as defined in Figure 7.2.4. Discuss the results predicted.

7.7 Determine the parameters for the Robinson model given in Problem 7.6 for the data given in Appendix 2.3 and compare the model predictions to experimental data.

7.8 Develop a three-dimensional representation for the Chaboche model described in Problem 6.8. What is the specific form of the equations for tension–torsion loading?

7.9 Make up your own list of the advantages and disadvantages of the state-variable approach to constitutive modeling. How does the state-variable approach compare to the classical methods of plasticity and creep in Chapter 5?

CHAPTER 8

Single-Crystal Superalloys

Nickel-base single-crystal alloys have attracted considerable attention for use in rocket and gas turbine engines because their high-temperature properties are superior to those of polycrystalline nickel-base superalloys. In high-temperature applications, grain boundaries in polycrystalline alloys provide passages for diffusion and oxidation. Eliminating grain boundaries and the grain-boundary strengthening elements produces materials with superior high-temperature fatigue and creep properties to those of conventional superalloys. However, the absence of grains make the alloys anisotropic or orientation dependent.

Components made of single-crystal superalloys are cast. The crystallographic orientation of the component is controlled during manufacture by the physical orientation of a seed crystal that is placed in the mold. The mold is cooled slowly from the bottom so that solidification grows from the seed and the crystallographic planes are continuous. Components can be cast in almost the net shape except for special features like holes and notches that must be machined.

Section 4.5.1 contains a description of the elastic properties of metals with cubic symmetry. This chapter contains a description of the inelastic response properties of single-crystal superalloys as a function of temperature, strain rate, and orientation. A description of the deformation mechanisms are included to provide a basis for the state variable approach to modeling and definition of the associated state variables. It was shown in Section 1.2.1 that face-centered cubic (FCC) crystals slip primarily on the octahedral and cube planes in specific slip directions. This information is used to construct a model for the global inelastic strain by summing the slip on individual slip planes. The constitutive model for the inelastic strain response is then constructed from a relationship between the stress and slip on the slip planes. Many experimental and numerical results are given for modern single-crystal superalloys at typical operating temperatures.

Finally, the reader should note that state variable constitutive modeling of single-crystal superalloys is a recent addition to the literature. Observations of the physical mechanisms of deformation and the evolution of physically based models were motivated by new alloys for high-temperature environments. At this time there is a reasonable understanding of the mechanisms associated with the yield, tensile, and cyclic response of nickel-base single-crystal alloys, and there is a reasonable consensus of a modeling approach. Creep modeling is not as well advanced and it is not included, but there is a discussion of some of the key experimental results and physical observations related to creep.

8.1 INTRODUCTION

Before beginning with a discussion of the mechanical properties of single-crystal alloys, it is useful to introduce the idea of a sterograph projection, a simple method to visualize the symmetry properties of crystals. Recall that planes are defined by their normal vector or pole. A sterographic sphere is a sphere showing the intersection of crystallographic planes and poles with the surface of the sphere. Poles intersect the sphere as points and planes intersect the sphere in great circles. A sterographic projection is the projection of a hemisphere of the sphere onto a plane. Figure 8.1.1*a* shows the sterographic projections of the [001], [100], and [010] planes and poles on a plane in front of the sphere. Observe that the traces of the [110] and [111] planes are great circles along the curved lines shown in Figure 8.1.1*b* and *c*. Any plane or pole can be represented by a similar sterographic projection.

Observe that the FCC crystal lattice has some symmetry. For example, the crystal can be rotated 90° about the [001] axis in either direction to an identical configuration. There are also some rotational symmetries for the face and cubic diagonals. A convenient way to view the symmetry is to plot all the points where the cubic edges (or material principal axes), face diagonals, and cubic diagonals intersect the sterographic sphere, as shown in Figure 8.1.2. The traces of the great circles from the planes similar to [001] and [011] generate a system of sterographic triangles that each contain a cubic axis, face diagonal, and cubic diagonal. There are 48 similar triangles on the front and reverse sides of the sphere. This system of triangles is important since each triangle, with a cubic axis, face diagonal, and cubic diagonal, corresponds to a similar region of the crystal lattice. Thus any three points such as *a*, *b*, and *c*, are all equivalent if they occupy the same relative position in these sterographic triangles. That is, if *a*, *b*, and *c* correspond to the orientation of three tensile specimens, each specimen would exhibit the same mechanical properties. Use of a sterographic triangle is a convenient way to show the orientations of a set of test samples.

The chemical compositions of superalloys vary some with the manufacturer, but they all have similar microstructural characteristics (see Sheh and Stouffer, 1988, for details). A typical microstructure of a two-phase single-crystal alloy is shown in Figure 8.1.3. Modern materials are two-phase alloys with a large volume fraction of the gamma prime (γ') phase, an ordered aluminum–nickel FCC crystal lattice structure, $L1_2$, with nickel at the center

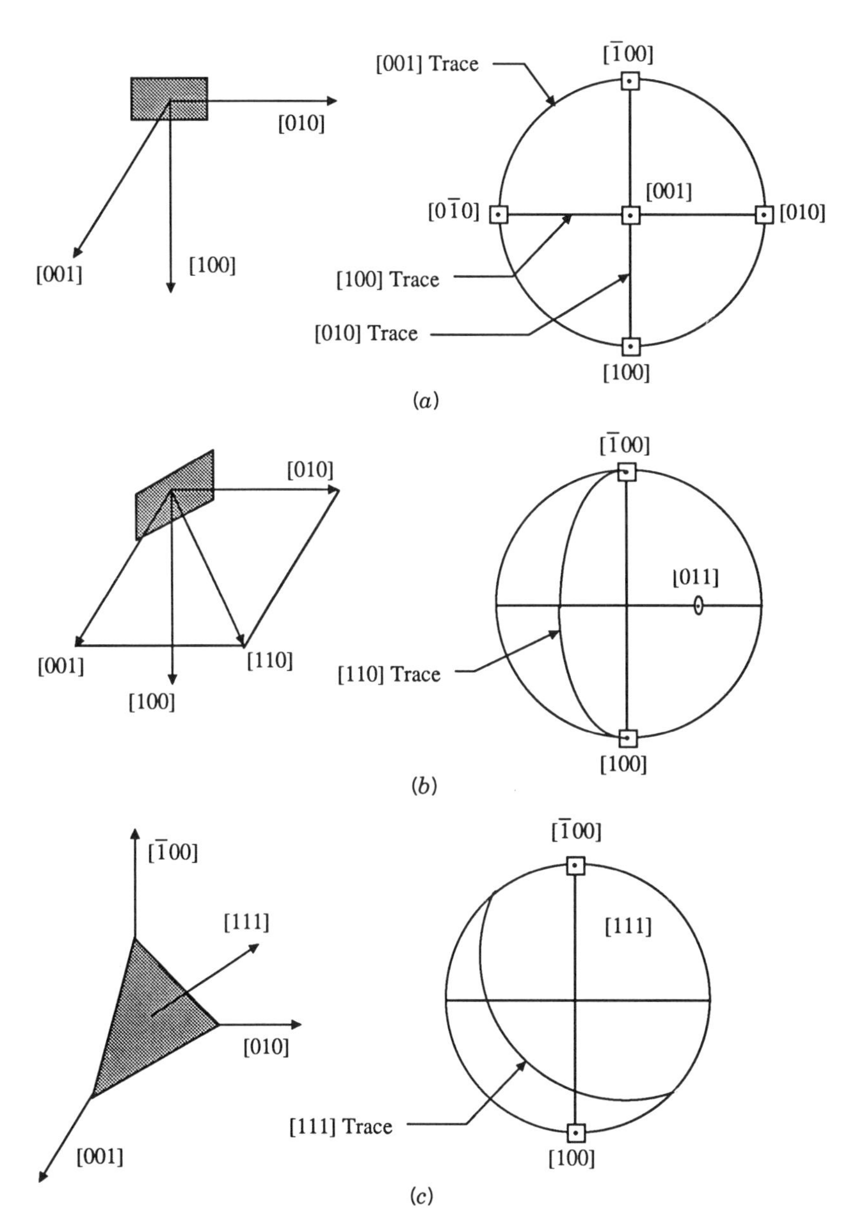

Figure 8.1.1 Standard stereographic projections of the (*a*) [001], [010], [100], (*b*) [101], and (*c*) [111] planes and poles. Squares represent the intersection with the principal axes or cubic edges, oblongs represent intersection with face diagonals, and triangles represent intersection with cubic diagonals.

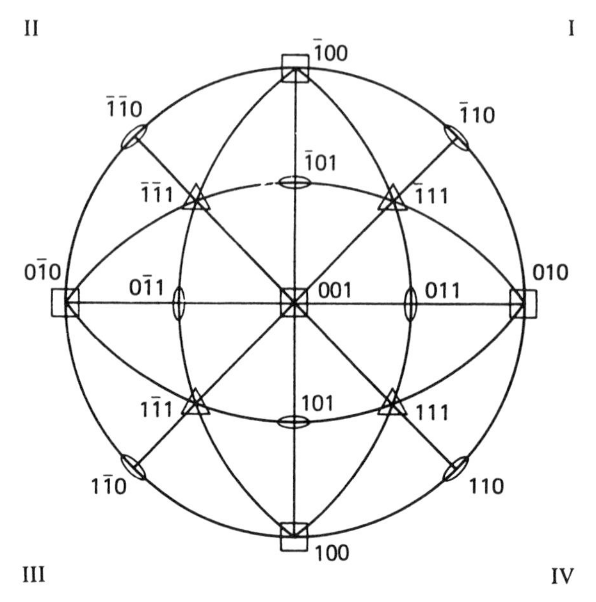

Figure 8.1.2 Standard [001] stereographic projection of the cubic edges (material principal axes), face diagonals, and cubic diagonals of an FCC or BCC cubic crystal. The great circles of the [001] and [101] planes create 24 similar stereographic triangles in each hemisphere. (After Verhoeven, Fundamentals of Physical Metallurgy. Copyright © 1975. Reprinted by permission of John Wiley & Sons, Inc.)

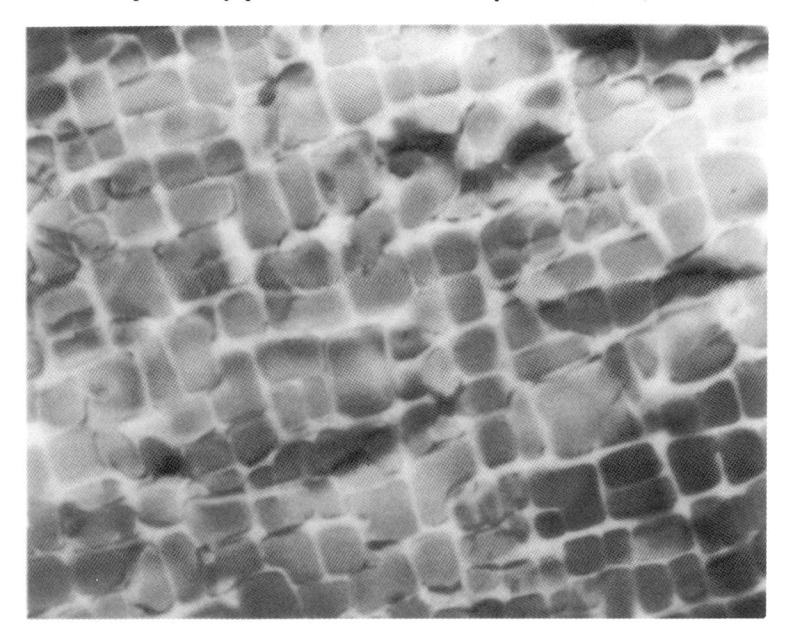

Figure 8.1.3 Bright-field transmission electron micrograph of the γ–γ' structure of PWA 1480 nickel-based single-crystal superalloy. (After Milligan and Antolovich, 1989.)

of the six faces and aluminum on the eight corners. The γ' is interspersed in a coherent FCC solid solution, the gamma phase (γ) with a random aluminum–nickel lattice structure. The lattice planes remain plane and are continuous across the interface, but the sequence of atoms in the lattice changes at the γ–γ' boundary. The interface between the γ and γ' phases is an antiphase boundary. Recall that lattice planes are discontinuous and change direction at a grain boundary; thus there are no grain boundaries in this alloy. The strength of the alloy is a function of the γ' size and volume fraction. Experiments have shown that the maximum creep resistance is achieved with a γ' volume fraction of about 60% (Shah and Duhl, 1984; Pope and Ezz, 1984). Even though the chemical compositions differ, commercial alloys are 60 to 65% γ' by volume, and the typical dimensions are around 1 μ and less.

8.2 TENSION, COMPRESSION, AND ORIENTATION PROPERTIES

The inelastic response of single-crystal alloys is quite different from that of polycrystalline alloys. The yield strength is a function of material orientation relative to the uniaxial load applied. The material also exhibits tension–compression asymmetry of the yield stress that is a function of orientation. An example of yield strength asymmetry for PWA 1480 is shown in Figure 8.2.1 for orientations between [001] and [011] in the [100] plane. The tension–compression asymmetry and orientation effects become negligible for loading near the [111] orientation. Above a critical temperature of approxi-

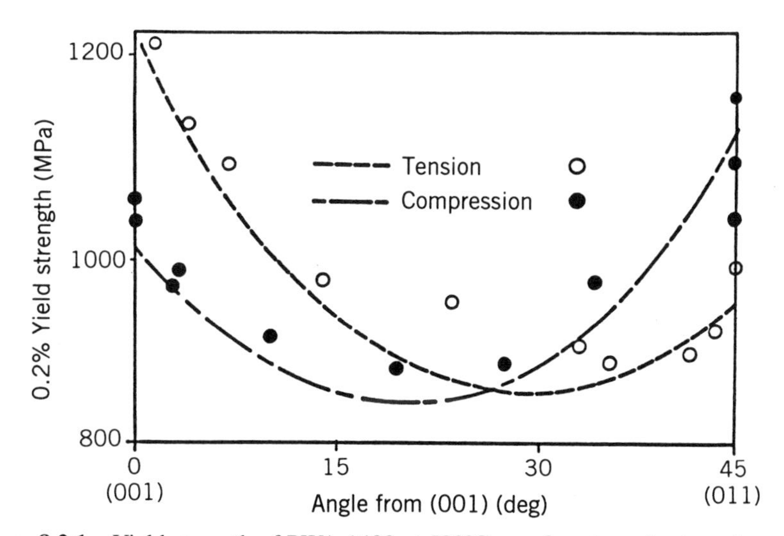

Figure 8.2.1 Yield strength of PWA 1480 at 593°C as a function of orientation along the [001]–[011] boundary of the standard stereographic triangle. (After Sheh and Stouffer, 1988; data from Shah and Duhl, 1984.)

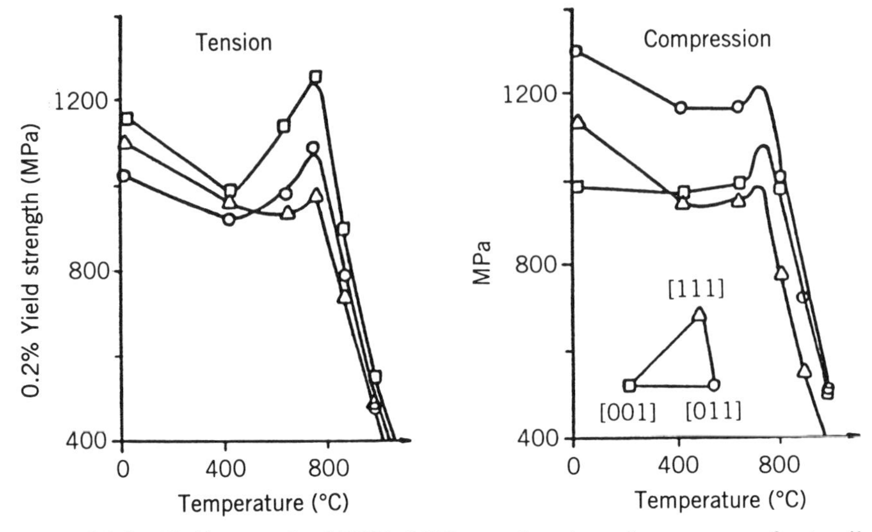

Figure 8.2.2 Yield strength of PWA 1480 as a function of temperature for tensile and compression tests in the [001], [011], and [111] orientations. (After Shah and Duhl, 1984.)

mately 700 to 750°C there is a sharp drop in the yield stress, and the tension–compression asymmetry and orientation effects disappear with increasing temperature, as shown in Figure 8.2.2. The yield strength of these alloys is also strain-rate dependent if the temperature is above the critical temperature (Figure 8.2.3), due to the balance between hardening and dynamic recovery as discussed in Section 2.4.

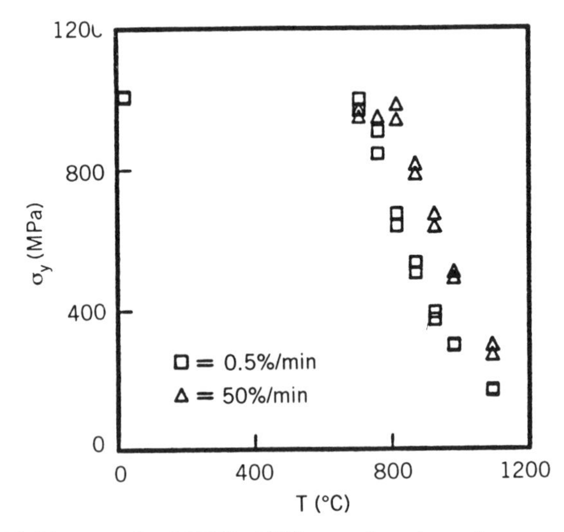

Figure 8.2.3 Yield strength of PWA 1480 as a function of temperature and strain rate for tensile tests in the [001] orientation. (After Milligan, 1986.)

The small strain, post yield response of nickel-base single-crystal alloys is not always smooth and continuous for temperatures near and below the critical temperature. Alden (1990) conducted a number of double tensile tests on René N4 at 760°C, where the specimen was initially loaded at a strain rate of 1×10^{-4} sec^{-1}, unloaded and held at zero stress for 2 min, and then reloaded at 6×10^{-4} sec^{-1}. Figure 8.2.4 shows the result of double tensile tests on specimens near the [001], [011], [012], [$\bar{1}$23] and [$\bar{1}$11] orientations as summarized in Figure 8.2.4*f*. Figure 8.2.4*a* and *e* for the [001] and

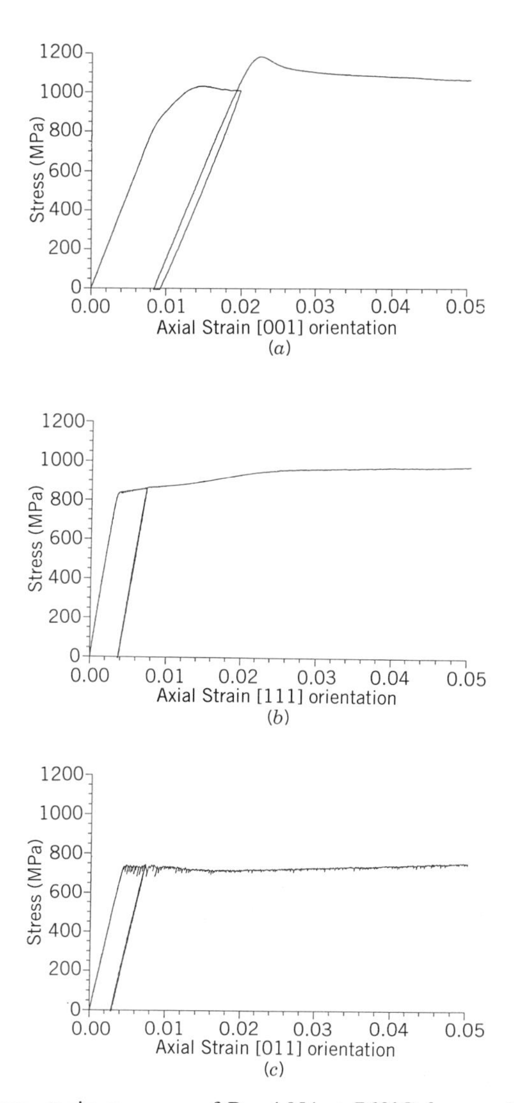

Figure 8.2.4 Stress–strain response of René N4 at 760°C from a double tensile test for specimens in the (*a*) [001], (*b*) [$\bar{1}$11], (*c*) [011], *d*) [012], (*e*) [$\bar{1}$23] orientations, and (*f*) a summary of orientations in a stereographic triangle. (After Alden, 1990.)

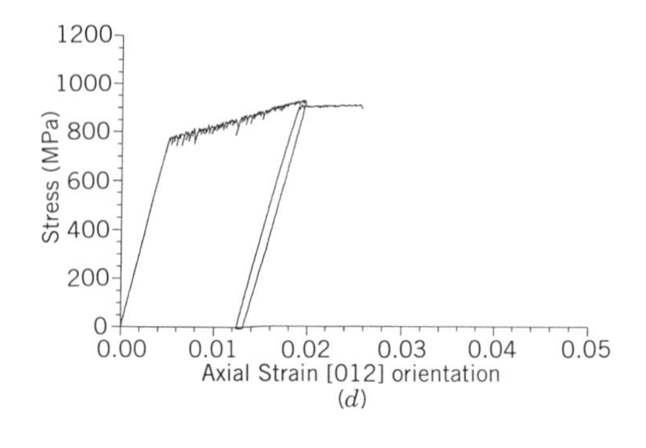

(*d*)

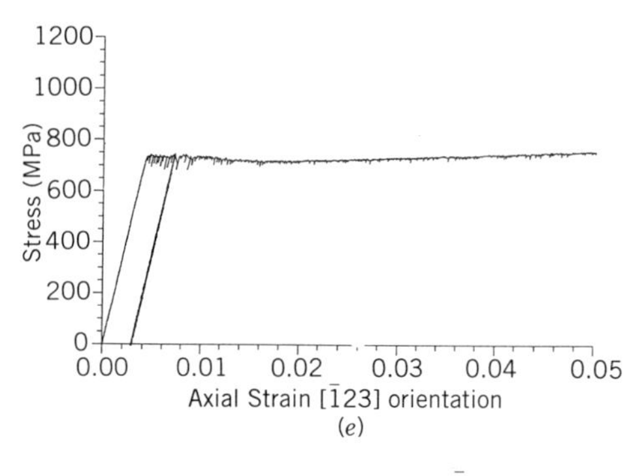

(*e*)

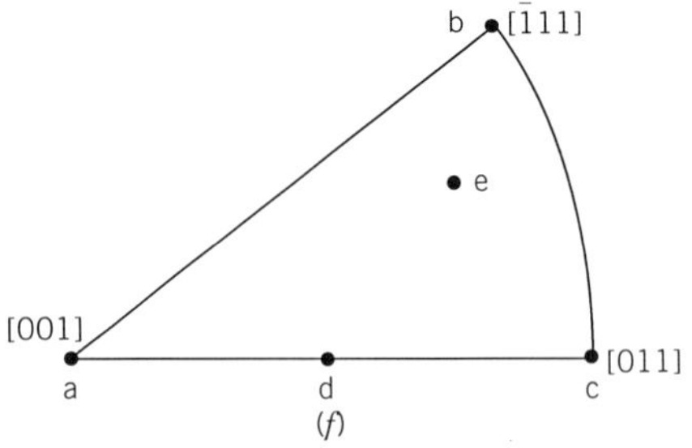

(*f*)

Figure 8.2.4 (*Continued*)

[$\bar{1}$23] orientations, respectively, show the range of response characteristics. The tensile response of a specimen in the [001] orientation produced the largest values of stress, there were no slip bursts, the response was strain-rate sensitive, and there was a significant amount of anelastic recovery during the 2-min hold period at zero stress. Conversely, the tensile response of the [$\bar{1}$23] specimen had audible slip bursts during the initial straining, the change in strain rate did not effect the response, and there was no recovery during the 2-min hold period at zero stress. There were also double tension tests at 982°C for specimens in the [100] and [$\bar{1}$11] orientations, as shown in Figures 1.3.3 and 1.3.4, respectively. The results at 982 and 760°C both suggest that the amount of recovery during unloading is a strong function of orientation. Tensile tests at 982°C in the [123] orientation at two strain rates (not shown) did not have slip bursts; thus slip bursts are a function of temperature and orientation but not strain rate. The results of the double tension tests show the broad effects of orientation, strain rate, and strain recovery on the stress response.

A study of the effect of orientation and temperature on the tensile response to failure of René N4 was reported by Miner et al. (1986a). Tests were run at a constant cross-head speed, giving an initial strain rate of about 2×10^{-4} sec^{-1}. Figure 8.2.5 shows the results of tensile tests to failure at room temperature at 760 and 980°C. Observe that the response at all temperatures is somewhat flat due to the initial hardness created by the large volume fraction of γ'. An interesting observation is that the yield stress in the [001] orientation is higher at 760°C than at room temperature. The orientations were selected to eliminate serrated yielding and crystallographic rotation (discussed in Section 8.4), A study of SC 7-14-6 by Dalah, Thomas, and Dardi (1984) presented in Figure 8.2.6 shows that the tensile strength or ultimate engineering stress above 760°C (1400°F) is substantially less anisotropic than the tensile strength at 760°C and below. Test specimens near the [011] orientation generally display the lowest tensile strength and the greatest ductility. Initially round specimens in the [011] orientation deform into cross sections with an elliptical shape. Specimens near the [111] orientation generally have the highest tensile strength.

The mechanical response properties observed in tension–compression testing are also present in cyclic loading. Figure 8.2.7 shows the initial cyclic response and stable response of PWA 1480 at 760°C for a specimen oriented in the [123] direction. Observe that the slip bursts disappear as cyclic hardening progresses. Cyclic hardening is observed by the decrease in the plastic strain range and increase in the stress range. There is also some tension–compression asymmetry in the response. Strain-rate sensitivity is present in cyclic tests above the critical temperature, as shown in Figure 8.2.8. However, it should be observed that variation in elastic modulus with orientation affects the balance between the elastic and plastic strain ranges in total strain range testing of crystals. Thus the inelastic (or plastic) strain range, rather than total strain range, is often used for cyclic hardening studies and low-cycle fatigue life testing.

8.3 DEFORMATION MECHANISMS

The orientation and tension–compression asymmetry properties can be explained from the active slip systems and dislocation mechanisms. Recall from Section 1.2.1 that there may be up to three families of active slip systems, depending on the orientation of the applied load and temperature. Milligan

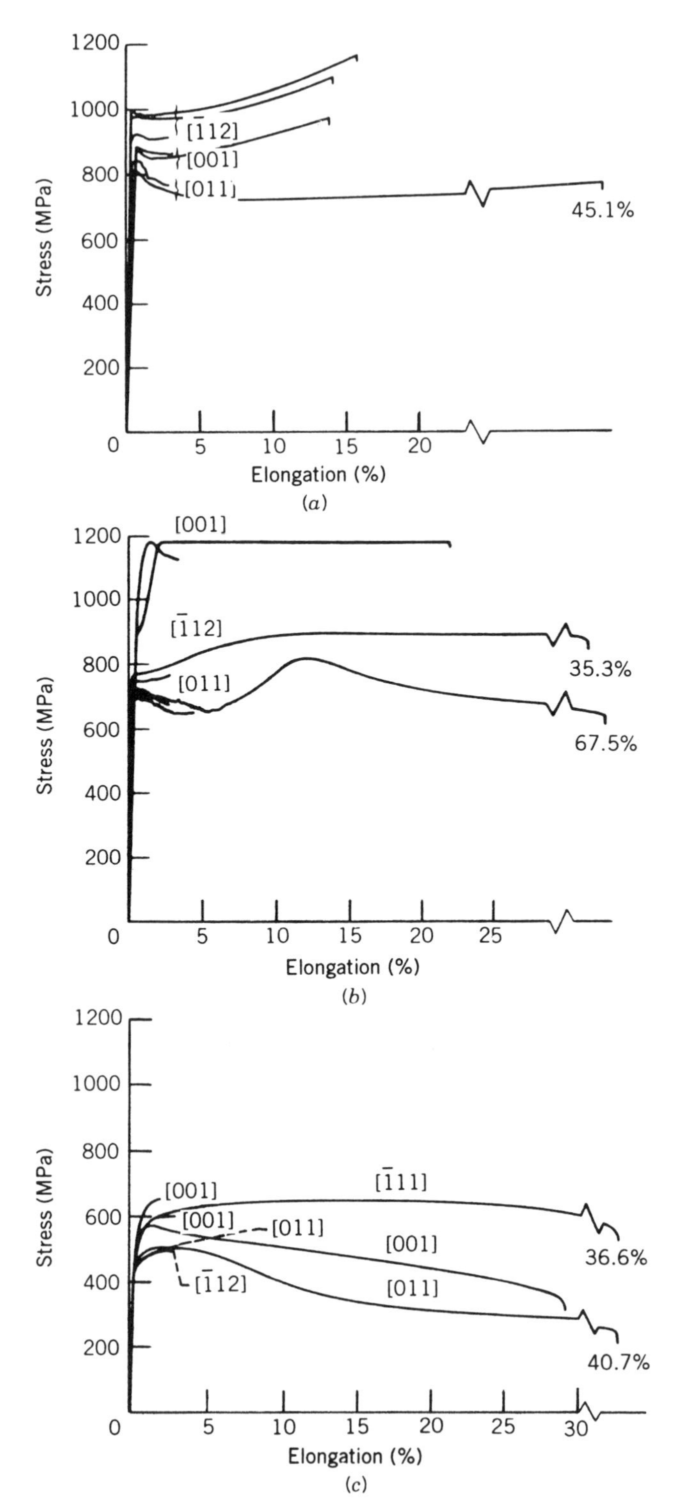

Figure 8.2.5 Tensile curves for René N4 specimens, including tests that were interrupted at 2% strain. The experiments were run at (*a*) room temperature, (*b*) 760°C, and (*c*) 980°C. (From Miner et al., 1986a; reproduced with permission of The Minerals, Metals & Materials Society.)

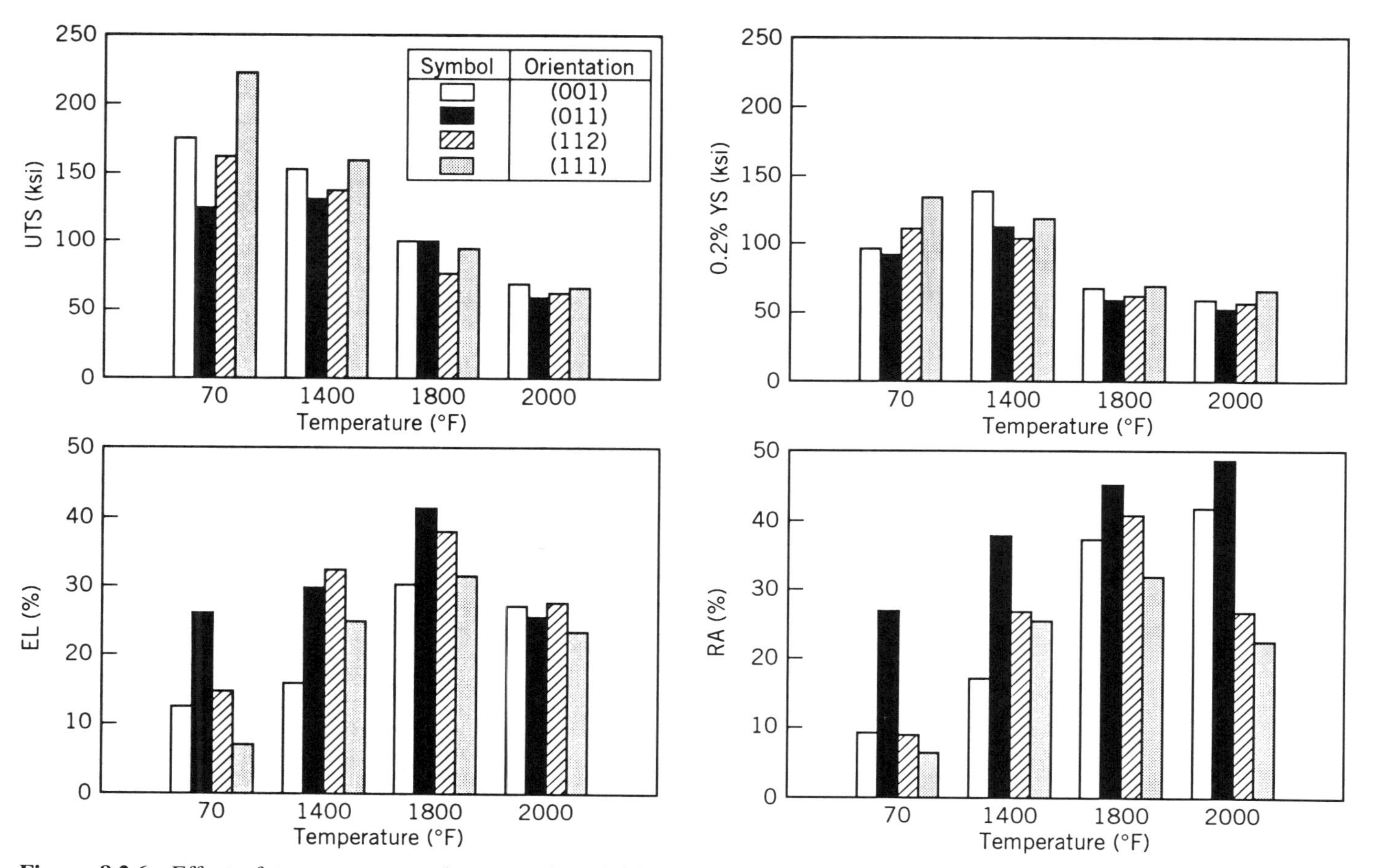

Figure 8.2.6 Effect of temperature on the properties of SC 7-14-6. (From Dalah, Thomas, and Dardi, 1984; reproduced with permission of the American Institute of Mining, Metallurgy and Petroleum Engineers.)

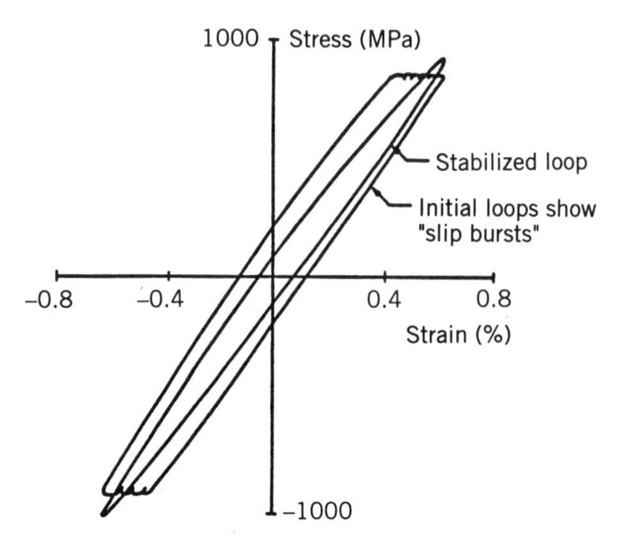

Figure 8.2.7 Cyclic response of PWA 1480 in the [321] orientation at 760°C. Hardening is observed by the reduction in the plastic strain range and increase in stress. (After Swanson, 1984b.)

and Antolovich (1989) reported for PWA 1480 for strain rates other than creep (typically, greater than 10^{-4} sec^{-1}; see Figure 3.5.2) that:

1. Below about 600°C deformation occurs due to slip on the octahedral planes, [111] planes, in the 12 directions similar to the [101] direction.

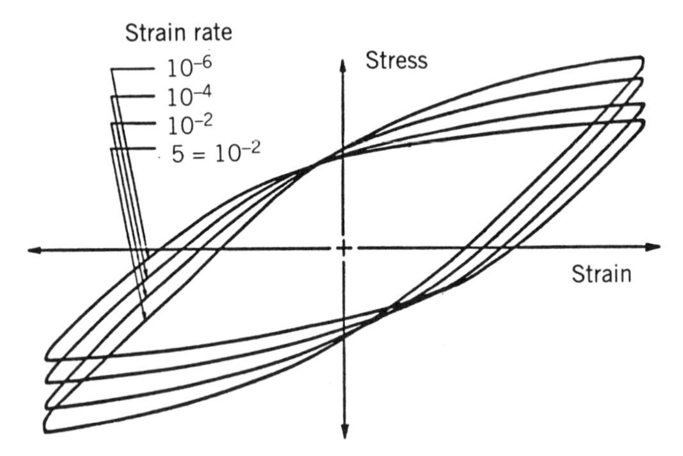

Figure 8.2.8 Strain-rate sensitivity of PWA 1480 in the [111] orientation at 871°C. (After Swanson, 1984b.)

Octahedral slip is shown in Figure 1.2.8 and corresponds to slip numbers 1 to 12 in Tables 1.2.1 and 1.2.2.

2. Between about 600°C and 850°C deformation occurs simultaneously by octahedral slip and cube slip. Cube slip occurs on the three planes similar to [001] in the two directions similar to the [011] direction (six slip systems) as defined in Figure 1.2.9 and slip numbers 25 to 30 in Tables 1.2.1 and 1.2.2.
3. Above about 850°C the strong effect of climb and cross slip reduces the dependence on orientation and the particular slip systems that are active.

These results are consistent with the observations of Miner (1986a, b) for René N4 at 760, 870, and 982°C. Miner found only cube slip for loading near the [111] orientation and only octahedral slip for loading near the [100] and [110] orientations. These results are summarized in the sterographic triangle in Figure 8.3.1, showing the specimen orientations that deform by octahedral, cube, and mixed slip.

The octahedral and cube slip systems can be displayed on the stereographic projection. Using Figure 8.1.1 as a guide, the trace of the four octahedral planes can be added to the standard stereographic projection in Figure 8.1.2 to get the diagram shown in Figure 8.3.2*b*. The three slip systems shown in Figure 8.3.2*a* can be identified in Figure 8.3.2*b* locating the [111] normal vector and the $[0\bar{1}1]$, $[\bar{1}01]$, and $[\bar{1}10]$ slip systems. Notice that the three slip systems must lie on the trace of the [111] plane, and the three vectors from [000] to $[0\bar{1}1]$, $[10\bar{1}]$ and $[\bar{1}10]$ in Figure 8.3.2*b* are parallel to the three slip directions shown in Figure 8.3.2*a*. The direction of slip along the line of action (the sign of the vector) is from the origin to the intersection with the

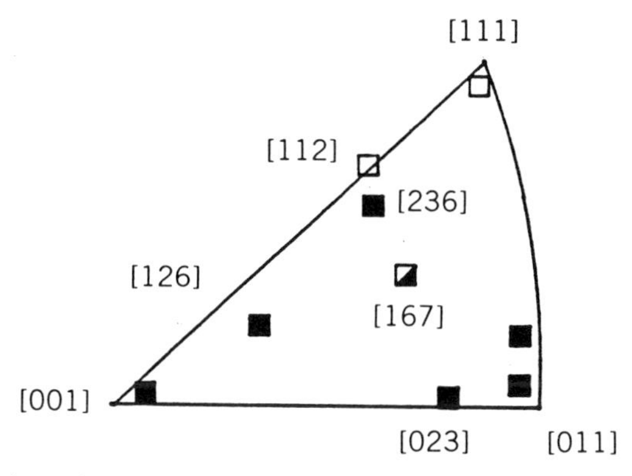

Figure 8.3.1 Orientations of the alloy René N4 that exhibited octahedral, cube, and mixed octahedral–cube slip. (From Sheh and Stouffer, 1988; data from Miner et al., 1986*a*, *b*; reproduced with permission of The Minerals, Metals & Materials Society.)

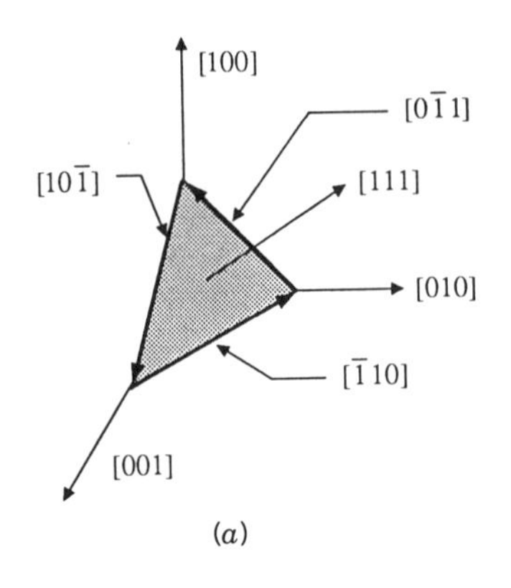

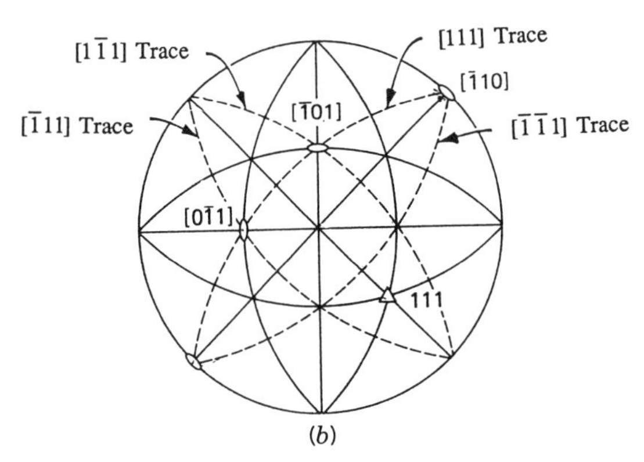

Figure 8.3.2 Stereographic projection showing the octahedral and cube slip systems in FCC crystals. (After Verhoeven, Fundamentals of Physical Metallurgy, Copyright © 1975. Reprinted by permission of John Wiley & Sons, Inc.)

sphere, as shown in Figure 8.3.2*b*. This process can be repeated to obtain the other nine slip systems similar to [111][101]. The locations of the cube planes are shown in Figure 8.1.1*a*. The slip directions on the [001] plane are along the two diagonal lines in Figure 8.3.2*b*. The other four slip directions are along the horizontal and vertical lines in Figure 8.3.2*b*.

The number of active slip systems similar to [111][101] depends on the orientation and magnitude of the applied load. The shear stress τ parallel to any slip system can be calculated from the applied tensile stress σ using Schmid's law (see Problem 1.6),

$$\tau = \sigma \cos \alpha \cos \beta \qquad (8.3.1)$$

where β is the angle between the tensile axis and the slip plane normal vector and α is the angle between the tensile axis and the slip direction. Equation 8.3.1 was used to determine the shear stresses in Table 1.2.1 for the three slip system types. For slip systems similar to [111][101], Table 1.2.1 shows that loading in the [001] direction will activate eight slip systems simultaneously if the stress is equivalent to the critical resolved shear stress (CRSS). Loading in the [110] will activate four simultaneous slip systems, whereas only one slip system will be activated for loading in the [123] direction. Figure 8.3.3 summarizes the number of slip systems similar to the [111][101] slip system with equal stresses (or slip rates) as a function of orientation.

Next, recall from Section 1.2.1 that slip on the octahedral plane occurs by the disassociation of a total dislocation into two partial dislocations. The atoms in the core of the extended dislocation have a different stacking sequence, the stacking fault energy controls the core width, and small core widths have a greater potential for cross slip to a cubic plane. In an ideal setting slip occurs by adding atoms to the core and removing atoms from the core as shown in Figure 1.2.10. In practice, extended dislocation movements are disrupted by other dislocations, antiphase boundaries, cross slip, and climb at high temperatures.

In single crystals, Lall, Chin, and Pope (1979) showed that extended dislocations create the tension–compression asymmetry shown in Figure 8.2.1 because they behave differently in tension and compression. Consider a tensile load that creates a shear stress on an octahedral plane that is not parallel to the slip direction. This shear stress can be resolved into two components as shown in Figure 8.3.4, with one component parallel to the slip direction τ and one component normal to the slip direction τ_1. During tensile loading in the [001] direction, for example, the normal component of the shear stress τ_1

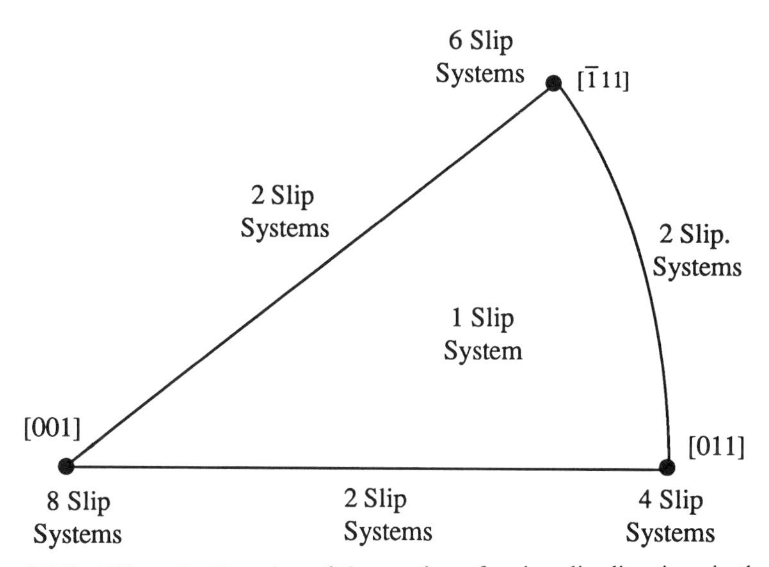

Figure 8.3.3 Effect of orientation of the number of active slip directions in the slip systems similar to [111][101] for the FCC crystal lattice structure.

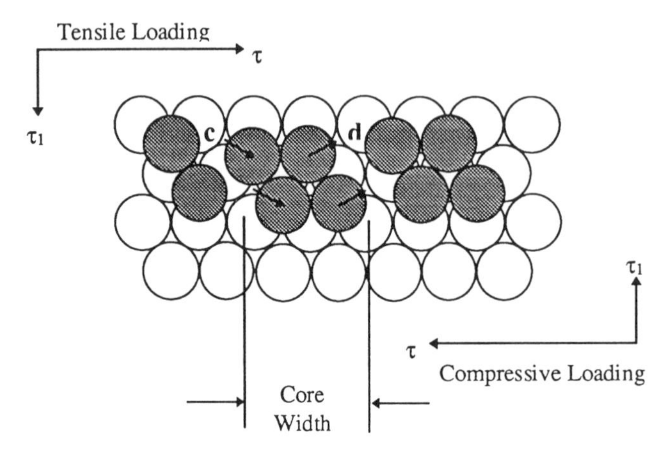

Figure 8.3.4 Effect of tensile and compressive stresses on the core width of an extended dislocation.

will increase the potential that atoms are added to the core, increase the resistance to cross slip, and increase the yield stress. When the applied stress is reversed, the shear stress $-\tau_1$ will favor the potential that atoms are removed from the core. Thus the core width is extended during tensile loading and reduced during compressive loading. Next recall that at temperatures below the critical temperature there is no slip in the cubic slip systems. Thus dislocations that cross slip from the octahedral planes to the cube planes during tensile loading in the [001] direction are pinned and their movement is restricted. At high temperatures the dislocations are not pinned on the cube planes and cross slip increases the potential for slip in the cube slip systems. Thus the core width controls the potential for cross slip, and the response in tension and compression will be different.

The movement of extended dislocations can be influenced by other stresses in the crystal lattice. Lall, Chin, and Pope (1979) also proposed that the stress on the cube plane, τ_2, parallel to τ on the octahedral plane will promote cross slip from the octahedral plane to the cube plane (Figure 8.3.5*a*). The potential for the cross slip is increased with increasing temperatures. The stress τ_2 is particularly important when the loading is near the [111] orientation, where the stresses on the cube plane may be larger than the octahedral stress. (Table 1.2.1 shows how the stresses in the cube system increase with increasing proximity to the [111] orientation.) Thus, τ_2 is the thermally aided contribution to cross slip that is important for loading near the [111] orientation.

Paidar, Pope, and Vitek (1984) later proposed that the stress normal to the slip direction on the intersecting octahedral plane (τ_3), as shown in Figure 8.3.5*b*, also influences the core width. The effect of τ_3 is similar to that of τ_1 on the core width, and increasing values of τ_3 increase the potential for cross slip of the dislocation.

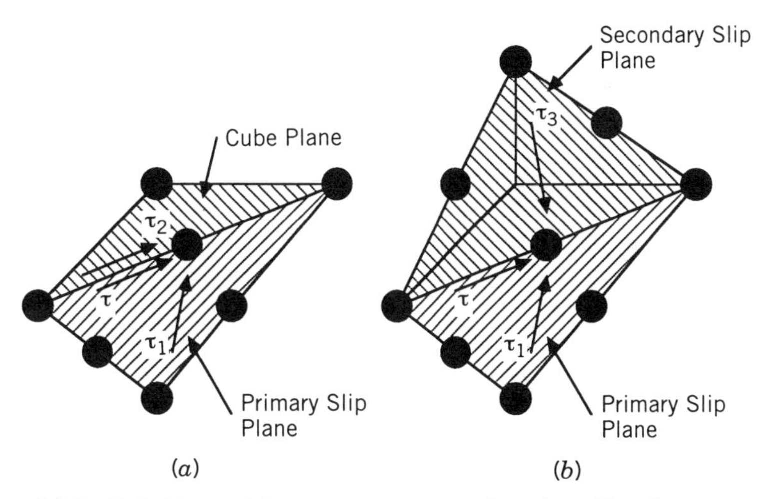

Figure 8.3.5 Definitions of the stresses τ_1, τ_2, and τ_3 that affect the movement of extended dislocations.

Milligan and Antolovich (1989) investigated the tensile and cyclic behavior of PWA 1480 in the temperature range 20 to 1093°C at strain rates of 0.50 and 50.0% min^{-1}. They found two types of dislocation networks at high and low temperatures. At temperatures between approximately 20 and 760°C and for loading in the [001] orientation with plastic strains up to 0.3%, the dislocation network was dominated by pairs of extended dislocations shearing the γ' phase as shown in Figure 8.3.6. In this temperature range dislocations traveled through the γ' as closely spaced pairs. In contrast to the shearing mechanism at low temperature, a γ' bypass mechanism was observed at temperatures of 982°C and higher. For specimens tested to failure at 982°C there were high dislocation densities in the γ phase, and that deformation was controlled by looping or bypass of the γ' particles. Shearing of the γ' phase was also observed after a large increase in the matrix dislocation density and significant strain hardening. In the intermediate temperature range from 760 to 927°C a transition from shearing to bypass was observed. Micrographs showing the bypass and shearing mechanisms at high temperature are given in Figures 8.3.7 and 8.3.8.

Consider next a few more single-crystal effects that are not seen in polycrystalline materials. The exact cause for the slip bursts reported in Figure 8.2.4 has not been documented. The slip bursts occurred only at the lower temperatures when climb is not significant and for orientations where the deformation occurred in one-, two-, or four-slip systems (see Figure 8.2.4*f*). Alden (1990) proposed that slip bursts result from dislocations cross slipping to a cube plane at the γ–γ' antiphase boundary and becoming pinned. The

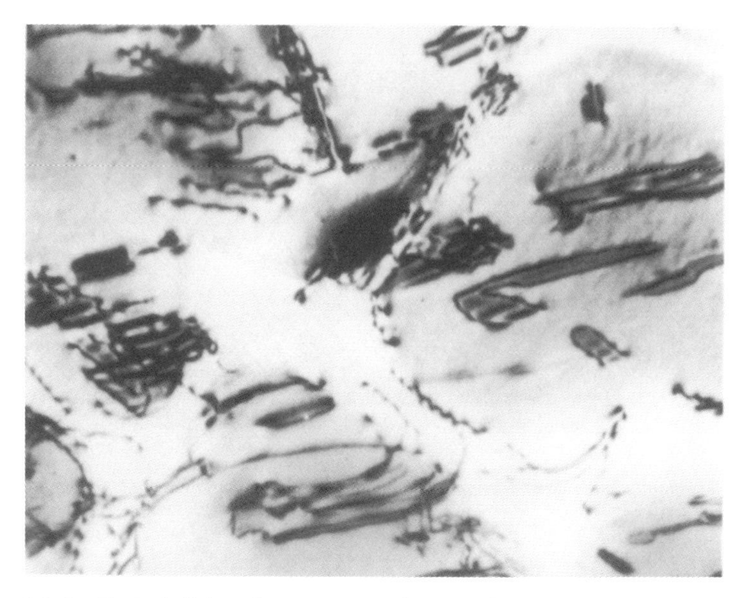

Figure 8.3.6 Typical dislocation structures from an interrupted fatigue test ($N = 6$) at 20°C. The test at 50% min^{-1} on a [100] specimen shows evidence of partial dislocation shearing and dislocation loops in the γ' phase. (After Milligan and Antolovich, 1989.)

stress increases as dislocations pile up until the resistance is overcome. Then a number of dislocations shear the γ' particle and produce a slip band in the γ' and audible sound. This proposal agrees with the slip bands in the γ' observed by Milligan and Antolovich (Figure 8.3.6).

The large amount of recovery observed in Figures 8.2.4a, d and 1.3.3 occurred only near the [001] orientation, where there is a potential for eight active slip systems. Dislocation tangles and pileups occur due to intersection of several slip planes. It is expected that these pileups relax during the hold period. High temperatures increase the rate of relaxation.

To summarize, deformation of single crystals occurs by slip in the octahedral and cube slip systems. The tension–compression asymmetry results from the propagation of partial dislocations that depend on four shear stress components: τ and τ_2 in the slip direction, and τ_1 and τ_3 normal to the slip direction. To use this information in constitutive modeling it is necessary to develop the equations for local stresses from the applied stress in a global or reference coordinate system. It will also be necessary to determine a representation for the inelastic strain resulting in the global coordinate system from the local slip in each slip system. This is the subject of Section 8.5.

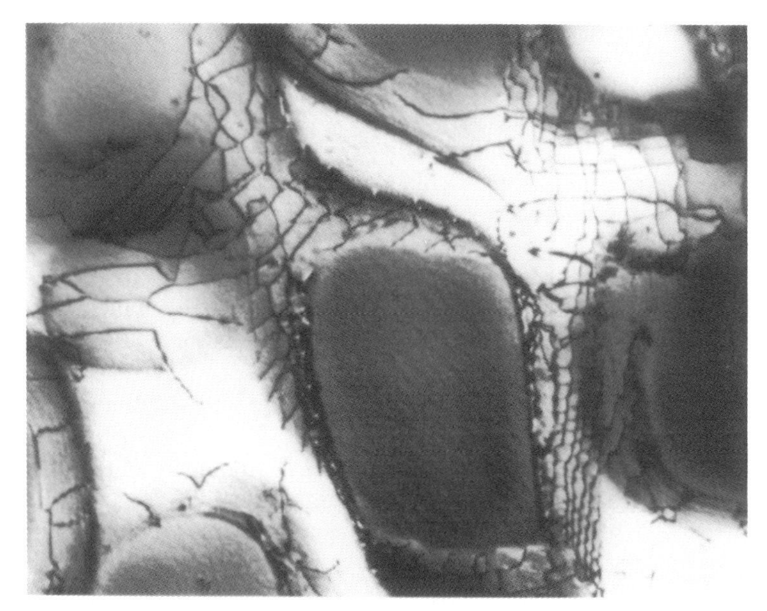

Figure 8.3.7 γ' bypass structures developed during a cyclic test ($N = 765$) at 1093°C on a [001] crystal 0.5% min^{-1} and a plastic strain range of 0.23%. (After Milligan and Antolovich, 1989.)

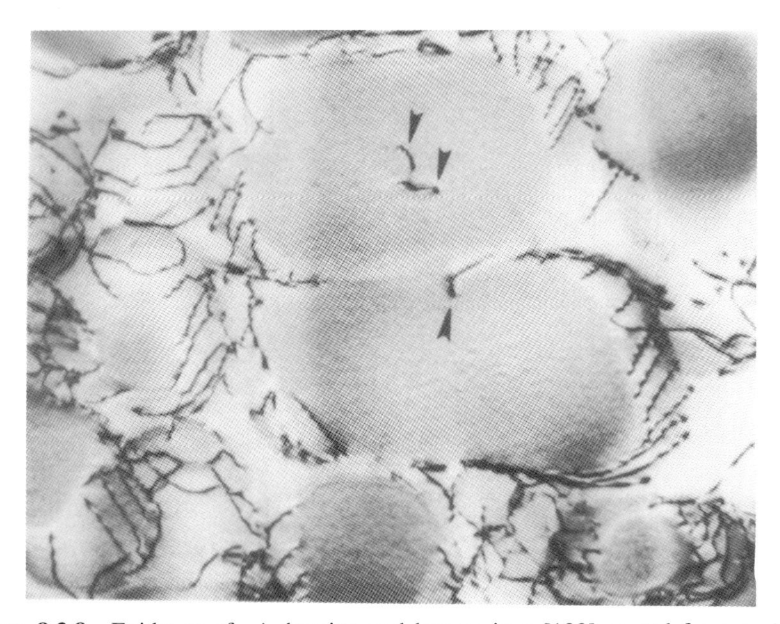

Figure 8.3.8 Evidence of γ' shearing and bypass in a [123] crystal from an interrupted tensile test at 927°C and 50% min^{-1}. (After Milligan and Antolovich, 1989.)

8.4 LATTICE ROTATION AND CREEP

The deformation discussed so far in this chapter has been primarily limited to small plastic deformations typical of low-cycle-fatigue applications. However, if the deformations are larger than 3 or 4%, as in creep and tensile loading, rotation of the crystallographic structure may occur during slip. The rotation occurs in single-slip orientations to accommodate the deformation and restraints at the boundary simultaneously. For example, consider the deformation of the single-crystal tensile specimen oriented in a single slip orientation as shown in Figure 8.4.1. The grips hold the ends of the specimen in a fixed alignment. During application of the load, slip will initiate on a number of parallel slip planes located close to the maximum shear stress, and slip bands will be observed on the surface of the specimen. As deformation progresses, continued slip would not be possible without rotation of the slip planes to accommodate the change in length of the specimen. Bending will occur near the grips to maintain continuity and alignment of the specimen. During elongation the slip planes rotate toward the tensile axis, and the angle between the slip plane and tensile axis decreases. As rotation continues a second slip plane will become oriented in a position to permit slip. When the stress on this plane reaches the critical resolved shear stress, slip will occur in a second slip system. Eventually, the slip on the primary and secondary planes will become equal and duplex slip will occur (identical slip in both systems).

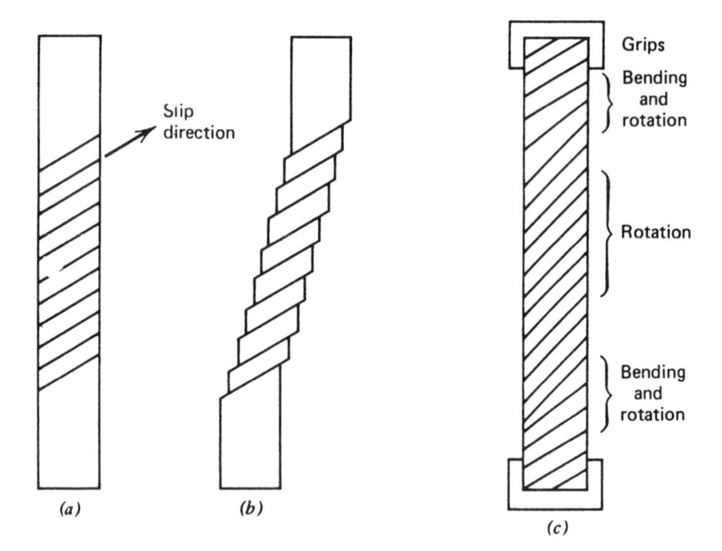

Figure 8.4.1 Schematic diagram of the deformation characteristics of a single-crystal alloy in a tensile testing machine. (After Verhoeven, Fundamentals of Physical Metallurgy, Copyright © 1975. Reprinted by permission of John Wiley & Sons, Inc.)

The rotation of a specimen in a single slip orientation can be plotted in a sterographic projection. Specimen D in Figure 8.4.2 has one active slip system in the $[\bar{1}01]$ direction for loading in the [123] as direction shown in Table 1.2.1 (note that orientation of the axes in Figure 8.4.2 and Table 1.2.1 are different). The tensile force produces a rotation that moves the $[\bar{1}01]$ slip system vector toward the tensile axis at point *D*, and the angle λ decreases in Figure 8.4.2*a*. However, instead of rotating the sterographic projection toward point *D*, it is more convenient to leave the sterographic projection fixed and rotate point *D* toward the $[\bar{1}01]$ slip system. Using this approach, when point *D* reaches point 2, two slip systems ($[111][\bar{1}01]$ and $[111][01\bar{1}]$) will have the same stress and duplex slip will result. From point *E* the specimen will rotate along the $[001]$–$[\bar{1}11]$ boundary to $[\bar{1}12]$, the point midway between $[\bar{1}01]$ and $[011]$ in the $[1\bar{1}1]$ plane (the vector sum of the two slip systems). The specimen will remain at this orientation until failure. Figure 8.4.3*a* shows the directions of rotation along the $[001]$–$[\bar{1}11]$ boundary for a specimen starting in any orientation within the sterographic triangle. Crystals with an initial orientation on the perimeter of the sterographic triangle will exhibit duplex slip upon loading and rotate initially to the poles shown in Figure 8.4.3*a*.

The classic tensile response of a pure single-crystal specimen (not a two-phase alloy) oriented within the sterographic triangle is very different from the deformation of a specimen oriented on the perimeter. Figure 8.4.4 shows the typical response of a pure (single-phase) single-crystal alloy loaded in four different orientations. Specimens A, B, and C deform with multiple intersecting slip systems that develop dislocation intersections and tangles. As

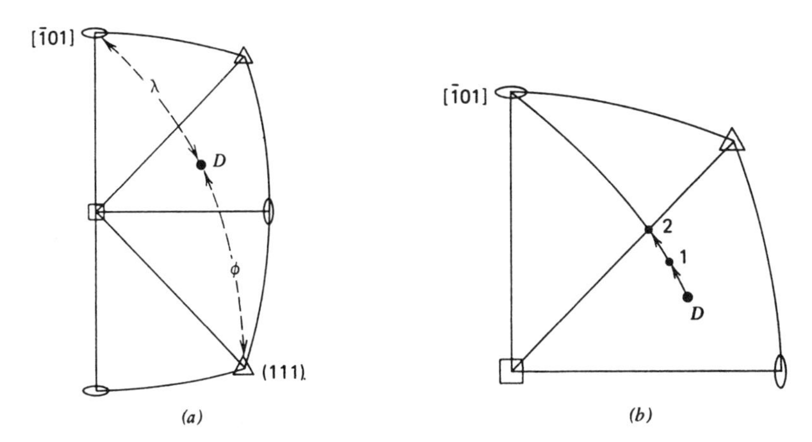

Figure 8.4.2 Stereographic projection showing the rotation of a tensile specimen loaded in a single slip orientation. The angles ϕ and λ define the slip plane and slip direction, respectively, in Schmid's law (equation 8.1.1). (After Verhoeven, Fundamentals of Physical Metallurgy, Copyright © 1975. Reprinted by permission of John Wiley & Sons, Inc.)

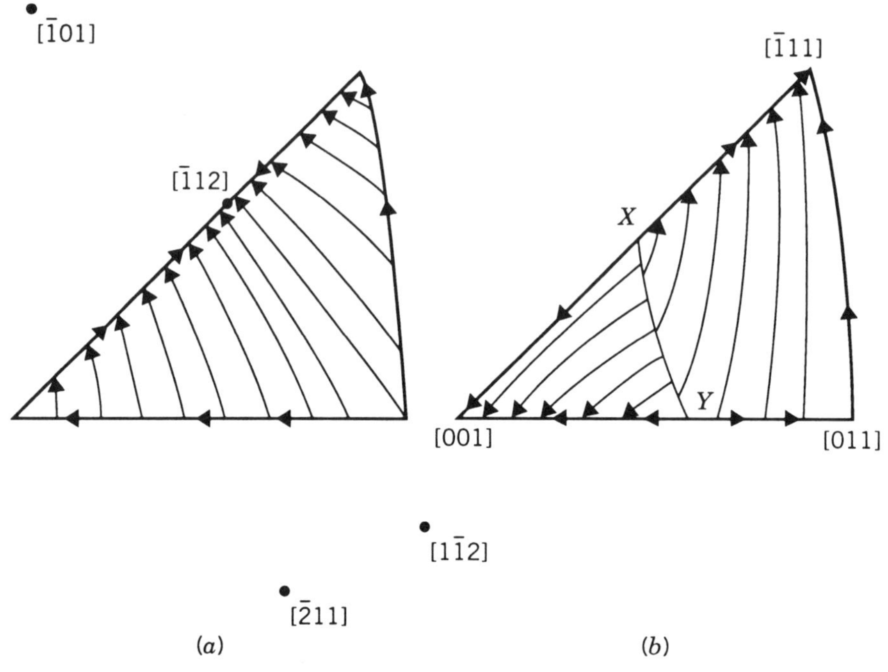

Figure 8.4.3 Directions of rotation for crystals load in single-slip orientations in the standard [001] stereographic triangle for the (*a*) [111][$\bar{1}$11] and (*b*) [111][112] slip systems. (From MacKay and Maier, 1982; reproduced with permission of The Minerals, Metals & Materials Society.)

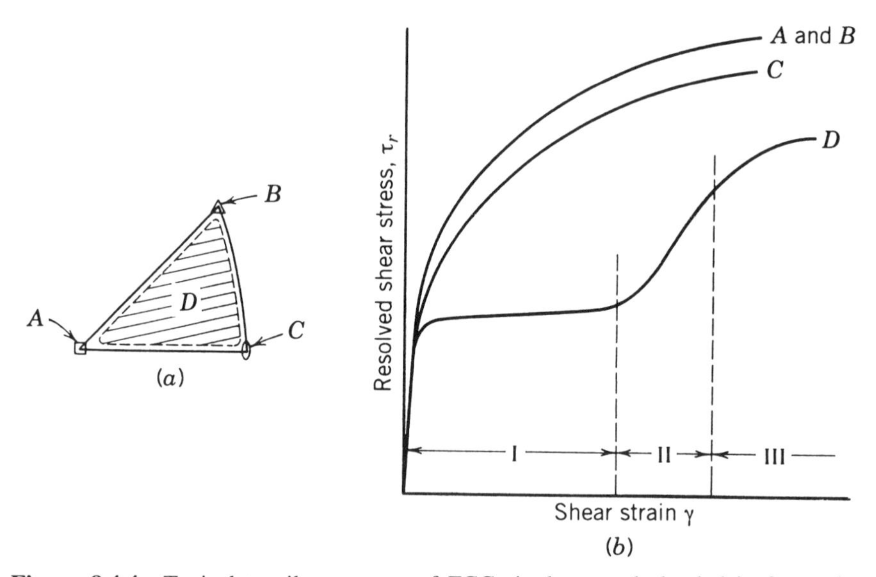

Figure 8.4.4 Typical tensile response of FCC single crystals loaded in four orientations. (After Verhoeven, Fundamentals of Physical Metallurgy, Copyright © 1975. Reprinted by permission of John Wiley and Sons, Inc.)

a result, the specimens experience considerable strain hardening. Specimen D, which is located within the sterographic triangle, deforms initially by single slip or "easy glide". There is very little hardening for specimen D until the end of stage I and the onset of a second slip system and rapid hardening (defined as stage II). Stage III corresponds to development of steady-state response and balance between the rates of strain hardening and dynamic recovery. It is important to note that the response of a pure metal is not the same as that of a nickel-base superalloy. Comparing the yield stresses of the specimens in the [011], [012], [$\bar{1}$23], and [$\bar{1}$11] orientations in Figure 8.2.4 shows that the classic single-slip response shown in Figure 8.4.4 is not present. Finally, notice how the choice of multiple slip orientations eliminated the effect of rotation in the tensile tests reported by Miner (1986a) (Figure 8.2.5).

The effect of rotation on the creep of single-crystal alloys makes the response much different from that of polycrystalline alloys presented in Chapter 3. The creep response is a strong function of the orientation, stress, and temperature. For example, Figure 8.4.5 shows the creep–rupture response of Mar M247 at 774°C and 724 MPa. Specimen I is near the [011] pole, and the other four specimens are near the [001] pole. All specimens are within the sterographic triangle and deform initially by a single-slip system. It is clear that the amount of primary strain decreases with increasing proximity to the [001] pole.

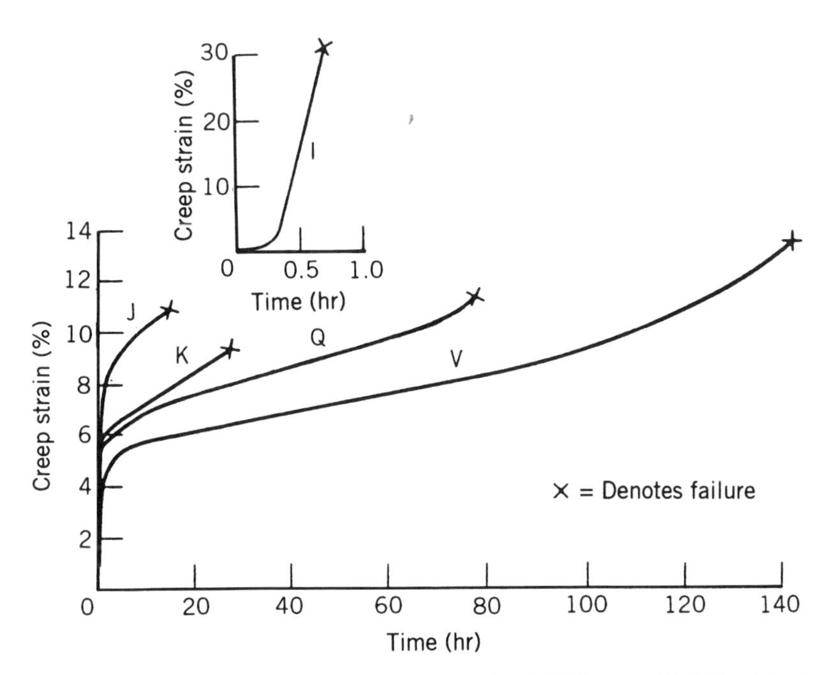

Figure 8.4.5 Creep curves for Mar M-247 tested at 774°C and 724 MPa. The letter associated with each curve is the specimen designation. (From MacKay and Maier, 1982; reproduced with permission of The Minerals, Metals & Materials Society.)

Recall that creep deformation at the higher stress levels but below the yield stress can occur by dislocation creep and dislocation glide. Bulk diffusion (Nabarro–Herring creep) and grain boundary diffusion (Coble creep) occur at lower values of stress creep, as shown in Section 3.2. Further, since there are no grain boundaries, Coble creep does not exist for single-crystal alloys and the deformation map is somewhat similar to Figure 3.2.6*b*. Single-crystal creep is more complicated because the active slip system for dislocation creep or dislocation glide can change during creep or be different for different alloys. Kear, Leverant and Olyak (1969), Leverant and Kear (1970), Leverant, Kear and Olyak (1973) and MacKay and Maier (1982) reported for Mar M200 and Mar M247 that primary creep and secondary creep occur in different slip systems. In particular, for temperatures near 750°C:

1. Primary creep occurs by deformation in the slip systems similar to [111][112]. This is the second octahedral slip shown in Figure 1.2.8 and corresponds to slip numbers 13 to 24 in Tables 1.2.1 and 1.2.2.
2. Secondary creep occurs by deformation in the slip systems similar [111][101].

However, Hopgood and Martin (1986) reported for SRR99 at 750 and 800°C that:

1. Both primary and secondary creep occurred by deformation in the slip systems similar to [111][112].
2. Deformation in the cube system has not been reported for creep near the [111] orientation.

Thus it appears that primary creep occurs in slip systems similar to [111][112], but the active slip system for secondary creep may be an alloy-specific property.

Lattice rotations for the [111][112] slip system are shown in Figure 8.4.3*b*. The boundary X–Y is defined by the Schmid factor $\cos\lambda \cos\phi = 0.427$ in equation 8.3.1. For the [111][112] slip system, duplex slip only occurs along the [001]–Y and X–[$\bar{1}$11] boundaries of the stereographic triangle. Lattice rotations were determined for selected specimens in the MacKay and Maier study. The solid dots in Figure 8.4.6 indicate the initial orientation, and the crosses represent the orientation at the end of the test. Their tests showed that in duplex creep the results followed the same order as the Schmid factor: stress in the slip plane in the slip direction. Specimens in single-slip creep exhibited the largest primary creep, and the amount of primary creep was ordered with the amount of rotation. Secondary creep did not occur until the onset of duplex deformation. In several studies on various nickel-base alloys, primary creep ranged from 6 to 10% for orientations within 5 to 10° of the [001] pole.

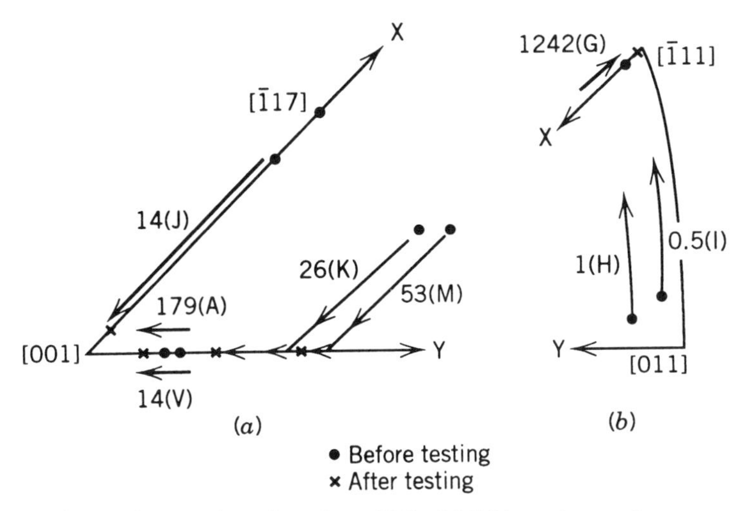

Figure 8.4.6 Lattice rotations for selected Mar M-247 specimens that were tested at 774°C and 724 MPa. Arrows show the directions of rotation, and the numbers and letters associated with each orientation are the rupture time in hours and specimen designation in Figure 8.4.5, respectively. (From MacKay and Maier, 1982; reproduced with permission of The Minerals, Metals & Materials Society.)

Nickel-base single crystals also exhibit a delay in the onset of creep for small values of stress. Pollock and Argon (1991) conducted a study of the creep resistance of CMSX3 in the [001] orientation at temperatures between 800 and 900°C and loads between 450 and 600 MPa. Figure 8.4.7 shows the primary creep transients of virgin crystals at three temperatures and one load. There was no instantaneous straining on loading (elastic strains are subtracted out), and delays of 1.6×10^4 and 600 sec were recorded for the tests at 800 and 850°C, respectively. A delay was reported for specimen I in Figure 8.4.5. No delay was observed at 900°C, and lower stresses were found to produce longer incubation periods in the onset of primary creep in Figure 8.4.7. To document the process during the incubation period, a creep test was terminated during the delay period at 4.5×10^3 sec. A typical micrograph (Figure 8.4.8) shows that dislocations spread out to dislocation-free regions from initial grown-in dislocation tangles that are formed during casting. Primary creep is assumed to be initiated when an active deformation system is achieved by the meeting of dislocation loops.

Before closing this section it is worthwhile briefly to discuss two closely related effects, overshooting and latent hardening, that are present in pure crystals but to our knowledge have not been reported for γ–γ' nickel-base alloys. An extensive study of single-slip system rotation was carried out by Bell and Green (1967) on aluminum crystals of 99.47% purity. They examined 168 specimens for the rotation of the tensile axis using X-ray diffraction. Their results showed overshooting of the tensile axis past the [001]–[$\bar{1}$11]

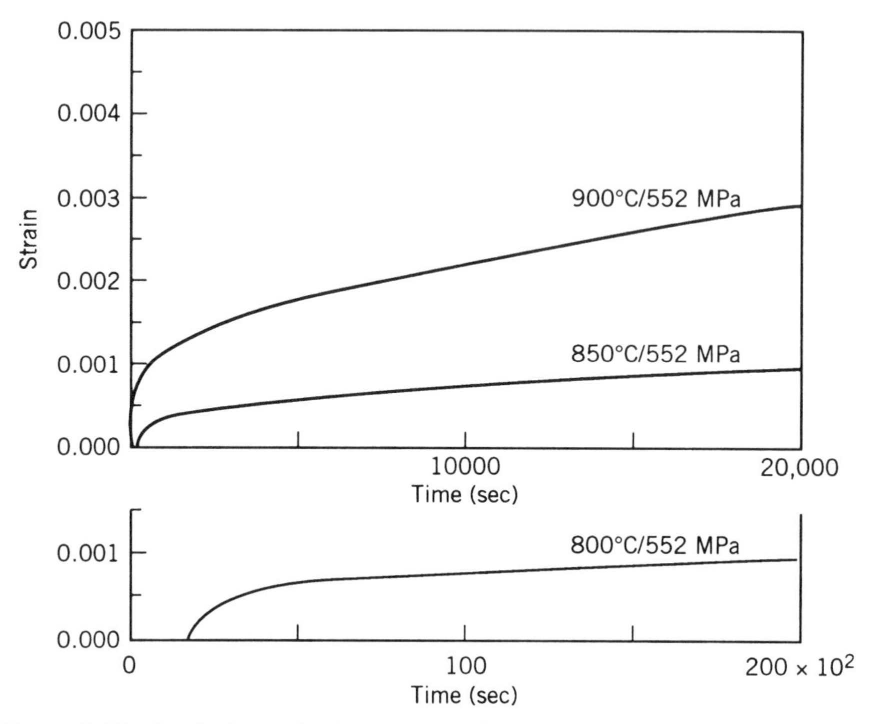

Figure 8.4.7 Incubation and primary creep of CMSX-3 in the [001] orientation for three temperatures. (From Pollock and Argon, 1991; reproduced with permission of The Minerals, Metals & Materials Society.)

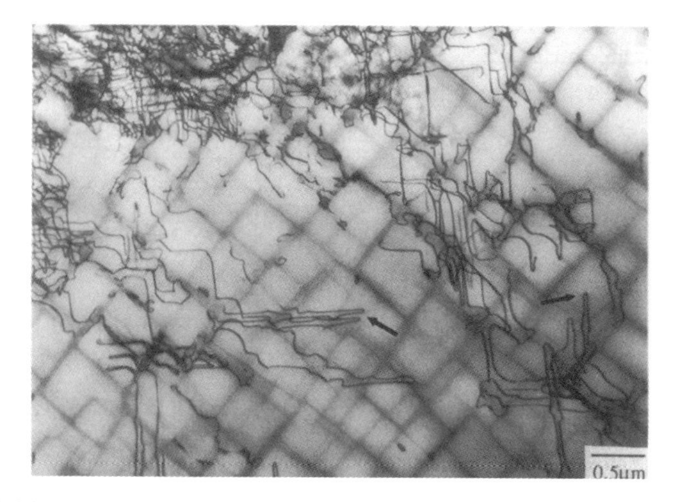

Figure 8.4.8 Dislocations spreading out from an initial grown-in tangle during the incubation period. The tangle is in the upper left corner and the arrows denote the expanding loop. (From Pollock and Argon, 1991; reproduced with permission of The Minerals, Metals & Materials Society.)

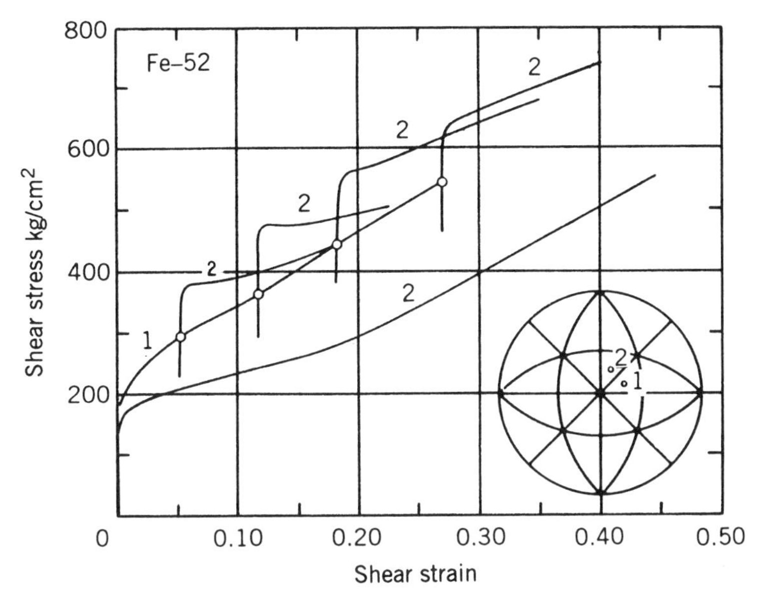

Figure 8.4.9 Effect of strain on the latent hardening in FE 52 at 77 K. The virgin response is shown in the long curves 1 and 2. The short curves 2 show the latent hardening after various amounts prestrain in the first orientation. (Reprinted from Acta Metallurgica, V 14.8, "Latent Hardening of Iron Single Crystals", pg. 961, Copyright 1966, with kind permission from Elsevier Science Ltd, The Boulevard, Langford Lane, Kidlington OX5 1GB, UK.)

duplex slip line in several tests. The generally accepted reason for the overshoot is latent hardening. To detect latent hardening, a large specimen is axially deformed in a single-slip orientation to a predetermined strain. After unloading, several small specimens were cut from the large prestrained crystal in other (latent) single-slip orientations. The specimens were then reloaded so that deformation occurs in the latent slip systems. Figure 8.4.9 shows the results of a well-known latent hardening test on Fe-52 at room temperature obtained by Nakada and Keh (1966). The experiments show that there is a significant increase in the resistance to slip in latent slip systems, due to the dislocation network in the original slip system. This creates slip systems with different properties so that duplex slip (equal slip rates in two slip systems) is not achieved on the $[001]$–$[\bar{1}11]$ boundary and overshoot follows. This effect is not present in γ–γ' alloys because the hardening created by the dislocation network is small compared to the initial hardening created by the γ' phase.

8.5 KINEMATIC EQUATIONS FOR CRYSTALLOGRAPHIC SLIP

It is necessary to develop the equations for local shear stresses and local slip in each slip system for use in constitutive modeling. Consider first a method to determine the shear stress components on octahedral and cube planes from the data given in Table 1.2.2. The stress vector **t** acting on some plane with an outward normal vector **n** is given by

$$\begin{bmatrix} t_1 \\ t_2 \\ t_3 \end{bmatrix} = \begin{bmatrix} \sigma_{11} & \sigma_{12} & \sigma_{13} \\ \sigma_{21} & \sigma_{22} & \sigma_{23} \\ \sigma_{31} & \sigma_{32} & \sigma_{33} \end{bmatrix} \begin{bmatrix} n_1 \\ n_2 \\ n_3 \end{bmatrix} \tag{8.5.1}$$

where σ_{ij} are the stress components in a principal coordinate of the material (parallel to the right-hand coordinate system in the [100], [010], and [001] orientations) at the point in question. The component of the stress, τ, in some direction **p** is then calculated from

$$\tau = [t_1 \quad t_2 \quad t_3] \cdot [p_1 \quad p_2 \quad p_3] \tag{8.5.2}$$

where **p** and **n** are unit vectors. The local shear stresses can be determined from equations 8.5.1 and 8.5.2 for each of the 30 slip numbers given in Table 1.2.2 and shown graphically in Figures 1.2.7 and 1.2.8. The result for the primary octahedral slip systems similar to [111][$\bar{1}$01] is

$$\begin{bmatrix} \tau_1^{11} \\ \tau_1^{12} \\ \tau_1^{13} \\ \tau_1^{21} \\ \tau_1^{22} \\ \tau_1^{23} \\ \tau_1^{31} \\ \tau_1^{32} \\ \tau_1^{33} \\ \tau_1^{41} \\ \tau_1^{42} \\ \tau_1^{43} \end{bmatrix} = \begin{bmatrix} \tau^{1} \\ \tau^{2} \\ \tau^{3} \\ \tau^{4} \\ \tau^{5} \\ \tau^{6} \\ \tau^{7} \\ \tau^{8} \\ \tau^{9} \\ \tau^{10} \\ \tau^{11} \\ \tau^{12} \end{bmatrix} = \frac{1}{\sqrt{6}} \begin{bmatrix} 1 & 0 & -1 & 1 & 0 & -1 \\ 0 & -1 & 1 & -1 & 1 & 0 \\ 1 & -1 & 0 & 0 & 1 & -1 \\ -1 & 0 & 1 & 1 & 0 & -1 \\ -1 & 1 & 0 & 0 & -1 & -1 \\ 0 & 1 & -1 & -1 & -1 & 0 \\ 1 & -1 & 0 & 0 & -1 & -1 \\ 0 & 1 & -1 & -1 & 1 & 0 \\ 1 & 0 & -1 & -1 & 0 & -1 \\ 0 & -1 & 1 & -1 & -1 & 0 \\ -1 & 0 & 1 & -1 & 0 & -1 \\ -1 & 1 & 0 & 0 & 1 & -1 \end{bmatrix} \begin{bmatrix} \sigma_{11} \\ \sigma_{22} \\ \sigma_{33} \\ \sigma_{12} \\ \sigma_{31} \\ \sigma_{23} \end{bmatrix} \tag{8.5.3}$$

The shear stresses in the slip systems similar to the secondary octahedral [111][112] are given by

$$\begin{bmatrix} \tau_2^{11} \\ \tau_2^{12} \\ \tau_2^{13} \\ \tau_2^{21} \\ \tau_2^{22} \\ \tau_2^{23} \\ \tau_2^{31} \\ \tau_2^{32} \\ \tau_2^{33} \\ \tau_2^{41} \\ \tau_2^{42} \\ \tau_2^{43} \end{bmatrix} = \begin{bmatrix} \tau^{13} \\ \tau^{14} \\ \tau^{15} \\ \tau^{16} \\ \tau^{17} \\ \tau^{18} \\ \tau^{19} \\ \tau^{20} \\ \tau^{21} \\ \tau^{22} \\ \tau^{23} \\ \tau^{24} \end{bmatrix} = \frac{1}{3\sqrt{2}} \begin{bmatrix} -1 & 2 & -1 & 1 & -2 & 1 \\ 2 & -1 & -1 & 1 & 1 & -2 \\ -1 & -1 & 2 & -2 & 1 & 1 \\ -1 & 2 & -1 & -1 & -2 & -1 \\ -1 & -1 & 2 & 2 & 1 & -1 \\ 2 & -1 & -1 & -1 & 1 & 2 \\ -1 & -1 & 2 & 2 & -1 & 1 \\ 2 & -1 & -1 & -1 & -1 & -2 \\ -1 & 2 & -1 & -1 & 2 & 1 \\ 2 & -1 & -1 & 1 & -1 & 2 \\ -1 & 2 & -1 & 1 & 2 & -1 \\ -1 & -1 & 2 & -2 & -1 & -1 \end{bmatrix} \begin{bmatrix} \sigma_{11} \\ \sigma_{22} \\ \sigma_{33} \\ \sigma_{12} \\ \sigma_{31} \\ \sigma_{23} \end{bmatrix} \tag{8.5.4}$$

and the stresses in the cube slip systems similar to [001][011] are

$$\begin{bmatrix} \tau_3^{11} \\ \tau_3^{12} \\ \tau_3^{21} \\ \tau_3^{22} \\ \tau_3^{31} \\ \tau_3^{32} \end{bmatrix} = \begin{bmatrix} \tau^{25} \\ \tau^{26} \\ \tau^{27} \\ \tau^{28} \\ \tau^{29} \\ \tau^{30} \end{bmatrix} = \begin{bmatrix} 0 & 0 & 0 & 1 & 1 & 0 \\ 0 & 0 & 0 & 1 & -1 & 0 \\ 0 & 0 & 0 & 1 & 0 & 1 \\ 0 & 0 & 0 & 1 & 0 & -1 \\ 0 & 0 & 0 & 0 & 1 & 1 \\ 0 & 0 & 0 & 0 & -1 & 1 \end{bmatrix} \begin{bmatrix} \sigma_{11} \\ \sigma_{22} \\ \sigma_{33} \\ \sigma_{12} \\ \sigma_{31} \\ \sigma_{23} \end{bmatrix} \tag{8.5.5}$$

The final step is to develop a correlation between stress in the slip direction, τ, and the stresses τ_1, τ_2, and τ_3 that control the core width of a partial dislocation in the octahedral slip system. Consider for example the stresses in the $[111][10\bar{1}]$ slip system or slip number 1 in Table 1.2.2. Recall that τ_1 is normal to τ in the slip plane; thus τ_1 can be identified as τ_{13} from Figure 1.2.7 and Table 1.2.2. The stress τ_2 is parallel to τ_1 or $[10\bar{1}]$ on an intersection cube plane; thus τ_2 is τ_{28} from Table 1.2.2. The stress τ_3 is normal to τ in the intersecting octahedral plane; thus from Figure 1.2.7, τ_3 is system 1 on plane 2 or τ_{16} in Table 1.2.2. The same approach is used to determine the remaining correlations given in Table 8.5.1.

The inelastic strain in the principal coordinates of the material will be developed next from the slip in each slip system. The local slip is equivalent to the local engineering shear strain. Let γ represent the local engineering shear strain in the right-hand coordinate system defined by the unit normal vector **n**, the unit slip direction vector, **p**, and $\mathbf{s} = \mathbf{n} \times \mathbf{p}$, where **s** is the vector parallel to the slip systems, similar to [111][112] defined in Table 1.2.2. Applying the second-order tensor transformation rule gives the global tensor strain as a function of the local slip in the form

Table 8.5.1 Correlation Among the 12 Shear Stresses τ in the Octahedral Slip System and the Stress Components τ_1, τ_2, and τ_3 That Control the Core Width of a Partial Dislocation

τ	τ_1	τ_2	τ_3
1	13	28	16
2	14	26	20
3	15	30	24
4	16	28	13
5	17	29	19
6	18	25	22
7	19	29	17
8	20	26	14
9	21	27	23
10	22	25	18
11	23	27	21
12	24	30	15

$$[\varepsilon] = \frac{1}{2}\lambda \begin{bmatrix} n_1 & n_2 & n_3 \\ p_1 & p_2 & p_3 \\ s_1 & s_2 & s_3 \end{bmatrix}^T \begin{bmatrix} 0 & 1 & 0 \\ 1 & 0 & 0 \\ 0 & 0 & 0 \end{bmatrix} \begin{bmatrix} n_1 & n_2 & n_3 \\ p_1 & p_2 & p_3 \\ s_1 & s_2 & s_3 \end{bmatrix} \tag{8.5.6}$$

Calculation of the inelastic strain, ε_{ij}^{11}, in the principal coordinates of the material due to the slip in the 12 primary octahedral slip directions gives

$$\frac{12}{\sqrt{6}} \begin{bmatrix} \varepsilon_{11}^{11} & \varepsilon_{12}^{11} & \varepsilon_{13}^{11} \\ \varepsilon_{21}^{11} & \varepsilon_{22}^{11} & \varepsilon_{23}^{11} \\ \varepsilon_{32}^{11} & \varepsilon_{32}^{11} & \varepsilon_{33}^{11} \end{bmatrix} =$$

$$+ \gamma_{11}^1 \begin{bmatrix} 2 & 1 & 0 \\ 1 & 0 & -1 \\ 1 & -1 & -2 \end{bmatrix} + \gamma_{12}^1 \begin{bmatrix} 0 & -1 & 1 \\ -1 & -2 & 0 \\ 1 & 0 & 2 \end{bmatrix} + \gamma_{13}^1 \begin{bmatrix} 2 & 0 & 1 \\ 0 & -2 & 1 \\ 1 & -1 & 0 \end{bmatrix}$$

$$+ \gamma_{21}^1 \begin{bmatrix} -2 & 1 & 0 \\ 1 & 0 & -1 \\ 0 & -1 & 2 \end{bmatrix} + \gamma_{22}^1 \begin{bmatrix} -2 & 0 & -1 \\ 0 & -2 & -1 \\ -1 & -1 & 0 \end{bmatrix} + \gamma_{23}^1 \begin{bmatrix} 0 & -1 & -1 \\ -1 & 2 & 0 \\ -1 & 0 & -2 \end{bmatrix}$$

$$+ \gamma_{31}^1 \begin{bmatrix} 2 & 0 & -1 \\ 0 & -2 & -1 \\ -1 & -1 & 0 \end{bmatrix} + \gamma_{32}^1 \begin{bmatrix} 0 & -1 & 1 \\ -1 & 2 & 0 \\ 1 & 0 & -2 \end{bmatrix} + \gamma_{33}^1 \begin{bmatrix} 2 & -1 & 0 \\ -1 & 0 & -1 \\ 0 & -1 & -2 \end{bmatrix}$$

$$+ \gamma_{41}^1 \begin{bmatrix} 0 & -1 & -1 \\ -1 & -2 & 0 \\ -1 & 0 & 2 \end{bmatrix} + \gamma_{42}^1 \begin{bmatrix} -2 & -1 & 0 \\ -1 & 0 & -1 \\ 0 & -1 & 2 \end{bmatrix} + \gamma_{43}^1 \begin{bmatrix} -2 & 0 & 1 \\ 0 & 2 & -1 \\ 1 & -1 & 0 \end{bmatrix} \tag{8.5.7}$$

and the strain, ε_{ij}^{13}, for the cube slip system can be written as

$$\frac{4}{\sqrt{2}} \begin{bmatrix} \varepsilon_{11}^{13} & \varepsilon_{12}^{13} & \varepsilon_{13}^{13} \\ \varepsilon_{21}^{13} & \varepsilon_{22}^{13} & \varepsilon_{23}^{13} \\ \varepsilon_{31}^{13} & \varepsilon_{32}^{13} & \varepsilon_{33}^{13} \end{bmatrix} =$$

$$\gamma_{11}^3 \begin{bmatrix} 0 & 1 & 1 \\ 1 & 0 & 0 \\ 1 & 0 & 0 \end{bmatrix} + \gamma_{12}^3 \begin{bmatrix} 0 & 1 & -1 \\ 1 & 0 & 0 \\ -1 & 0 & 0 \end{bmatrix} + \gamma_{21}^3 \begin{bmatrix} 0 & 1 & 0 \\ 1 & 0 & 1 \\ 0 & 1 & 0 \end{bmatrix}$$

$$+ \gamma_{22}^3 \begin{bmatrix} 0 & 1 & 0 \\ 1 & 0 & -1 \\ 0 & -1 & 0 \end{bmatrix} + \gamma_{31}^3 \begin{bmatrix} 0 & 0 & 1 \\ 0 & 0 & 1 \\ 1 & 1 & 0 \end{bmatrix} + \gamma_{32}^3 \begin{bmatrix} 0 & 0 & -1 \\ 0 & 0 & 1 \\ -1 & 1 & 0 \end{bmatrix} \tag{8.5.8}$$

As a check of the calculations, notice that the inelastic strains given by equations 8.5.7 and 8.5.8 are symmetric.

8.6 ASSUMPTIONS FOR MODELING

Before proceeding to develop a constitutive equation for the mechanical response of nickel-base single-crystal superalloys, it is important to restate that the use of these alloys in structural applications is relatively new. As a result, modeling is a new, contemporary topic. There is no consensus on the best approach or even all the physical effects that may be present under different loading conditions. The underlying deformation mechanisms are not fully understood. Thus the following results should be viewed only as an interim report on an evolving technology.

As a first step, consider a review of the key ideas that will used to develop a model for René N4, a two phase nickel base alloy:

1. Only octahedral slip is observed at temperatures below the critical temperature, about 700°C, for loading above 10^{-4} sec. At temperatures above the critical temperature deformation occurs by both octahedral and cube slip. Cube slip dominates near the [111] orientation and octahedral slip dominates near the [001] and [011] orientations.
2. Slip, or plastic flow, occurs when the shear stress on the slip plane in the slip direction τ exceeds the critical shear stress. The tension–compression anisotropy occurs because the stresses normal to the slip direction on the slip plane and intersecting slip plane, τ_1 and τ_2, respectively, extend and contract the core width of partial dislocations. Small core widths increase the potential for cross slip. The shear stress on the cube plane parallel to an active slip system, τ_3, also promotes cross slip at higher temperatures.
3. The small strain tensile response in curves in Figure 8.2.4 shows the presence of strain recovery; thus a back-stress variable should be included as part of the model. The variation in recovery with orientation shows that the effect of back stress will be different in the octahedral and cube slip systems.
4. The flat tensile curves show that there must be a balance between strain hardening and dynamic recovery. The strain-rate sensitivity occurs from a change in the balance between the hardening rate and recovery rate.
5. The cyclic hardening observed in Figure 8.2.7 suggests the need for a drag stress variable to model cyclic hardening that is different from strain hardening. However, some two-phase superalloys such as René N4 are nearly stable in cyclic testing, so only the initial value of the drag stress may be required.
6. Crystal lattice rotation occurs above a few percent strain in single-slip-system orientations. Crystal lattice rotation effects are expected to be present in tensile and creep response beyond 3 to 5% strain.
7. The inelastic or plastic strain or strain rate is obtained by summing the slip or slip rate of the individual slip systems. The strain in the principal

coordinates of the material for the octahedral and cubic slip systems is given in Section 8.5.

In addition to the foregoing observations, a constitutive model is constructed with the following limitations:

1. No attempt is made to model the slip bursts observed in Figure 8.2.4. Only the shape of the response curve will be modeled.
2. The strain range, consistent with low cycle fatigue, is limited to a few percent; therefore, lattice rotation effects are neglected. The orientation constants in the equations for the local stresses and inelastic strains (Section 8.5) are not valid if the crystallographic planes rotate.
3. Creep modeling is not included. At present there is no consensus of what the active slips systems are for secondary creep; thus it is difficult to establish a physically based model for creep.

The constitutive model is based on a unified strain approach where the total strain is decomposed into elastic, inelastic, and thermal components (equation 6.1.1). The coordinate system in equation 6.1.1 is assumed to be in the principal coordinates of the material. For FCC crystals the principal directions are along the cubic edges, or in the [100], [010], and [001] directions. Thus the load or strain at a point in a structure must first be transformed to the principal coordinates of the material before using the equations developed in this section and Section 8.5. The elastic strain in the principal directions is given by equation 4.5.10. Recall for cubic symmetry that the elastic modulus, shear modulus, and Poisson's ratio are independent. A method for determining the elastic constants was presented in Section 4.5.1. The thermal strain in the principal-coordinate system is isotropic; therefore, the thermal strain components can be determined from the thermal strain terms in equation 4.7.8. Equations 8.5.7 and 8.5.8 can be used to determine the total inelastic strain rate as

$$\dot{\varepsilon}_{ij}^{\mathrm{I}} = \dot{\varepsilon}_{ij}^{\mathrm{I1}} + \dot{\varepsilon}_{ij}^{\mathrm{I3}} \tag{8.6.1}$$

for the octahedral and cube slip systems acting simultaneously. This approach will require two separate flow equations, one for each type of slip system. Recall that partial dislocations are present in octahedral slip but not in cube slip. The parameters must be determined such that the correct flow equation will be operative under the correct conditions, as summarized above. The accumulated inelastic strain is determined by integrating the inelastic strain rate. The modeling task therefore reduces to determining the flow and state variable evolution equations for the octahedral and cube slip systems, incorporating the tension–compression asymmetry, and evaluating the material parameters.

8.7 MODELING OF SINGLE CRYSTAL ALLOYS

Several of the polycrystalline materials discussed in Chapters 6 and 7 have chemical compositions rather similar to the γ–γ' single crystal superalloys alloys and exhibit many similar time-dependent features. The state-variable equations discussed in Chapter 7 were successful in modeling the multiaxial response of nickel-base superalloys. Thus it is reasonable to use the same type of equation for the slip rate in the octahedral and cube slip systems. However, recall that the stresses in all three slip systems will be required to model the core width effect.

Let us define the following indical notation for the three slip systems:

$k = 1$ defines slip systems similar to [111][101] and $\alpha = 1,4$ and $\beta = 1,3$.
$k = 2$ defines slip systems similar to [111][112] and $\alpha = 1,4$ and $\beta = 1,3$.
$k = 3$ defines slip systems similar to [001][011] and $\alpha = 1,3$ and $\beta = 1,2$.

The flow equations for local slip in the octahedral and cube slip systems $\dot{\gamma}^{\mathrm{I}k}_{\alpha\beta}$ $(k \neq 2)$ are assumed to be similar to equation 6.2.13. Rewriting this result in the notation above gives

$$\dot{\gamma}^{\mathrm{I}k}_{\alpha\beta} = D_k \exp\left[-\frac{1}{2}\left(\frac{Z_k^{\alpha\beta}}{|\tau_k^{\alpha\beta} - \Omega_k^{\alpha\beta}|}\right)^{n_{k1}}\right]\frac{\tau_k^{\alpha\beta} - \Omega_k^{\alpha\beta}}{|\tau_k^{\alpha\beta} - \Omega_k^{\alpha\beta}|} \tag{8.7.1}$$

where $Z_k^{\alpha\beta}$ and $\Omega_k^{\alpha\beta}$ are the drag stress and back stress, respectively, associated with the slip rate in each slip system. The shear stresses in two active slip systems are denoted by $\tau_k^{\alpha\beta}$, and the difference between the shear stress and back stress, $\tau_k^{\alpha\beta} - \Omega_k^{\alpha\beta}$, is the net stress required to produce slip. The parameters D_k are scale factors that will be used to separate the high- and low-strain-rate response, and n_{k1} controls the strain-rate sensitivity. Finally the factor $2/\sqrt{3}$ is not necessary for this application.

The back-stress evolution equations discussed in Chapter 6 included strain-hardening and dynamic recovery terms. Sheh and Stouffer (1988) found that equation 6.3.7 satisfied the saturation condition and was more flexible than equation 6.3.6 for modeling the response of the three types of slip systems. Rewriting equation 6.3.7 in the notation above gives

$$\dot{\Omega}_k^{\alpha\beta} = F_k|\dot{\gamma}_k^{\alpha\beta}|^{n_{k3}}\left[\mathrm{sgn}(\dot{\gamma}_k^{\alpha\beta}) - \frac{\Omega_k^{\alpha\beta}}{\Omega_{mk}}\right] + G_k\left(\frac{\tau_k^{\alpha\beta}}{\tau_0}\right)^{n_{k2}}\dot{\tau}_k^{\alpha\beta} \qquad (k \neq 2) \tag{8.7.2}$$

where F_k, G_k, n_{k2}, n_{k3}, are material parameters and Ω_{mk} is the maximum value of the back stress. The anelastic recovery is modeled primarily by the strain-rate term. If the G_k term is small relative to the F_k term, changes in the value of the back stress will be small during elastic loading and unloading. The recovery and softening were both significant during the 120-sec hold period for the [001] orientation in Figure 1.3.3. Thus there was a large change in the back stress during the inelastic strain recovery, and the contribution of the second term is relatively small in the [001] orientation. There is almost no softening and recovery in the [111] orientation in Figure 1.3.4. The exponents in equation 8.7.2 give more flexibility, to model the amount of recovery. The parameter $\tau_0 = 1$ MPa or ksi is included to normalize units of stress. The saturated back stress, Ω_{mk}, is assumed to be a constant for tensile and cyclic loading.

The drag stress is used to model the initial hardness through Z_0 and the cyclic hardening through equation 6.6.3. The initial value of the drag stress, Z_0, was chosen to model the tension compression anisotropy because the yield stress is a measure of the initial resistance to slip. Let the cyclic hardening component of the drag stress, designated as $H_k^{\alpha\beta}$, have the same form as equation 6.6.3. Using the slip system notation, the cyclic growth in drag stress can be rewritten as

$$H_k^{\alpha\beta} = H_{k2} - (H_{k1} - H_{k2}) \exp(-m_k W_k^{\alpha\beta}) \qquad (k \neq 2) \qquad (8.7.3)$$

where $W_k^{\alpha\beta} = \int \tau_k^{\alpha\beta} \, d\gamma_k^{\alpha\beta}$ is the accumulated inelastic work in each slip system. The parameters H_{k1} and H_{k2} are the initial and saturated values of the drag stress, respectively. Equation 8.7.3 is a measure of the resistance to slip in each slip system due to the initial hardness and the development of a dislocation network. Since the tension–compression asymmetry occurs in slip systems similar to [111][101], only the $H_1^{\alpha\beta}$ needs to be modified. The simplest approach is to assume a linear relation between $Z_1^{\alpha\beta}$ and $\tau_1^{\alpha\beta}$, $\tau_2^{\alpha\beta}$, and $\tau_3^{\alpha\beta}$; that is, let

$$\begin{aligned} Z_1^{\alpha\beta} &= H_1^{\alpha\beta} + V_1 \tau_1^{\alpha\beta} + V_2 |\tau_2^{\alpha\beta}| + V_3 \tau_3^{\alpha\beta} \\ Z_3^{\alpha\beta} &= H_3^{\alpha\beta} \end{aligned} \qquad (8.7.4)$$

The material parameters V_1, V_2, and V_3 must be determined to give the required asymmetry. From their definitions (Section 8.3), $\tau_1^{\alpha\beta}$ and $\tau_3^{\alpha\beta}$ extend or constrict the core width and control the ability of the dislocation to cross slip from the octahedral to cube plane. Thus when $\tau_1^{\alpha\beta}$ and $\tau_3^{\alpha\beta}$ are positive, the resistance to slip, $Z_1^{\alpha\beta}$, is increased, but the resistance to slip is decreased when $\tau_1^{\alpha\beta}$ and $\tau_3^{\alpha\beta}$ are negative. Thus the resistance to inelastic flow depends on both the magnitude and sign of $\tau_1^{\alpha\beta}$ and $\tau_3^{\alpha\beta}$. Recalling that $\tau_2^{\alpha\beta}$ is the thermally aided contribution of the cross slip, and the sign of $\tau_2^{\alpha\beta}$ is not relevant since the potential to cross slip is the same in tension and compression (see Figures 8.3.4 and 8.3.5). The parameters V_1, V_2, and V_3 become less important with increasing temperature above the critical temperature. At very high temperatures $Z_1^{\alpha\beta} \rightarrow H_1^{\alpha\beta}$ and the orientation dependence disappears as shown in Figure 8.2.3. The stresses $\tau_1^{\alpha\beta}$, $\tau_2^{\alpha\beta}$, and $\tau_3^{\alpha\beta}$ are defined in Table 8.5.1.

8.8 MATERIAL PARAMETERS AND CALCULATED RESULTS

The following general assumptions and limitiations are made to guide development of the material parameters:

1. The material constants are determined under isothermal conditions. Extension to thermomechanical response has not been attempted for single-crystal alloys.
2. The material parameter evaluation procedure is for strain-rate-controlled tensile and cyclic data.
3. Single-crystal response depends on strain rate and temperature. At a high strain rate below about 600°C deformation occurs in the 12 slip systems similar to [111][101], between about 600 and 850°C, deformation occurs simultaneously by octahedral slip and cube slip (six slip systems similar to [001][011]), and above 850°C orientation dependence becomes negligible with increasing temperature.
4. The octahedral slip constants are determined from tests were there is no cube slip. For example, data from tests in the [100] and [110] orientations can be used at any temperature to determine the octahedral constants. Conversely, the cube slip constants were determined from tests in the [111] orientation. The local stresses are given in Table 1.2.1 for a 1000-lb tensile load in the [100], [110], and [111] orientations.

Table 1.2.1 shows that tensile loading in the [100] orientation will produce equal stresses in eight slip systems. The shear stress in the slip plane producing slip is 0.408 of the applied tensile load. Similarly, there are assumed to be eight equal slip rates, $\dot{\gamma}$, that add up to the inelastic strain rate, $\dot{\varepsilon}^{\mathrm{I}}$, in the [100] orientation. The slip system indicies are dropped since there are eight equal shear stresses and slip rates in the [100] orientation. In this case equation 8.5.7 can be used to show that $\dot{\gamma} = (12/16\sqrt{6})\dot{\varepsilon}_{11}^{\mathrm{I}} \equiv 0.306\dot{\varepsilon}^{\mathrm{I}}$ after noting that stresses corresponding to the $\dot{\gamma}_{21}^{1}$, $\dot{\gamma}_{22}^{1}$, $\dot{\gamma}_{42}^{1}$, and $\dot{\gamma}_{43}^{1}$ slip systems are negative (see Table 1.2.1). Similar results can be established for the [110] (or [111]) orientation since there are four (or three) equal shear stresses and slip rates. Thus the local shear stresses and local slip rates for tensile, cyclic and creep tests in the [100], [110], and [111] orientations can be obtained from experimental data.

Since the equations are the same in both octahedral and cube slip systems, the discussion of determining parameters will be limited to the octahedral system as an example. This system includes the tension–compression asymmetry terms. The local slip equations are the same as the uniaxial equations in Chapter 6, and many of the results from Section 6.4 for strain rate sensitive materials can be applied since the temperatures in this example are 760 and 982°C. The method of determining parameters is divided into the following three sequential steps for evaluation of the (1) flow equation parameters, (2) back-stress parameters, and (3) drag stress parameters. The order of the steps must be preserved since the information flow in sequential.

Consider first the flow equation parameters. For uniaxial loading, the mapped local slip rate, $X(\dot{\gamma})$, in the saturated state can be determined from equation 6.4.2. Equation 6.4.2 can be rewritten as

$$\tau_s X^{1/2n} = H_{11} + V_1 \tau_{1s} + V_2 |\tau_{2s}| + V_3 \tau_{3s} + \Omega_m X_s^{1/2n} \tag{8.8.1}$$

after using equation 8.7.4a in Equation 8.7.1 with $k = 1$. Recall that the initial value of the drag stress in the octahedral slip system is H_{11}, and Ω_m is one of the eight equal values of the maximum saturated back stress in the [100] orientation. The choice of the parameter $D_1 = 10^4 \text{ sec}^{-1}$ is explained in Section 6.4. Assuming a value for n_{11} (the strain-rate-sensitivity exponent), equation 8.8.1 becomes a linear equation with five unknown parameters, H_{11}, V_1, V_2, V_3, and Ω_s. At least five sets of data pairs ($X_s^{1/2n}$, τ_s) are required to evaluate these parameters in a least-squares routine. These data can come from tension and compression tests at different strain rates in the [100] and [110] orientations. The iteration procedure in Section 6.4 can be used to determine the parameters. To summarize briefly, start by assuming a value for n_{11} ($0.1 > n_{11} > 3.0$) and calculate H_{11}, V_1, V_2, V_3, and Ω_{sat} from the data. Iterate until the correlation coefficient is close to 1.000.

The back-stress parameters in equation 8.7.2 are determined next. The parameters n_{12} and G_1 for the elastic component of the back-stress rate can be determined from equation 6.4.6 (and Figure 6.4.2) except that the plot is log-linear rather than linear. The parameters n_{13} and F_1 for the inelastic term can be determined from the back-stress rate in a tensile test. The flow equation parameters can be used with the stress–strain data from a test to calculate the back-stress history from equation 8.8.1 rewritten in the form

$$\Omega(t) = \tau(t) - \frac{1}{X^{1/2n}(t)} [H_{11} + V_1 \tau_1(t) + V_2 |\tau_2(t)| + V_3 \tau_3(t)] \tag{8.8.2}$$

as discussed in Section 6.4. The back-stress rate is then determined numerically using the postyield uniaxial test data. Rearranging equation 8.7.2 for $k = 1$ and $\tau_0 = 1$ MPa or ksi gives

$$\frac{\dot{\Omega} - G_1(\tau)^{n_{k2}}\, \dot{\tau}}{1 - \Omega/\Omega_{m1}} = F_1 |\dot{\gamma}|^{n_{13}} \tag{8.8.3}$$

The equation can be used with tensile data and/or recovery to determine n_{13} and F_1. An example of log-linear recovery during a 2-min hold period is shown in Figure 8.8.1. The parameters n_{13} and F_1 can be picked to model the data.

The response for René N4 was calculated using parameters that were determined as described above. All parameters are given in Appendix 8.1. In the course of the calculations it was found that the elastic back-stress term was negligible; thus G_1 was defined as zero, $G_1 \equiv 0$. Figure 8.8.2*a* and *b* show the experimental and calculated response for the [111] orientation. The back-stress parameters that were determined from the tensile response underpredicted the initial strain-hardening rate, whereas the parameters that were determined from the recovery data overpredicted the initial strain hardening, The conditions for determining n_{11} were approximate. Sheh and Stouffer (1988) proposed optimizing n_{11}, n_{12}, and F_1 together; thus n_{11} was iteratively adjusted to give the best values for n_{12}, and F_1. The optimized results are shown in Figure 8.8.3 for loading in the [111] and [001] orientations. Figure

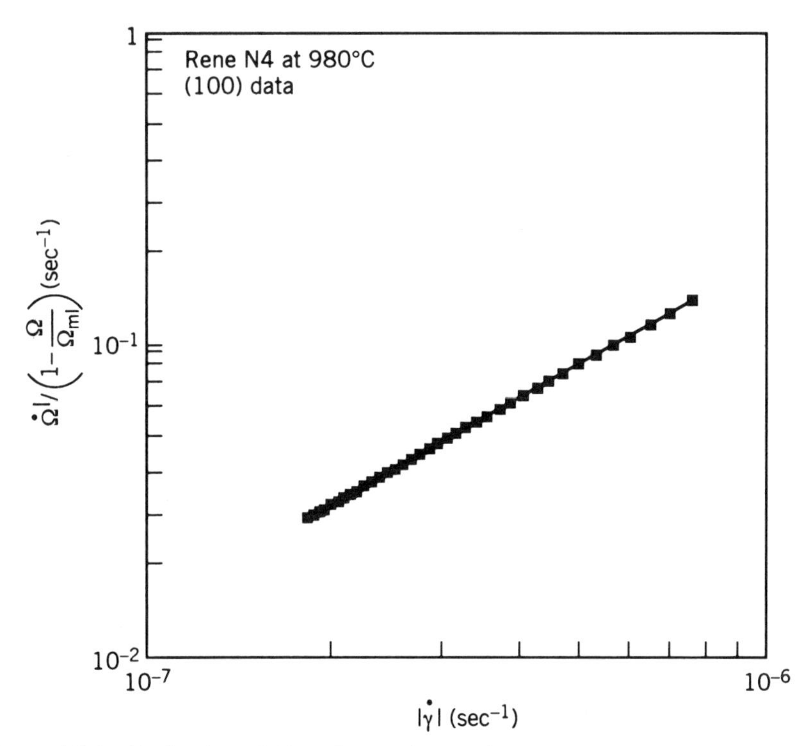

Figure 8.8.1 Evaluation of F_1 and n_{13} using strain recovery data from a the double tensile test of René N4 at 982°C. The double tensile test is shown in Figure 1.3.3 (and $\dot{\Omega}^I = \dot{\Omega} - G_1(\tau)^{n_{k2}}\, \dot{\gamma}$). (After Sheh and Stouffer, 1988.)

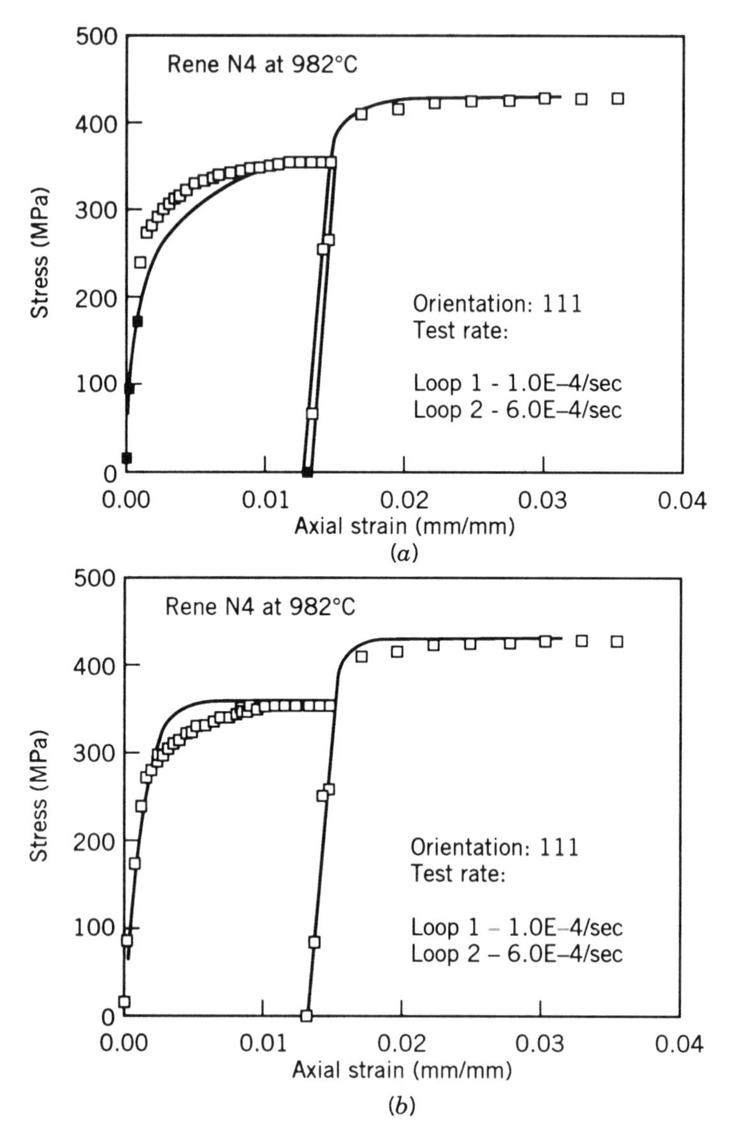

Figure 8.8.2 Predicted double tensile test in the [111] orientation using F_1 and n_{13} as determined from the (*a*) tensile curve and (*b*) inelastic recovery data. (After Sheh and Stouffer, 1988.)

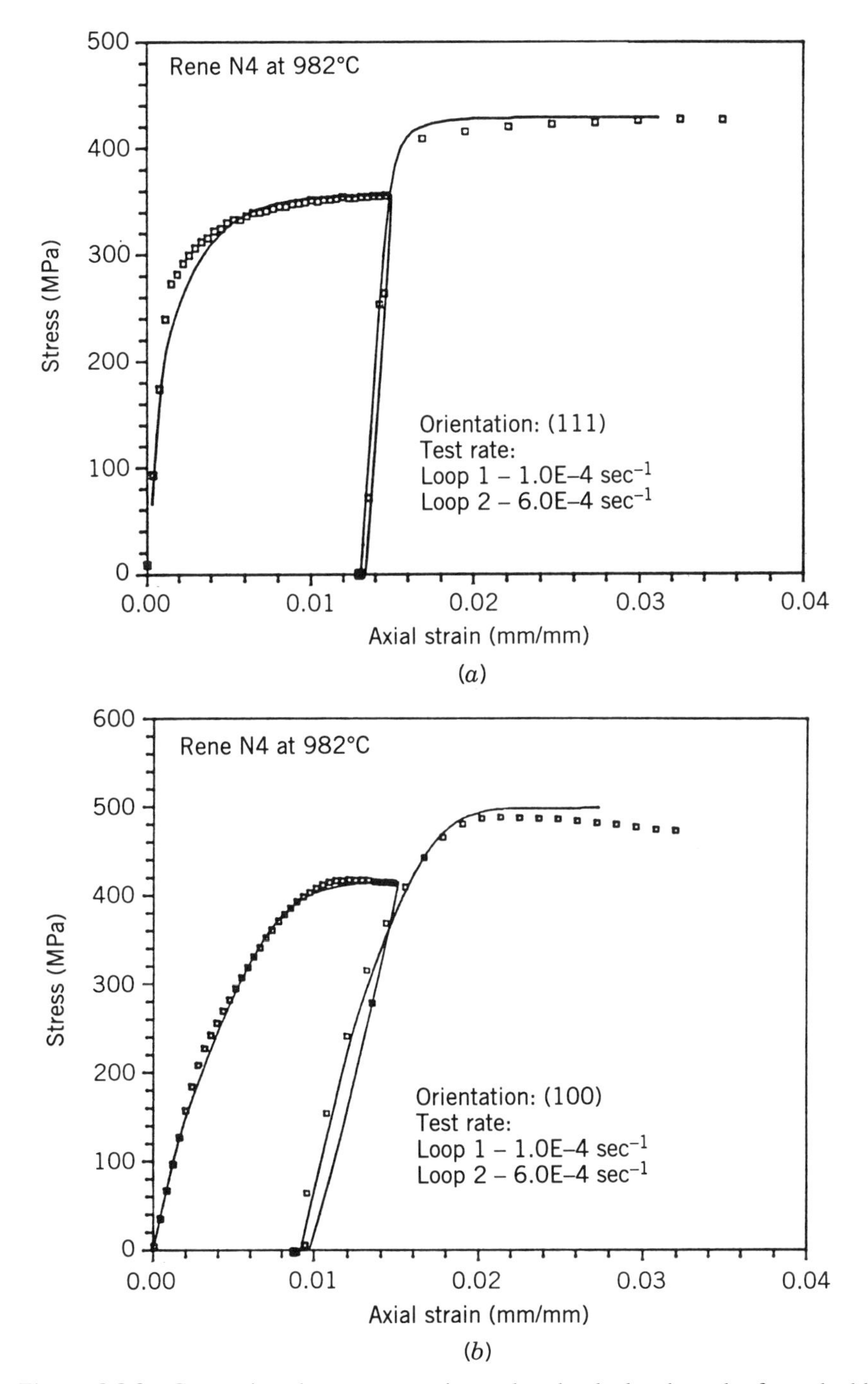

Figure 8.8.3 Comparison between experimental and calculated results for a double tensile test using the optimized values of F_1 and n_{13} in the (*a*) [111] orientation and (*b*) [001] orientation. (After Sheh and Stouffer, 1988.)

8.8.4 shows the tensile results as a function of strain rate for the [011] orientation and the [123] orientation, a single-slip-system test involving the simultaneous effect of octahedral and cube slip.

The drag stress equations are identical to the results in Section 6.6. Thus these parameters can be determined by the methods discussed earlier. However, the fatigue data for the particular case of René N4 at 982°C exhibited no cyclic hardening. Thus the initial values of H_{11} and H_{31} determined earlier for the octahedral and cube slip, respectively, were used for all cyclic calculations. Equation 8.7.3 was not used. To check the model without the use of the drag stress equation, the cyclic response in the [100] and [111] orientations was determined and compared to the experimental data. Figure 8.8.5 shows that the comparison, based on parameters from tensile data, is almost perfect. Figure 8.8.6 shows the results of two cyclic tests in the [321] orientation. Figure 8.8.6*b* shows the results for a test with a 30-sec hold period in compression.

There is one further refinement that may be necessary to obtain an acceptable set of parameters. Recall that the initial hardness H_{k1} is assumed to remain constant during tensile loading. This assumption is not strictly valid. The drag stress is a function of the plastic work; thus a more accurate value

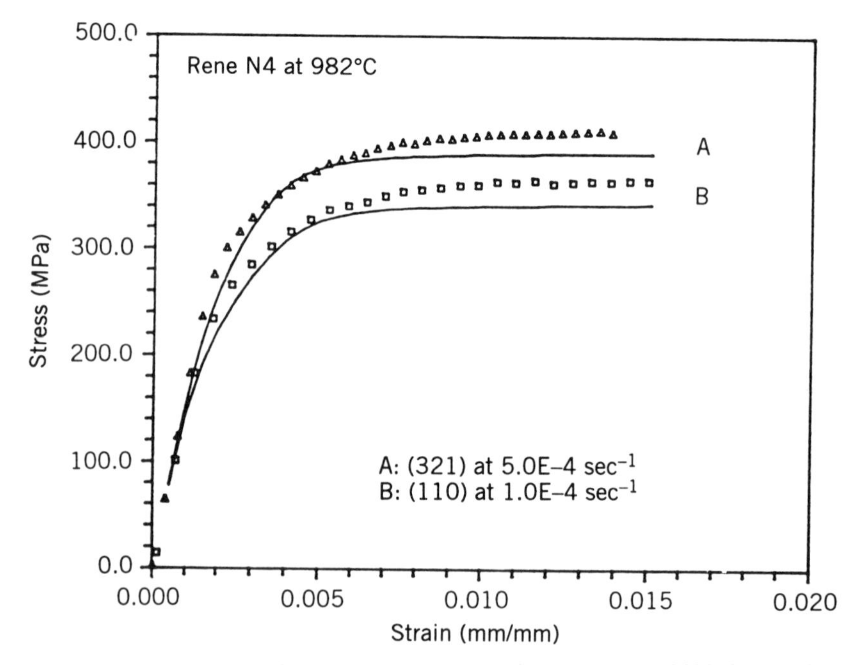

Figure 8.8.4 Prediction of the tensile response of René N4 at 982°C for loading in the [110] and [123] orientations at two different strain rates. (After Sheh and Stouffer, 1988.)

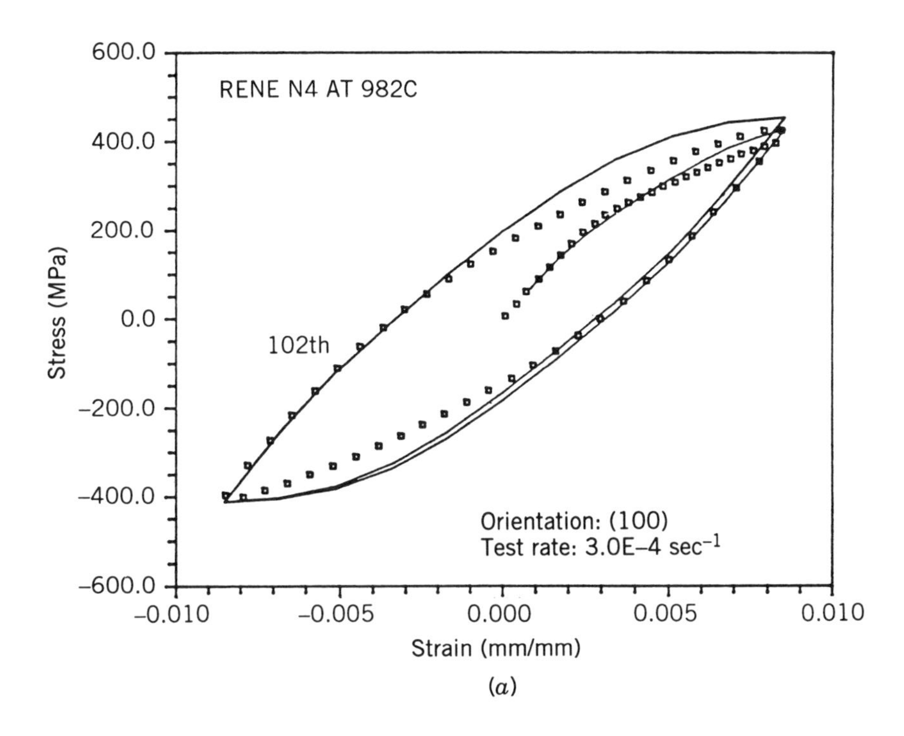

(*a*)

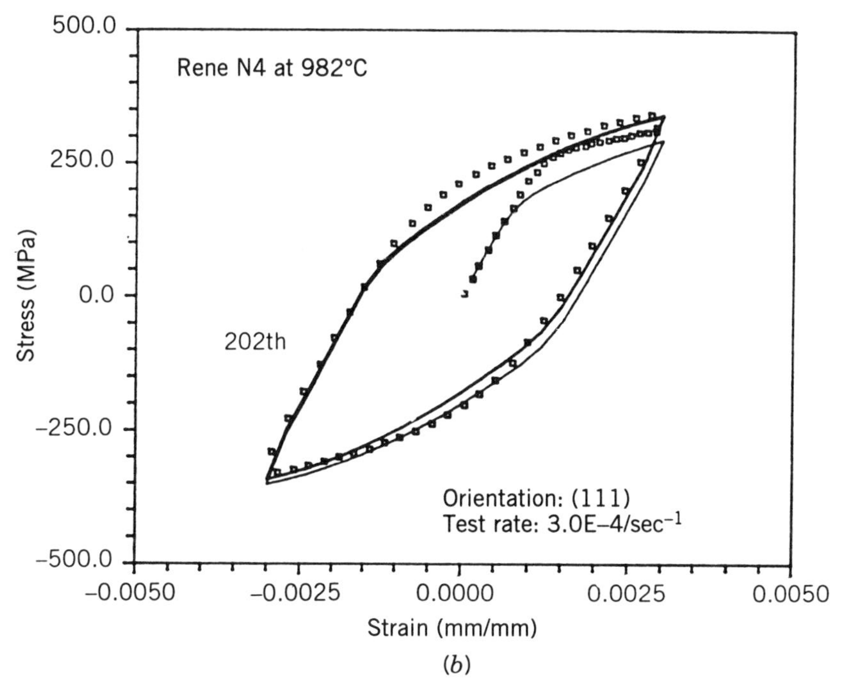

(*b*)

Figure 8.8.5 Comparison of the experimental and calculated response of René N4 at 982°C for loading in the (*a*) [001] and (*b*) [111] orientations. (After Sheh and Stouffer, 1988.)

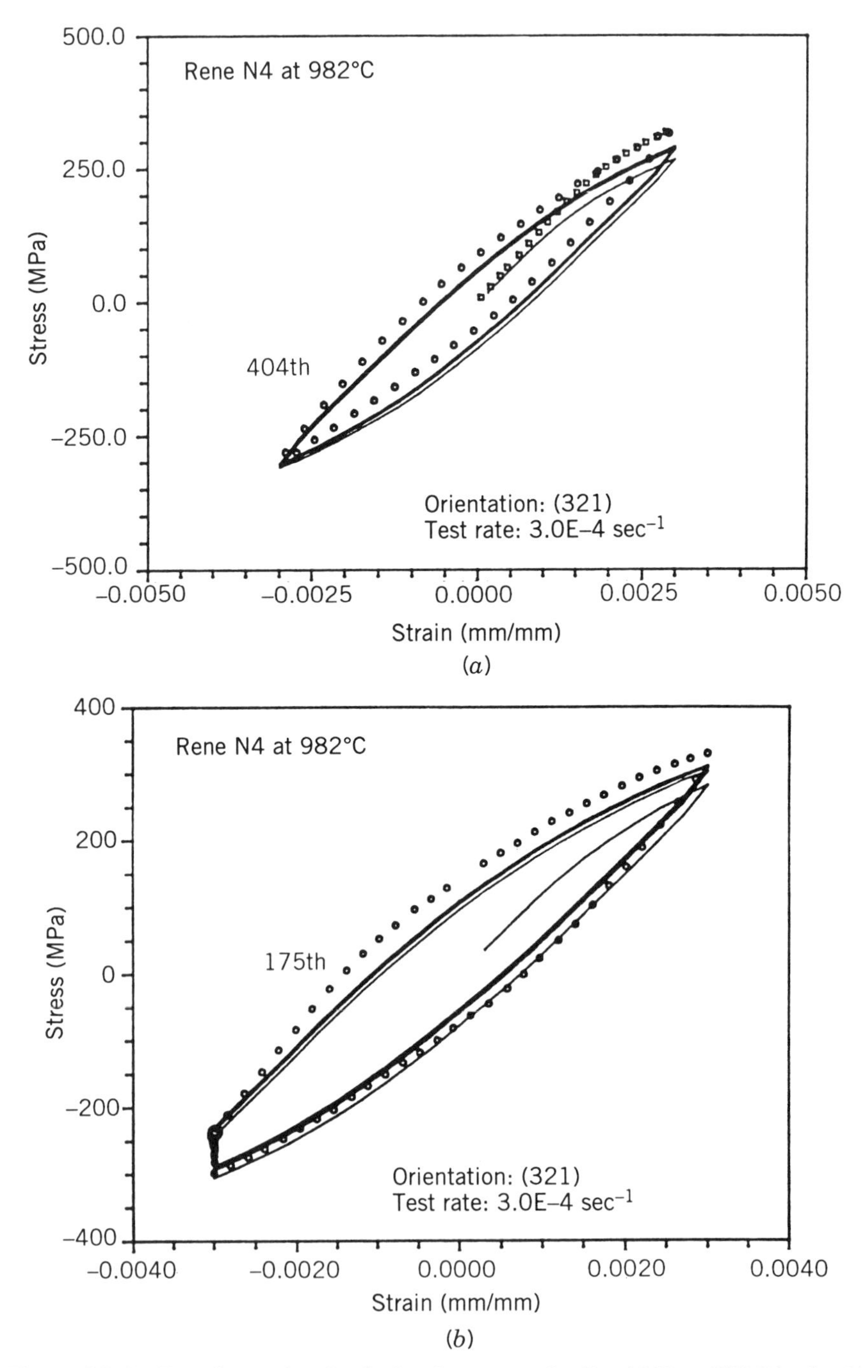

Figure 8.8.6 Experimental and calculated response for René N4 at 982°C in the (*a*) [321] orientation and (*b*) [321] orientation with a strain hold at the peak compressive stress. (After Sheh and Stouffer, 1988).

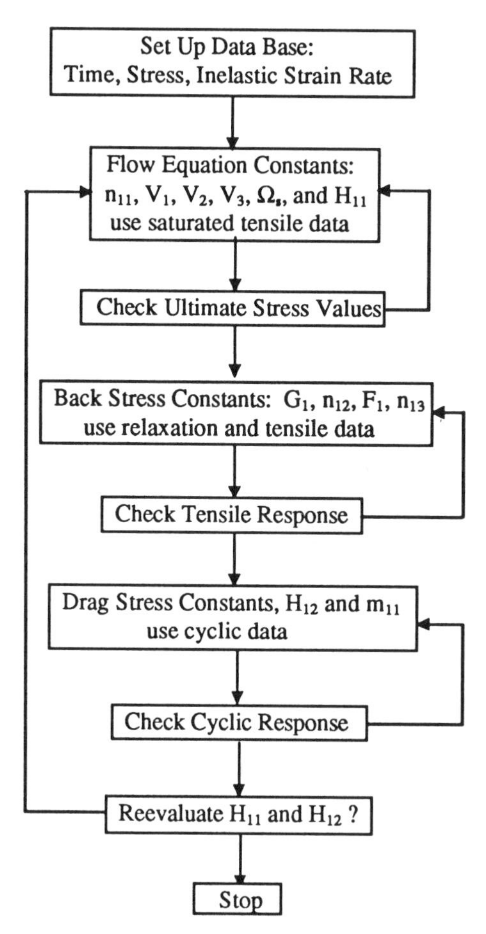

Figure 8.8.7 Material parameter evaluation procedure. (After Sheh and Stouffer, 1988.)

than H_{k1} for $H_1^{\alpha\beta}$ can now be obtained by using the drag stress equation to update $H_1^{\alpha\beta}$ at saturation. This will change the values of the other parameters that are determined from equations of 8.8.1 to 8.8.3. A diagram of the material parameter iteration procedure is outlined in Figure 8.8.7.

8.9 SUMMARY

The understanding and modeling of single-crystal superalloys is far from being a mature topic. These alloys are relatively new, and although much has been achieved in the past decade, there is still much to be done. A summary of the results of this chapter will be given for the numerical efficiency, evaluation of material parameters, and accuracy.

The constitutive equations above have been implemented in stand-alone constitutive equation codes and finite element analysis codes following the procedures in Section 6.7. The CPU time required to produce the calculated results in this chapter were about 10 to 15% longer than the isotropic constitutive equation results in Chapters 6 and 7. The computation efficiency in finite element design codes is acceptable.

Evaluation of the material parameters requires a well designed system of experiments as described in Section 8.8 and knowledge of the active slip system and deformation mechanism. The tests must be chosen to activate the octahedral and cube slip systems separately; thus most of the tests should be in the [100] and [111] orientations, but some tests should be in the [110] orientation to determine tension–compression asymmetry. Verification tests should also be run in other orientations to confirm the model. If more than isothermal results are required, a set of tests should be run above and below the critical temperature of about 800°C for γ–γ' alloys since the deformation mechanisms are different, and the flow equation should be modified to include temperature as shown in Sections 7.4 and 7.5. The methods of Sections 6.4, 6.6, and 8.8 can be used to determined the material parameters. The elastic response can be evaluated as shown in Section 4.5.1.

The model results are reasonably accurate for tensile and fatigue applications with time and rate effects. The model has been applied to creep of René N4 but without verification of the active slip system and without including the incubation period at low stress levels. These aspects of creep need further study. The model has been applied at only two isothermal temperatures, and thermomechanical cycling has not been considered.

REFERENCES

Alden, D. (1990). An analysis of the yield phenomena in René N4+ single crystals with respect to orientation and temperature, Ph.D. dissertation, University of Cincinnati.

Asaro, R. J. (1983). Crystal plasticity *Journal of Applied Mechanics,* Vol. 50, pp. 921–934.

Asaro, R. J. (1985). Material modeling and failure modes in metal plasticity, *Mechanics of Materials,* Vol. 4, pp. 343–373.

Asaro, R. J., and J. R. Rice (1987). Strain localization in ductile single crystals, *Journal of the Mechanics and Physics of Solids,* Vol. 25, p. 309.

Bacroix, B., J. J. Jones, F. Montheillet, and A. Skalli (1986). Grain reorientation during plastic deformation of FCC crystals, *Metallurgica,* Vol. 34, pp. 937–950.

Bell, J. F., and R. E. Green (1967). An experimental study of the double slip deformation hypothesis in face centered cubic crystals, *Philosophical Magazine,* Series 8, Vol. 15, pp. 469–476.

Courtney, T. H. (1990). *Mechanical Behavior of Materials,* McGraw-Hill, New York.

Dalah, R. D., C. R. Thomas, and L. E. Dardi (1984). The effect of crystallographic orientation on the physical and mechanical properties of investment cast single crystal nickel base superalloy, *Superalloy 1984, Proceedings of the 5th International Symposium on Superalloys,* Metals Park, OH, p. 185.

Dame, L. T., and D. C. Stouffer (1985). Anisotropic constitutive model for nickel base single crystal alloys: model development and finite element implementation, *NASA CR-175015.*

Dame, L. T., and D. C. Stouffer (1988). A crystallographic model for nickel base single crystal alloys, *Journal of Applied Mechanics,* Vol. 55, pp. 325–331.

Ezz, S., D. Pope, and V. Paidar (1982). The tension/compression flow stress asymmetry in Ni_3(Al, Nb) single crystals, *Acta Metallurgica,* Vol. 330, p. 921.

Gabb, T., J. Gayda, and R. Miner (1986). Orientation and temperature dependence of some mechanical properties of single crystal nickel base superalloy René N4, Part 2: Low cycle fatigue behavior, *Metallurgical Transactions,* Vol. 17A, p. 497.

Gell, M., and G. Leverant (1968). The fatigue of nickel base superalloy Mar M200, in single crystal and columnar forms at room temperature, *Transactions of AIME,* Vol. 22, p. 242.

Havner, K. S. (1992). *Finite Plastic Deformation of Crystalline Solids,* Cambridge Monographs on Mechanics and Applied Mathematics, Cambridge University Press, New York.

Heredia, F., and D. Pope (1986). The tension/compression flow asymmetry in high γ' volume fraction nickel base alloys, *Acta Metallurgica,* Vol. 34, No. 2, pp. 279–285.

Hill, R., and K. S. Havner (1982). Perspectives in the mechanics of elastoplastic crystals, *Journal of the Mechanics and Physics of Solids,* Vol. 30, p. 5.

Hopgood, A. A., and J. W. Martin (1986). The creep behavior of a nickel based crystal superalloy, *Materials Science and Engineering,* Vol. 82, pp. 27–36.

Jackson, P. J. (1985). Dislocation modeling of shear in F.C.C. crystals, *Progress in Materials Science,* Vol. 29, pp. 139–175.

Jackson, P. J. (1986). The mechanisms of plastic relaxation in single crystal deformation, *Materials Science and Engineering,* Vol. 81, pp. 169–174.

Kear, H. B., G. R. Leverant, and J. M. Objak (1969). An analysis of creep induced intrinsic/extrinsic fault pairs in a precipitation hardened nickel-base alloy, *Transactions American Society of Metals,* Vol. 62, pp. 639–650.

Lall, C., S. Chin, and D. Pope (1979). The orientation and temperature dependence of the yield stress of Ni_3(Al, Nb) single crystals, *Metallurgical Transactions,* Vol. 10A, p. 1323.

Leverant, G. R., and H. B. Kear (1970). The mechanism of creep in a γ' precipitation hardened nickel-base superalloy at intermediate temperatures, *Metallurgical Transactions,* Vol. 1, pp. 491–498.

Leverant, G. R., H. B. Kear, and J. M. Objak (1973). Creep of precipitation hardened nickel-base superalloy single crystals at high temperature, *Metallurgical Transactions,* Vol. 4, pp. 355–362.

MacKay, R. A., and R. D. Maier (1980). Anisotropy of nickel base superalloy single crystals, *Superalloys,* pp. 385–394.

MacKay, R. A., and R. D. Maier (1982). The influence of orientation on the stress rupture properties of nickel base superalloy single crystals, *Metallurgical Transactions,* Vol. 13A, pp. 1747–1754.

Milligan, W. W. (1986). Yielding and deformation behavior of the single crystal nickel base superalloy PWA 1480, *NASA CR 175100.*

Milligan, W. W., and S. D. Antolovich (1987). Yielding and deformation of single crystal superalloy PWA 1480, *Metallurgical Transactions,* Vol. 18A, p. 85.

Milligan, W. W., and S. D. Antolovich (1989). Deformation modeling and constitutive modeling of anisotropic superalloys, *NASA CR 4215.*

Milligan, W. W., N. Jayarmann, and R. C. Bill (1984). Low cycle fatigue of Mar M-200 single crystals at 760 and 970°C, *NASA TM 86993.*

Miner, R., R. Voigt, J. Gayda, and T. Gabb (1986). Orientation and temperature dependence of some mechanical properties of the nickel base single crystal alloy René N4, Part 1: Tensile behavior, *Metallurgical Transactions,* Vol. 17A, p. 491.

Miner, R., R. Voigt, J. Gayda, and T. Gabb (1986). Orientation and temperature dependence of some mechanical properties of the nickel base single crystal alloy René N4, Part 3: Tension/compression anisotropy, *Metallurgical Transactions,* Vol. 17A, p. 507.

Nakada, Y., and A. S. Keh (1966). Latent hardening in iron single crystals, *Acta Metallurgica,* Vol 23, No. 2, pp. 436–448.

Paidar, V., D. P. Pope, and V. Vitek (1984). A theory of the anomalous yield behavior in LI_2 ordered crystals, *Acta Metallurgica,* Vol. 23, No. 3, pp. 435–448.

Pollock, T., and A. Argon (1991). Creep resistance of superalloy single crystals: experiments and modeling, in *Modeling the Deformation of Cystalline Solids,* T. C. Lowe, A. D. Rollet, P. S. Follansbee, and G. S. Daehn, eds., The Minerals, Metals & Materials Society, Warrendale, PA.

Pope, D., and S. Ezz (1984). Mechanical properties of nickel base alloys with a high volume fraction of γ', *International Metals Reviews,* Vol. 29, No. 3, pp. 136–167.

Shah, D., and D. Duhl (1984). The effect of orientation, temperature and γ' size on the yield strength of a single crystal nickel base superalloy, *Proceedings of the 5th International Symposium on Superalloys,* ASM, Metals Park, OH.

Sheh, M. Y., and D. C. Stouffer (1988). Anisotropic constitutive modeling of nickel base single crystal superalloy, *NASA CR 182157.*

Sheh, M. Y., and D. C. Stouffer (1990). A crystallographic model for tensile and fatigue response of René N4 at 982°C, *Journal of Applied Mechanics,* Vol. 57, pp. 25–31.

Swanson, G. (1984). Life prediction and constitutive model for hot section anisotropic materials program, *Contract NAS3-23939, Monthly Report PWA-5968-9,* August.

Swanson, G. (1984). Life prediction and constitutive model for hot section anisotropic materials program, *Contract NAS3-23939, Monthly Report PWA-5968-10,* September.

Umakoshi, Y., D. Pope, and V. Vitek (1984). The asymmetry of the flow stress in Ni_3(Al, Ta) single crystals, *Acta Metallurgica,* Vol. 32, No.3, p. 449.

Verhoeven, J. D. (1975). *Fundamentals of Physical Metallurgy,* Wiley, New York.

Ver Snyder, F., and B. Piearcey (1966). Single crystal alloy extend turbine blade service life four times, *SAE Journal,* Vol. 74, p. 36.

Yang, S. (1984). Elastic constants of a monocrystalline nickel base superalloy, *Metallurgical Transactions,* Vol. 16A, No. 4, p. 661.

PROBLEMS

8.1 Sketch the standard [110] sterographic projections of the front and back hemispheres for the [110][111] cubic slip system described in Table 1.2.2.

8.2 Repeat Problem 8.1 except for the [110][111] octahedral slip system described in Table 1.2.2.

8.3 Figure P8.3 shows the Thompson tetrahedron that is formed from the four planes in the octahedral slip system. Identify the Miller indices for the four planes and the slip directions on the two front surfaces. Show the results in a sketch.

8.4 Verify the first few rows of the stress matrices given in equations 8.5.3 to 8.5.5.

8.5 Establish the first few terms of the inelastic strain equations 8.5.7 and 8.5.8.

8.6 An FCC crystal is subjected to a tensile test in the $[\bar{1}12]$ orientation. Determine the shear stresses in the octahedral slip system for an applied stress of 1000 psi. Repeat the calculation for another orientation on the $[001]$–$[\bar{1}11]$ boundary, say $[\bar{1}13]$. Why do crystals rotate toward $[\bar{1}12]$ along the $[001]$–$[\bar{1}11]$ boundary?

8.7 Verify the correlation between the shear stresses in the octahedral slip system and the stress components τ_1, τ_2, and τ_3 in Table 8.5.1.

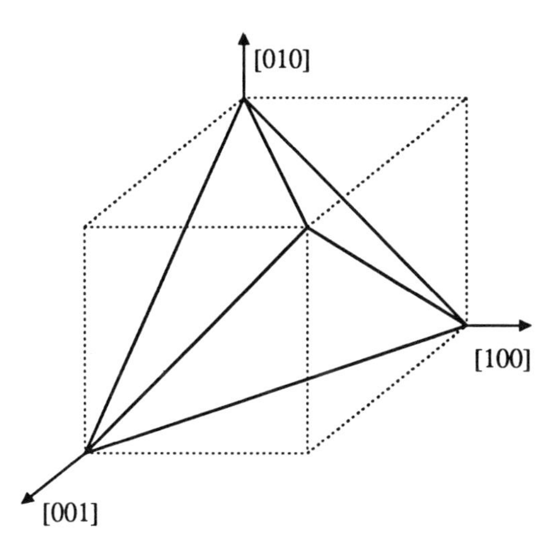

Figure P8.3

8.8 Determine a system that will exhibit the equivalent slip component for each of the six stereographic triangles in the [001]–[010]–[$\bar{1}$00] quadrant of the standard [001] stereographic triangle shown in Figure P8.8. Sketch the direction of rotation for specimen with each of the orientations above.

8.9 The data in Figure P8.9 shows the critical resolved shear stress for René N4 as a function of temperature and strain rate. Estimate the yield stress of a specimen oriented in the [011] direction at strain rates of 0.5 and 50% min^{-1}.

8.10 Determine the stress and inelastic strain rate scale factors between the active slip systems for loading in the [101] and [111] orientations. What assumptions or restrictions are necessary to determine material parameters from a single-slip-system test? Determine the scale factors for a specimen in the [225] orientation.

8.11 Why is yield stress in the [001] orientation higher at 760°C than at room temperature? (See Figure 8.2.5.)

8.12 What is the inelastic strain-rate tensor for primary creep in the principal-coordinate system of the crystal (similar to equations 8.5.7 and 8.5.8) due to slip in the 12 systems similar to [111][112]?

8.13 Why is the small strain response of René N4 in Figure 8.2.4 different from the classic single-slip response for pure metals in Figure 8.4.4?

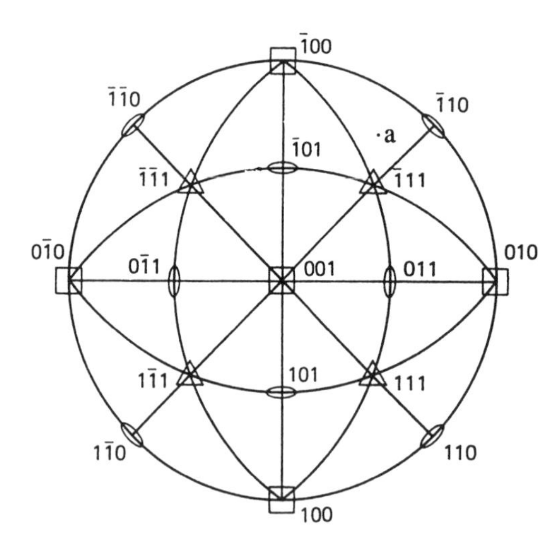

Figure P8.8 (After Verhoeven, Fundamentals of Physical Metallurgy, Copyright © 1975. Reprinted by permission of John Wiley & Sons, Inc.)

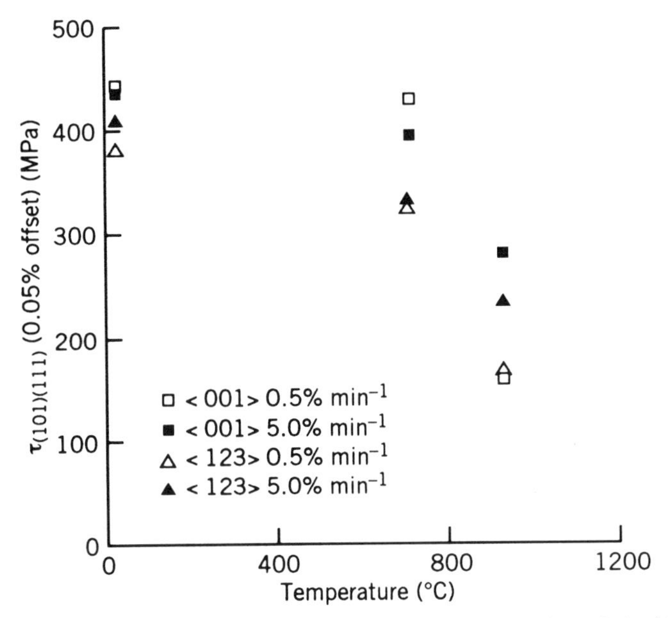

Figure P8.9 Average critical resolved shear stress determined from Schmid's law for René N4. (After Milligan and Antolovich, 1989.)

CHAPTER 9

Finite Element Methods

To be of practical use, the material models discussed in this book must ultimately be incorporated into structural finite element codes. Although many excellent books exist that develop the finite element method in detail, the authors believe that any book on nonlinear material modeling would be incomplete without an overview as to how these models are incorporated in finite element codes.

Linear structural analysis is based on the assumptions that (1) the infinitesimal strain measure is valid, (2) boundary conditions are independent of displacements, (3) structural stiffness is independent of displacement and loads, and (4) the relationship between stress and strain is linear. The first three assumptions can be broadly classified as geometric linearity, and the last is material linearity. If any of these assumptions are violated, a valid solution will require the use of nonlinear finite element methods. Often, more than one type of nonlinearity may occur in the same problem. There are various finite element formulations that address some or all of these nonlinearities. Since the focus is on the nonlinear behavior of metals, the discussion is limited to only material nonlinear finite element formulations. The discussion is also limited to quasi-static problems where loads are applied slowly and inertial terms are neglected.

9.1 FINITE ELEMENT EQUATIONS WITH MATERIAL NONLINEARITIES

The purpose of this section is to develop the finite element equations from the principles of mechanics and to show how nonlinear material models are included in the global equations.

9.1.1 Finite Element Equilibrium Equation

Although the equations for the displacement based finite element method may be derived in various ways, one of the simplest approaches is from the principle of virtual work, which states that a loaded deformable body is in equilibrium if the virtual work of the real internal stresses is equal to the virtual work of the real external forces and moments. Virtual work is the work done by virtual displacements acting on the real stresses, forces, and moments acting on the body. Virtual displacements are any arbitrary displacements consistent with continuity and displacement boundary conditions on the body. Mathematically, the principle of virtual work may be expressed as

$$\int_V \sigma_{ij}\hat{\varepsilon}_{ij}\, dV = \int_V b_i\hat{d}_i\, dV + \int_S t_i\hat{d}_i\, dS + f_i\hat{d}_i \tag{9.1.1}$$

The left-hand side of the equation is the internal virtual work that is a result of the strain due to the virtual displacements, $\hat{\varepsilon}_{ij}$, acting on the real stresses, σ_{ij}. The first term on the right-hand side is the virtual work due to the virtual displacements $\hat{d}_i$ acting on the real body forces, b_i. The second term on the right-hand side is the virtual work due to the virtual displacements $\hat{d}_i$ acting on the real surface pressures or tractions t_i. The third term on the right-hand side is the virtual work due to the virtual displacements $\hat{d}_i$ acting on real point loads. For developing the finite element equations it is more convenient to express equation 9.1.1 as a matrix equation:

$$\int_V \{\hat{\varepsilon}\}^T \{\sigma\}\, dV = \int_V \{\hat{d}\}^T \{b\}\, dV + \int_S \{\hat{d}\}^T \{t\}\, dS + \{\hat{d}\}^T \{f\} \tag{9.1.2}$$

In the finite element method the body is subdivided into a set of *elements* that are connected on their boundaries at discrete points or *nodes*. This allows the volume and surface integrals in equation 9.1.2 to be computed over each element individually and summed over the entire structure. Assuming that point loads are applied at nodal points, integrating over each of the element with volume, V_e, and surface, S_e, and summing, equation 9.1.2 may be written

$$\sum \int_{V_e} \{\hat{\varepsilon}\}^T \{\sigma\}\, dV_e = \sum \int_{V_e} \{\hat{d}\}^T \{b\}\, dV_e + \sum \int_{S_e} \{\hat{d}\}^T \{t\}\, dS_e + \{\hat{d}\}^T \{F_n\} \tag{9.1.3}$$

The displacements at any interior point of an element, $\{d\}$, are completely defined as a function of the nodal displacements on the boundary of the element, $\{D\}$, by using a matrix of interpolating functions, $[N]$:

$$\{d\} = [N]\{D\} \tag{9.1.4}$$

The interpolating functions, usually polynomials, are assumed a priori, depending on the specific type of element. For most element types there is an interpolating function associated with each node. The values of the interpolating functions at a point depend on the location of the point. Interpolating functions are generally defined so that the sum of the interpolating functions at any location in the element is 1. Furthermore, the interpolating function associated with a particular node has a value of 1 at that node and a value of zero at all other nodes. The interior displacements therefore depend on the assumed form of the interpolating functions and the nodal displacements that are the principal unknowns in the problem.

Strains are determined from spatial derivatives of displacements as defined in equation 4.3.13. Within an element the strains may be written as a function of the nodal displacements,

$$\{\varepsilon\} = [B]\{D\} \tag{9.1.5}$$

The strain displacement matrix $[B]$ consists of spatial derivatives of the interpolating functions, $[N]$. An example of these matrices for a typical element is given is Section 9.1.5. Substituting equations 9.1.4 and 9.1.5 into equation 9.1.3 gives

$$\{\hat{D}\}^T \left[\sum \int_{V_e} [B]^T \{\sigma\} \, dV_e \right] = \{\hat{D}\}^T \left[\sum \int_{V_e} [N]^T \{b\} \, dV_e + \sum \int_{S_e} [N]^T \{t\} \, dS_e + \{F_n\} \right] \tag{9.1.6}$$

The virtual displacements may be eliminated from equation 9.1.6 to obtain the finite element equilibrium equation

$$\sum \int_{V_e} [B]^T \{\sigma\} \, dV_e = \sum \int_{V_e} [N]^T \{b\} \, dV_e + \sum \int_{S_e} [N]^T \{T\} \, dS_e + \{F_n\} \tag{9.1.7}$$

The first term on the right-hand side of equation 9.1.7 is the vector of equivalent nodal body forces, $\{F_b\}$, and the second term is the vector of equivalent nodal surface traction loads, $\{F_s\}$. The elemental volume and surface integrals in equation 9.1.7 are nearly always evaluated numerically. Efficient integration algorithms have been developed for the various element types. All numerical integration procedures evaluate the integral by computing the integrand at discrete integration points within each element, multiplying by a weighting factor, and summing.

9.1.2 Linear Finite Element Equilibrium Equation

For the geometrically linear deformations the change in the shape of the body from the unloaded to the loaded configuration is considered to be negligible, and the surface and volume integrals in equation 9.1.7 are evaluated over the undeformed configuration. The small strain assumption (equation 4.3.13) is used and the total strain is assumed to be the sum of the elastic strain and the thermal strain components in equation 6.1.2:

$$\{\varepsilon\} = \{\varepsilon^e\} + \{\varepsilon^{th}\} \tag{9.1.8}$$

Substituting equations 9.1.8 and 4.5.3 (the matrix form of Hooke's law for linear elastic materials) into equation 9.1.7 produces

$$\sum \int_{V_e} [B]^T [C][B]\, dV_e \{D\} = -\sum \int_{V_e} [B]^T \{\varepsilon^{th}\}\, dV_e + \{F_b\} + \{F_s\} + \{F_n\} \tag{9.1.9}$$

where $[C]$ is the 6×6 matrix of elastic constants. The left-hand side of the equation is the sum of the products of the elemental stiffness matrices and nodal displacements. The first term on the right-hand side arises from the thermal strain and is usually referred to as the thermal load vector, $\{F_t\}$.

Finally, for the linear finite element formulation (linear material, small strain, and displacement independent stiffness and boundary conditions), the global equilibrium equation may be rewritten

$$[K]\{D\} = \{F_t\} + \{F_b\} + \{F_s\} + \{F_n\} \tag{9.1.10}$$

where the global stiffness and thermal force vectors are defined by

$$[K] = \sum \int_{V_e} [B]^T[C][B]\, dV_e \quad \text{and} \quad \{F_t\} = -\sum \int_{V_e} [B]^T\{\varepsilon^{th}\}\, dV_e$$

Symbolically, the solution of the equilibrium equation for nodal displacements can be written as

$$\{D\} = [K]^{-1}(\{F_t\} + \{F_b\} + \{F_s\} + \{F_n\}) \tag{9.1.11}$$

In practice the inverse of the global stiffness matrix is never calculated. The set of linear equations 9.1.10 is usually solved by Gauss elimination or a closely related technique. Once nodal displacements are obtained, strains within each element are determined from equation 9.1.5 and stresses are calculated from equations 9.1.8 and 4.5.3.

It is important to note that in the linear finite element formulation the stiffness matrix K plays a dual role. It is part of the equilibrium equation (9.1.10) and it is also used to relate forces to displacements, equation (9.1.11). For nonlinear finite element formulations the stiffness matrix relates incremental forces to incremental displacements, but it does not appear in the equilibrium equation 9.1.7.

9.1.3 Material Nonlinear Finite Element Equilibrium Equation

When inelastic deformations are present, the equilibrium equation 9.1.7 is not satisfied unless the correct stresses are identified. The stresses are no longer a linear function of displacements but depend on the inelastic strains. At a particular point in the solution process, equilibrium is not satisfied and the equilibrium imbalance is put in the form of a residual force; that is,

$$\begin{aligned}\{R\} &= \sum \int_{V_e} [N]^T \{b\}\, dV_e + \sum \int_{S_e} [N]^T \{t\}\, dS_e \\ &\quad + \{F_n\} - \sum \int_{V_e} [B]^T \{\sigma\}\, dV_e \end{aligned} \qquad (9.1.12)$$

Equilibrium is satisfied when the residual force vanishes. The first three terms on the right-hand side are applied loads, and the fourth term is the "internal" force resulting from stresses in the body. The residual force can therefore be thought of as the difference between the external and internal forces or an out-of-balance force.

The nonlinear problem is solved iteratively and the residual force is reduced successively in each iteration. During an iteration i, an incremental displacement can be calculated from the residual force and the stiffness matrix, and then summed to the displacement from the prior iteration.

$$\{\Delta D\}_i = [K]^{-1} \{R\}_i \qquad (9.1.13)$$

$$\{D\}_i = \{D\}_{i-1} + \{\Delta D\}_i \qquad (9.1.14)$$

To converge in the minimum number of iterations the stiffness matrix should be a tangent matrix that is updated every iteration and reflects the current configuration of the body, the current state of the material, and the algorithm used to integrate the inelastic strain increments. The tangent stiffness matrix is defined as the change in the residual force due to an incremental change in displacement $\{\Delta D\}$ and can be written

$$[K^T] = \frac{\partial \{R\}}{\partial \{\Delta D\}}$$

A solution strategy where the stiffness matrix is updated every iteration is called a full Newton–Raphson procedure and is illustrated in Figure 9.1.1 for a problem with a single displacement unknown (degree of freedom). However, the calculation and decomposition (inversion) of the stiffness matrix are computationally very expensive. This means that the minimum-cost solution strategy is not necessarily the one with the fewest global iterations. A solution procedure that does not reform the stiffness at all during the iteration procedure is called an initial stiffness method or initial stress method and is illustrated in Figure 9.1.2 for a single-degree-of-freedom problem. For highly nonlinear problems the initial stiffness method may require an excessive number of iterations to converge or may not converge at all. Therefore, adaptive strategies that monitor stiffness changes and update the stiffness only when necessary have been developed (see, e.g., Crisfield, 1991, or Hinton, 1992).

Often, for creep or unified constitutive models an initial stiffness method is used. Since most rate integration procedures for creep and unified models require relatively small steps to integrate the rate equations accurately, the initial stiffness calculated with the elastic material matrix is often adequate. Usually, for plasticity the tangent material matrix is used in calculation of the stiffness. The tangent material matrix, C^T, (Section 9.2) is defined as the

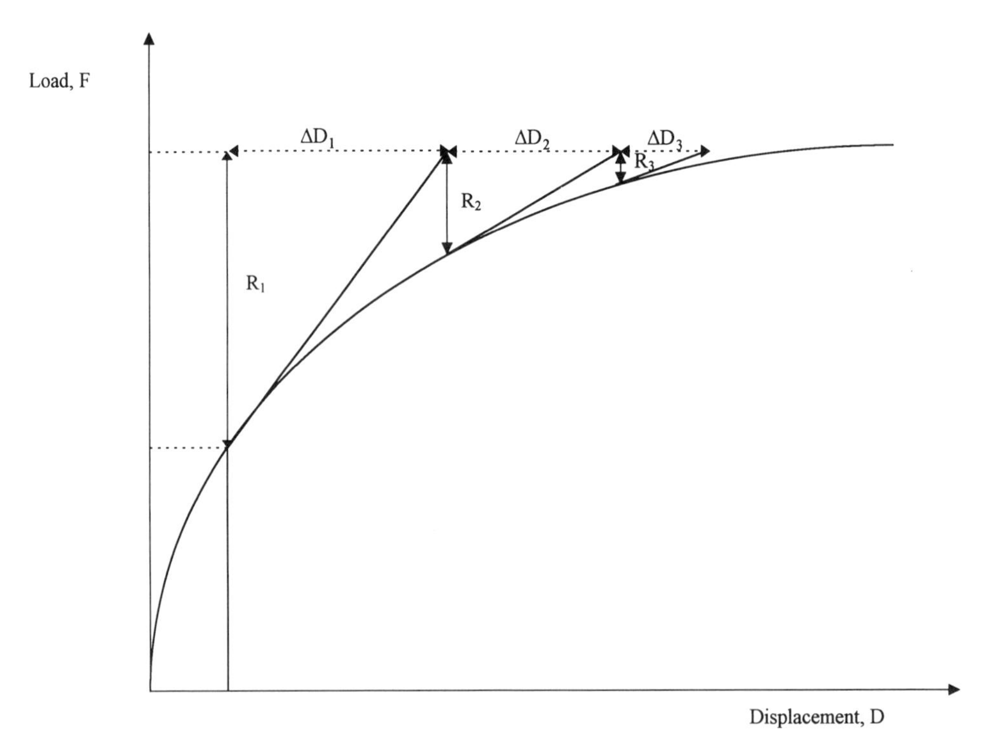

Figure 9.1.1 Newton–Raphson iteration method for elastic–plastic material response.

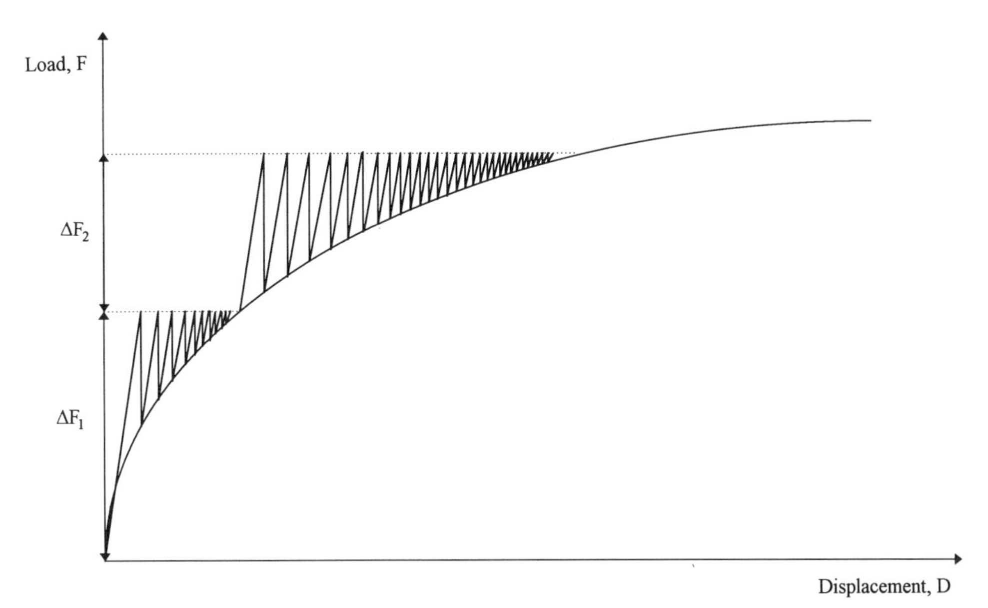

Figure 9.1.2 Initial stiffness iteration method for elastic–plastic material response.

matrix relating incremental stresses to incremental strains. When using the tangent material matrix the tangent stiffness matrix is given by

$$[K^T] = \sum \int_V [B]^T[C^T][B]\, dV \tag{9.1.15}$$

Since the stiffness is calculated numerically, the tangent material stiffness, $[C^T]$, must be calculated at each discrete integration point throughout the body. From a coding perspective this is accomplished by computing and using the tangent material stiffness instead of the elastic stiffness within the element stiffness routines. As the body continues to deform with changing loads, the tangent material stiffness will also continue to change.

9.1.4 Typical Solution Procedure

A typical solution sequence for a material nonlinear finite element code would contain the following main steps.

1. Increment applied loads and applied displacements. If the constitutive model is rate dependent, time must also be incremented.

$$\{F_{\text{applied}}\}^n = \{F_{\text{applied}}\}^{n-1} + \{\Delta F_{\text{applied}}\}^n$$

$$t^n = t^{n-1} + \Delta t$$

2. Compute internal forces for the current load increment and time step. At each integration point within each element, compute stress, accounting for prior inelasticity and current inelastic strain increments.

$$\{\varepsilon\} = [B]\{d\}$$

$$\{\varepsilon^e\} = \{\varepsilon\} - \{\varepsilon^{\text{th}}\} - \{\varepsilon^I\} - \{\Delta\varepsilon^I\}$$

$$\{\sigma\} = [C]\{\varepsilon^e\}$$

Assemble internal forces for all elements.

$$\{F_{\text{internal}}\} = \sum \int_V [B]^T \{\sigma\}\, dV$$

3. Compute the force residual.

$$\{R\} = \{F_{\text{applied}}\} - \{F_{\text{internal}}\}$$

4. Compute the stiffness matrix if necessary (stress stiffening effects similar to classical "beam-column" effects, are also usually included).

$$[K^T] = \int_V [B]^T[C^T][B]\, dV$$

5. Compute the displacement increment and update displacements for the current step n and current iteration i.

$$\{\Delta D\}_i = [K^T]^{-1}\,\{R\}_i$$

$$\{D\}_i^n = \{D\}_{i-1}^n + \{\Delta D\}_i$$

6. Compute inelastic strain increments at each element integration point. Total strains are computed from the current displacements. The inelastic strain increments are a function of the current stress, current strain, and material history or material state variables.
7. Check convergence. If not converged, go to step 2 and proceed with the next iteration. If converged, update inelastic strains and go to step 1 for the next load increment, or exit if the end of the loading sequence has been reached.

This particular solution sequence is not unique and many variations and enhancements of this basic scheme are possible. For further details and a discussion of geometric nonlinear effects, see Owen and Hinton (1980), Bathe (1982), Crisfield (1991), Zienkiewicz and Taylor (1991), and Hinton (1992).

9.1.5 Example Element

It is the purpose of this section to demonstrate by example how material nonlinearities are included in the element tangent stiffness matrix and the element internal force matrix. Other computed element quantities, such as body force, surface force, and mass matrix, are not affected by the presence of material nonlinearities and are not discussed. It is assumed that the reader has some familiarity with the linear finite element theory. Required preliminaries such as the concept of mapped elements and numerical integration are presented but not developed in detail.

The 10-node quadratic tetrahedral element that is widely used in finite element codes will be used as an example. It is a displacement-based finite element that is developed from the assumption that displacements within the element are a quadratic function of the spatial coordinates and of nodal displacements. The functions used to determine the displacement field within the element from nodal values are the interpolation functions. There are three degrees of freedom per node, the translational displacements u, v, and w, in the x, y, and z directions.

The quadratic tetrahedral element is a mapped element. Element interpolation functions and numerical integration points and weights are defined for a parent element in a local coordinate system. The parent element is mapped to a distorted but topologically similar shape in a global Cartesian coordinate system as shown in Figure 9.1.3. By using mapped elements it is possible to mesh very general shapes. The same interpolation functions used for displacements are used to map the geometry, which by definition makes this an isoparametric element. Since the interpolation functions are used to map the geometry, they are also called shape functions. Virtually all elements in use today are mapped elements.

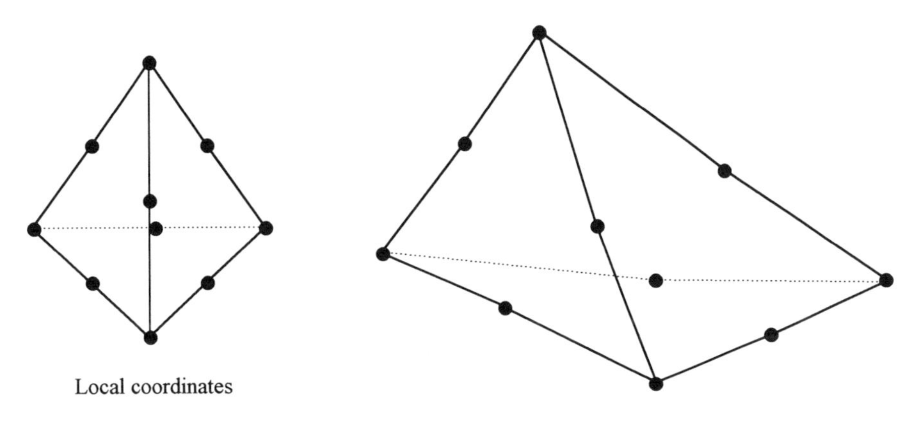

Figure 9.1.3 Three-dimensional mapping of a tetrahedral element between the local volume coordinates and the global Cartesian coordinates.

The interpolation functions for the tetrahedral element are written in a special local coordinate system identified as the volume coordinates. Volume coordinates, V_1, V_2, V_3, V_4, are defined in terms of a global Cartesian coordinate, x, y, z system by

$$\begin{Bmatrix} 1 \\ x \\ y \\ z \end{Bmatrix} = \begin{bmatrix} 1 & 1 & 1 & 1 \\ x_1 & x_2 & x_3 & x_4 \\ y_1 & y_2 & y_3 & y_4 \\ z_1 & z_2 & z_3 & z_4 \end{bmatrix} \begin{Bmatrix} V_1 \\ V_2 \\ V_3 \\ V_4 \end{Bmatrix} \tag{9.1.16}$$

where the Cartesian coordinates of corner node i are (x_i, y_i, z_i). This system of linear equations may be inverted to obtain the volume coordinates as a function of the Cartesian coordinates. Physically, the volume coordinates of a point P inside the tetrahedron may be interpreted as the ratio of a tetrahedral subvolume divided by the total volume of the tetrahedron, as shown in Figure 9.1.4; for example,

$$V_1 = \frac{\text{volume P234}}{\text{volume 1234}} \tag{9.1.17}$$

The volume coordinate V_i will vary linearly from a value of 1 at node i to zero on the face opposite node i. Since there are four volume coordinates describing a point in three-dimensional space, one of the volume coordinates

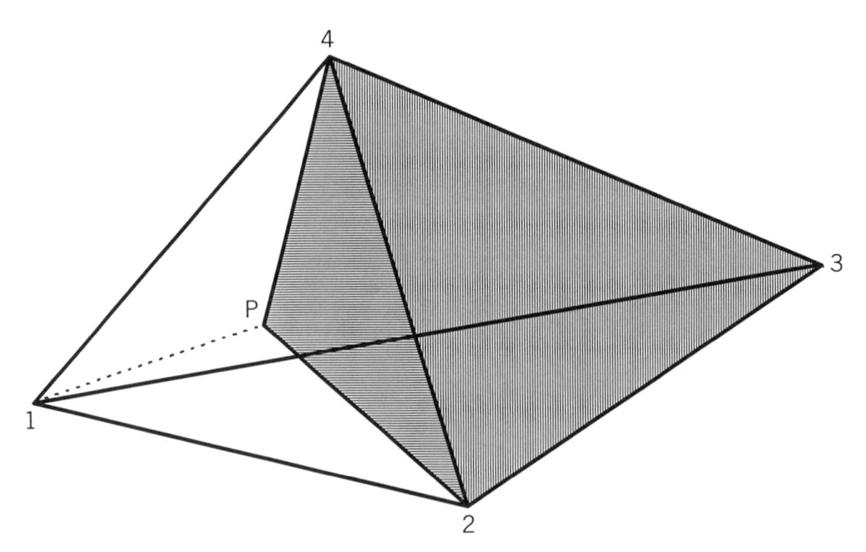

Figure 9.1.4 Tetrahedron defined by nodes 1–2–3–4 and volume V_1 of the subtetrahedron defined by nodes 2–3–4 and point P.

is redundant. The sum of the volume coordinates must be equal to 1, so the redundant volume coordinate may be eliminated using the equation

$$V_1 + V_2 + V_3 + V_4 = 1 \tag{9.1.18}$$

For the isoparametric quadratic tetrahedral element with nodes numbered as shown in Figure 9.1.5, the interpolation functions in volume coordinates can be established as

$$\begin{aligned}
N_1 &= V_1(2V_1 - 1) & N_6 &= 4V_2V_3 \\
N_2 &= V_2(2V_2 - 1) & N_7 &= 4V_1V_3 \\
N_3 &= V_3(2V_3 - 1) & N_8 &= 4V_1V_4 \\
N_4 &= V_4(2V_4 - 1) & N_9 &= 4V_2V_4 \\
N_5 &= 4V_1V_2 & N_{10} &= 4V_3V_4
\end{aligned} \tag{9.1.19}$$

The displacements can be computed at any location inside the element by using the shape functions and the nodal displacement values; that is,

$$u = \sum_{i=1}^{10} N_i u_i, \qquad v = \sum_{i=1}^{10} N_i v_i, \qquad w = \sum_{i=1}^{10} N_i w_i$$

or in matrix form,

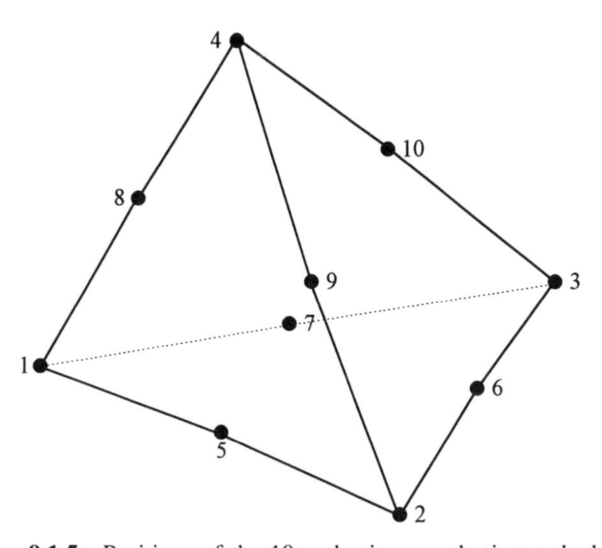

Figure 9.1.5 Positions of the 10 nodes in a quadratic tetrahedron.

$$\begin{Bmatrix} u \\ v \\ w \end{Bmatrix} = \begin{bmatrix} N_1 & 0 & 0 & N_2 & \cdots & N_{10} & 0 & 0 \\ 0 & N_1 & 0 & 0 & \cdots & 0 & N_{10} & 0 \\ 0 & 0 & N_1 & 0 & \cdots & 0 & 0 & N_{10} \end{bmatrix} \begin{bmatrix} u_1 \\ v_1 \\ w_1 \\ \vdots \\ u_{10} \\ v_{10} \\ w_{10} \end{bmatrix} \tag{9.1.20}$$

The calculation of strains within the element requires the spatial derivatives of the displacements with respect to the global Cartesian coordinate system (equation 4.3.13). Using the element interpolation functions and the nodal displacements, the strains are computed by

$$\begin{Bmatrix} \varepsilon_x \\ \varepsilon_y \\ \varepsilon_z \\ \gamma_{xy} \\ \gamma_{xz} \\ \gamma_{yz} \end{Bmatrix} = \begin{bmatrix} \frac{\partial N_1}{\partial x} & 0 & 0 & \frac{\partial N_2}{\partial x} & \cdots & \frac{\partial N_{10}}{\partial x} & 0 & 0 \\ 0 & \frac{\partial N_1}{\partial y} & 0 & 0 & \cdots & 0 & \frac{\partial N_{10}}{\partial y} & 0 \\ 0 & 0 & \frac{\partial N_1}{\partial z} & 0 & \cdots & 0 & 0 & \frac{\partial N_{10}}{\partial z} \\ \frac{\partial N_1}{\partial y} & \frac{\partial N_1}{\partial x} & 0 & 0 & \cdots & \frac{\partial N_{10}}{\partial y} & \frac{\partial N_{10}}{\partial x} & 0 \\ \frac{\partial N_1}{\partial z} & 0 & \frac{\partial N_1}{\partial x} & 0 & \cdots & \frac{\partial N_{10}}{\partial z} & 0 & \frac{\partial N_{10}}{\partial x} \\ 0 & \frac{\partial N_1}{\partial z} & \frac{\partial N_1}{\partial y} & 0 & \cdots & 0 & \frac{\partial N_{10}}{\partial z} & \frac{\partial N_{10}}{\partial y} \end{bmatrix} \begin{bmatrix} u_1 \\ v_1 \\ w_1 \\ \vdots \\ u_{10} \\ v_{10} \\ w_{10} \end{bmatrix} \tag{9.1.21}$$

or

$$\{\varepsilon\} = [B]\{d\} \tag{9.1.22}$$

Since the interpolation functions are defined in terms of volume coordinates, the spatial derivatives with respect to global coordinates required in equation 9.1.21 cannot be calculated directly. The relationship between the shape function derivatives with respect to the independent volume coordinates (V_4 is eliminated by using equation 9.1.18) and the shape function derivatives with respect to global coordinates can be written as

$$\begin{Bmatrix} \frac{\partial N_i}{\partial V_1} \\ \frac{\partial N_i}{\partial V_2} \\ \frac{\partial N_i}{\partial V_3} \end{Bmatrix} = \begin{bmatrix} \frac{\partial x}{\partial V_1} & \frac{\partial y}{\partial V_1} & \frac{\partial z}{\partial V_1} \\ \frac{\partial x}{\partial V_2} & \frac{\partial y}{\partial V_2} & \frac{\partial z}{\partial V_2} \\ \frac{\partial x}{\partial V_3} & \frac{\partial y}{\partial V_3} & \frac{\partial z}{\partial V_3} \end{bmatrix} \begin{Bmatrix} \frac{\partial N_i}{\partial x} \\ \frac{\partial N_i}{\partial y} \\ \frac{\partial N_i}{\partial z} \end{Bmatrix} = \mathbf{J} \begin{Bmatrix} \frac{\partial N_i}{\partial x} \\ \frac{\partial N_i}{\partial y} \\ \frac{\partial N_i}{\partial z} \end{Bmatrix} \tag{9.1.23}$$

where **J** is the Jacobian matrix of the mapping from local to global coordinates. For this element the interpolation functions used to map the geometry are the same as the interpolation functions used to map the displacements. The Jacobian matrix in terms of the local shape function derivatives and nodal coordinates is given by

$$\mathbf{J} = \begin{bmatrix} \sum_{i=1}^{10} \frac{N_i}{\partial V_1} x_i & \sum_{i=1}^{10} \frac{N_i}{\partial V_1} y_i & \sum_{i=1}^{10} \frac{N_i}{\partial V_1} z_i \\ \sum_{i=1}^{10} \frac{N_i}{\partial V_2} x_i & \sum_{i=1}^{10} \frac{N_i}{\partial V_2} y_i & \sum_{i=1}^{10} \frac{N_i}{\partial V_2} z_i \\ \sum_{i=1}^{10} \frac{N_i}{\partial V_3} x_i & \sum_{i=1}^{10} \frac{N_i}{\partial V_3} y_i & \sum_{i=1}^{10} \frac{N_i}{\partial V_3} z_i \end{bmatrix} \tag{9.1.24}$$

Equation 9.1.24 may be inverted to obtain the derivatives of the interpolation functions with respect to the global coordinates so that

$$\begin{Bmatrix} \frac{\partial N_i}{\partial x} \\ \frac{\partial N_i}{\partial y} \\ \frac{\partial N_i}{\partial z} \end{Bmatrix} = \mathbf{J}^{-1} \begin{Bmatrix} \frac{\partial N_i}{\partial V_1} \\ \frac{\partial N_i}{\partial V_2} \\ \frac{\partial N_i}{\partial V_3} \end{Bmatrix} \tag{9.1.25}$$

The determinant of the Jacobian also relates the differential volume in the parent space of the element, dV_p, to the differential volume in the global space of the element, dV_e.

$$dV_e = \det \mathbf{J} \, dV_p \tag{9.1.26}$$

The volume integrals used to compute element mass and stiffness matrices and element force vectors can now be transformed into integrals in the parent space. The element stiffness matrix, $[K_e]$ for example, may be calculated using

$$[K_e] = \int_{V_e} [B]^T[C][B] \, dV_e = \int_{V_p} [B]^T[C][B] \det \mathbf{J} \, dV_p \tag{9.1.27}$$

The integral in equation 9.1.27 is computed using numerical integration. The integral is evaluated by computing the integrand at discrete points within the element, multiplying by a weighting factor and summing, that is,

$$\int_{V_p} [B]^T[C][B] \det \mathbf{J}\, dV_p = \sum_{i=1}^{n} [B_i]^T[C_i][B_i] \det \mathbf{J}_i\, W_i \qquad (9.1.28)$$

where n is the number of integration points. The weights and integration point locations required to evaluate the integrand depend on the polynomial order of the integrand. The integration point locations and weights typically used for evaluation of the stiffness matrix of the quadratic tetrahedral element are shown in Figure 9.1.6. It is important to remember that the determinant of the Jacobian and the matrix $[B_i]$ depend on location. When material nonlinearities are present a tangent material matrix is often used in the stiffness calculation. The tangent material matrix (discussed in the next section) depends on the material state at that particular location. Therefore, it is necessary to compute and store the inelastic strains, inelastic strain rates or increments, state variables, and so on, at the element integration points for every element in the model.

The element internal force vector defined in equation 9.1.12 is numerically integrated with the same rule as that used for the element stiffness calculation. The strain displacement matrix $[B]$ and stress vector $\{\sigma\}$ vary with location and must be evaluated at each integration point. To compute stress, the total strain is first recovered using the strain displacement matrix and the displacements (equation 9.1.5). The elastic strain is then recovered from the total strain by subtracting out the thermal and inelastic components:

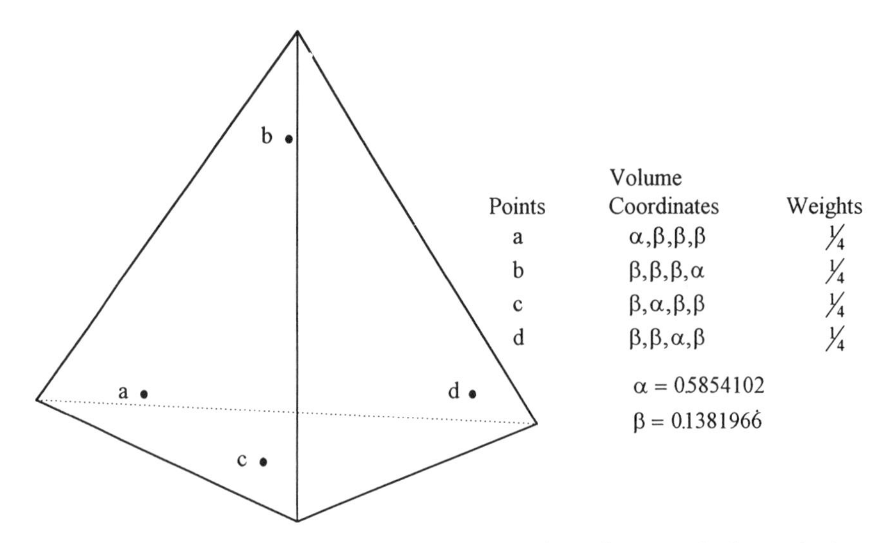

Figure 9.1.6 Numerical integration points and weights for a quadratic tetrahedron.

$$\{\varepsilon^e\} = \{\varepsilon\} - \{\varepsilon^{th}\} - \{\varepsilon^I\}$$

The stress is computed from Hooke's law using the matrix of elastic constants and the elastic strains, equation 4.5.3.

The inelastic strain and the inelastic strain increment are required at the element integration points for both the element internal force vector and the element tangent stiffness matrix calculations. Therefore, the inelastic strains and material history variables unique to the particular constitutive model are calculated and stored for element integration points throughout the model.

9.2 TANGENT MATERIAL MATRIX

When using the finite element method it may be necessary to compute a tangent material modulus $[C^T]$ that relates the stress increment to the total strain increment. The tangent material modulus is used to compute the tangent stiffness matrix in Equation 9.1.15 and is defined by

$$\{d\sigma\} = [C^T]\{d\varepsilon\} \tag{9.2.1}$$

in matrix notation. The tangent material matrix will vary from point to point in a body, depending on the material state at that particular location and the applied loads. The equation for the tangent material modulus will also depend on the constitutive model and on the algorithm used to integrate the incremental or rate equations.

9.2.1 Strain Rate Independent Materials

Strain rate independent models include yield surface plasticity presented in Chapter 5. First a continuum tangent modulus will be developed for infinitesimal stress and strain increments. Then an algorithmic or consistent tangent modulus will be developed for finite stress and strain increments.

The stress–elastic strain relationship is defined using Hooke's Law (equation 4.5.1), and the decomposition of total strain into elastic and plastic components is given by equation 5.2.1. Combining these equations produce an expression for the stress increment as a function of the elastic constants, the total strain increment, and the plastic strain increment:

$$d\sigma_{ij} = C_{ijkl}(d\varepsilon_{kl} - d\varepsilon_{kl}^p) \tag{9.2.2}$$

The tangent material modulus C_{ijkl}^t relates the stress increment to the total strain increment and is defined by

$$d\sigma_{ij} = C_{ijkl}^t\, d\varepsilon_{kl} = C_{ijkl}^t(d\varepsilon_{kl}^e + d\varepsilon_{kl}^p) \tag{9.2.3}$$

The tangent material modulus may be found as a function of the elastic material properties, the current stress state, and the plastic potential function. First, using the general expression for the plastic strain increment in terms of a plastic potential function (equation 5.2.12) gives

$$d\sigma_{ij} = C^t_{ijkl}\, d\varepsilon_{kl} = C^t_{ijkl}\left(d\varepsilon^e_{kl} + d\lambda\,\frac{\partial g}{\partial \sigma_{kl}}\right) \tag{9.2.4}$$

The consistency condition (equation 5.4.5) requires that during plastic flow the stress increment remains on the yield surface. Substituting equations 9.2.2 and 5.2.12 into equation 5.4.5 produces

$$\frac{\partial F}{\partial \sigma_{ij}} C_{ijkl}\left(d\varepsilon_{kl} - d\lambda\,\frac{\partial g}{\partial \sigma_{kl}}\right) + \frac{\partial F}{\partial \varepsilon^p_{ij}}\, d\lambda\,\frac{\partial g}{\partial \sigma_{ij}} = 0 \tag{9.2.5}$$

and solving for $d\lambda$ gives

$$d\lambda = \frac{(\partial F/\partial \sigma_{ij})C_{ijkl}\, d\varepsilon_{kl}}{C_{ijkl}(\partial F/\partial \sigma_{ij})(\partial g/\partial \sigma_{kl}) - (\partial F/\partial \varepsilon^p_{ij})(\partial g/\partial \sigma_{ij})} \tag{9.2.6}$$

Substituting equation 9.2.6 into equation 9.2.4 gives

$$d\sigma_{ij} = C^t_{ijkl} d\varepsilon_{kl}$$

where

$$C^t_{ijkl} = C_{ijkl} - \frac{C_{ijrs}(\partial F/\partial \sigma_{rs})(\partial g/\partial \sigma_{mn})C_{mnkl}}{(\partial F/\partial \sigma_{qr})C_{qrst}(\partial g/\partial \sigma_{st}) - (\partial F/\partial \varepsilon^p_{uv})(\partial g/\partial \sigma_{uv})} \tag{9.2.7}$$

Equation 9.2.7 may be expressed in matrix form as

$$[C^t] = [C] - \frac{[C]\{a\}\{b\}^T[C]}{A + \{a\}^T[C]\{b\}} \tag{9.2.8}$$

The column vector $\{b\}$ is the plastic flow vector,

$$\{b\} = \frac{\partial g}{\partial \{\sigma\}} \tag{9.2.9}$$

the column vector $\{a\}$ is defined by

$$\{a\} = \frac{\partial F}{\partial \{\sigma\}} \tag{9.2.10}$$

and the scalar A is

$$A = -\left\{\frac{\partial F}{\partial\{\varepsilon^p\}}\right\}^T \left\{\frac{\partial g}{\partial\{\sigma\}}\right\} \tag{9.2.11}$$

Recall that F is the yield function (equation 5.1.1) and g is a plastic potential function used to derive the plastic strain increments (equation 5.2.12). For the special case of an associated flow rule, which is typical for metals, the yield function and the plastic potential function are the same and $\{b\} = \{a\}$. Equation 9.2.8 may be simplified further to

$$[C^T] = [C] - \frac{[C]\{a\}\{a\}^T[C]}{A + \{a\}^T[C]\{a\}} \tag{9.2.12}$$

The tangent material moduli defined by equations 9.2.8 and 9.2.12 are for infinitesimal strain steps and are usually called continuum tangent moduli.

Finite length steps are required in a numerical procedure, and an algorithmic tangent modulus or *consistent tangent modulus* is often used. A Newton–Raphson solution procedure that uses a tangent stiffness based on the consistent tangent modulus will converge more rapidly than a procedure that uses a tangent stiffness based on the continuum tangent modulus. The consistent tangent modulus is derived from the derivatives of the incremental equations. Using matrix notation the incremental form of equation 9.2.2 is

$$\{\Delta\sigma\} = [C](\{\Delta\varepsilon\} - \Delta\lambda\{b\}) \tag{9.2.13}$$

Taking the differential of the incremental equation produces

$$d\{\Delta\sigma\} = [C](d\{\Delta\varepsilon\}) - d(\Delta\lambda)[C]\{b\} - \Delta\lambda[C]\frac{\partial\{b\}}{\partial\{\Delta\sigma\}}d\{\Delta\sigma\} \tag{9.2.14}$$

The matrix $\partial\{b\}/\partial\{\Delta\sigma\}$ is the change in the flow vector over the finite stress increment $\Delta\sigma$ and leads to the difference between the continuum tangent modulus and the consistent tangent modulus. Rearranging equation 9.2.14 gives

$$\left([I] + \Delta\lambda[C]\frac{\partial\{b\}}{\partial\{\Delta\sigma\}}\right)d\{\Delta\sigma\} = [C](d\{\Delta\varepsilon\}) - d(\Delta\lambda)[C]\{b\} \tag{9.2.15}$$

Making the substitution

$$[\hat{C}] = \left([I] + \Delta\lambda[C]\frac{\partial\{b\}}{\partial\{\Delta\sigma\}}\right)^{-1}[C] \tag{9.2.16}$$

equation 9.2.15 may be written as

$$d\{\Delta\sigma\} = [\hat{C}](d\{\Delta\varepsilon\} - d(\Delta\lambda)\{b\}) \tag{9.2.17}$$

The consistency condition (equation 5.4.5) written in matrix form is

$$\{a\}^T d\{\Delta\sigma\} - d(\Delta\lambda)A = 0 \tag{9.2.18}$$

where the vector $\{a\}$ is defined by equation 9.2.10 and the scalar A is given by equation 9.2.11. Substituting the expression for $d\{\Delta\sigma\}$ in equation 9.2.17 into equation 9.2.18 and solving for $d(\Delta\lambda)$ gives

$$d(\Delta\lambda) = \frac{\{a\}^T[\hat{C}]d\{\Delta\varepsilon\}}{A + \{a\}^T[\hat{C}]\{b\}} \tag{9.2.19}$$

Substituting equation 9.2.19 into equation 9.2.17 results in

$$d\{\Delta\sigma\} = [C^{ct}]d\{\Delta\varepsilon\} \tag{9.2.20}$$

where the consistent tangent modulus is given by

$$[C^{ct}] = [\hat{C}] - \frac{[\hat{C}]\{b\}\{a\}^T}{A + \{a\}^T[\hat{C}]\{b\}}[\hat{C}] \tag{9.2.21}$$

For an associated flow rule (i.e., $\{b\} = \{a\}$), the expression for the consistent tangent modulus reduces further to

$$[C^{ct}] = [\hat{C}] - \frac{[\hat{C}]\{a\}\{a\}^T}{A + \{a\}^T[\hat{C}]\{a\}}[\hat{C}] \tag{9.2.22}$$

The consistent moduli in equations 9.2.21 and 9.2.22 are mathematically similar to the continuum moduli in equations 9.2.8 and 9.2.11, except that $[\hat{C}]$ is used rather than $[C]$. Further details of the derivation of a consistent tangent modulus for rate-independent plasticity models may be found in Simo and Taylor (1985), Crisfleld (1991), and Hinton (1992).

9.2.2 Strain-Rate-Sensitive Materials

Experience has shown that finite element procedures for models with a creep or inelastic strain rate that require frequent reforming of the stiffness matrix can be prohibitively expensive. When the inelastic strain increments are small, it is usually more efficient to use the elastic material properties to form an elastic stiffness matrix once and use this stiffness matrix throughout the solution procedure. However, when the inelastic strain increments are large, it may be better to use a tangent stiffness matrix based on an incremental tangent material modulus.

For constant temperature the incremental stress change over a time increment is given by

$$\{\Delta\sigma\} = [C](\{\Delta\varepsilon\} - \{\Delta\varepsilon^{\mathrm{I}}\}) \tag{9.2.23}$$

The creep or inelastic strain increment may be written as

$$\{\Delta\varepsilon^{\mathrm{I}}\} = \Delta t\{\dot{\hat{\varepsilon}}^{\mathrm{I}}\} \tag{9.2.24}$$

where the average inelastic strain rate or creep rate vector, $\{\dot{\hat{\varepsilon}}^{\mathrm{I}}\}$ is defined for the generalized trapezoidal rule as

$$\{\dot{\hat{\varepsilon}}^{\mathrm{I}}\} = (1 - \beta)\{\dot{\varepsilon}^{\mathrm{I}}\}_t + \beta\{\dot{\varepsilon}^{\mathrm{I}}\}_{t+\Delta t} \tag{9.2.25}$$

and for the generalized midpoint rule as

$$\{\dot{\hat{\varepsilon}}^{\mathrm{I}}\} = \{\dot{\varepsilon}^{\mathrm{I}}\}_{t+\beta\Delta t} \tag{9.2.26}$$

The incremental stress equation 9.2.23, can be written as

$$\{\Delta\sigma\} - [C]\{\Delta\varepsilon\} + \Delta t[C]\{\dot{\hat{\varepsilon}}^{\mathrm{I}}\} = \{0\} \tag{9.2.27}$$

To determine a relationship between the stress and inelastic strain increments, it is convenient to consider only the differential of $\{\dot{\hat{\varepsilon}}^{\mathrm{I}}\}$ with respect to stress and assume that the other components of the differential may be neglected. This approximation leads to

$$d\{\Delta\sigma\} - [C]d\{\Delta\varepsilon\} + \Delta t[C]\left\{\frac{\partial\{\dot{\hat{\varepsilon}}^{\mathrm{I}}\}}{\partial\{\sigma\}}\right\} d\{\Delta\sigma\} = 0 \tag{9.2.28}$$

Solving equation 9.2.28 for the differential of the stress increment produces

$$d\{\Delta\sigma\} = \left([I] + \Delta t[C]\left\{\frac{\partial\{\dot{\hat{\varepsilon}}^{\mathrm{I}}\}}{\partial\{\sigma\}}\right\}\right)^{-1} [C]d\{\Delta\varepsilon\} \tag{9.2.29}$$

The tangent material modulus for a strain-rate-sensitive material is therefore given by

$$[C'] = \left([I] + \Delta t[C]\left\{\frac{\partial\{\hat{\dot{\varepsilon}}^{\mathrm{I}}\}}{\partial\{\sigma\}}\right\}\right)^{-1}[C] \tag{9.2.30}$$

When using the generalized trapezoidal rule, the inelastic strain rate at the beginning of the time step is fixed and not a function of the stress change during the time step; therefore, using equation 9.2.25, the derivative of the average inelastic strain rate with respect to stress becomes

$$\left\{\frac{\partial\{\hat{\dot{\varepsilon}}^{\mathrm{I}}\}}{\partial\{\sigma\}}\right\} = \beta\left\{\frac{\partial\{\dot{\varepsilon}^{\mathrm{I}}\}_{t+\Delta t}}{\partial\{\sigma\}}\right\} \tag{9.2.31}$$

The derivation of a specific tangent modulus for unified state variable model or a creep model requires using the appropriate representation in equation 9.2.30. Specific representations for the Miller model and the Bodner model using a generalized trapezoidal rule integrator appear in Dombrovsky (1992). Further details of the derivation of a consistent tangent modulus for the classical creep models may be found in Zienkiewicz and Taylor (1991).

9.3 APPLICATION OF AN ASYMPTOTIC INTEGRATOR

The unified strain constitutive models using state variables have been found to represent the mechanical response of metals fairly effectively. However, these constitutive models are generally defined by a set of coupled, mathematically stiff, nonlinear first-order differential equations, which are difficult to integrate numerically. The integration often requires relatively small time steps (Gear, 1971), which can lead to numerical instability or extensive computing times, especially in nonlinear finite element programs. An asymptotic integration algorithm was recently proposed by Walker and Freed (1991) for state-variable constitutive models. This algorithm involves integrating the first-order differential equations analytically to obtain a recursive form. The formulation is unconditionally stable when applied to viscoplasticity.

The section begins by recasting the constitutive equations into first-order differential equations, followed by the asymptotic algorithm. Implementation of the iteration procedure and update schemes are discussed. Numerical results are compared in terms of accuracy and time-step sizes to those obtained from a predictor–corrector trapezoidal rule. For ease of comparison, only fixed time steps were used in both calculations.

The state variable model in this development is defined by inelastic strain rate (equation 6.2.13) and back stress (equation 6.3.6). This application is directed toward fatigue calculations for cyclically stable materials, so drag stress evolution and the effect of static thermal recovery associated with creep are not included for convenience. Therefore, Ω_m and Z are constant. Introducing the total deviatoric strain rate, $\dot{E}_{ij}$, equation 4.5.7 can be rewritten as

$$\dot{S}_{ij} = 2\mu(\dot{E}_{ij} - \dot{\varepsilon}^{I}_{ij}) \tag{9.3.1}$$

Equation 9.3.1 can then be used with equations 6.2.3, 7.2.5 and 7.5.1 to establish a set of first-order differential equations in the form

$$\dot{S}_{ij} + 2\mu g_1 S_{ij} = 2\mu \dot{E}_{ij} + 2\mu g_1 \Omega_{ij}$$

and (9.3.2)

$$\dot{\Omega}_{ij} + g_2 \Omega_{ij} = 2\mu f_2 \dot{E}_{ij} + (\tfrac{2}{3} f_1 - 2\mu f_2) g_1 S_{ij}$$

where g_1 and g_2 are defined as

$$g_1 = D_0 \exp\left[-\left(\frac{Z^2}{3T^2 K_2}\right)^n\right] \frac{1}{\sqrt{K_2}} \quad \text{and}$$

$$g_2 = \left(\frac{2f_1\sqrt{K_2}}{\sqrt{3}\Omega_m} + \frac{2}{3} f_1 - 2\mu f_2\right) g_1$$

Equations 9.3.2 are a system of 12 independent first-order nonlinear coupled differential equations for the deviatoric stress and back stress.

The properties of the asymptotic integration algorithm proposed by Walker and Freed (1991) can be illustrated through the use of a single first-order differential equation

$$\dot{X} + U_1 X = V_1 \tag{9.3.3}$$

The parameters U_1 and V_1 are functions of the variable X and time and are defined such that

$$U(t) - U(0) = \int_0^t U_1(\tau)\, d\tau \qquad \text{and} \qquad V(t) - V(0) = \int_0^t V_1(\tau)\, d\tau$$

Introducing the integrating factor λ as given by Boyce and DiPrima (1969), equation 9.3.3 can be rewritten in the form

$$\lambda(\dot{X} + U_1 X) \equiv \frac{d}{dt}\lambda X = \dot{\lambda} X + \lambda \dot{X}$$

Upon simplifying the equation above, it follows that $\dot{\lambda} = \lambda U_1$; thus the integrating factor is defined as $\lambda = e^U$. Multiplying equation 9.3.3 by $\lambda = e^U$ gives

$$\frac{d}{dt}(Xe^U) = V_1 e^U$$

Integrating equation 9.3.3 over the interval $[t, t + \Delta t]$ and solving for $X(t + \Delta t)$ gives the general recursive integration solution

$$X(t + \Delta t) = X(t)e^{-[U(t+\Delta t)-U(t)]} + \int_t^{t+\Delta t} e^{-[U(t+\Delta t)-U(\tau)]} \frac{\partial V(\tau)}{\partial \tau} d\tau \quad (9.3.4)$$

The integral is the contribution of the nonhomogeneous term to the solution. Equation 9.3.4 is practical only when a solution exists for the integral

$$I(\Delta \tau) = \int_t^{t+\Delta t} e^{-[U(t+\Delta t)-U(\tau)]} \frac{\partial V(\tau)}{\partial \tau} d\tau \quad (9.3.5)$$

Walker and Freed found that the integral can be approximated by expanding the integrand in a Taylor series and then integrating term by term. The implicit solution is found by expanding about the upper limit, $t + \Delta t$, while the explicit solution is obtained by expanding about the lower limit, t. Since it was found that the function U_1 is greater than zero in viscoplasticity, U is a monotonically increasing function, and the exponential term in the integral will make the largest contribution to the solution at the upper limit. Further, the solution possesses a fading memory of the forcing function V_1, so the solution will depend more strongly on the most recent values of V_1. Thus an implicit Taylor series expansion about the upper limit is more appropriate for viscoplasticity and the expansion will require a smaller number of terms.

The next step is to expand the functions U and V in a Taylor series to obtain an implicit representation of the integral. By making a change of variable in equation 9.3.5 and setting $\tau = t + \Delta t + z$, the integral becomes

$$I(\Delta \tau) = -\int_0^{\Delta t} e^{-[U(t+\Delta t)-U(t+\Delta t-z)]} \frac{\partial V(t + \Delta t - z)}{\partial z} dz$$

Expanding the function U and V in a Taylor series about the upper limit $t + \Delta t$ and neglecting the higher-order terms gives

$$U(t + \Delta t - z) \approx U(t + \Delta t) - U_1(t + \Delta t)z$$

$$V(t + \Delta t - z) \approx V(t + \Delta t) - V_1(t + \Delta t)z$$

Evaluating the integral with the use of the Taylor expansions gives an implicit asymptotic solution as

$$X(t + \Delta t) \approx X(t)e^{-U_1(t+\Delta t)\Delta t} + \frac{V_1(t + \Delta t)}{U_1(t + \Delta t)}(1 - e^{-U_1(t+\Delta t)\Delta t}) \quad (9.3.6)$$

This is a recursive solution of the differential equation. Since this is an implicit expansion, the parameters U_1 and V_1 are evaluated at $t + \Delta t$ and are unknown a priori. For large time steps when $U_1(t + \Delta t)\Delta t \to \infty$, the solution becomes

$$\lim_{U_1(t+\Delta t)\Delta t \to \infty} X(t + \Delta t) \approx \frac{V_1(t + \Delta t)}{U_1(t + \Delta t)}$$

which is asymptotically accurate and stable for large time steps.

Let us summarize by recasting the solution of the differential equations for the deviatoric stress and back stress (equations 9.3.2) in the form of equation 9.3.6. Define

$$\begin{aligned} X_{ij}^{(1)} &= S_{ij} \qquad U^{(1)} = 2\mu g_1 \qquad V_{ij}^{(1)} = 2\mu \dot{E}_{ij}^t + 2\mu g_1 X_{ij}^{(2)} \\ X_{ij}^{(2)} &= \Omega_{ij} \qquad U^{(2)} = g_2 \qquad V_{ij}^{(2)} = 2\mu f_2 \dot{E}_{ij}^t + (\tfrac{2}{3} f_1 - 2\mu f_2) g_1 X_{ij}^{(1)} \end{aligned} \quad (9.3.7)$$

Then for $\alpha = 1, 2$, the solution for the constitutive model becomes

$$X_{ij}^{\alpha}(t + \Delta t) \approx X_{ij}^{\alpha}(t)e^{-U^{\alpha}(t+\Delta t)\Delta t} + \frac{V_{ij}^{\alpha}(t + \Delta t)}{U^{\alpha}(t + \Delta t)}(1 - e^{-U^{\alpha}(t+\Delta t)\Delta t}) \quad (9.3.8)$$

Note that the inelastic strain rate is not explicitly determined using the asymptotic integral. The accumulated inelastic strain is determined from the integral of equation 9.3.1, $S_{ij} = 2\mu(E_{ij} - \varepsilon_{ij}^I)$ once the deviatoric stress has been obtained. The total deviatoric strain history $E_{ij}(t)$ is the independent variable in the solution procedure.

Since equation 9.3.8 is implicit in nature, an iteration and update procedure is required for each time step. An algorithm for these equations is summarized by the following steps:

1. Determine the input loading ε_{ij}, $\dot{\varepsilon}_{ij}$, E_{ij}, and $\dot{E}_{ij}$ for the current time step.

2. Estimate or assume the initial values for S_{ij}, Ω_{ij}, U^1, and U^2.
3. Compute the forcing terms V_{ij}^1, V_{ij}^2 from S_{ij}, Ω_{ij} and equation 9.3.7.
4. Solve for S_{ij} and Ω_{ij} using equation 9.3.8.
5. Check for convergence: If satisfied, update S_{ij} and Ω_{ij} and go to next time step; if not satisfied, update U^1 and U^2 and return to step 3.

Several update procedures have been suggested in the literature, but experience has shown that a simple method of successive substitution is superior to the other update procedures. Convergence is checked for a_α, variables (α = 1, 4): stress and inelastic strain, U^1 and U^2, respectively. The variables a_α in two successive iterations are required to be less than a specified tolerance b_α, in the form

$$\left| \frac{a_\alpha^i - a_\alpha^{i-1}}{a_\alpha^i} \right| > b_\alpha \tag{9.3.9}$$

where the superscript denotes the iteration number.

To benchmark and demonstrate the properties of the asymptotic integration algorithm, the results of the algorithm are compared to results of the constitutive model integrated with an implicit predictor–corrector trapezoidal rule algorithm. The exercises were run for René 80 at 871°C under strain controlled uniaxial cyclic loading. Applications of the method in a P version finite element model of a two material strip under mechanical cycling are given by Chen (1993).

A René 80 tensile specimen was loaded through $1\frac{1}{4}$ strain controlled cycles at 871°C. The piecewise linear strain history is defined by the strain and time coordinates (0, 0.005, −0.005, 0.005) and (0, 30, 90, 150 sec), respectively. The material parameters are listed in Appendix 6.1. The constitutive equation was implemented using 500 fixed time steps to serve as the baseline for comparison. The implicit asymptotic algorithm (equation 9.3.8) was implemented using a successive substitution update procedure.

The algorithm was tested first for a strain amplitude of 0.005, which corresponds to an inelastic strain amplitude of approximately 0.0015. The results obtained from the asymptotic algorithm with 50 time steps are practically identical to the baseline using the trapezoidal rule as shown in Figure 9.3.1*a*. Larger time step sizes were then tested. The trapezoidal rule failed to converge with 25 time steps, while the asymptotic method had reasonable results (about 2% error). Figure 9.3.1*b* shows the results of the asymptotic method using 500, 25, and 5 time steps. The error in stress was larger when only 5 time steps were used. Note that these results were computed by using only the linear terms in a Taylor series expansion. Table 9.3.1 shows the total number of iterations required to produce the results.

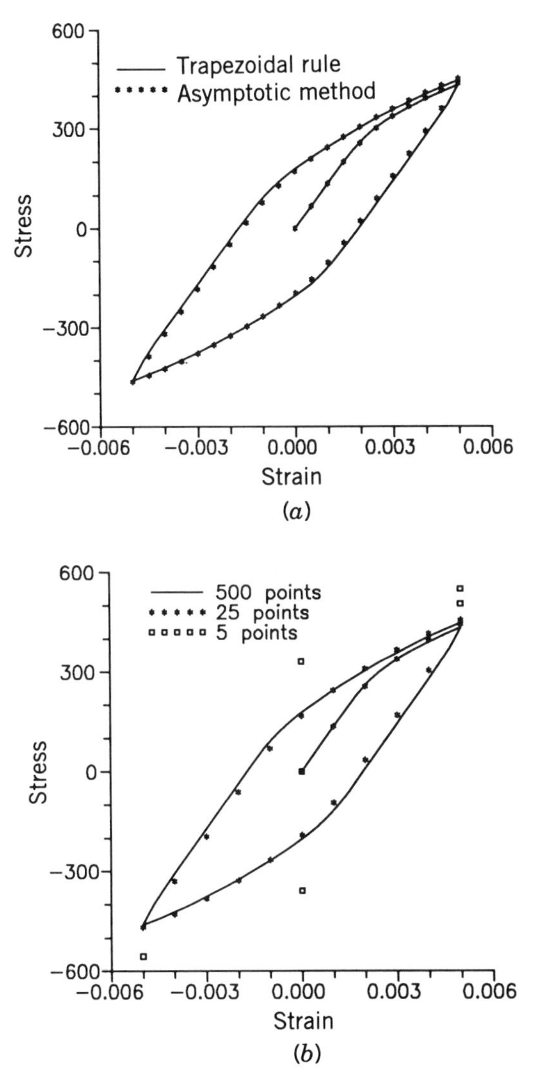

Figure 9.3.1 Fatigue calculations for a strain amplitude of 0.005: (*a*) comparison of the trapezoid rule of integration with the asymptotic integration algorithm for 500 and 50 time steps; (*b*) asymptotic integration algorithm for 500, 25, and 5 time steps. (After Chen, 1993.)

The method is next examined when the inelastic strain amplitude is large. The strain amplitude was increased from 0.005 to 0.02, increasing the inelastic strain amplitude to about 0.015. The results in Figure 9.3.2 show the error was less than 1% for five time steps with the larger strain range. It appears that the implicit asymptotic algorithm permits the use of very large time steps and is increasingly accurate for larger inelastic strain cycles. The successive

Table 9.3.1 Total Number of Iterations for Two Methods of Integration

Number of Time Steps	Trapezoid Algorithm[a]	Asymptotic Algorithm
500	1003	1740
50	146	341
25	NA	225
10	NA	207
5	NA	140

Source: After Chen (1993).

[a]NA, not applicable; the trapezoid method failed to converge for 5, 10, and 25 time steps.

substitution update method is simple and effective and results can be fiirther improved by improving the initial values of the variables.

Chang, Saleeb, and Iskovitz (1993) recently implemented Walker's model (Section 7.9) and Robinson's model (Problem 7.6) in a finite element code using the asymptotic integration algorithm. Results indicate that the asymptotic integration algorithm is more eflicient than a trapezoidal integration rule.

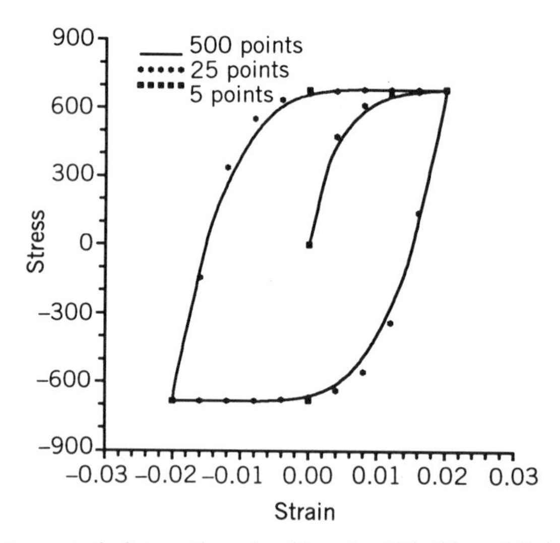

Figure 9.3.2 Asymptotic integration algorithm for 500, 25, and 5 time steps applied to a fatigue loop with a total strain amplitude of 0.02. (After Chen, 1993.)

Iskovitz, Chang, and Saleeb (1994) applied the technique to Robinson's model for transversely isotropic materials with similar success.

9.4 FINITE ELEMENT ANALYSIS OF NOTCH BEHAVIOR

This section presents the results of a finite element study comparing the predictions of the Bodner–Partom state-variable model (Section 7.7) implemented in a finite element code and experimentally obtained data for a notched specimen made of the high-temperature nickel-based superalloy Inconel 718 (Figure 9.4.1). The purpose of the study was to test a state-variable model for multiaxial stress conditions (Dame, Stouffer, and Abuelfoutouh, 1984).

Two forms of the Bodner–Partom state-variable equations were used in this study. The first is one of the earlier published versions of the model (Bodner and Partom, 1972, 1975). The inelastic strain rate is given by

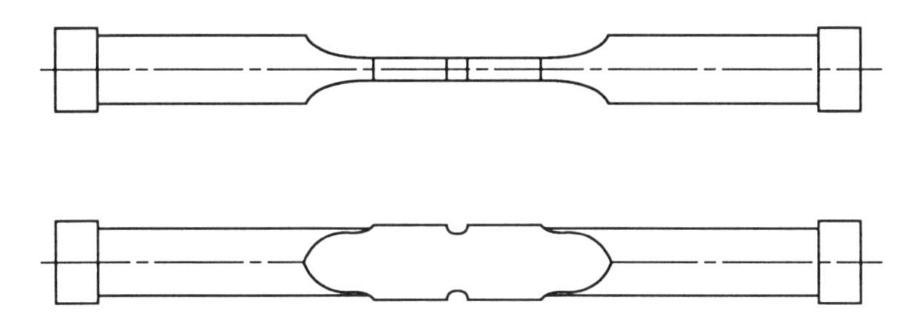

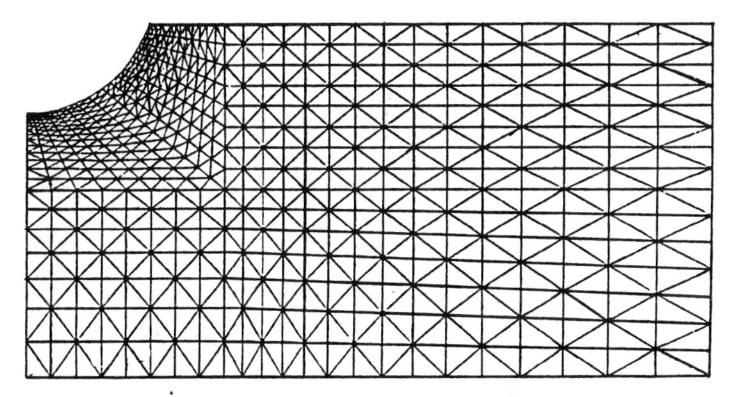

Figure 9.4.1 Benchmark notch specimen and finite element mesh at the root of the notch. (From Dame, Stouffer, and Abuelfoutouh, 1984.)

$$\dot{\varepsilon}_{ij}^{I} = D_0 \exp\left[-\frac{n+1}{2n}\left(\frac{Z^2}{3J_2}\right)^n\right]\frac{S_{ij}}{\sqrt{J_2}} \tag{9.4.1}$$

rather than equation 6.2.12. The only difference is the presence of $n + 1/n$ in the flow law. The quantities D_0 and n are material parameters, S_{ij} the deviatoric stress tensor, J_2 the second invariant of the deviatoric stress tensor, and Z a scalar state variable. The evolution equation for Z is given by equation 7.7.2. The second model is a state variable model of Abuelfoutouh (1983) that is derived from a potential function that was established from thermodynamics. One of the main results from the study was that the flow and evolution equations should not be picked independently. The potential function concept was used to derive a second evolution equation for the Bodner–Partom flow law. The material constants used in both models were derived from uniaxial smooth bar tensile and creep test data. The material constants for Inconel 718 at 649°C are listed in Appendix 7.2.

The Bodner model was implemented in a nonlinear finite element code that used an initial stiffness iteration procedure. The inelastic strain rate and state variable rate equations were integrated using a predictor–corrector method with iterative corrections. The predictor is a forward Euler integrator and the corrector is a trapezoidal rule (see Section 5.6). Plane stress linear triangular elements were used in the finite element model. Since these elements are constant stress elements, a fine mesh was required to obtain reasonable results. The quality of the mesh in Figure 9.4.1 was verified by comparing elastic finite element results with the theoretical stress concentration factor.

A comparison of experimental and calculated response for uniaxial tensile and creep tests is shown in Figure 9.4.2. This comparison validates the solution procedure and material constants for uniaxial conditions. The experimental Inconel 718 notch results were obtained for a variety of loading conditions using a laser interferometric strain displacement gage (Domas et al., 1982). The gage was 100 μm in length and was placed at the root of the notch. Measurements were made for six load histories, including continuous cycling and cycling with hold-time periods in tension, tension and compression, and compression. The specimen was the thin flat double-notched bar in Figure 9.4.1 with an elastic stress concentration factor of 1.9. The experimental study also included a number of smooth bar tests that were used to calculate material parameters.

Typical strain response results at the root of the notch for three tests are shown in Figures 9.4.3 to 9.4.5. The load is held constant for 2 min in compression, tension and compression, and tension, respectively. The finite element and experimental results are for the first cycle. Although the inelastic strain levels are not large, the analytical predictions can be seen to be fairly accurate.

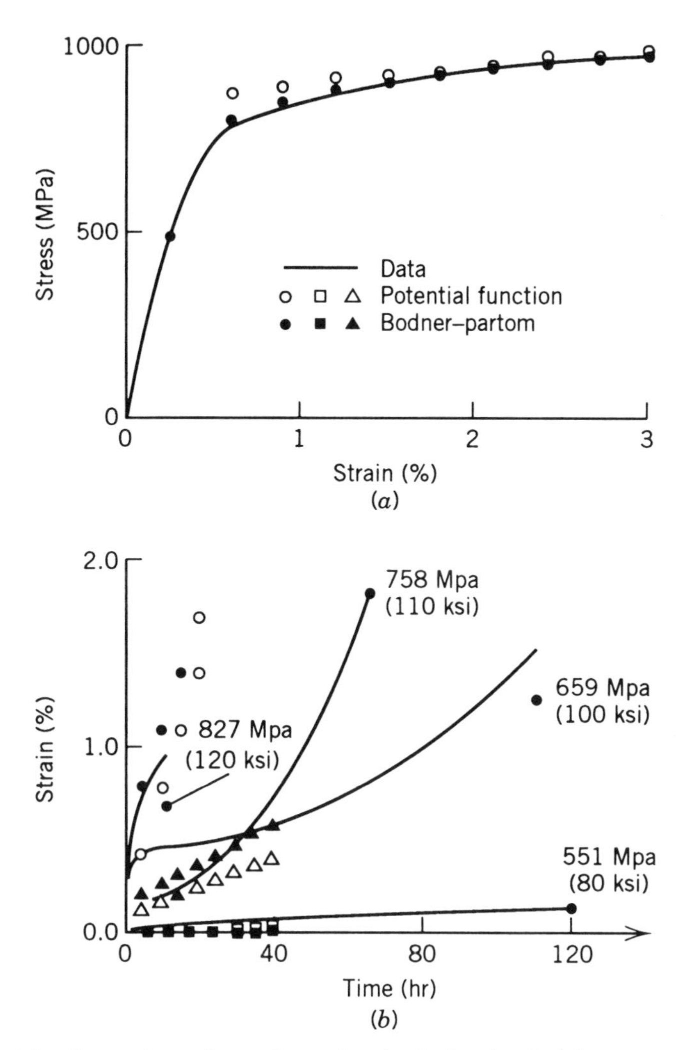

Figure 9.4.2 Comparison of experimental and calculated uniaxial response of Inconel 718 at 649°C using the Bodner–Partom model and a potential function formulation that resulted in a modified form of the Bodner–Partom evolution equation: (*a*) tensile response at 1% min^{-1}; (*b*) creep response at 689, 758, and 827 MPa. (From Dame, Stouffer, and Abuelfoutouh, 1984.)

9.5 SUMMARY

The finite element formulation in this chapter is limited to small strains, the stiffness matrix and boundary conditions are independent of the displacements, and the motions are quasi-static. These assumptions simplify the for-

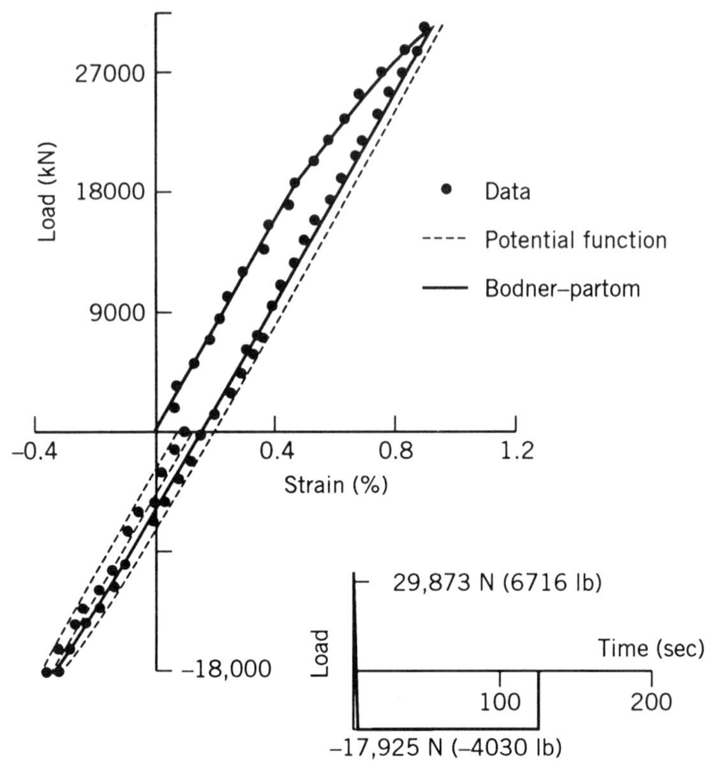

Figure 9.4.3 Comparison of experimental and calculated load-strain response at the root of the notch for Benchmark notch test 6 with a hold in compression. (From Dame, Stouffer, and Abuelfoutouh, 1984.)

mulation to focus on the implementation on nonlinear material response. Extensions of the finite element method to include dynamic and nonlinear geometric effects may be found in the references cited.

The essential difference between linear and nonlinear finite element formulations is that the nonlinear solution is iterative, and it is obtained as a sequence of linear solutions. The stiffness matrix relates the incremental displacements to forces, but it does not appear in the equilibrium equation. The most popular methods for solving the nonlinear equations involve some form of the Newton–Raphson iteration procedure.

Calculation of the equilibrium imbalance or residual force requires stresses that depend on the inelastic strains. The tangent stiffness matrix is a function of the tangent material matrix, which depends on the material model for the inelastic strains. Since the residual force vector and tangent stiffness matrix are integrated numerically, the constitutive equation must be evaluated at the element integration points.

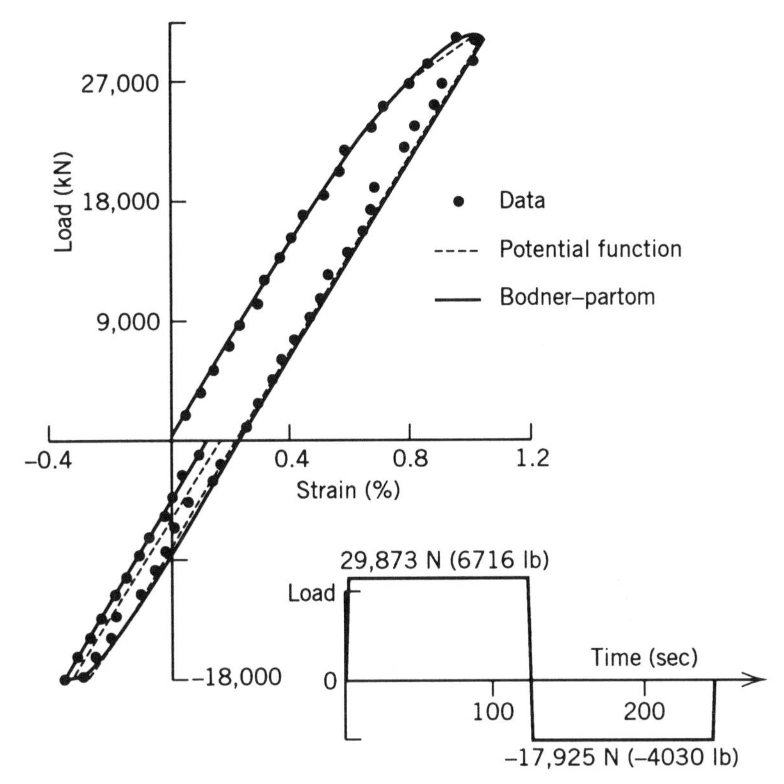

Figure 9.4.4 Comparison of experimental and calculated load-strain response at the root of the notch for Benchmark notch test 9 with a hold in tension and compression. (From Dame, Stouffer, and Abuelfoutouh, 1984.)

Nonlinear finite element solutions require significantly more computer resources than the linear solution of the same problem. Each iteration involving an update of the global stiffness matrix requires approximately the same computer time as a complete linear solution. Strain-rate-dependent constitutive models that must be integrated over time are significantly more costly to run than strain-rate-independent models. A small amount of additional disk space is required for storage of the inelastic strains, inelastic strain increments, state variables, and so on, at each integration point.

Implementation of rate-independent models, the classical creep models, and unified state variables is very similar. Plasticity and creep formulations have been available in commercial finite element codes for several years, but unified state-variable models have not yet appeared in commercial finite element codes.

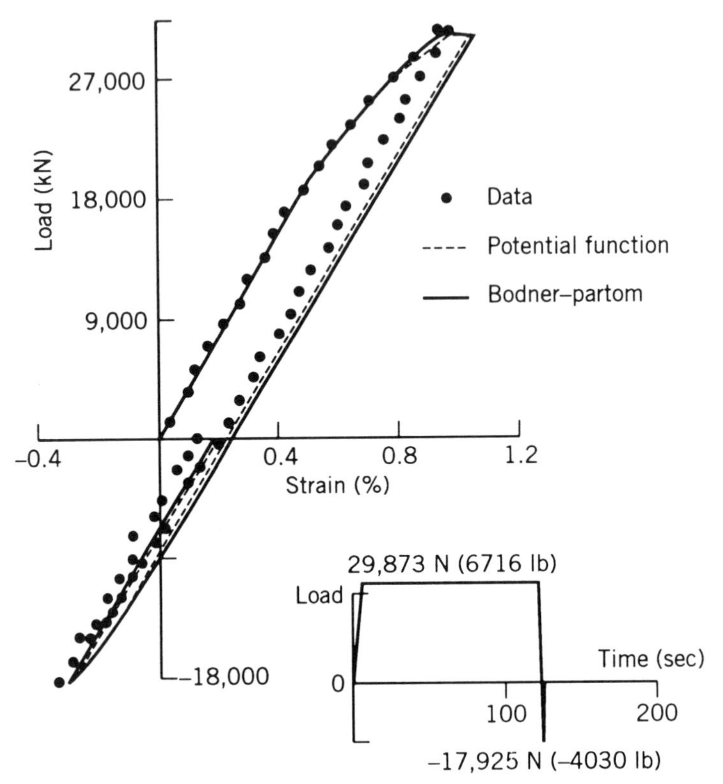

Figure 9.4.5 Comparison of experimental and calculated load-strain response at the root of the notch for Benchmark notch test 8 with a hold in tension. (From Dame, Stouffer, and Abuelfoutouh, 1984.)

REFERENCES

Abuelfoutouh, N. (1983). A thermodynamically consistent constitutive model for inelastic flow of materials, Ph.D. dissertation, University of Cincinnati.

Bathe, K. J. (1982). *Finite Element Procedures in Engineering Analysis,* Prentice Hall, Englewood Cliffs, NJ.

Bhattachar, V., and D. Stouffer (1993a). Constitutive equations for the response of René 80, Part 1: Development from isothermal data, *Journal of Engineering Materials and Technology,* Vol. 115, No. 4, pp. 351–357.

Bhattachar, V., and D. Stouffer (1993b). Constitutive equations for the response of René 80, Part 2: Effects of temperature history, *Journal of Engineering Materials and Technology,* Vol. 115, No. 4, pp. 358–364.

Bodner, S., and Y. Partom (1972). A large deformation elastic viscoplastic analysis of thick walled spherical shells, *Journal of Applied Mechanics,* Vol. 39, pp. 751–757.

Bodner, S., and Y. Partom (1975). Constitutive equations for elastic viscoplastic strain hardening materials, *Journal of Applied Mechanics,* Vol. 42, pp. 385–389.

Boyce, W., and R. DiPrima (1969). *Elementary Differential Equations,* 2nd ed., Wiley, New York.

Chang, T., A. Saleeb, and I. Iskovitz (1993). Finite element implementation of state variable based viscoplasticity models, *Computers and Structures,* Vol. 46, pp. 33–45.

Chen, C. (1993). A nonlinear P-version finite element program with a state variable constitutive equation and asymptotic integrator, Ph.D. dissertation, University of Cincinnati.

Chula, A., and K. Walker (1991). A new uniformly valid asymptotic integration algorithm for elasto-plastic-creep and unified viscoplastic theories including continuum damage, *International Journal of Numerical Methods in Engineering,* Vol. 32, pp. 385–418.

Crisfield, M.(1991). *Nonlinear Finite Element Analysis of Solids and Structures, Vol. 1: Essentials,* Wiley, Chichester, West Sussex, England.

Dame, L., D. Stouffer, and N. Abuelfoutouh (1984). Finite element analysis of notch behavior using a state variable constitutive equation, *Proceedings of the 2nd Symposium on Nonlinear Constitutive Equations for High Temperature Application,* NASA Lewis Research Center, Cleveland, OH, June 13–15.

Domas, P., W. Sharpe, M. Ward, and J. Yau (1982). Benchmark notch test for life prediction, *NASA CR-165571,* NASA Lewis Research Center, Cleveland, OH.

Dombrovsky, L. (1992). Incremental constitutive equations for Miller and Bodner–Partom viscoplastic models, *Computers and Structures,* Vol. 44, pp. 1065–1072.

Freed, A., and K. Walker (1992). Exponential integration algorithms applied to viscoplasticity, *NASA TM 104461.*

Gear W. (1971). *Numerical Initial Value Problems in Ordinary Differential Equations,* Prentice Hall, Englewood Cliffs, NJ.

Hinton, E., ed.(1992). *NAFEMS Introduction to Nonlinear Finite Element Analysis,* NAFEMS, Glasgow.

Hughes, J. (1984). Numerical implementation of constitutive models, in *Theoretical Foundation for Large Scale Computations in Nonlinear Material Behavior,* S. Nemat-Nasser, R. J. Asaro, and G. A. Hegemier, eds., Martinus Nijhoff, The Hague, The Netherlands.

Iskovitz, I., T. Chang, and A. Saleeb (1994). Extension of an asymptotic algorithm to orthotropic viscoplastic structural analysis, *Computers and Structures,* Vol. 52, pp. 667–668.

Owen, D., and E. Hinton (1980). *Finite Elements in Plasticity: Theory and Practice,* Pineridge Press, Swansea, Wales.

Simo, J., and R. Taylor (1985). Consistent tangent operators for rate-independent elastoplasticity, *Computer Methods in Applied Mechanics and Engineering,* Vol. 48, pp. 101–118.

Walker, K. P., and A. D. Freed (1991). Asymptotic integration algorithms for nonhomogeneous, nonlinear, first order, ordinary differential equations, *NASA TM 103793.*

Zienkiewicz, 0., and R. Taylor (1991). *The Finite Element Method, Vol. 2: Solid and Fluid Mechanics, Dynamics and Non-linearity,* 4th ed., McGraw-Hill, London.

PROBLEMS

9.1 Use the definition of the deviatoric stress and strain tensors to recast the linear elastic stress strain relationship as equation 9.3.1.

9.2 Consider a yield surface plasticity model obeys the von Mises yield criteria and has an associated flow rule. The model has a bilinear stress–strain curve and the slope of the curve up to the yield point is the elastic modulus E, and the slope of the curve after the yield point is E_T. Compute the continuum tangent modulus **(a)** assuming isotropic hardening; **(b)** assuming Ziegler kinematic hardening; **(c)** assuming mixed hardening using the Ziegler model for the kinematic component of the hardening.

9.3 Determine the tangent material modulus for the Bailey–Norton creep law (equation 5.8.9) when using a trapezoidal integration rule.

9.4 Sketch a full Newton–Raphson iteration procedure and an initial stiffness iteration procedure for a hardening structure (load displacement curve concave up).

9.5 The response of a material is represented by a bilinear stress–strain curve with an elastic modulus of 30 GPa, an initial yield point of 30 MPa, and a tangent modulus of 10 GPa after yielding. A bar of this material is subjected to a load that produces a uniaxial stress of 50 MPa. Sketch two iterations of a full Newton–Raphson procedure to predict the strain in the bar.

9.6 Repeat Problem 9.5 using two iterations of an initial stiffness method.

9.7 Compute the continuum tangent modulus for the material in Problem 9.5 for an applied stress of 50 MPa. Assume a von Mises material model with isotropic hardening and an associated flow rule.

Appendix: Parameters And Properties

This appendix contains material parameters for most of the constitutive equations given earlier. As a result, there are also parameters for many engineering materials. The tables are identified by appropriate name and are listed in approximately the order in which the models were introduced. Cross reference to a particular material is given in the Index.

Since many of the models evolved over several years, earlier forms of the equations are given as necessary. Further, the method of determining parameters has evolved from very primitive approaches such as guessing to become the methods presented in Section 6.4. As a result, the methods presented in Section 6.4 will not always give the parameters listed in the following tables.

APPENDIX 2.1: RELATIONSHIP BETWEEN HARDNESS AND STRENGTH

Following are the approximate equivalent Vickers, Brinell, and Rockwell hardness numbers with the tensile strength for carbon and alloy steels. The values are abstracted from the *Metals Handbook*, desk ed., Howard E. Boyer and Timothy L. Gall, eds., American Society for Metals, Metals Park, OH, 1985, page 1–60.

Vickers Hardness Number	Brinell 3000-kg Load, 10-mm Tungsten Carbide Ball	Rockwell B 100-kg Load, $\frac{1}{16}$-inch Diameter Indenter	Rockwell D 150-kg Load, Brale Indenter	Tensile Strength (1000 psi)
670	—	—	58.8	348
630	591	—	56.8	323
600	564	—	55.2	303
570	535	—	53.6	288
530	497	—	51.1	265
500	471	—	49.1	247
470	442	—	46.9	228
430	405	—	43.6	205
400	379	—	40.8	187
380	360	—	38.8	175
360	341	—	36.6	164
340	322	—	34.4	155
320	303	—	32.2	146
300	284	—	29.8	138
280	265	—	27.1	129
260	247	—	24.0	120
240	228	98.1	20.3	111
220	209	95.0	—	101
200	190	91.5	—	92
180	171	87.1	—	84
160	152	81.7	—	75
140	133	75.0	—	66
120	114	66.7	—	57

APPENDIX 2.2: PARAMETERS FOR THE UNIAXIAL RAMBERG–OSGOOD PLASTICITY EQUATION

Following are the typical values for the strain-hardening exponent (n), strength coefficient (K) in Equation 2.3.8, and the cyclic strain hardening exponent (n') and cyclic strength coefficient (K') in equations 2.5.2 and 2.5.3. *References*: (a) D. C. Stouffer, original data; (b) monotonic data from J. R. Low and F. Garofalo, *Proceedings of the Society for Experimental Stress Analysis*, Vol. 4, No. 2, 1947; (c) cyclic data from N. E. Dowling, *Mechanical Behavior of Materials*, Prentice Hall, Englewood Cliffs, NJ, 1993; (d) N. E. Dowling and A. K. Khosrovaneh, Simplified analysis of helicopter fatigue spectra, in *Development of Fatigue Loading Spectra*, ASTM STP 1006, J. M. Potter and R. T. Watanabe, eds., 1989; (e) F. A. Conle, R. W. Landgraf, and F. D. Richards, *Materials Data Handbook: Monotonic and Cyclic Properties of Engineering Materials*, Ford Motor Co. Science and Research Staff, 1984; (f) *AISI Automotive Steel Design Manual*, Revision 4, AISI, Southfield, MI, 1993.

Material	Ref.	n	K (ksi)	K (MPa)	n'	K' (ksi)	K' (MPa)
Al 1145–H16	(a)	0.042–0.056	18.2	126	0.100–0.200	25–34	172–234
Al 2024–T351	(d)	—	—	—	0.070	96	662
Al 7075-T6	(e)	—	—	—	0.106	142	977
AISI 950X, BHN = 124	(f)	0.22	117	810	0.17	128	883
AISI 950X, BHN = 187	(f)	0.10	116	800	0.09	130	900
AISI 1045, BHN = 500	(f)	0.04	341	2351	0.20	672	4634
SAE 4340 steel	(b, c)	0.15	93	641	0.131	240	1655
Copper, annealed	(b)	0.54	46	320	—	—	—
Inconel X	(e)	0.12	269	1855	—	—	—

APPENDIX 2.3: TENSILE AND FATIGUE RESPONSE OF INCONEL 100 AT 1350°F

The tables contain data for three tensile curves and one fatigue loop with stress relaxation in compression for IN 100 at 1350°F. The three tensile tests were run at constant strain rates of 1.43×10^{-3} sec^{-1}, 6.33×10^{-5} sec^{-1}, and 6.67×10^{-6} sec^{-1}. The fatigue test was run with the following the strain rate history: $+8.33 \times 10^{-4}$ sec^{-1} for 25 sec, -8.33×10^{-4} sec^{-1} for the next 25 sec, then held at zero strain for the next 26 sec, and then loaded at $+8.33 \times 10^{-4}$ sec^{-1} for 50 sec.

Specimen #1		Specimen #2		Specimen #3		Specimen #4		Specimen #4	
Rate: 1.43 10^{-3} sec^{-1}		Rate: 6.33 10^{-5} sec^{-1}		Rate: 6.67 10^{-6} sec^{-1}		Rate: 6.67 × E-6		Rate: Continued	
Strain	Stress-ksi	Strain	Stress-ksi	Strain	Stress-ksi	Strain	Stress-ksi	Strain	Stress-ksi
0.0000	0.0	0.0000	0.0	0.0000	0.0	−0.00024	0.0	0.01336	−30.0
0.0003	6.0	0.0005	6.0	0.0005	10.2	0.00000	6.0	0.01272	−40.0
0.0010	20.0	0.0010	20.0	0.0009	20.4	0.00038	16.0	0.01188	−50.0
0.0015	30.0	0.0015	30.0	0.0011	25.5	0.00056	20.0	0.01136	−60.0
0.0020	40.0	0.0020	40.0	0.0013	30.6	0.00096	30.0	0.01076	−70.0
0.0028	56.0	0.0025	56.0	0.0015	35.7	0.00136	40.0	0.00960	−80.0
0.0032	66.0	0.0030	66.0	0.0018	40.3	0.00180	50.0	0.00888	−90.0
0.0036	74.0	0.0032	74.0	0.0021	45.9	0.00220	60.0	0.00808	−100.0
0.0039	80.0	0.0035	80.0	0.0023	50.9	0.00260	70.0	0.00672	−110.0
0.0044	90.0	0.0040	90.0	0.0024	56.1	0.00300	80.0	0.00572	−120.0
0.0047	96.0	0.0045	96.0	0.0026	59.1	0.00340	90.0	0.00440	−130.0
0.0053	108.0	0.0050	108.0	0.0032	69.3	0.00392	100.0	0.00248	−140.0
0.0056	114.0	0.0055	114.0	0.0038	81.5	0.00432	110.0	0.00032	−150.0
0.0059	120.0	0.0060	120.0	0.0043	91.7	0.00476	120.0	0.00000	−152.8
0.0062	126.0	0.0065	126.0	0.0049	101.9	0.00520	130.0	−0.00020	−153.0
0.0066	134.0	0.0070	129.0	0.0054	108.0	0.00572	140.0	−0.00020	−140.0
0.0069	140.0	0.0076	132.0	0.0059	112.1	0.00592	142.0	−0.00020	−130.0
0.0070	142.0	0.0080	133.0	0.0066	116.2	0.06080	144.0	−0.00020	−120.0
0.0071	145.0	0.0085	134.0	0.0073	118.2	0.00640	146.0	0.00036	−110.0
0.0075	148.0	0.0095	135.0	0.0084	120.3	0.00672	148.0	0.00076	−100.0
0.0085	152.0	0.0110	135.0	0.0099	122.3	0.00720	150.0	0.00112	−90.0
0.0100	156.0	0.0130	135.0	0.0101	122.3	0.00790	150.0	0.00152	−80.0
0.0115	158.0	0.0150	134.0	0.0126	122.3	0.00900	154.0	0.00192	−70.0
0.0140	160.0			0.0144	120.3	0.01120	156.0	0.00232	−60.0
0.0160	161.0					0.01240	156.4	0.00268	−50.0
0.0200	162.0					0.01360	156.6	0.00290	−40.0
0.0245	162.0					0.01920	156.4	0.00348	−30.0
0.0300	162.0					0.02080	156.0	0.00388	−20.0

Specimen #1 Rate: 1.43 10^{-3} sec^{-1}		Specimen #2 Rate: 6.33 10^{-5} sec^{-1}		Specimen #3 Rate: 6.67 10^{-6} sec^{-1}		Specimen #4 Rate: 6.67 × E-6		Specimen #4 Rate: Continued	
Strain	Stress-ksi	Strain	Stress-ksi	Strain	Stress-ksi	Strain	Stress-ksi	Strain	Stress-ksi
						0.02144	155.6	0.00420	−10.0
						0.02140	152.0	0.00460	0.0
						0.02138	150.0	0.00500	10.0
						0.02136	140.0	0.00552	20.0
						0.02120	136.0	0.00608	30.0
						0.02100	130.0	0.00656	40.0
						0.02052	120.0	0.00716	50.0
						0.02008	110.0	0.00776	60.0
						0.01960	100.0	0.00848	70.0
						0.01920	90.0	0.00920	80.0
						0.01880	80.0	0.01000	90.0
						0.01800	70.0	0.01080	100.0
						0.01780	60.0	0.01152	110.0
						0.01720	50.0	0.01228	120.0
						0.01640	40.0	0.01320	130.0
						0.01636	30.0	0.01440	140.0
						0.01592	20.0	0.01592	50.0
						0.01548	10.0	0.01720	154.0
						0.01500	0.0	0.01760	154.4
						0.01436	−10.0	0.03280	154.4
						0.01388	−20.0		

APPENDIX 3.1: PARAMETERS FOR THE ARRHENIUS CREEP EQUATION FOR NARloy Z

Following are the parameters for equations 3.4.1 and 3.4.3 to 3.4.5 with the modification proposed by A. K. Miller in equations 7.8.8*a* and 7.8.8*b* for steady-state creep above and below a critical temperature $T_{cr} = 0.5T_{melt}$ for NARloy Z. *Reference*: A. D. Freed and K. P. Walker, Viscoplastic model development with an eye towards characterization, in *Material Parameter Estimation for Modern Constitutive Equations*, L. A. Bertram, S. B. Brown, and A. D. Freed, eds., MD Vol. 43, and AMD Vol. 168, ASME, New York, 1993, p. 57.

Parameter	Units	Value
A	s^{-1}	8×10^{18}
D	MPa	14
n	—	4
Q	J/mol	450,000
T_{melt}	K	$\approx$1350

APPENDIX 3.2: PARAMETERS FOR THE BAILEY–NORTON CREEP EQUATION FOR RENÉ 80

The following parameters were reported for the Bailey–Norton creep law (equation 3.5.3) with $\sigma_0 = 10$ ksi and $t_0 = 1$ hr. *Reference*: R. L. McKnight, J. H. Laflen, and G. T. Spamer, Turbine blade tip durability analysis, *Contractor Report 165268*, NASA Lewis Research Center, Cleveland, OH, 1982.

T (°F)	1200	1600	1700	1800	1900	2000	2100
A (ksi)	2.69×10^{-7}	1.12×10^{-6}	9.24×10^{-6}	9.25×10^{-5}	5.40×10^{-4}	1.23×10^{-3}	2.36×10^{-3}
m	2.54	5.19	4.60	3.65	2.65	2.00	1.43
n	0.452	0.685	0.813	0.890	0.920	0.923	0.924

APPENDIX 3.3: PARAMETERS FOR THE MARIN–PAO CREEP EQUATION FOR UDIMENT 700W

The following representation was used for the creep of Udiment 700W (wrought) at 1700°F as a function of stress as shown in Figure 3.3.6. The parameters A, $B = \dot{\varepsilon}^{C}_{min}$, the minimum creep rate, and k are defined in equation 3.5.5. The units of stress are ksi. *Reference*: J. H. Laflen and D. C. Stouffer, An analysis of high temperature metal creep, Part 1: Experimental definition of an alloy, *Journal of Engineering Materials and Technology*, Vol. 100, 1978, pp. 363–370.

$$A(\sigma) = \frac{\dot{\varepsilon}^{C}_{max}(\sigma) - \dot{\varepsilon}^{C}_{min}(\sigma)}{\beta(\sigma)}$$

$$\dot{\varepsilon}^{C}_{min} = 1.89 \times 10^{-4} \left(\frac{|\sigma|}{50}\right)^{10.64} \quad \text{min}^{-1}$$

$$\dot{\varepsilon}^{C}_{max} = 8.4 \times 10^{-4} \left(\frac{|\sigma|}{50}\right)^{8.808} \quad \text{min}^{-1}$$

$$k = 3.8 \times 10^{-3} \exp(9.12 \times 10^{-2}|\sigma|) \quad \text{min}^{-1}$$

APPENDIX 5.1: PARAMETERS FOR THE PRANDTL–REUSS EQUATION FOR IN 100

The following parameters were used to plot Figure 5.5.1 with equation 5.5.11, a modified form of the Prandtl–Reuss equation with kinematic hardening in the flow law. The material is Inconel 100 at 1300°F described in Appendix 2.3. The tensile curve was modeled with

$$\sigma = \begin{cases} E\varepsilon & \text{for } \varepsilon < \varepsilon_y \\ \sigma_y + A\{1 - \exp[-B(\varepsilon - \varepsilon_y)]\} & \text{for } \varepsilon > \varepsilon_y \end{cases}$$

so the plastic tangent modulus can be established as

$$\frac{1}{\frac{4}{3}E^p} = \frac{1}{AB[1 - (\sigma - \sigma_y)/A]} - \frac{1}{E}$$

The tensile parameters below were determined from the data in Appendix 2.3 except for the duplicate test at $5.50 \times 10^{-5}/\text{sec}^{-1}$. (Courtesy of Binyu Tian, 1992.)

Strain Rate (sec^{-1})	A (ksi)	B	E (ksi)	Yield Stress σ_y (ksi)	Yield Strain, ε_y
1.43×10^{-3}	20.0	3.1723	19,862	142	0.0070
6.33×10^{-5}	28.0	8.5506	22,574	114	0.0055
5.50×10^{-5}	27.4	9.4044	22,395	116	0.0054
6.67×10^{-6}	18.4	6.5448	21,044	102	0.0049
8.33×10^{-4}[a]	23.5	5.52	21,050	130	0.0061

[a]Parameters determined by interpolation.

APPENDIX 6.1: PARAMETERS FOR THE ISOTHERMAL RESPONSE OF RENÉ 80

The parameters below were used with the following flow equation:

$$\dot{\varepsilon}^{\mathrm{I}}_{ij} = D_0 \exp\left[-\frac{B}{2}\left(\frac{Z^2}{3T^2K_2}\right)^n \right] \frac{S_{ij} - \Omega_{ij}}{\sqrt{K_2}}$$

and equations 6.6.3, 7.2.5, 7.2.9, and 7.2.10 for three-dimensional response. The corresponding uniaxial equations 6.5.3, 6.5.4, and 6.5.8 are used with the reduced uniaxial flow equations 6.2.13 with the extra parameter B that was later combined with Z_0 and Z_1. The drag stress equation is the same for uniaxial and multiaxial loading. The elastic moduli, Poisson ratio, and coefficient of thermal expansion as a function of temperature are given in Figure 4.5.1. The extra hardening parameter $\alpha_m = 1.0$. *References*: V. G. Ramaswamy, R. H. Van Stone, L. T. Dame, and J. H. Laflen, Constitutive modeling for isotropic materials, *NASA Conference Publication 2339*, 1984; V. G. Ramaswamy, A constitutive model for the inelastic multiaxial cyclic response of a nickel base superalloy René 80, *NASA CR 3998*, 1986.

Parameter	982°C	871°C	760°C	538°C
n	0.2418	0.3005	0.6	1.0
D (sec^{-1})	1.0	10.0	10,000	10,000
B	1.0	1.0	0.0609	0.000916
f_1 (MPa)	2.88×10^4	4.65×10^4	4.97×10^4	4.86×10^4
f_2	0.3005	0.2926	0.4944	0.5058
Ω_m (MPa)	283	384	551	551
b_1	0.4772	0.4566	0.5018	0
b_2 (MPa)	10.3	−29.5	−28.3	0
b_3 (MPa)	91.8	91.8	229.6	367.4
A (sec^{-1})	4.526×10^{-4}	2.126×10^{-3}	7.625×10^{-7}	0
p	8.458	3.0609	13.7	0
σ_2 (MPa)	188.4	256.0	367.4	367.4
m (MPa^{-1})	0.103	0.0126	0	0
Z_0 (MPa)	5.1×10^4	5.1×10^4	5.1×10^4	5.1×10^4
Z_1 (MPa)	4.0×10^4	2.1×10^4	2.1×10^4	2.1×10^4

APPENDIX 6.2: PARAMETERS FOR TITANIUM ALUMINIDE UP TO 871°C

The tensile parameters for equations 6.3.5 and 6.2.13 are given below for Ti–14Al–21Nb. The value of σ_S is the saturated tensile stress, and D_0 = 10,000 sec^{-1}. *Reference*: D. C. Stouffer and B. Kane, *Inelastic Deformation Modeling Titanium Aluminide Matrix Material*, GE Aircraft Engines, Lynn MA, 1988.

Temp. (°C)	E (MPa)	σ_S (MPa)	Ω_S (MPa)	f_1 (MPa)	f_2	Z_0 (MPa)	n
21	94,065	693	476	144,375	0.7926	484	1.30
93	91,238	562	435	65,317	0.780	484	1.30
204	88,817	500	377	36,437	0.7665	491	1.25
316	89,328	450	341	5,500	0.7740	517	1.12
427	73,165	425	327	68,750	0.7680	551	1.00
538	70,333	381	211	139,500	0.5547	1,258	0.87
649	44,035	351	214	88,000	0.5000	1,649	0.70
760	25,012	277	123	17,187	0.3500	4,348	0.52
871	18,953	174	25	7,562	0.0010	49,012	0.30

APPENDIX 6.3: PARAMETERS FOR Al 7050–T7451 AT ROOM TEMPERATURE

The tensile response of the alloy was almost elastic–perfectly plastic, but after cyclic the response gradual elastic–plastic transition. This behavior was modeled by allowing f_1 to change in equation 6.3.6. Thus

$$f_1 = f_{1f} - (f_1 - f_{1f}) \exp(-m_2 W^{\mathrm{I}})$$

where f_1 and f_{1f} are the initial and final values of f_1, respectively, and m_2 controls the rate of yield softening. Equations 6.2.13 and 6.6.3 remain the same. Equations 6.4.4, 6.4.9, 6.4.13, and 6.4.19 were used to find four of the parameters. *Reference*: M. D. Kuruppu, J. F. Williams, N. Bridgeford, R. Jones, and D. C. Stouffer, Constitutive modeling of the elastic plastic behavior of 7050-T7459 aluminum alloy, *Journal of Strain Analysis*, Vol. 27, No. 2, 1992, pp. 85–92.

$E = 71{,}000$ MPa	$\Omega_m = 414$ MPa	$m = 0.6$
$f_1 = 132{,}000$ MPa	$Z_1 = 182$ MPa	$f_{1f} = 25{,}000$ MPa
$f_2 = 0.816$	$m_2 = 30$	$Z_0 = 40$ MPa
$D_0 = 10{,}000\ \text{sec}^{-1}$	$n = 3$	

APPENDIX 7.1: PARAMETERS FOR THE THERMOMECHANICAL RESPONSE OF RENÉ 80

The revised parameters for the thermomechanical model (equations 7.5.3, 6.3.5, 7.6.1, and 7.5.4) are given below except for the value of Z_0, Z_1, and n in Figure 7.5.3*a* and *b* which are used between the test temperatures because interpolation is not adequate. The parameters for equation 6.5.2 are given in Appendix 6.1. The parameters for the extra hardening are $m_2 = 1.0$ for $T < T_C$, and $m_2 = 0.5$ for $T > T_C$, where $T_C = 700°C$. The parameter β was assumed to be a stiff exponential function as shown in Figure 7.6.2. *Reference*: V. S. Bhattachar and D. C. Stouffer, Constitutive equations for the thermomechanical response of René 80, Part 1: Development from isothermal data; Part 2: Effects of temperature history, *Journal of Engineering Materials and Technology*, Vol. 115, 1993, pp. 351–364. Copyright © The American Society of Mechanical Engineers, 345 East 47th St., N.Y., N.Y. 10017. Used with permission.

Parameter	811 K	1033 K	1144 K	1255 K	1366 K
E (MPa)	153,940	152,000	134,807	110,900	100,000
ν	0.38	0.38	0.38	0.38	0.38
n	1.0	0.6	0.16	0.15	0.14
D (sec^{-1})	10,000	10,000	10,000	10,000	10,000
f_1 (MPa)	161,010	150,000	119,010	61,500	37,500
f_2	0.465	0.46	0.358	0.33	0.30
Ω_m (MPa)	500	490	420	228	100
A (sec^{-1})	0.004857	0.004857	0.005	0.006841	0.025
p	6.5	5.499	3.695	1.88	1.70
σ_0 (MPa)	780	750	540	330	70
m (MPa^{-1})	0	0	0.087	0.09	0.10
Z_0' (MPa)	0.9×10^6	0.3×10^7	0.2×10^{10}	0.302×10^{10}	0.316×10^{10}
Z_1 (MPa)	—	—	0.1183×10^{10}	0.195×10^{10}	0.26×10^{10}

APPENDIX 7.2: PARAMETERS FOR THE BODNER EQUATION FOR SEVERAL MATERIALS

There are two forms of the Bodner Partom flow equation, equation 9.4.1 denoted as form A and equation 6.2.7 denoted as form B. The evolution of Z is defined in equations 7.7.1 to 7.7.5. There are two sets of material parameters for Inconel 718 at 649°C/650°C. Bodner's parameters are designated as form A-1 and those used in Section 9.4 are denoted as form A-2. *Reference*: S. R. Bodner, Review of a unified elastic–viscoplastic theory, in *Unified Constitutive Equations for Creep and Plasticity*, A. K. Miller, ed., Elsevier Applied Science, Barking, Essex, England, 1987.

	Material					
	Titanium	Copper	Copper	Aluminum	Al 2024-0	René 95
Temp. (°C)	Room temp.	Room temp.	550	20	Room temp.	650
Form	C	B	B	B	C	C
D_0 (sec^{-1})	10,000	10,000	10,000	10,000	10,000	10,000
n	1.0	7.5	1.2	1.4	10.0	3.2
Z_0 (MPa)	1150	63	16	39	90	1680
Z_1 (MPa)	1400	250	250	220	20	2200
Z_2 (MPa)	NA	63	16	31	NA	1680
Z_3 (MPa)	NA	NA	NA	NA	NA	NA
m_1 (MPa^{-1})	0.087	0.13	2.5	5.0	0.22	0.37
m_2 (MPa^{-1})	NA	NA	NA	NA	NA	NA
m_3 (MPa^{-1})	NA	NA	NA	NA	NA	NA
A_1 (sec^{-1})	NA	NA	4.3×10^{-3}	3.4×10^{-1}	NA	5.3
r_1	NA	NA	3.6	3.25	NA	3.0
E (GPa)	118	120	$\approx$100		73	177

Appendix 7.2 Continued

Material				
Inconel 100	Inconel 718	Inconel 718	Hastelloy X	B1900+Hf
732	650	649	RT	RT
C	A	C	B	B
10,000	10,000	10,000	10,000	10,000
0.7	1.17	1.954	1.0	1.05
6300	3130	1805	1200	2700
7000	4140	2253	2000	3000
4130	2760	1805	1200	2700
NA	NA	NA	1200	1150
0.37	0.024	0.16	0.02	0.27
NA	NA	NA	0.9	1.52
NA	NA	NA	0.001	0
1.9×10^{-3}	1.1×10^{-3}	5.6×10^{-5}	NA	NA
2.66	2.89	1.37	NA	NA
179	165	—	207	200

APPENDIX 7.3: PARAMETERS FOR THE MILLER EQUATION

Following are the type 304 stainless steel parameters for equations 7.8.7, 7.8.8*a*, 7.8.8*b*, 7.8.12, and 7.8.15 except for the initial value of the drag stress, D_0, which is given in Figure A7.3b. The parameters were proposed for the conditions that (1) $T_{cr} = 0.6\, T_m$, and (2) linear interpolation or extrapolation of the elastic moduli between the values of 28×10^6 ksi at 23°C and 22.5×10^5 at 538°C. *Reference*: A. Miller, An inelastic constitutive equation for monotonic, cyclic and creep deformation, Part 1: Equations development and analytical procedures; Part 2: Application to 304 stainless steel, *Journal of Engineering Materials and Technology*, Vol. 98H, 1976, pp. 97–113.

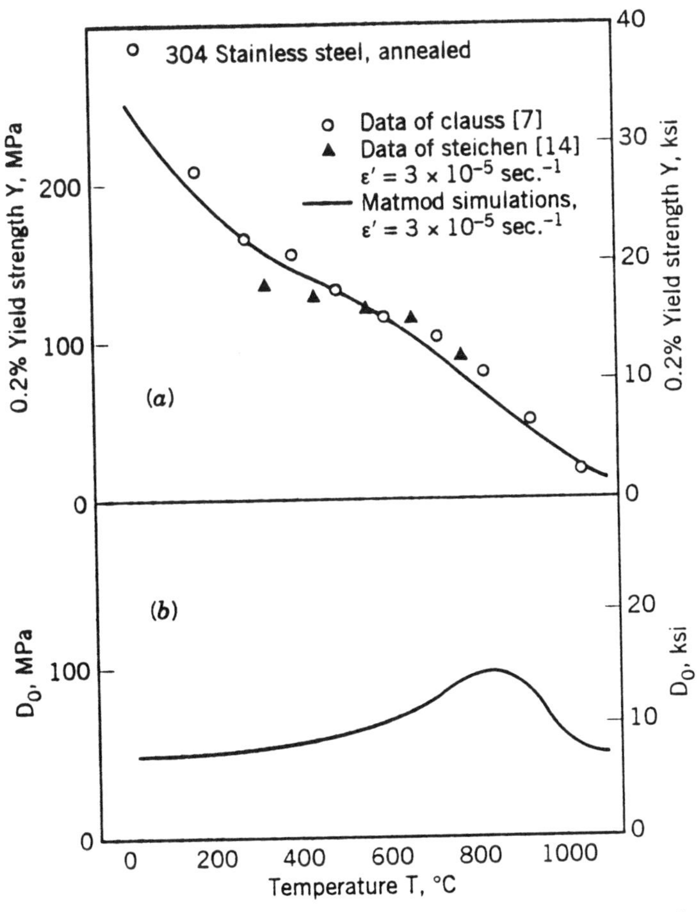

Figure A7.3 Yield strength (a) and initial value of drag stress (b) as a function of temperature for type 304 stainless steel. (After Miller, Journal of Engineering Materials and Technology, Vol. 98, pp 109, 1976. Copyright © The American Society of Mechanical Engineers, 345 East 47th St., N.Y., N.Y. 10017. Used with permission.)

$A_1 = 0.8 \text{ ksi}^{-1}$	$C_2 = 0.1 \text{ ksi}$	$n = 5.8$
$A_2 = 7.42 \times 10^{-5} \text{ ksi}^{-3}$	$f_1 = 280 \text{ ksi}$	$Q = 91{,}000 \text{ cal/mol}$
$D_0' = 1.0 \times 10^{15} \text{ sec}^{-1}$	$g_1' = 100$	$T_m = 1800 \text{ K}$

The Hastelloy X parameters for the Miller model with the flow rule as

$$\dot{\varepsilon}^{\mathrm{I}}_{ij} = D_0 \left[\sinh \left(\frac{|S_{kl} - \Omega_{kl}|}{Z} \right)^3 \right]^n \frac{\frac{3}{2} S_{ij} - \Omega_{ij}}{|S_{kl} - \Omega_{kl}|}$$

and $T_{cr} = 0.6T_m$ are given below. The elastic moduli and Poisson ratio as a function of temperature are given in Appendix 7.4 for Hastelloy X. *Reference*: A. Chula and K. P. Walker, *NASA TM 102344*, 1989.

$A_1 = 9.3 \times 10^{-4} \text{ ksi}^{-1}$	$C_2 = 50{,}000 \text{ ksi}$	$n = 1.598$
$A_2 = 5.943 \times 10^{-12} \text{ ksi}^{-3}$	$f_1 = 1.0 \times 10^7 \text{ ksi}$	$Q = 104{,}600 \text{ cal/mol}$
$D_0' = 1.0293 \times 10^{14} \text{ sec}^{-1}$	$g_1' = 100$	$T_m = 1588 \text{ K}$

APPENDIX 7.4: HASTELLOY X PARAMETERS FOR THE WALKER EQUATION

Below are the parameters for equations 6.2.9, 6.2.10, 7.9.8, and 7.9.9 with $Z = Z_0$, which give the results in Figure 7.9.1. *Reference*: A. Chula and K. P. Walker, A new uniformly valid asymptotic integration algorithm for elastic–plastic-creep and unified viscoplastic theories including continuum damage, *NASA TM 102344*, NASA Lewis Research Center, Cleveland, OH, 1989.

T (°F)	800	1000	1200	1400	1600	1800
E (psi)	26×10^6	24×10^6	24×10^6	22×10^6	18.6×10^6	13.2×10^6
ν	0.322	0.328	0.334	0.339	0.345	0.351
Z_0 (psi)	50,931	75,631	95,631	251,866	91,505	59,292
n	0.059	0.059	0.079	0.224	0.195	0.223
m	1.158	1.158	1.158	1.158	1.158	1.158
n_2 (psi)	30×10^7	6×10^7	1.5×10^7	2×10^7	5×10^6	1.5×10^6
n_3 (psi)	8000	1000	781.2	1178.6	672.6	312.5
n_6 (psi)	0	0	0	0	8.977×10^{-4}	2.733×10^{-3}
$\mathring{\Omega}_{11}$ (psi)	0	0	−2000	−2000	−1434	−1200

APPENDIX 7.5: HASTELLOY X PARAMETERS FOR THE KREIG, SWEARENGEN, AND ROHDE EQUATION

An earlier form of the Kreig, Swearengen, and Rohde equation is defined by equations 6.2.9 and 6.2.10 and the back-stress evolution equation

$$\dot{\Omega}_{ij} = A_1\dot{\varepsilon}_{ij} - A_2\Omega_{ij}\sqrt{\tfrac{2}{3}\Omega_{pq}\Omega_{pq}}\,[\exp(A_3\,\tfrac{2}{3}\Omega_{pq}\Omega_{pq}) - 1]$$

with drag stress $Z = Z_0$. These results of are reported in Figure 7.9.2. *References*: A. Chula and K. P. Walker, A new uniformly valid asymptotic integration algorithm for elastic–plastic-creep and unified viscoplastic theories including continuum damage, *NASA TM 102344*, NASA Lewis Research Center, Cleveland OH, 1989; R. D. Krieg, J. C. Swearengen, and R. W. Rohde, A physically based internal variable model for rate dependent plasticity, in *Inelastic Behavior of Pressure Vessel and Piping Components*, ASME PVP-PB-028, T. Y. Chang and E. Krempl, eds., ASME, New York.

T (°F)	800	1000	1200	1400	1600	1800
E (psi)	26×10^{6}	24×10^{6}	24×10^{6}	22×10^{6}	18.6×10^{6}	13.2×10^{6}
ν	0.322	0.328	0.334	0.339	0.345	0.351
Z_0 (psi)	50,931	75,631	95,631	251,866	91,505	59,292
n	0.059	0.059	0.079	0.224	0.195	0.223
A_1 (psi)	3.0×10^{8}	6.0×10^{7}	1.5×10^{7}	2.0×10^{7}	5.0×10^{6}	1.0×10^{6}
A_2 (psi^{-1} sec^{-1})	0.59	0.00179	0.66	1.54	14.96	243.0
A_3 (psi^{-1})	1.0×10^{-12}	1.0×10^{-12}	1.0×10^{-12}	1.0×10^{-12}	1.0×10^{-12}	1.0×10^{-12}

APPENDIX 8.1: ISOTHERMAL PARAMETERS FOR RENÉ N4

The parameters below are for René N4 at 982°C and René N4 VF317 at 760°C, an earlier heat of the alloy. No cube slip constants were evaluated for René N4 VF317 at 760°C. The parameters are for use with equations 8.7.1, 8.7.2, 8.7.3, and 8.7.5. The elastic constants for René N4 at 982°C are E = 81.3 GPa, G = 90.0 GPa, and ν = 0.398. The elastic constants for René N4 VF 317 at 760°C are E = 100.0 GPa, G = 96.5 GPa, and ν = 0.380. *Reference*: M. Y. Sheh and D. C. Stouffer, Anisotropic constitutive modeling of nickel-base single crystal superalloy, *NASA CR 182157*, 1988.

	René N4 at 982°C		VF317 at 760°C
	Octagonal Slip, $k = 1$	Cube Slip, $k = 3$	Octagonal Slip, $k = 1$
D_k (sec^{-1})	1.0	1.0	1.0
n_{k1}	0.53	0.70	1.40
H_{k1} (MPa)	20,510	7184	3926
H_{k2} (MPa)	0	0	0
m_k	0	0	0
V_1	−40.415	—	−2.7662
V_2	−28.231	—	−5.3904
V_3	17.661	—	0.98201
G_k	0	0	0.10
n_{k2}	0	0	0
F_k	0.21619×10^7	0.1600×10^6	0.40974×10^6
n_{k3}	1.2108	1.1202	0.80567
Ω_k^{sat} (MPa)	95.293	61.838	208.46

INDEX